15th International Conference on Thin Films 2011

(ICTF-15)

AA008881

Journal of Physics: Conference Series Volume 417

Kyoto, Japan
8-11 November 2011

ISBN: 978-1-62276-994-0
ISSN: 1742-6588

Printed from e-media with permission by:

Curran Associates, Inc.
57 Morehouse Lane
Red Hook, NY 12571

CURRAN ASSOCIATES INC. proceedings.com

Some format issues inherent in the e-media version may also appear in this print version.

Printed by Curran Associates, Inc. (2013)

Institute of Physics
Dirac House, Temple Back
Bristol BS1 6BE UK

Phone: 44 1 17 929 7481
Fax: 44 1 17 920 0979

techtracking@iop.org

Additional copies of this publication are available from:

Curran Associates, Inc.
57 Morehouse Lane
Red Hook, NY 12571 USA
Phone: 845-758-0400
Fax: 845-758-2634
Email: curran@proceedings.com
Web: www.proceedings.com

TABLE OF CONTENTS

Preface

The International Conference on Thin Films is the most established conference for all researchers and persons interested in thin films and coatings. It is one of the tri-annual conference series endorsed and co-organized by the Thin Film Division of the International Union for Vacuum Science, Technique and Applications (IUVSTA), a union of national member societies whose role is to stimulate international collaboration in the fields of vacuum science, techniques and applications and related multi-disciplinary topics including solid-vacuum and other interfaces.

The 15th International Conference on Thin Films (ICTF-15) is organized by The Vacuum Society of Japan (VSJ) and held at Kyoto TERRSA in Kyoto, Japan on 8–11 November 2011, following the 14th International Conference on Thin Films (ICTF-14), which was held in Ghent, Belgium in 2008.

Thin films and coatings are daily becoming increasingly important in the fields of various industries. This International Conference provides a multi-disciplinary forum for recent advances in basic research, development and applications of thin films and coatings. This conference will present a unique opportunity for researchers, engineers and managers to acquire new knowledge of thin films and coatings. We hope that our understanding on thin films and coatings will be deepened through this conference.

The conference site, 'Kyoto TERRSA' is located in the historical heart of the old capital Kyoto. Kyoto is an ancient city with a 1200-year history. It was established as Japan's capital under the name 'Heian-kyo' in the year 794. Although many transformations have taken place over the years, Kyoto has always embraced the most advanced standards of the times. It has greatly contributed to the nation's industrial, economic and cultural development. The dauntless spirit of leadership of Kyoto's past as a capital city is still felt here today. Kyoto also preserves the beloved examples of its culture as testimonials of time. This is shown in the ancient temples and shrines, as well as private houses, which are built in styles unique to Kyoto. Furthermore, many festivals, ceremonies and traditional activities reveal the will of this city to convey and develop its 1200-year-old culture. Participants of the conference will also be able to see many world heritage sites in the city. Moreover, November is the best time of year to visit Kyoto. We hope you will enjoy Kyoto very much.

We would like to offer our thanks for all the contributions from the members of the International Advisory Committee and Organizing Committee, Symposium Chairs, the Secretary General, the Thin Film Division of IUVSTA (chair: Professor Dr Alberto Tagliaferro), IUVSTA, VSJ and other cooperating societies, and to all of the supporting organizations and enterprises. We would also like to express our thanks to all of the participants, secretariat members and members of the supporting staff.

I am very pleased to welcome you to ICTF-15 and Kyoto!

Director Professor Dr Osamu Takai
Chairperson of ICTF-15
EcoTopia Science Institute, and Department of Materials, Physics and Energy Engineering, Graduate School of Engineering, Nagoya University, Japan

Peer review statement

All papers published in this volume of *Journal of Physics: Conference Series* have been peer reviewed through processes administered by the proceedings Editors. Reviews were conducted by expert referees to the professional and scientific standards expected of a proceedings journal published by IOP Publishing.

Journal of Physics: Conference Series **417** (2013) 012001 doi:10.1088/1742-6596/417/1/012001

Effect of Interfacial Reactions in Radical Process on Electrical Properties of Al_2O_3/Ge Gate Stack Structure

Kimihiko Kato[1,2], Mitsuo Sakashita[1], Wakana Takeuchi[1], Osamu Nakatsuka[1], and Shigeaki Zaima[1]
[1] Graduate School of Engineering, Nagoya University
Furo-cho, Chikusa-ku, Nagoya 464-8603, Japan
[2] Research Fellow of Japan Society for the Promotion of Science

E-mail: kkato@alice.xtal.nagoya-u.ac.jp

Abstract. We have investigated the interfacial reactions of Al_2O_3/Ge structure with nitrogen (N) radical process, and the effect of the chemical bonding states on electrical properties. The angle resolved hard X-ray photoelectron spectroscopy measurement revealed that oxygen is supplied from the Al_2O_3 film to the interface with Ge substrate during N radical process and Ge atoms diffuse into the Al_2O_3 film. As a result, AlGeON/Ge suboxide/Ge stacked structure is formed by N radical process. From capacitance-voltage characteristics of Al/AlGeON/Ge suboxide/Ge MOS capacitors, both the interface state density at the interface and the carrier trap density in the Al_2O_3 film are effectively decreased by the formation of AlGeON/Ge suboxide/Ge structure. The interface state density decreases with increasing in the thickness of Ge suboxide at the interface.

1. Introduction

Germanium (Ge) has much attention for the further increase in the operation speed of the metal-oxide-semiconductor field effect transistor (MOSFET), because of high mobilities for both electron and hole. Also, gate dielectrics with high dielectric constant (high-k) film become essential materials to reduce the equivalent oxide thickness (EOT) with suppressing the increase in the leakage current. To control the high-k/Ge interfacial structure is an important key technique for realizing of high-performance Ge based MOSFETs. For example, the interface state density at a high-k/Ge interface should be decreased to less than 10^{11} $eV^{-1}cm^{-2}$, and a high thermal stability of interfacial structure is also required. Focusing on the interface state density, there are many reports about realizing the low interface state density as low as 5×10^{10} $eV^{-1}cm^{-2}$ with the GeO_2/Ge structure formed by the thermal or plasma oxidation [1-4]. However, it is difficult to realize the GeO_2/Ge structure with keeping the thin EOT and the high thermal stability. In particular, the poor thermal stability is reported in which the GeO desorption from the GeO_2/Ge interface easily occurs [5,6]. On the other hand, Al_2O_3/Ge structure has a superior thermal stability. The interfacial mixing between Al_2O_3 and Ge does not occur even after annealing in N_2 ambient at as high as 550°C [7-9]. We recently reported that AlGeO/Ge structure can be formed by Ge diffusion from a Ge substrate into an Al_2O_3 film with annealing to the 1 nm-thick-Al_2O_3/Ge structure in O_2 ambient at from 400 to 550°C [9]. The interface state density of this AlGeO/Ge interfacial structure can be decreased by O_2 annealing with keeping thin EOT. Therefore, the AlGeO film is a new attractive candidate as an interfacial control layer with the thin EOT and the high thermal stability.

The most important issue for the formation of the AlGeO/Ge structure is control of the oxygen (O) supply and reaction at the the Al_2O_3/Ge interface. Excess O supply to the interface causes the segregation of GeO_2 on the surface of the Al_2O_3 film, although O supply is essential to cause the Ge diffusion from the Ge substrate [9]. Also, Zhang *et al.* reported that the exposure of O radicals to the Al_2O_3/Ge structure leads to the formation of the GeO_x interfacial layer [10]. The GeO_x formation immediately causes the increase in the EOT, although the interface state density effectively decreases. In this study, we focus on the O supply from an Al_2O_3 film by the N radical irradiation process. We can lead to the chemical reaction in the films and at the interface at wide process temperature by using the radical process since radicals have the high internal energy. In addition, we have proposed the N radical irradiation process to Al_2O_3/Ge and HfO_2/Si structures to improve on the interfacial properties [11,12]. N radicals break the O-Al or O-Hf bond and the O atoms are supplied from the oxide film to the interface. As a result, the interface state density of the Al_2O_3/Ge structure decreases after the N radical process [11]. However, the interfacial reactions at Al_2O_3/Ge interface with N radical process have not clarified yet. It is necessary to understand the crystalline structure and electrical properties of the Al_2O_3/Ge interface after N radical process in detail.

In this study, we investigated the reaction at the Al_2O_3/Ge interface with N radical process by using the angle resolved hard X-ray photoelectron spectroscopy (AR-HAXPES) method. Also, we clarified the effect of the chemical bonding state on electrical properties of the Al_2O_3/Ge structure after N radical process.

2. Experimental procedure

We used a Ga-doped p-type Ge(001) substrate with a resistivity of 1~5 Ω·cm. After the chemical cleaning of the substrate by using diluted-HF and deionized-H_2O to remove native oxide, a 1 nm-thick Al_2O_3 film (1st-Al_2O_3 film) was formed with ALD method (Picosun SUNALE R-150B). We used precursors of trimethylaluminium ($Al(CH_3)_3$, TMA) and H_2O with a pulse time are 0.1 and 1.0 s, respectively. The formation temperature of the Al_2O_3 film was 300°C. We confirmed that the thickness of the Ge oxide interlayer between the Al_2O_3 film and Ge substrate was as thin as 0.1 nm after the Al_2O_3 film deposition at 300°C [11]. Then, we performed the N radical process to the 1st-Al_2O_3/Ge structure in an ultra high vacuum chamber with a base pressure of less than 7×10^{-7} Pa. Nitrogen radicals were formed by the inductively coupled plasma (ICP) source with an RF power of 500 W and a N_2 partial pressure of 4×10^{-3} Pa. The heater temperature for the 1st-Al_2O_3/Ge sample was set at 50°C during the N radical process. The process time was ranging from 1 to 20 min. The distance from the Al_2O_3/Ge sample and ICP source was 5 or 18 cm. Then, we additionally deposited a 3 nm-thick-Al_2O_3 film (2nd-Al_2O_3 film) on some samples at 300 °C to suppress the leakage current in the electrical property measurement. Gate electrodes with an area of $1.3\sim1.5\times10^{-4}$ cm^2 were fabricated with the vacuum deposition of Al. Finally, some samples were annealed at 300 °C for 30 min in H_2 ambient at an atmospheric pressure (H_2 annealing).

The chemical bonding states were investigated from the X-ray photoelectron spectroscopy (XPS) with Al$K\alpha$ line ($h\nu$=1486.6 eV). HAXPES was also performed at the beamline BL46XU in the super photon ring-8 GeV (SPring-8). The monochromated X-ray with the photon energy of 7939 eV was used in the HAXPES measurement. Photoelectron take-off angle was ranging from 8 to 90°. Capacitance-voltage (*C-V*) properties were measured at room temperature with a high-frequency of 1 MHz. Interface state density was evaluated from the peak value of conductance-voltage properties [13].

3. Results and discussion

Figure 1 shows the *C-V* characteristics of Al/2nd-Al_2O_3/1st-Al_2O_3/Ge MOS capacitors with N and O radical process. The increase in the capacitance from inversion to accumulation conditions in the *C-V* curve becomes steep after N radical processes, indicating the decrease in the interface state density. Also, the hysteresis voltage width decreases with the N radical process. The counterclockwise hysteresis in *C-V* characteristics is probably related to the carrier traps from the Ge substrate to the defects formed in the Al_2O_3 film near the Al_2O_3/Ge interface. On the other hand, the hysteresis voltage width increases with the O radical process and it is much larger than that of the sample with the N radical process. Although, the interface state density decreases in the sample with the O radical process.

Figure2 shows the Ge3d photoelectron core spectra measured with XPS of 1st-Al_2O_3/Ge samples with (a) N and (b) O radical processes. The intensity of these peaks is normalized by the Al_2O_3 peak intensity in the Al2p spectra. The binding energy is corrected by that of the Ge substrate peak in

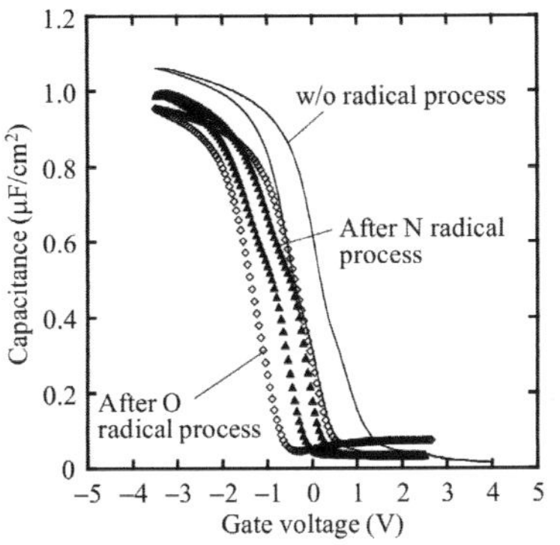

Figure 1 *C-V* characteristics of Al/2nd-Al_2O_3/1st-Al_2O_3/Ge MOS capacitor after N and O radical process without H_2 annealing. Distance from Al_2O_3/Ge and ICP source is 5 cm.

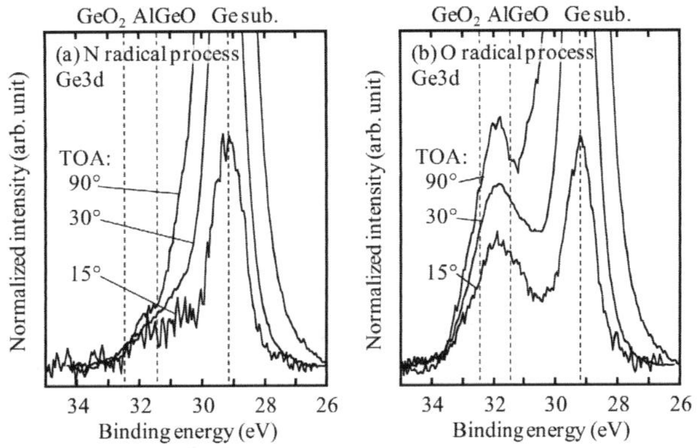

Figure 2 Ge3d photoelectron core spectra of 1st-Al_2O_3/Ge structures after (a) N and (b) O radical process measured by XPS with TOA of 90, 30, and 15°. Distance from Al_2O_3/Ge and ICP source is 5 cm. Binding energy is corrected by that of the Ge substrate peak and intensity is normalized by area intensity of the Ge substrate peak.

$Ge3d_{5/2}$ spectra (29.2 eV). The take-off angles are 90, 30, and 15°. The area intensity of the Ge oxide peak at a binding energy of about 32 eV in the sample with the N radical process is much smaller than that of the sample with the O radical process. The area densities of Ge atoms bonding with O atoms are evaluated by the area intensity ratio of the Ge oxide peak to the Ge substrate peak. Those for the sample with N radical and O radical process are 9.8×10^{14} and 3.0×10^{15} cm^{-2}, respectively. We can also observe the TOA dependence of the normalized area intensity of the Ge oxide peak for the sample after the O radical process. The normalized area intensity of the Ge oxide peak decreases with the TOA. This result indicates that a Ge oxide layer formed at the Al_2O_3/Ge interface with O radical process, since XPS with a small TOA enables surface sensitive measurement. On the other hand, we can find almost no TOA dependence of the normalized intensity of the Ge oxide peak in the sample with the N radical process. It is found that N radical process forms an AlGeON film on a Ge substrate due to the uniform diffusion of Ge atoms into the 1st-Al_2O_3 film from the Ge substrate.

Next, we performed the AR-HAXPES measurement in order to investigate the interfacial chemical bonding state in detail and the interfacial reaction at Al_2O_3/Ge structure with N radical process.

Figure 3 shows (a) $Ge2p_{3/2}$, (b) Al1s, and (c) N1s photoelectron core spectra of the 1st-Al_2O_3/Ge samples with the N radical process for from 1 to 10 min. The TOA of the HAXPES measurement was 8°. The intensity of $Ge2p_{3/2}$ spectra is normalized by the area intensity of the Ge substrate peak and that of Al1s and N1s spectra is normalized by the area intensity of the Al_2O_3 peak in Al1s spectra. The binding energy is corrected by that of the Ge substrate peak of 1217.0 eV. We can find the peak related to AlGeON in $Ge2p_{3/2}$ spectra at a binding energy of 1219.5 eV [14,15]. We can also find

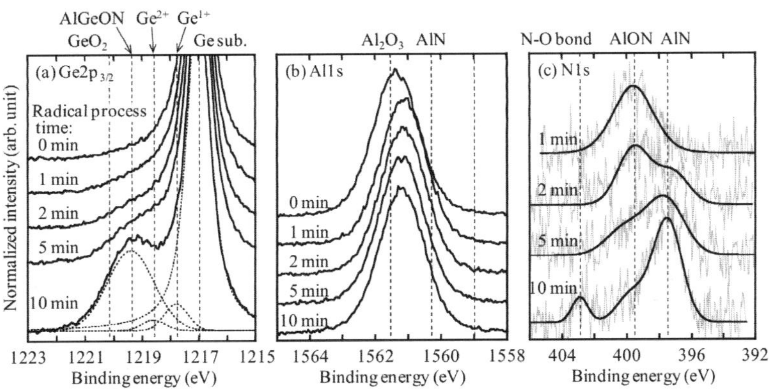

Figure 3 (a) $Ge2p_{3/2}$, (b) Al1s, and (c) N1s photoelectron core spectra of 1st-Al_2O_3/Ge structures after N radical process for from 1 to 10 min measured by HAXPES. Distance from Al_2O_3/Ge and ICP source is 18 cm. Hard X-ray energy is 7939 eV and TOA is 8°. Intensity of $Ge2p_{3/2}$ spectra is normalized by area intensity of the Ge substrate peak and that of Al1s and N1s spectra is normalized by area intensity of the Al_2O_3 peak in Al1s spectra. Binding energy is corrected by that of the Ge substrate peak of 1217.0 eV.

peaks related to Ge suboxides of Ge^{1+} and Ge^{2+} in $Ge2p_{3/2}$ spectra at a binding energy of 1217.8 and 1218.6 eV, respectively, as shown in Figure 3. The radical process time dependence of the area intensity ratio of the Ge oxide peak to the Ge substrate peak is shown in Figure 4. In this figure, we represent the sum of Ge^{1+} and Ge^{2+} as Ge suboxide since both these components exist at the AlGeON/Ge interface as discussed later. The intensity ratio of the AlGeON peak increases with the radical process time, indicating the increase in the composition ratio of Ge in the AlGeON film. The intensity ratio of Ge suboxide increases with the N radical process time of shorter than 2 min, and it hardly changes with the N radical process for longer than 2 min.

Figure 5 shows the area intensity ratio of the Ge oxide peak in the $Ge2p_{3/2}$ spectra to the Al_2O_3 peak in Al1s spectra as shown in Figure 3 for the sample with the N radical process as a function of TOA. The N radical process time was 10 min. We can find almost no TOA dependence of the area intensity ratio of the AlGeON peak to the Al_2O_3 peak, indicating that Ge atoms are uniformly distributed in the 1st-Al_2O_3 film. This result supports the formation of the AlGeON film after the N radical process to the 1st-Al_2O_3/Ge structure. The atomic density ratio of Ge atoms to Al atoms in the AlGeON film is evaluated to be 16% by the photoelectron intensity ratio of the AlGeON peak in $Ge2p_{3/2}$ spectra to Al1s spectra. On the other hand, the area intensity ratios of the Ge suboxide peaks to the Al_2O_3 peak decreases with TOA. This result indicates that Ge suboxide exists at the AlGeON/Ge interface. The area intensity ratio of the Ge suboxide peaks to the Ge substrate peak corresponds to a thickness of 0.22 nm, which is less than a thickness of 1 monolayer of a quartz-GeO_2 (0.32 nm [16,17]). Here, we assumed the formation of the ideal GeO_2 film as a Ge oxide interlayer. The atomic densities used to evaluate the thickness of Ge suboxide are 2.53×10^{22} and 4.42×10^{22} cm^{-3} for GeO_2 and bulk Ge, respectively [17, 18].

Next, we will discuss the chemical reaction in the film and at the interface with N radical process to the Al_2O_3/Ge structure using Al1s and N1s photoelectron core spectra as shown Figure 3(b) and 3(c). We can find the peak related to the 1st-Al_2O_3 film at a binding energy of 1561.5 eV in the Al1s spectra before the N radical process [7,19]. In the sample after the N radical process, the binding energy of the peak in the Al1s spectra decreases compared with that of the sample before the N radical process. The decrease in the binding energy of the peak indicates the formation of Al-N bonds in the 1st-Al_2O_3 film [20-22]. In the N1s spectra, we can find three peaks at binding energies of 398, 400, and 403 eV for the sample after the N radical process. These peaks relates to the AlON (~400 eV [21]), GeON (at a higher binding energy of AlN (~398 eV [21,22]), or Ge_3N_4 (~397.5 eV [23,24])). Also, the peak at a binding energy of 403 eV is probably attributed to the N-O bond.

Figure 6 shows the TOA dependence of the area intensity ratios of three peaks in N1s spectra to AlGeON peak in Al1s spectra for the sample after the N radical process for 10 min. The area intensity ratio for all peaks in N1s spectra increases with TOA. This result indicates that the N atoms exist

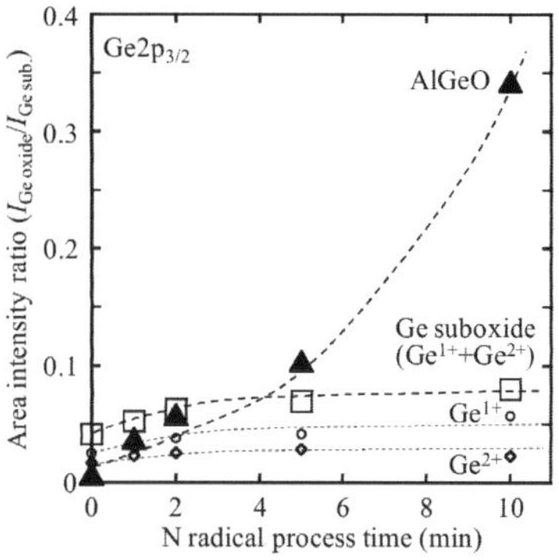

Figure 4 Area intensity ratio of Ge oxide peak to Ge substrate peak in $Ge2p_{3/2}$ photoelectron core spectra of 1st-Al_2O_3/Ge structures as a function of N radical process time measured by HAXPES. TOA is 8°.

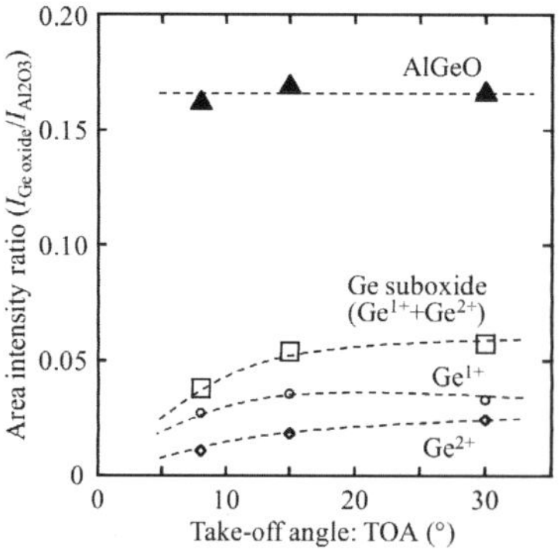

Figure 5 TOA dependence of area intensity ratio of Ge oxide peak in $Ge2p_{3/2}$ photoelectron core spectra to Al_2O_3 peak in Al1s spectra of 1st-Al_2O_3/Ge structures measured by HAXPES. N radical process time is 10 min.

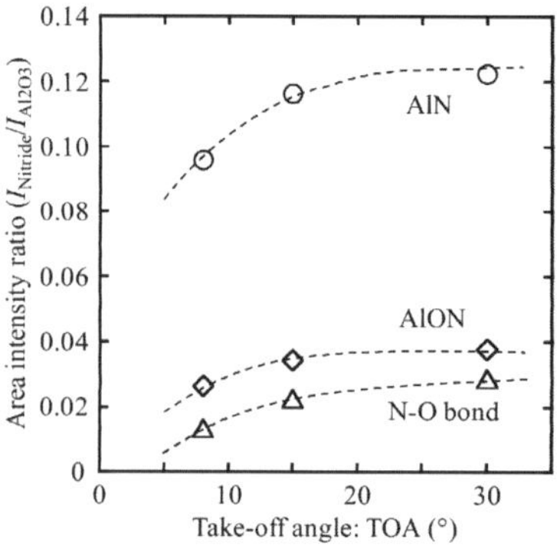

Figure 6 TOA dependence of area intensity ratio of peaks in N1s photoelectron core spectra to Al_2O_3 peak in Al1s spectra of 1st-Al_2O_3/Ge structures measured by HAXPES. N radical process time is 10 min.

Journal of Physics: Conference Series **417** (2013) 012001 doi:10.1088/1742-6596/417/1/012001

mainly near the AlGeON/Ge interface. Here, considering the TOA dependence of the area intensity ratio of Ge oxide peak to Al oxide peak as discussed with Figure 5, Ge atoms in the AlGeON film bind with O or N atoms and uniformly distribute in the AlGeON film. In addition, the composition ratio of Ge in the AlGeON film is not so large compared to that of N. Therefore, it is difficult to explain the depth distribution and composition ratio of N by the formation of Ge nitride and oxynitride in the Al_2O_3 film, and N atoms mainly bonds with Al atoms. At the same time, the N1s spectra shown in Figure 3(c) support the result of the formation of Al-N bonds, and it is considered that AlON is formed in the 1st-Al_2O_3 film. Here, in Figure 3(c), the intensity of the peak at a binding energy of 400 eV decreases with the N radical process time. At the same time, the intensity of the peak at a binding energy of 398 eV increases with the N radical process time. These results indicate that the composition ratio of N in the AlGeON film increases with the N radical process time.

From the above discussion, the interfacial reactions and interfacial structure for the sample after the N radical process are considered as follows. At first, Al-O bonds in the Al_2O_3 film are broken with the N radical irradiation to the Al_2O_3/Ge structure. At the same time, Al-N bonds are formed in the Al_2O_3 films with the N radical process to the Al_2O_3/Ge structure. Following that, O atoms diffuse from the Al_2O_3 film to the Al_2O_3/Ge interface, and the Ge of the substrate at the interface is oxidized. The oxidation of the Ge becomes a trigger for the diffusion of Ge atoms from the Ge substrate to the Al_2O_3 film. As a result, the AlGeON film is formed on the Ge substrate. The Ge suboxide with the thickness of 0.22 nm corresponds to the Ge atoms bonding with O atoms at the AlGeON/Ge interface.

Finally, we investigated the N radical process time dependence on the electrical properties of the Al_2O_3/Ge sample with the N radical process. Figure 7 shows the *C-V* characteristics of MOS capacitors after H_2 annealing. The N radical process is performed for from 1 to 20 min to the 1st-Al_2O_3/Ge structure. Results of the *C-V* measurement indicate the decrease in the interface state density with increasing in the N radical process time as well as result shown in Figure 1. The interface state density for Al/Al_2O_3/Ge MOS capacitors decreases with increasing in the N radical process time. Here, we will show the relationship between the interface state density and the thickness of Ge suboxide at the AlGeON/Ge interface in Figure 8. The interface state density also tends to decrease with increasing in the thickness of the Ge suboxide. This trend implies that we can realize the low interface state density less than 10^{11} $eV^{-1}cm^{-2}$ with the formation of very thin Ge suboxide film less than 1 monolayer. At the same time, we can find the decrease in the hysteresis voltage width in *C-V* characteristics with increasing in the N radical process time as shown in Figure 7. The hysteresis in *C-V* curve suggests that existence of defects near the Al_2O_3/Ge interface and in the Al_2O_3 film. This result indicates that the decrease in the oxide trap density with increasing in the N radical process time, and that is probably due to the incorporation of N or Ge atoms in the Al_2O_3 film.

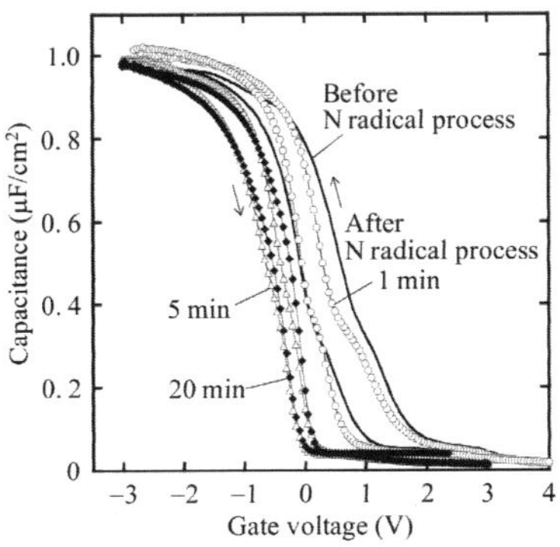

Figure 7 *C-V* characteristics of Al/2nd-Al_2O_3/1st-Al_2O_3/Ge MOS capacitor after N radical process and H_2 annealing.

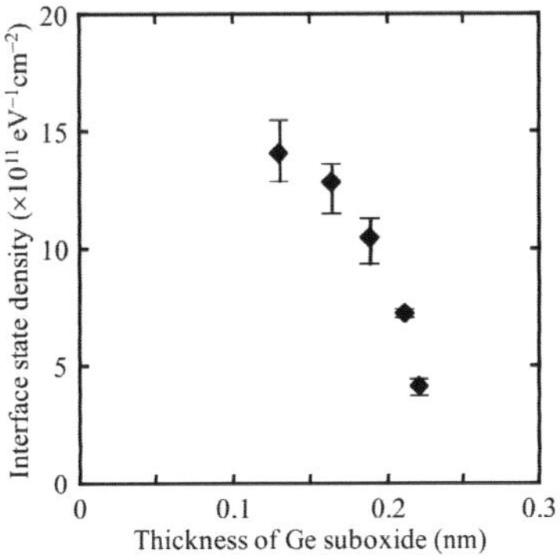

Figure 8 Relationship between interface state density and thickness of Ge suboxide for samples Al/2nd-Al_2O_3/1st-Al_2O_3/Ge MOS capacitor after N radical process and H_2 annealing.

4. Conclusions

We have investigated the interfacial reactions of the Al_2O_3/Ge gate stack structure with N radical process, and the effect of the chemical bonding states on electrical properties. In the AR-HAXPES results, we found that the uniform distribution of Ge atoms in the Al_2O_3 film and the formation of the AlGeON films. Also, the intensity ratio of the Ge suboxide peak to the Ge substrate peak hardly changes after the N radical process for more than 2 min, and it corresponds to the Ge suboxide thickness of 0.22 nm for the sample after the N radical process for 10 min. We can observe the steep increase in the capacitance from inversion to accumulation conditions and the decrease in the hysteresis voltage width in the *C-V* characteristics of the Al/Al_2O_3/Ge MOS capacitors. The interface state density decreases with increasing in the N radical process time. Also, that tends to decrease with increasing in the thickness of Ge suboxide. The interface state density of the sample after the N radical process for more than 10 min is as low as 4×10^{11} $eV^{-1}cm^{-2}$. The relationship between the interface state density and the thickness of Ge suboxide implies that we can realize the interface state density less than 10^{11} $eV^{-1}cm^{-2}$ with the very thin Ge suboxide less than 1 monolayer. The oxide trap density in the oxide film also decreases with N radical process to Al_2O_3/Ge structure. Therefore, it is found that N radical process is an attractive process to control of the amount of O atoms from an Al_2O_3 film to the Al_2O_3/Ge interface. We can achieve the decrease in both the interface state density and oxide trap density with the formation of AlGeON/Ge suboxide/Ge structure.

Acknowledgements

This work was partly supported by a Grant-in-Aid for Specially Promoted Research (No. 22000011) from the Ministry of Education, Culture, Sports, Science and Technology of Japan. The hard X-ray photoelectron spectroscopy experiments were performed at the BL46XU in SPring-8 with the approval of the Japan Synchrotron Radiation Research Institute (JASRI, Proposal No. 2010B0026).

References

[1] Matsubara H, Sasada T, Takenaka M and Takagi S 2008 *Appl. Phys. Lett.* **93** 032104-1-3.
[2] Tabata T, Lee C H, Kita K and Toriumi A 2008 *ECS Trans.* **16** 479-86.
[3] Hosoi T, Kutsuki K, Okamoto G, Saito M, Shimura T and Watanabe H 2009 *Appl. Phys. Lett.* **94** 202112-1-3.
[4] Kobayashi M, Thareja G, Ishibashi M, Sun Y, Griffin P, McVittie J, Pianetta P, Saraswat K and Nishi Y 2009 *J. Appl. Phys.* **106** 104117-1-7.
[5] Kita K, Suzuki S, Nomura H, Takahashi T, Nishimura T and Toriumi A 2008 *Jpn. J. Appl. Phys.* **47** 2349-53.
[6] Prabhakaran K and Ogino T 1995 *Surf. Sci.* **325** 263-71.
[7] Kato R, Kyogoku S, Sakashita M, Kondo H and Zaima S 2009 *Jpn. J. Appl. Phys.* **48** 05DA04-1-4.
[8] Wilk G D, Wallace R M and Anthony J M 2001 *J. Appl. Phys* **89** 5243-75.
[9] Shibayama S, Kato K, Sakashita M, Takeuchi W, Nakatsuka O and Zaima S 2011 *Thin Solid Films, in press.*
[10] Zhang R, Iwasaki T, Taoka N, Takenaka M and Takagi S 2011 *Appl. Phys. Lett.* **98** 112902-1-3.
[11] Kato K, Kyogoku S, Sakashita M, Takeuchi S, Kondo H, Takeuchi S, Nakatsuka O and Zaima S 2011 *Jpn. J. Appl. Phys.* **50** 10PE02-1-7.
[12] Takahashi R, Sakashita M, Sakai A, Zaima S and Yasuda Y 2004 *Jpn. J. Appl. Phys.* **43** 7821-5.
[13] Fukuda Y, Otani Y, Itayama Y and Ono T 2007 *IEEE Trans. Electron Dev.* **54** 2878-83.
[14] Merckling C, Penaud J, Kohen D, Bellenger F, Alian A, Brammertz G, El-Kazzi M, Houssa M, Dekoster J, Caymax M, Meuris M and Heyns M M 2009 *Microelectron. Eng.* **86** 1592-5.
[15] Bellenger F, De Jaeger B, Merckling C, Houssa M, Penaud J, Nyns L, Vrancken E, Caymax M, Meuris M, Hoffmann T, De Meyer K and Heyns M 2010 *IEEE Electron Dev. Lett.* 31 402-4.
[16] Seifert K J, Howotny H and Hauser E 1971 *Monatshefte für Chemie* **102** 1006-9.
[17] Yamanaka T 1991 *J. Appl. Crystallography* **24** 111-8.
[18] Sze S M 1981 *Physics of Semiconductor Devices* (Hoboken NJ: J. Wiley and Sons, Inc.).
[19] Tsuchiya Y, Yoshiki M, Koga J, Nishiyama A, Koyama M, Ogawa M and Zaima S 2008 *IEEE Trans. Electron Dev.* **55** 2648-56.
[20] Schosera S, Bräuchlea G, Forgeta J, Kohlhofa K, Webera T, Voigta J and Rauschenbachb B 1998 *Surf. Coatings Technol.* **103-104** 222-6.
[21] Sohal R, Torche M, Henkel K, Hoffmann P, Tallarida M, and Schmeißer D 2006 *Mater. Sci. Semicond. Process.* **9** 945-8.
[22] Kum D and Byun D 1997 *J. Electron. Mater.* **26** 1098-102.
[23] Maeda T, Yasuda T, Nishizawa M, Miyata N and Morita Y 2006 *J. Appl. Phys* **100** 014101-1-7.
[24] Kato K, Kondo J, Sakashita M and Zaima S 2010 *Thin Solid Films* **518** S226-30.

Journal of Physics: Conference Series **417** (2013) 012002 doi:10.1088/1742-6596/417/1/012002

Fabrication of Oriented Thin Films Composed of Polyfluorene and Oligothiophene, and Application for Polarized White Light Emitting Devices

C Heck, T Mizokuro, and N Tanigaki*

Research Institute for Ubiquitous Energy Devices, National Institute of Advanced Industrial Science and Technology AIST, Ikeda, Osaka 563-8577, Japan

*E-mail: no.tanigaki@aist.go.jp

Abstract. We have been studying the production of oriented films of various conjugated polymers by means of the friction transfer technique. The method has been applied to produce oriented poly(9,9-dioctylfluorene) (PFO) films which were then used to produce devices that emit polarized blue light. We fabricate oriented films composed of PFO and sexithiophene (6T), which is an orange emitting dye, for production of white polarized light emitting devices. We make a comparative study of the characteristics of devices produced with different doping techniques and discuss the advantages of each method.

1. Introduction

Light emitting organic materials based on oriented layered structures have been studied for their applications in electroluminescent devices that emit polarized light. There are various possible applications of these polarized light emitting devices in the field of optoelectronics like, for example, as a backlight source of liquid crystal displays or as flat-area light sources that can be produced with low cost. Various techniques, such as blending of polymers of copolymerization[1-3], have been employed for the whitening of the emission of devices based on oriented polymeric layers but most of the successful techniques employ rather complex compounds.

We have been producing oriented polymer films [4,5], by means of the friction transfer technique [6], for the fabrication of polarized electroluminescence (EL) devices. Oriented poly(9,9-dioctylfluorene) (PFO) films were produced with this technique to fabricate polarized blue light emitting EL devices. PFO is a simple structured conjugated polymer that, besides showing efficient blue electroluminescence [4, 5, and 7] and good charge transport, is also a material with easy processability. We have been studying the application of these oriented PFO films combined with some processes that allow the change of the colour of the device in a controlled manner, with the ultimate purpose of producing devices that emit white light [8, 9].

One of the techniques for the whitening of the emitted light is doping α-sexithiophene (6T) molecules into the oriented PFO layers by means of the vapour transportation method [9, 10]. This method is suitable for oriented layers because it allows doping after film production without loss of molecular orientation. With this method the whitening of the emission of a device based on 6T-doped oriented PFO films [9] was possible, but intrinsically it does not allow a good control of the amount of dye doped into the matrix film.

Another method employed to produce white emitting devices is the deposition of 6T molecules, directly on top of the oriented PFO films, before device production. With this method we tried to better understand the light emission mechanisms of the devices by analysing the effect of the amount and morphology of the dye at the interface region. Devices' colour change and light polarization were analyzed and related to 6T film thickness and morphology.

The photoelectrical characteristics of devices produced with each doping method, i.e., vapour transportation and vacuum deposition, are also analyzed and discussed.

2. Experimental

Oriented PFO films were produced on clean substrates by means of the friction transfer method. This method consists of sliding the PFO polymer block directly onto the substrate by controlling the temperature (95°C) of the substrate, and the sliding speed (0.5 m/min) and the load (2.45 MPa) of the block onto the substrate. With this process the PFO molecules naturally orient themselves in the direction of the sliding movement, but further thermal treatment is performed (at 180°C, in vacuum, during 1 h) to perfect orientation and surface smoothness [6]. The resulting PFO film is about 50 nm thick.

To change the colour of the blue emitting PFO, two methods were employed for comparison. With one method 6T was doped into the oriented PFO films and with the other one 6T was deposited onto the oriented PFO films, as described in the following. For doping 6T into the PFO film the vapour transportation method was employed. With this method the dye and the oriented PFO film are loaded together, without contact, into a glass tube. The whole system is evacuated and then the glass tube is closed/sealed forming an ampoule. Finally this sealed system is heated in an oven at 180°C for about 13 hours, causing the 6T dye to sublimed and diffuse into the PFO film. The other method, for deposition of 6T onto the PFO films, is the conventional vacuum deposition technique, where deposition is performed at room temperature in a vacuum better than $2x10^{-4}$ Pa and 6T deposition rate is about 0.02 nm/s. Films of 6T with various thicknesses, between 0.5 and 5.0 nm, were prepared. For morphological comparison, 6T was also deposited on silicon substrates, which native oxide layer is left on top, degreased with acetone prior to deposition.

EL devices were produced in the following way. Indium thin oxide (ITO) coated glasses were used as substrates, which were initially cleaned in ultrasonic baths of ethanol and acetone, prior to a 30 minutes oxygen plasma treatment. To improve the efficiency of the device a poly(3,4-ethylenedioxythiophene):poly(styrene sulfonate) (PEDOT:PSS) hole injection layer was spin coated on top of the clean ITO substrate. The PFO layer was then produced, by the friction transfer method, directly on this PEDOT:PSS/ITO substrate. For the case of 6T doped PFO layers, they were inserted into a vacuum chamber and a 50-nm-thick 2,2',2"-(1,3,5-benzinetryl)-tris(1-phenyl-1-*H*-benzimidazole) (TPBi), which is an electron transport and hole-blocking layer, was evaporated on top at a vacuum better than $2x10^{-4}$ Pa. Finally, a 90-nm-thick magnesium-silver (Mg:Ag, 10:1) cathode was deposited. For the oriented PFO layers were 6T was to be deposited, the PFO layers were inserted into the chamber and 6T deposition was performed prior to the deposition, but without removing from the chamber, of TPBi and the Mg:Ag layers. The active area of the devices was 4 mm^2.

The PFO used in this work had molecular weight ranging from 20000 to 120000 and was purchased from American Dye Source, catalogue number ADS 129BE. A sublimated grade of 6T was purchased from Tokyo Chemical Industry, and used without further purification. The PEDOT:PSS (CLEVIOS P VP.AI4083) was purchased from H. C. Starck and TPBi from Luminescence Technology.

For the polarized electroluminescence (EL) measurements the light is collected with a glass fiber, dispersed in a spectrograph and detected with a charge coupled device camera (BWTEK BTC112). The Commission Internationale d'Énclairage (CIE) chromaticity coordinates and the luminance are measured by a luminance-colorimeter (Konica-Minolta CS-100A). Atomic Force Microscopy (AFM) measurements are performed by means of a commercial Seiko Instrument SPI-3700. All measurements were performed at room temperature, without sample encapsulation.

3. Doping of 6T into PFO by vapour transportation

The vapour transportation method was employed to dope 6T into oriented PFO films. These oriented PFO/6T doped films were then used for the production of EL devices. Figure 1 (a) shows the polarized EL spectra of the devices where it can be seen that, for both PFO (peaks around 430 and 450 nm) and 6T (broad peak between 510 and 550 nm), the electroluminescence with parallel polarization to the friction direction is much stronger than that with orthogonal polarization. The polarization of PFO originates from the uniaxially oriented PFO main chain along the friction direction. Considering that the peak from 6T also exhibits high polarization we suggest that the 6T molecules are also aligned along the polymer chain direction in the oriented film.

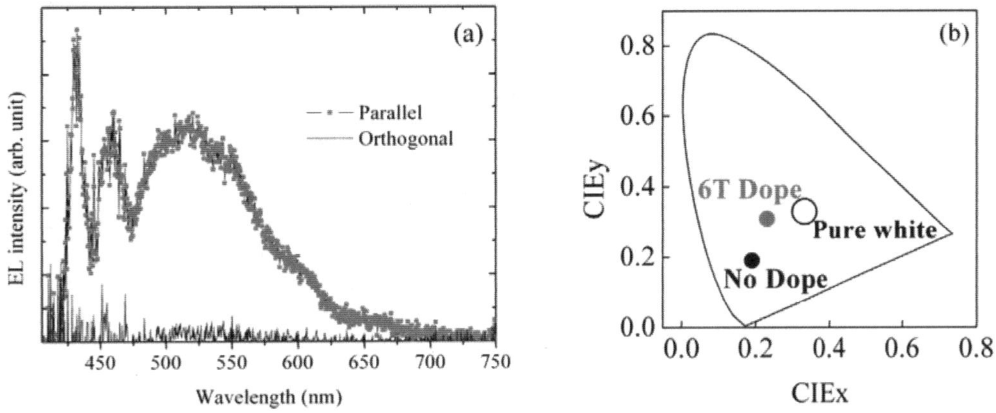

Figure 1. (a) Polarized EL spectra of the 6T-doped PFO oriented film. The EL device was based on PFO film deposited directly on ITO substrates, without the intermediate PEDOT:PSS film. (b) CIE chromaticity diagram of EL devices based on PFO films with and without 6T-doping.

The degree of orientation was evaluated semi quantitatively using the polarized EL spectra. The orientation of the entire PFO plus 6T structure was evaluated by calculating the dichroic ratio, *i.e.*, the parallel component to the orthogonal component ($EL_{\parallel}/EL_{\perp}$), of the integrated intensities for polarized EL spectra. The average dichroic ratio results were about 25 for both doped and non doped samples. The EL results for non doped samples show that there is a large dichroism due to an excellent uniaxial alignment of the polymer backbone. On the other hand, the results for doped samples show the same or better dichroism, compared to the non doped samples, which indicates that there is also an excellent alignment of the dye molecules parallel to the polymer orientation direction. CIE chromaticity diagram coordinates show the whitening of the device, as can be seen in figure 1 (b), where the blue PFO (0.19, 0.19) coordinates at 11 V bias voltage changed to (0.23, 0.31) for bias voltage 11 V after doping with 6T. These CIE coordinates are very stable with applied voltage up to the break down voltage. These results indicate that by doping the 6T (orange emitting material) into PFO (blue emitting material) the whitening of the EL device is possible with excellent polarization of the emitted light.

To improve the efficiency of the device a PEDOT:PSS layer is inserted between the ITO substrate and the PFO film. The properties of the EL device were then analyzed. For the 6T doped device luminance reached 3800 cd/m^2 at 10 V and the device without 6T doping reached ~2900 cd/m^2 at 9 V. Turn on voltage (the bias required to give a luminance of 1 cd/m^2) of the device is about 4 V for both doped and non doped devices. Efficiency results show that the doping of 6T leads to about 65% increase in efficiency of the device (from 1.48 to 2.28 cd/A). CIE chromaticity diagram coordinates are (0.19, 0.18) for PFO at 11 V bias voltage and change to (0.22, 0.24) for bias voltage 11 V after doping with 6T. These CIE coordinates are very stable with applied voltage up to the break down voltage. Dichroic ratio was about 16 and 13 for non-doped and doped EL devices, respectively. We observe the reduction of dichroic ratio and of the whitening, probably due to the difference in film thickness when PEDOT:PSS is included, but considering that there is a good improvement of efficiency and luminance, we are continuing the research for optimization of these of PEDOT:PSS based devices.

4. Deposition of 6T on PFO

For a better understanding of the light emission mechanisms of the 6T-oriented PFO based devices, we analyze the influence of the amount of 6T at the interface region, by producing vacuum deposited 6T layers on top of oriented PFO films before device production. Figure 2 (a) shows the normalized EL spectra of these devices with 6T thickness being varied from 2.0 to 4.0 nm (for 5.0 nm see the inset of figure 2(a)). The peaks derived from PFO can be seen at 430 and 450 nm while the broad region seen between 500 and 650 nm derives from 6T. The relative intensity of this region derived from 6T is seen to increases with 6T film thickness. CIE chromaticity diagram coordinates from these devices with 6T thickness between 0.5 and 5.0 nm plus the one for the device without 6T are shown in figure 2(b). A linear variation of the devices' coordinates with the thickness of the deposited 6T film can be seen. Emitted colour changes from blue for the device without 6T, (0.19, 0.19) coordinates at 11 V bias voltage, to white for the device with 6T around 4.0 nm thick, (0.30, 0.32) coordinates at 10 V bias voltage. These coordinates are stable around the maximum luminance voltage.

These results show that, by varying the thickness of deposited 6T film onto a PFO layer, an excellent control of the emitted colour is possible. A white emitting device can be produced by depositing about 4.0 nm thick 6T (orange emitting) onto the oriented PFO (blue emitting) layer.

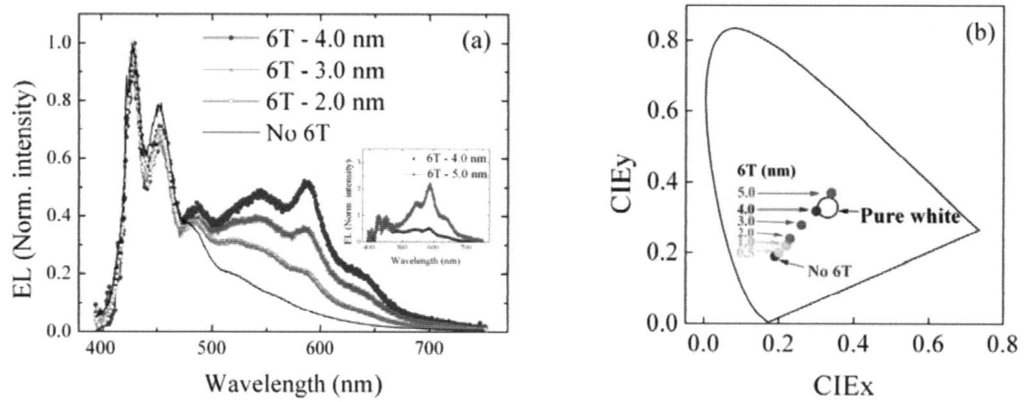

Figure 2. (a) EL spectra of devices based on pure PFO layers and on PFO with deposited 6T films of thickness from 2.0 to 4.0 nm (and of 5.0 nm shown in the inset). (b) CIE chromaticity diagram for these devices plus for those for 0.5 and 1.0 nm thick 6T.

The EL spectra, with polarization parallel and orthogonal to the friction direction, of the device produced with 4.0 nm thick 6T on the oriented PFO layer are shown, as a typical example, in figure 3 (a). One can see that the emission from PFO, at 430 and 450 nm, with parallel polarization to the friction direction is much stronger than that with orthogonal polarization. This polarization originates from the uniaxially oriented PFO main chain along the friction direction. In a similar way, the emission region derived from 6T (between 500 and 650 nm) also exhibits high polarization, indicating that the 6T molecules might be aligned along the polymer chain direction in the oriented film. The degree of this orientation was evaluated by calculating the dichroic ratio ($EL_{\parallel}/EL_{\perp}$), in the same way as described in the previous section. Average dichroic ratio results were around 25 for pure PFO samples (without 6T deposition), and varied from about 8 to 12 (see figure 3(b)) for 6T deposited samples. These results are interpreted as follows. For devices based on oriented pure PFO layers (without 6T), there is a large dichroism due to an excellent uniaxial alignment of the PFO polymer backbones. On the other hand, 6T deposited samples show a comparatively smaller but still large dichroism, which suggests some level of alignment of the dye molecules parallel to the polymer orientation direction. Furthermore, despite the fact that the emitted colour changes with 6T thickness, the dichroic ratio does not change significantly, which indicates that the orientation of 6T is not significantly affected by film growth, for the analyzed thickness range of this study.

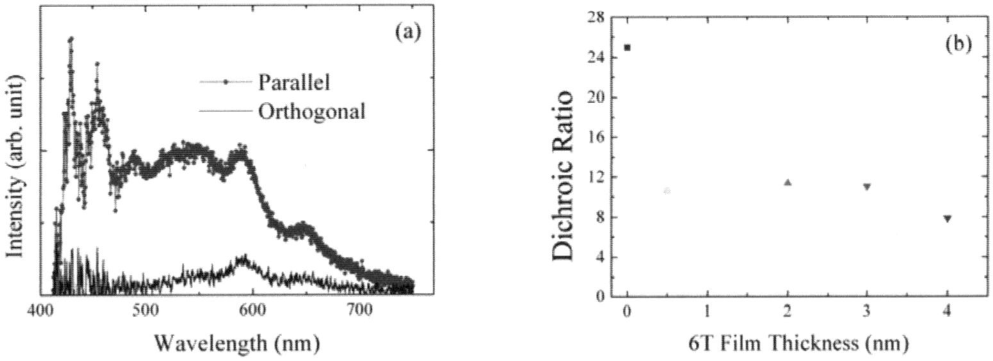

Figure 3. (a) Polarized spectra of a white EL device based on a 4.0 nm thick 6T layer deposited on top of and oriented PFO film. (b) Dichroic ratio determined from the polarized EL spectra of PFO and 6T deposited /oriented PFO based devices.

5. Morphology studies

Morphology studies of 6T/PFO films prepared with both vapour transportation and vacuum deposition methods were performed with AFM.

For PFO films doped with 6T by vapour transportation, no significant surface structure could be observed, and morphology was very similar to the oriented PFO film without 6T (data not shown). This is probably due to the fact that 6T molecules are evenly distributed inside the film and that the molecules that rest on the surface do not form aggregates large enough to be clearly observed.

For the case of vacuum deposition a comparative analysis has been performed by depositing 6T films on the oriented PFO and on a Si substrate, whose native oxide layer was kept on top. Figure 4 shows that 30 nm thick 6T films deposited on these substrates form very different structures, namely, drop-like islands are formed on oriented PFO layers while flat-plateaux like islands grow on top of Si. Figure 4(a) shows that on Si the flat islands are about 200 nm wide and 5.0 nm high. This height corresponds to approximately the 6T unit cell parameter a ($a = 4.4708$ nm) [11] indicating that the 6T molecules are aligned with their molecular axis perpendicular or almost perpendicular to the Si substrate. This result is in accordance with previously reported studies for 6T grown on top of flat surfaces [12, 13]. On the other hand, on top of the oriented PFO layer, figure 4 (b), 6T grows forming drop-shaped islands, about 100 nm wide and between 20 and 30 nm high, while flat regions are not noticeable. We suggest that these islands are composed of 6T molecules piled up in a way that their long axis is parallel or almost parallel to the friction direction, *i.e.*, to the orientation direction of the underneath PFO molecules which are parallel to the substrate's surface. These islands were observed to grow in width and height with the increase of 6T films thickness (not shown here). Similar results have been reported for oriented polytetrafluoroethylene (PTFE) [14] and TiO_2 [15] substrates, where 6T molecules were also observed to orient parallel to the orientation of these substrates. These results suggest that for 6T deposited thin films, substrate orientation acts as a driving force to preferentially orient the deposited 6T molecules with their long axis parallel to the underneath molecular orientation. This assumption of molecular orientation between 6T and PFO is also consistent with the present results that show that the EL devices produced with 6T on oriented PFO films produce polarized light, from both PFO and 6T emissions, with high dichroic ratio(figure 3(a)).

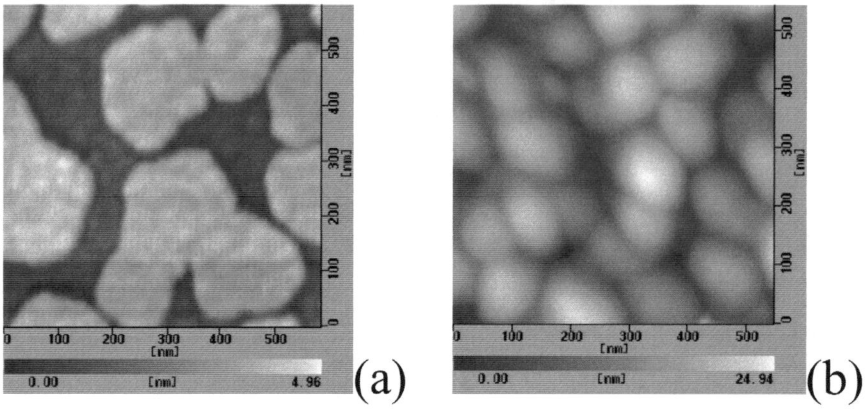

Figure 4. AFM images of 6T films (30 nm thick) deposited on (a) Si substrate (with native oxide kept on top) and on (b) oriented PFO substrate.

6. Discussion

For devices produced with 6T deposited/PFO the excellent control of the emission colour by changing the 6T film thickness without loosing molecular orientation makes it a promising candidate for application. An interesting feature is the fact that by increasing 6T film thickness, the current density does not change significantly while luminance decreases. Another peculiar result is that the efficiency increases slightly after deposition of 0.5 nm thick 6T (figure 5), compared to the pure PFO based device, and then decreases with further increase of 6T thickness. The mechanism of the increase of efficiency at 0.5 nm 6T deposition is not yet well understood.

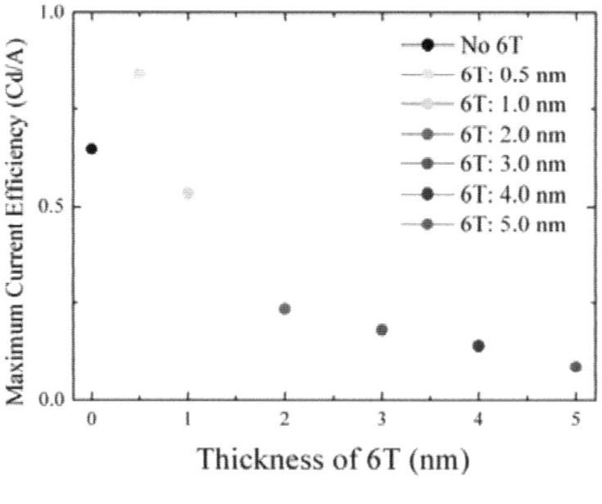

Figure 5. Variation of the maximum current efficiency with deposited 6T film thickness. The highest efficiency is observed for devices produced with a 0.5 nm thick 6T layer deposited on top of oriented PFO.

We compared the photoelectric results of these devices with those for the devices we produced by doping 6T into the oriented PFO. Results show that for the best colour change result for each production method, *i.e.*, the whitest CIE chromaticity diagram coordinates for devices produced with 6T deposition (CIE (0.30, 0.32)), the maximum luminance was about 380 Cd/m^2 and efficiency 0.1 Cd/A. This is much lower than the results for the 6T doped devices (CIE (0.22, 0.24)) where maximum luminance was 3850 Cd/m^2 and efficiency 0.75 Cd/A, despite a lack of significant colour change in the doped devices because of the inclusion of a PEDOT:PSS layer. This difference in luminance and efficiency can be explained as follows. Considering that for the deposited samples, the 6T layer is oriented on top of the PFO surface and might act as a quenching interface layer for the emitted light, as suggested by the decrease of luminance despite of constant current density with increase of 6T thickness results. On the other hand, vapour transportation doping leads to a homogenously distributed 6T into the PFO film, and because no quenching layer is formed on top of PFO the device shows a better performance. These results show that while the deposition process allows an excellent control of the colour as well as a better understanding of the emission mechanism of 6T/oriented PFO devices, the doped 6T/oriented PFO devices, despite of a lack of control of doping amount and consequently of the colour of the device, show better photoelectrical characteristics due to the oriented and homogenously distributed 6T in the oriented PFO matrix. Studies are under way to combine both methods, *i.e.*, deposition of 6T on top of a 6T doped PFO layer, to improve the whitening of the highly efficient doped devices.

7. Conclusions

Whitening of the emission color of EL devices was performed by depositing about 4.0 nm thick orange emitting 6T onto blue emitting oriented PFO films. The orientation of deposited 6T molecules parallel to the oriented PFO molecules leads to very good polarization of the light emitted by these EL devices. Excellent control of the color of the emitted light is achieved by controlling the thickness of the 6T film. On the other hand, high luminance and efficiency of devices prepared by doping 6T into oriented PFO layers by means of the vapor transportation process was observed, despite of a lack of control of the color of the device.

References

[1] Chen A C A, Culligan S W, Geng Y, Chen S H, Klubek K P, Vaeth K M and Tang C W 2004 *Adv. Mater.* **16** 783
[2] Yao Y-H, Kung L-R and Hsu C-S *Jap. J. Appl. Phys.* 2005 **44** 7648
[3] Zhu D X, Zhen H Y, Ye H and Liu X 2008 *Appl. Phys. Lett.* **93** 163309
[4] Misaki M, Ueda Y, Nagamatsu S, Chikamatsu M, Yoshida Y, Tanigaki N and Yase K 2005 *Appl. Phys. Lett.* **87** 243503
[5] Misaki M, Chikamatsu M, Yoshida Y, Azumi R, Tanigaki N, Yase K, Nagamatsu S and Ueda Y 2008 *Appl. Phys. Lett.* **93** 023304
[6] Misaki M, Ueda Y, Nagamatsu S, Yoshida Y, Tanigaki N and Yase K 2004 *Macromolecules* **37** 6926
[7] Grell M, Knoll W, Lupo D, Meisel A, Miteva T, Neher D, Nothofer H-G, Scherf U and Yasuda A 1999 *Adv. Mater.* **11** 671
[8] Mizokuro T, Heck C, Tanigaki N, Hiraga T and Tanaka N 2008 *Applied Physics Express* **1** 021804
[9] Heck C, Mizokuro T, Misaki M, Azumi R and Tanigaki N 2011 *Jpn. J. Appl. Phys.* **50** 04DK20
[10] Tanigaki N, Heck C, Mizokuro T, Minato H, Misaki M, Yoshida Y and Azumi R 2008 *Jpn. J. Appl. Phys.* **47** 416
[11] Servet B, Horowitz G, Reis S, Lagorsse O, Alnot P, Yassar A, Deloffre F, Srivastava P, Hajlaoui R, Lang P and Garnier F 1994 *Chem. Mater.* **6** 1809
[12] Okajima T, Narioka S, Tanimura S, Hamano K, Kurata T, Uehara Y, Araki T, Ishii H, Ouchi Y, Seki K, Nakahara H, Ogama T and Koezuka H 1996 *Jpn. J. Appl. Phys.* **35** 2822
[13] Granstrom E L and Frisbie D 1999 *J. Phys. Chem.* B **103** 8842
[14] Wittmann J C, Straupe C, Meyer S, Lotz B, Lang P, Horowitz G and Garnier F 1998 *Thin Solid Films* **333** 272
[15] Ivanco J, Haber T, Krenn J R, Netzer F P, Resel R and Ramsey M G 2007 *Surface Science* **601** 178

Journal of Physics: Conference Series **417** (2013) 012003 doi:10.1088/1742-6596/417/1/012003

Investigation of Mn adatoms adsorption on graphene using first-principles calculation

K Nakada[12], **K Mishima**[1], **A Ishii**[12]

[1]Department of Applied Mathematics and Physics, Tottori University
[2]JST-CREST, 5 Sanbancho, Chiyoda-ku, Tokyo 102-0075, Japan

E-mail: nakada@damp.tottori-u.ac.jp

Abstract. The electronic state of the magnetism in a coverage change by multiple Mn atoms and various layout was calculated. When we adsorbed two Mn atoms on graphene, a structure to locate vertical alignment is the most stable. Because difference produces for the large charge transfer between two Mn, large difference of the magnetic moment is created.

1. Introduction

Graphene is a flat monolayer of carbon atoms tightly packed into a two-dimensional honeycomb lattice, and is a basic building block for graphitic materials of all other dimensionalities [1]. It can be wrapped up into fullerenes C60, rolled into carbon nanotubes or stacked into graphite. The surface layer of graphite can be considered as the most popular graphene. Nowadays, many researchers are attempting to grow graphene on other surfaces, such as SiC(0001). However, after success in the development of technologies to grow graphene on the surface, the next required technology is to grow the same nano structures on the graphene on the substrate. Thus, graphene will be the focus of attention as a substrate for new nano structure devices using its two-dimensional property. For the functionalization of graphene or the construction of nano structures on graphene, it is necessary to investigate the bonding of materials on graphene. For example, as an important application of the construction of a device on graphene, the growth of GaN thin film on graphite using pulsed laser deposition has been achieved [2, 3, 4, 5, 6]. The structure of the grown GaN on graphite is calculated using the density functional eheory[8, 9]. In a theoretical study[7], the adsorption sites and adsorption energies for some atomic species have been reported. However, in general, the mechanism of the growth of three-dimensional materials or nano structures on the two-dimensional material is not yet understood. For the first step, it is necessary to investigate the adsorption of atoms on graphene. In a recent theoretical study[10, 11], we investigate computationally the adsorption energies and migration energies in adsorption sites on graphene sheet for many atomic species, including transition metals, noble metals, nitrogen and oxygen, from atomic number 1 to 83, using the DFT (Density Functional Theory) calculation. We discuss the adsorption and migration energy when an adatom is adsorbed onto graphene. There has been discussion about the adsorption energy in many previous studies[12, 13, 14], if we design a semiconductor device on graphene or a nano device using graphene, the adsorption and the migration energy from an adatom adsorbed onto the graphene are very important. For example, the issue of a metalgraphene junction is an important problem in device design. Therefore, we investigate the stabilities and electronic structures of adatoms of almost all atomic species to understand the three-dimensional material on graphene, which is itself a typical two-dimensional material. In this study, we perform multiple atomic adsorptions on the graphene. As for the next discussion of the single atom adsorption, it should be discussed by the adsorption of multiple atoms. We chose a Mn atoms as adatom in this sresearch. Because d-orbital is half occupied for the Mn atom, various magnetic order is expected. The electronic state of the magnetism in a coverage change by multiple Mn atom and various layout was calculated.

2. Calculation Method

In this work, we used a first-principles band calculation technique based on density functional theory. We used VASP (The Vienna Ab initio Simulation Package) [15, 16, 17, 18], which is a first-principles calculation code with high precision using the PAW (Projector Augmented Wave) method. We adopted LDA (Local Density Approximation) as the term exchange correlation with a cutoff energy of 500 eV and all calculations performed nonmagnetically. The unit cell for the graphene sheet adopted a 3 $\times$ 3 structure. The lattice constant of the graphene used the value optimized by calculation. The distance between graphene sheets is about 14.7 $\mathring{A}$ and the distance between adatoms is about 7.3 $\mathring{A}$. The final potential is constructed self-consistently from eigenstates at 24 sampling k-points in the irreducible Brillouin zone (IBZ). To obtain a final band structure, we selected 68 sampling k-points in the IBZ. Two Mn atoms are adsorbed to the 3$\times$3 graphene. For the calculation of adatoms at certain sites, the coordinate normal to the surface is fully relaxed. One atom of the edge of the 3$\times$3 structure of the graphene sheet is fixed during the relaxation of the other carbon atoms of the sheet. The calculation was carried out at many configuration sites. As a result of relaxation calculation, the construction converges in seven pattern shown in Fig.1,2.

3. Result and Discussion

The result of the monoatom adsorption is shown by previous study[10, 11]. When we adsorbed Mn atom on the 3$\times$3 graphene, H-site is more stable and migration energy is 0.14eV. In this study, we adsorbed multiple Mn atom to the 3$\times$3 graphene. We assumed that two Mn atoms has moment of the same direction for DFT calculation. In other words, this would assume ferromagnetism. Fig.1 is local stable structure when two Mn atoms adsorbed it on 3$\times$3 graphene. Fig.1(a) and fig.1(b) are placed in the hexagonal Hollow site (H-site). Fig.1(c) shows that two Mn atom is adsorbed in a six-membered ring. In fig.1(d), two Mn atom is adsorbed each on the direct top of neighboring C atom. As for the configuration of fig.1, two Mn atom is adsorbed each on H-site and the T-site. Fig.1(f) shows that two Mn atom is adsorbed each in H-site and the B-site. Fig.2 shows configuration to locate vertical alignment on H-site. As a result of relaxation calculation, the construction converges in seven pattern shown in Fig.1,2. The configuration that two Mn formed vertical alignment becomes most stable. The energy and the local magnetic moment in the local stable structure were shown in table 1. We show the Δ energy in difference from most stable structure (Fig.2). The d in Table1 shows distance between graphene and the adatom. The m is magnetic moment per Mn atom. The most stability is a fig.2, and configuration most stable next is fig.1(c). Furthermore, the next stable configuration is fig.1(a) and fig.1(d). These stable structure has the special feature that an adjacent distance of two Mn atom is close in. In other words, the dimer structure is stabilized. The fig.1(a) structure of the dimer configuration is slightly unstable to insert C-C atom bond between Mn-Mn dimer. When we located two Mn atom, there are some structures that got different magnetic moment in two Mn atom. In fig.1(e) and fig.1(f) and fig.2 structures, two Mn atoms get different moment. As for the other configuration, two Mn atoms have the same magnetic moment. Firstly we discuss on fig.1(e) and fig.1(f) structure. Atomic magnetic moments of the H-site adsorption side decrease in these structures. It is discussed by Mn monoatom adsorption in the graphene.

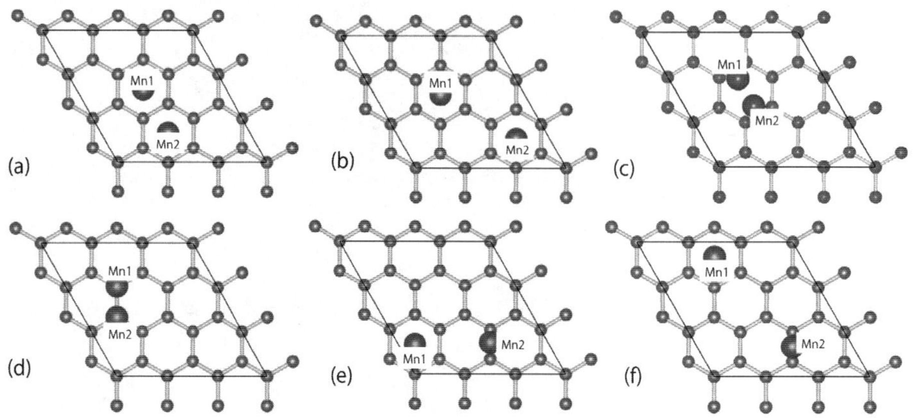

Figure 1. Local stable structure when two Mn atom adsorbed to the 3×3 graphenet

configuration	ΔE (eV)	d_{Mn1} (Å)	d_{Mn2} (Å)	m_{Mn1} (μ_B)	m_{Mn2} (μ_B)
fig.1(a)	0.81	1.74	1.74	3.51	3.52
fig.1(b)	1.91	1.64	1.64	3.56	3.56
fig.1(c)	0.25	2.03	2.03	0.55	0.55
fig.1(d)	0.72	1.92	1.91	0.63	0.64
fig.1(e)	2.54	1.51	2.12	0.25	4.16
fig.1(f)	1.48	1.57	2.36	1.33	3.93
fig.2	0.00	1.67	3.23	1.65	-0.04

Table 1. Calculation result in the local stable structure.: configuration:The stable coordinate that is shown in fig.1,2. ΔE:Energy difference on the basis of the most stable. d: distance between graphene and the adatom. m: magnetic momentt.

A spin magnetic moment of two Mn atom shows large polarization[11]. The magnetic moment of the Mn is made by remaining d-electron after charge transfer occurred from Mn atom to graphene. When the Mn atom adsorbs in the adsorption site except most stable H-site, the magnetic produces in the Mn atom. In this calculation result, tendency of the magnetic moment in the H,B and T-site adsorption does not have the change either. However, unlike monoatom adsorption, bond distance with the graphene is opened by T-site and the B-site adsorption with H-site adsorption. Therefore, electrons of Mn atom bonding to graphene decrease, and electrons to take magnetic of the Mn atom on increase. The magnetic configuration in fig.2 of the most stable structure is characteristic. When we adsorbed two Mn atoms on graphene, a structure to locate vertical alignment is the most stable. Magnetic moment of the Mn1 of the graphene side is 1.65 μ_B, and the Mn2 of the Mn1 direct top is -0.04 μ_B. Fig.2(c) shows a spin map at this configuration. A spin map is a real space map of magnetization density. As for the yellow in fig.2(c), magnetization density shows a positive region. The negative region is light blue. An interesting magnetic structure is shown in fig.2. Furthermore, we analyze it about charge transfer at this configuration. The charge density analysis checked integral value in the cut-off radius 1.323 Å in the Mn atom. We performed integral in the cut-off radius 0.863Å about the C atom. With the Mn of the graphene side, charge is transferred from Mn1 atom to the graphene side widely then. Furthermore, the Mn atomic charge of the graphene side is transferred to the Mn2 atom of the direct top. In other words, spin magnetic moments decrease because electron to

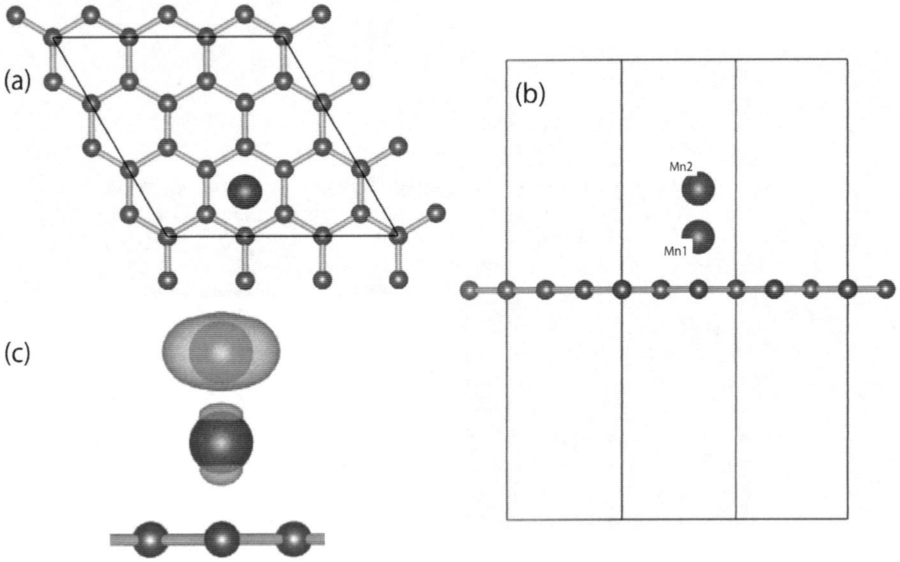

Figure 2. The most stable structure when two Mn atom adsorbed it.: (a) top view, (b) side view ,(c) sipn density map

take magnetic on decreases in the Mn1 atom. It is shown that the d-orbital to take magnetic on Mn1 atom decreased remarkably. Furthermore, we calculated the antiferromagnetism and the nonmagnetism status in a structure of fig.2. Because there is five d-orbital, as for the Mn atom, the magnetism with the variety has configuration. As a result, the solution of the ferromagnetic state and the solution of the antiferromagnetic state were degenerated. The antiferromagnetic state of the Mn atom is not kept, because charge is transferred widely in a structure of fig.2. For the calculation that assumed nonmagnetism, fig.2A structure is the most stable. However, a few local stable structure got different result. The structure that trimer of the Mn atoms forms vertical alignment is most stable.

4. Conclusion

We perform multiple atomic adsorptions on the graphene. We found a few local stable structure when two Mn atoms adsorbed to the graphene. When two Mn atoms adsorbed to the graphene top, the Mn atom tends to make a dimer on the graphene. When a Mn atom adsorbed to two sites of H-site and the B-site, magnetic occurred widely in the B-site. Furthermore, magnetic occurred in the T-site when a Mn atom adsorbed to two sites of H-site and the T-site. When we adsorbed two Mn atoms on graphene, a structure to locate vertical alignment is the most stable. In the vertical alignment construction which is the most stable structure, two Mn atomic magnetic moments are different. The magnetic of the Mn was not almost created in the H-site adsorption which was the most stable adsorption site in the monoatom adsorption. However, we could produce magnetic while keeping stable structure because a product displayed two dimers perpendicularly on H site. Because difference produces for the large charge transfer between two Mn, large difference of the magnetic moment is created.

5. Acknowledgement
This research was supported as the CREST project of Japan Science and Technology Agency (JST). Figures 1 and 2 were created using VESTA package by K. Momma and F. Izumi [19].

References
[1] A.K. Geim, K.S. Novoselov, Nature Materials 6 (3) (2007) 183191.
[2] A. Kobayashi, Y. Kawaguchi, J. Ohta, H. Fujioka, Appl. Phys. Lett. 89 (2006) 111918.
[3] G. Li, J. Ohta, K. Okamoto, A. Kobayashi, H. Fujioka, Appl. Phys. Lett. 89 (2006) 182104.
[4] M.-H. Kim, M. Oshima, H. Kinoshita, Y. Shirakura, K. Miyayama, J. Ohta, A. Kobayashi, H. Fujioka, Appl. Phys. Lett. 89 (2006) 031916.
[5] A. Kobayashi, S. Kawano, Y. Kawaguchi, J. Ohta, H. Fujioka, Appl. Phys. Lett. 90 (2007) 041908.
[6] M.-H. Kim, J. Ohta, A. Kobayashi, H. Fujioka, M. Oshima, Appl. Phys. Lett. 91 (2007) 151903.
[7] A. Ishii, M. Yamamoto, H. Asano, K. Fujiwara, J. Phys. Conf. Ser. 100 (2008) 052087.
[8] A. Ishii, T. Tatani , H. Asano and K. Nakada: Pnys.Sate.Soli C7 (2010) 347.
[9] A. Ishii, T. Tatani and K. Nakada: Phys. Status Solidi C 8 (2011) 1585.
[10] K. Nakada and A. Ishii, Solid State Commun. 151 (2011) 13
[11] K. Nakada, A. Ishii, Jian Ru Gong (Ed.), ISBN: 978-953-307-556-3, InTech
[12] M. Wu, C. Cao, J. Z. Jiang:New J.Phys 12 (2010) 063020
[13] H. Sevincli, M. Topsakal, E. Durgun, and S. Ciraci: Phys. Rev. B 77 (2008) 195434
[14] P. A. Khomyakov, G. Giovannetti, P. C. Rusu, G. Brocks, J. van den Brink, P. J. Kelly, Phys. Rev. B 79 (2009) 195425
[15] G. Kresse, J. Hafner, Phys. Rev. B 47 (1993) RC558.
[16] G. Kresse, Thesis, Technische Universitat Wien, 1993.
[17] G. Kresse, J. Furthmuller, Comput. Math. Sci. 6 (1996) 1550.
[18] G. Kresse, J. Furthmuller, Phys. Rev. B 54 (1996) 11169.
[19] K. Momma and F. Izumi, J. Appl. Crystallogr., 41 (2008) 653-658.

Growth of Ag thin films on glass substrates with a 3-mercaptopropyltrimethoxysilane (MPTMS) interlayer

M Kawamura, T Fudei, and Y Abe
Department of Materials Science and Engineering, Kitami Institute of Technology, Kitami, Hokkaido, Japan

E-mail: kawamumd@mail.kitami-it.ac.jp

Abstract. We report that very thin Ag films on glass substrates with a 3-mercaptopropyl trimethoxysilane (MPTMS) interlayer grow relatively flatter and become electrically continuous with a smaller thickness due to an interaction between Ag atoms and the molecule. Because optical transmittance of Ag films with and without the MPTMS layer is almost the same, it is found out that the Ag film with the interlayer is more suitable for the use as a semi-transparent electrode for display devices.

1. Introduction

Thin metal films with high transmittance and low resistivity can be used as transparent electrodes in various electronic devices such as solar cells or organic light-emitting diodes [1-3]. Though a thinner film shows a higher transmittance, it becomes discontinuous and has a high electrical resistance. Therefore, it is preferable to develop a process for thinner continuous films. It is well known that film growth is affected by various deposition parameters, such as deposition rate, temperature, substrate surface condition and so on. For Cu sputtered film, reports have been made that organic silane molecule layers, such as 3-mercaptopropyltrimethoxy silane (MPTMS) or 3-aminopropyltrimethoxysilane (APTMS) gave influences on nucleation of Cu films due to a chemical interaction between Cu atoms and the molecules[4, 5].

Previously, we have investigated properties of Ag thin films with an MPTMS interlayer on SiO_2/Si substrates. It was found that Ag thin films with the interlayer were electrically continuous compared to those without the interlayer [6]. For the use as transparent electrode, optical property of the films is an important factor. In the present paper, we prepare the Ag thin films on glass substrates with and without an MPTMS interlayer and investigate their electrical and optical properties.

2. Experiment

A Corning #1737 glass substrate was cleaned ultrasonically in acetone, 2-propanol, and de-ionized water, followed by treatment in a UV/ozone cleaner. The substrate was transferred to a nitrogen-filled glove box, immersed in 5 mM MPTMS toluene solution for 30 min, and then rinsed in toluene and dried overnight to provide the substrate with an MPTMS interlayer. The deposition of Ag films on the substrates, with and without the interlayer was carried out by vacuum evaporation below

3.0×10^{-4} Pa using an Ag wire (99.99% purity) evaporation source. The thickness and deposition rate (0.40 nm/s) of the films were controlled by a quartz crystal monitor.

Formation of MPTMS layer was confirmed by contact angle measurement using 10 μl water droplets, and the layer thickness was estimated by ellipsometry, as described somewhere else[7]. After deposition of 60-nm-thick Ag thin film, adhesion strength of the film on the substrate was examined by microscratch tester with a diamond stylus of R25μm. The surface morphology of the films was observed using an atomic force microscope (AFM). The average rms roughness (Rrms) was calculated based on images (5 μm × 5 μm) from three different locations on the sample surface. Electrical resistivity was measured at room temperature using a four-point probe method. Optical transmittance was investigated with UV/Vis spectrophotometer.

3. Results and Discussion

3.1. Formation of MPTMS layer on glass substrate

To confirm formation of MPTMS layer on glass substrate, we measured a static water contact angle, and the measured value was 64°, which is close to a value ($66\ ^{O}$) for MPTMS layer on SiO_2/Si substrate in our previous study [7]. As a thickness, we obtained a value of 1.4 nm +/- 0.6 nm which is slightly larger than that in previous study. As a result, we consider that the substrate surface is covered with MPTMS layer.

Next, we investigated adhesion strength at the interface of 60-nm-thick Ag thin film and the glass substrate. Figure 1 shows photos of representative samples after the scratch test. It is recognized that a Ag film without an MPTMS interlayer delaminated easily. Figure 2 shows average critical load for delamination of Ag films. It is found that the critical load becomes three times larger for the sample with MPTMS interlayer than that without the interlayer, which was also observed for Ag films on SiO_2/Si substrates [6]. As the reason, we consider that the interlayer enhanced the adhesion strength by anchoring effect of the MPTMS molecule. Therefore we think this enhanced adhesion is one of the proofs of the formation of the MPTMS layer.

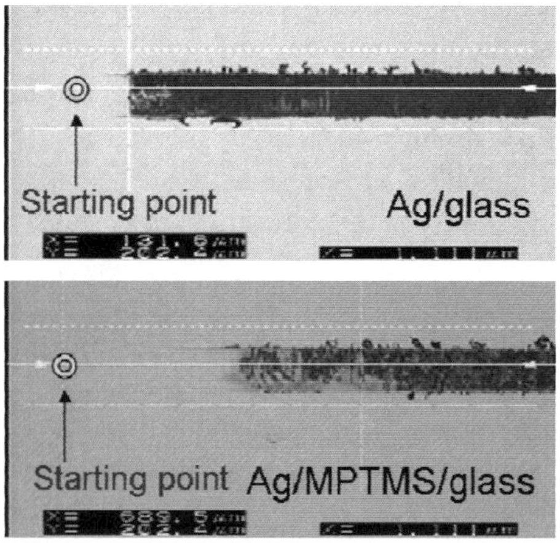

Figure 1. Sample surfaces after the scratch test.

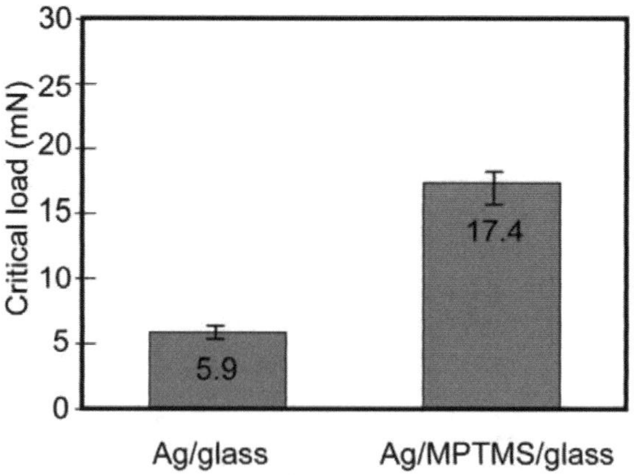

Figure 2. Critical load for delamination of Ag films on investigation of adhesion strength.

3.2. Characterization of Ag thin films on glass substrate with MPTMS interlayer

First, morphological analysis was carried out for the deposited samples. Figure 3 shows AFM images of 6-nm-thick Ag thin films with and without MPTMS interlayer. It is clearly observed that Ag grains are grown larger in round shape and look isolated each other on glass substrate without the interlayer.

However, they are smaller in peanut shape being connected each other on the MPTMS interlayer. This suggests that initial nucleation density was higher on MPTMS interlayer. At 8-nm-thick Ag films, grains were connected on both surfaces, but height from the substrate was smaller for Ag film with the interlayer. Therefore, it is suggested that growth of Ag thin films seems to be flatter by MPTMS interlayer, which was found on SiO_2/Si substrates, too.

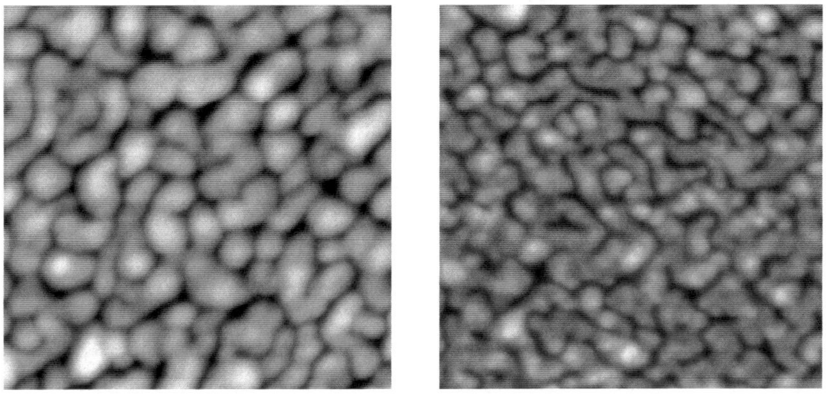

Figure 3. AFM images of 6-nm-thick Ag thin films on glass substrates with (right) and without (left) the MPTMS interlayer. Image size is 0.5μm x 0.5μm.

Further result of AFM analysis is shown in Figure 4. It is shown that the root-mean-square roughness of Ag film with the MPTMS interlayer is lower than that of Ag film without the interlayer at the same thickness. Then, the values of roughness decrease with increase of the film thickness because the films become more continuous. Consequently, we confirm that a Ag film grows flatter on a glass substrate with the MPTMS interlayer. As the reason, we consider that Ag atoms arrived on the substrate are less mobile due to a strong interaction with mercapto moiety of MPTMS, which causes a higher nuclei density.

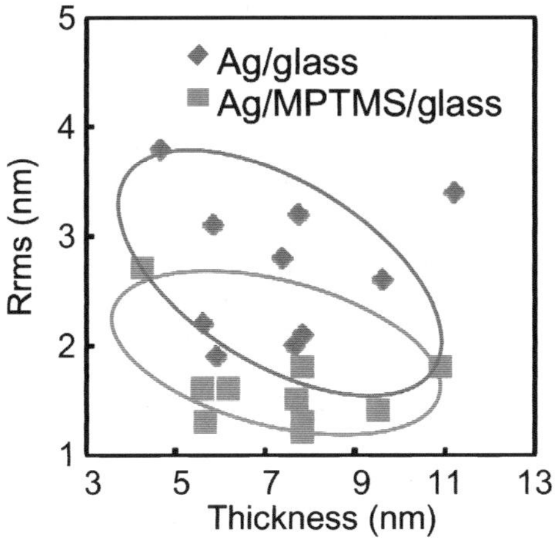

Figure 4. Root-mean-square roughness of the samples at each Ag film thickness.

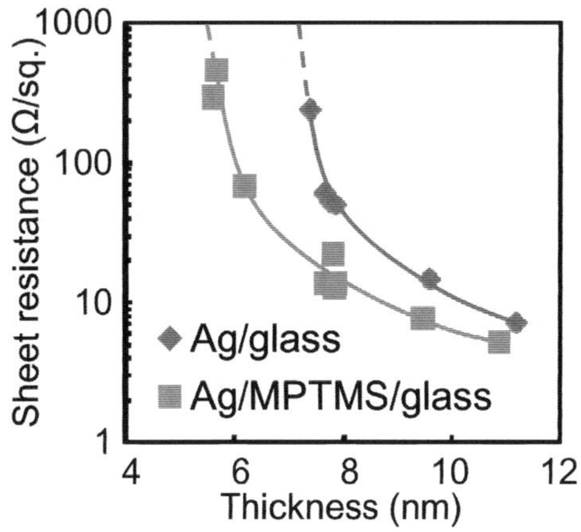

Figure 5. Sheet resistance of the samples as a function of Ag film thickness.

Then we compare an electrical property of both samples. Figure 5 shows the sheet resistance of Ag films with and without the MPTMS interlayer. In general, very thin films are discontinuous, and their resistance is too high to be measured. Above a critical thickness, films become continuous and their sheet resistances decrease with increasing thickness. It is found that the Ag film with the interlayer shows lower sheet resistance in any thickness. The figure shows that the Ag film without the interlayer becomes continuous at a thickness of 7.4 nm, on the other hand, that with the interlayer becomes continuous at a thickness of 5.6 nm. Namely, critical thickness was found to be 2 nm smaller. Thus, we confirmed that a thinner Ag film becomes continuous film by the presence of the MPTMS interlayer.

The sheet resistance of 6 and 8-nm-thick Ag films are summarized in Table 1 where the average values are shown with the highest and the lowest values in parenthesis. At 6 nm, a sheet resistance of Ag films without the MPTMS interlayer was too high to be measured, but that with the interlayer was relatively low. This difference coincides with microstructure of the films shown in Figure 3; Ag grains were isolated on glass substrate but connected on the interlayer.

As a result, we confirm the usefulness of MPTMS interlayer on glass substrate to obtain flat Ag film with a smaller critical thickness. We consider the following mechanism for this smaller critical thickness. By interaction with mercapto group of MPTMS layer, migration of arrived Ag atoms is suppressed. This causes formation of more nuclei on the substrate, and a continuous film can be obtained with fewer atoms.

Table 1. Sheet resistance of the samples

Sample	Sheet resistance (Ω/sq)
Ag(6nm)/MPTMS/glass	67 (63 – 73)
Ag(6nm)/ glass	-
Ag(8nm)/MPTMS/glass	13 (13 – 14)
Ag(8nm)/ glass	54 (50 – 60)

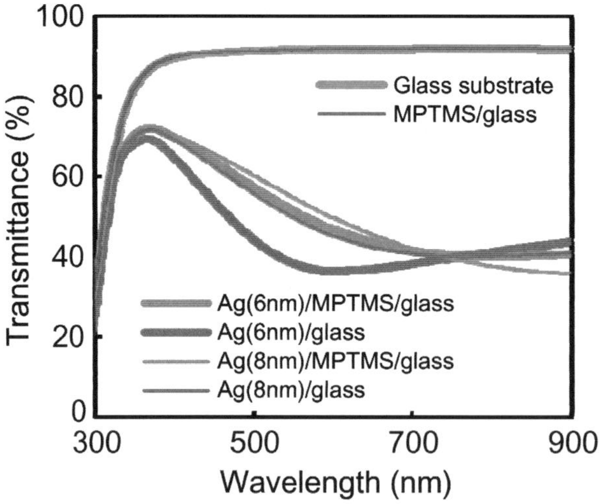

Figure 6. Transmittance spectra of the samples.

Next, transmittance of the films was measured. The transmission spectra of glass substrate and glass substrate with the MPTMS interlayer were the same in a region from 300 to 900 nm. Then, Ag films with a thickness of 8 nm with and without the interlayer were compared. As a result, there was no remarkable difference on their spectra and transmittance was higher than 70% at 400nm, and about 60% at 500nm, including absorption by the glass substrate. The transmittance of a 6-nm-thick Ag film does not show any increase compared to that of an 8-nm-thick film. Because the sheet resistance of the Ag film (8 nm) with the interlayer is much lower than that of the others as shown in Table 1, it is concluded that introduction of the MPTMS interlayer is very useful to prepare the Ag thin film with properties of a low sheet resistance and a high transmittance. And the best sample structure in the present study was found to be Ag (8 nm) film with MPTMS interlayer. According to a reference where Ag film was deposited on ITO film [8], sheet resistance and transmittance were reported to be 27 Ω/sq. and about 60% (λ=400 nm), respectively. Therefore, the sheet resistance of Ag film (8 nm) with the interlayer in the present study was found to be half with similar transmittance value. We now proceed to apply this film as electrode of an organic light emitting diode.

4. Conclusions

We have found flatter growth of Ag thin films on glass substrates with MPTMS interlayer due to the suppressed surface migration of Ag atoms by an interaction with mercapto moiety of MPTMS. As a result, Ag film with MPTMS interlayer showed a lower sheet resistance compared to a Ag film without the interlayer when the Ag thickness was the same. In addition, optical transmittance does not change by the presence of the MPTMS layer. Therefore, with this interlayer, we can prepare Ag thin film with a lower sheet resistance for the same optical transmittance. This is believed to be preferable for the use as semi-transparent electrode in display devices.

Acknowledgments

This work was supported in part by a Grant-in-Aid for Scientific Research (C) (No. 22560712) from the Japan Society for the Promotion of Science.

References

[1] Guillen C and Herrero J 2011 Thin Solid Films **520** 1
[2] Zhang X W, Li J, Zhang L, Jiang X Y, Haq K, Zhu W Q and Zhang Z L 2010 Thin Solid Films **518** 1756
[3] Lee C J, Pode R B, Han J I and Moon D G 2006 Appl. Phys. Lett. **89** 253508
[4] Hu M, Noda S, Tsuji Y, Okubo T, Yamaguchi Y and Komiyama H 2002 J. Vac. Sci. Technol. A **20** 589
[5] Liu Z, Wang Q, Liu X and Bao J 2008 Thin Solid Films **517** 653
[6] Fudei T, Kawamura M, Abe Y and Sasaki K 2012 J. Nanosci. Nanotechnol. **12** 1188
[7] Kawamura M, Fudei T, Abe Y and Sasaki K 2009 Jpn. J. Appl. Phys. **48** 118002
[8] Guillen C and Herrero J 2008 Sol. Energy Mater. Sol. Cells. **92** 938

ZnTe Amorphous Semiconductor Nanowires Array Electrodeposited into Polycarbonate Membrane Thin Films

T. Ohgai[1], T. Ikeda[2] and J. Ohta[3]

[1]Division of Chemistry and Materials Science, Nagasaki University, 1-14 Bunkyo-machi, Nagasaki 852-8521, JAPAN
[2]Graduate School of Science & Technology, Nagasaki University, 1-14 Bunkyo-machi, Nagasaki 852-8521, JAPAN
[3]Department of Materials Science and Engineering, Nagasaki University, 1-14 Bunkyo-machi, Nagasaki 852-8521, JAPAN

E-mail: ohgai@nagasaki-u.ac.jp

Abstract. ZnTe amorphous semiconductor nanowires array was electrodeposited into the nanochannels of ion-track etched polycarbonate membrane thin films from acidic aqueous solution at 313 K. ZnTe electrodeposits with Zn-rich composition was obtained over the wide range of cathode potential from -0.8 V to -1.1 V and the growth rate of ZnTe amorphous nanowires was around 3 nm•sec^{-1} at the cathode potential of -0.8 V. Cylindrical shape of the nanowires was precisely transferred from the nanochannels and the aspect ratio reached up to ca. 40. ZnTe amorphous phase electrodeposited at 313 K was crystallized by annealing at 683 K and the band gap energy of ZnTe crystalline phase reached up to ca. 2.13 eV.

1. Introduction

II-VI compound semiconductors (CdSe, CdTe, ZnSe, ZnTe, *etc.*[1, 2]) with wide band gap energy can be applied to the opto-electronic devices. Among of the II-VI compound semiconductors, ZnTe can be applied to green light emission diodes and photovoltaic solar cells because the band gap energy is 2.26 eV, which corresponds to around 550 nm. ZnTe thin films can be prepared using several techniques such as, thermal evaporation [3], sputtering [4], chemical vapor deposition [5] and electrodeposition form aqueous [6-9] or non-aqueous [10-13] solution. By the way, one-dimensional semiconductor nanowires with a large aspect ratio have received much attention due to their shape anisotropy and extremely large surface area [14, 15]. Nanowires can be fabricated by electrochemically depositing metallic atoms into a nano-well template with numerous nanochannels such as anodized aluminum oxide films or polycarbonate membrane films [16-18]. T. Gandhi *et al.* [19] and Y. Liu *et al.* [20] reported that ZnTe nanowires or nanoparticles can be electrodeposited into an anodized titanium oxide nanotubes array. However, for the opto-electronic device application, it is quite difficult to obtain the uniform structure of electrodeposited ZnTe nanowires array using a conductive titanium oxide template. In this paper, ZnTe amorphous semiconductor nanowires array was synthesized into the nanochannels of non-conductive ion-track etched polycarbonate membrane thin films to obtain a novel semiconductor device with extremely large surface area. Crystallization process and optical properties of the electrodeposited ZnTe were also investigated.

2. Experimental Procedures

Figure 1 shows experimental apparatus for electrodeposition of ZnTe thin films and nanowires. Bath composition and electrolysis condition are also shown in this figure. Aqueous electrolytic solutions were synthesized from $ZnSO_4$•$7H_2O$ (0.1 M), TeO_2 (0.002 M), Na_2SO_4 (0.5 M), H_3Cit (citrate acid: 0.05 M) and Na_3Cit•$2H_2O$ (sodium citrate: 0.05 M). The solution pH was adjusted to 4.0 by adding sulfuric acid and sodium hydroxide and the solution temperature was kept to 313K. Glass sheets coated with ITO were used as a cathode for growing ZnTe thin films, while ion track-etched polycarbonate membrane films were used as a template for growing ZnTe nanowires. Optimum deposition potential for growing ZnTe was determined by a cathodic polarization curve measured over the wide range of potential window. ZnTe were electrodeposited potentio-statically keeping the cathode potential to -0.8 V. Crystal phase, structure and chemical composition of electrodeposited ZnTe was investigated by using XRD and EDX. Band gap energy of ZnTe was estimated by using UV-VIS absorption spectroscopy.

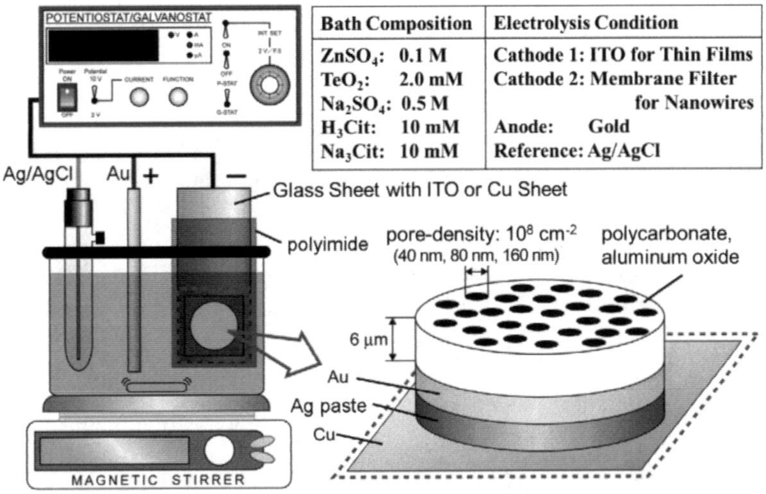

Figure 1. Experimental apparatus for electrodeposition of ZnTe thin films and nanowires (Bath composition and electrolysis condition are also shown in this figure).

3. Results and Discussion

3-1. Electrodeposition process of ZnTe nanowires

Cathodic polarization curve for electrodeposition of ZnTe from an aqueous solution containing Zn^{2+} and $TeCit^+$ ions is shown in **Fig.2**. At the potential of ca. -0.2 V, $TeCit^+$ ions began to be reduced. With increasing the current up to 10^{-2} A/m^2, the potential polarized to around -0.4 V and H^+ ions began to deposit. With increase in the current over 10 A/m^2, the potential polarized from -0.4 V to -1.0 V. In this potential range, the pH in the vicinity of the cathode can increase up to around 6 due to the diffusion limit of H^+ ions and $TeCit^+$ ions, then $Zn(OH)_2$ forms in the vicinity of cathode. Electrodeposition of ZnTe will proceed by Zn UPD due to the formation of $Zn(OH)_2$ by the following reaction.

$$Zn^{2+} + 2OH^- \rightarrow Zn(OH)_2$$
$$Zn(OH)_2 + TeCit^+ + 6e^- \rightarrow ZnTe + 2OH^- + Cit^{3-}$$

Zn-Te films deposited in the potential range from -0.8 V to -0.9 V mainly contained ZnTe phase with stoichiometric composition. Furthermore, with increase in the current more than 50 A/m^2, the potential polarized to less-noble than -1.0 V. In the potential range, massive metallic Zn began to deposit from $Zn(OH)_2$.

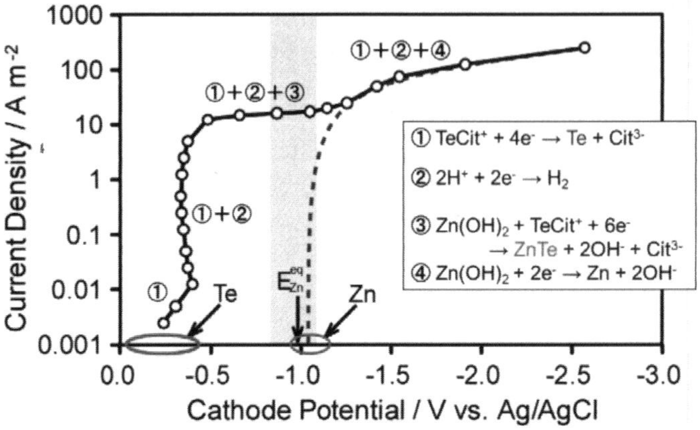

Figure 2. Cathodic polarization curve for electrodeposition of ZnTe from an aqueous solution containing Zn^{2+} and $TeCit^{+}$ ions.

Figure 3 shows the effect of cathode potential on Zn content in electrodeposits obtained from aqueous solutions containing Zn^{2+} and $TeCit^{+}$ ions. In the cathode potential range less-nobler than -0.9 V, Zn content in electrodeposit was more than 60%. On the other hand, in the cathode potential range nobler than -0.8 V, Zn content in electrodeposit was close to 50%. T. Ishizaki *et al.* reported that the potential-pH diagram at pH4.7 of Zn-Te-Cit-H_2O system and predicted that ZnTe can be electrodeposited over the wide range of cathode potential from -0.45 V to -1.1 V [21]. Therefore, in this study, ZnTe electrodeposits with Zn-rich composition could be obtained over the wide range of cathode potential from -0.8 V to -1.1 V.

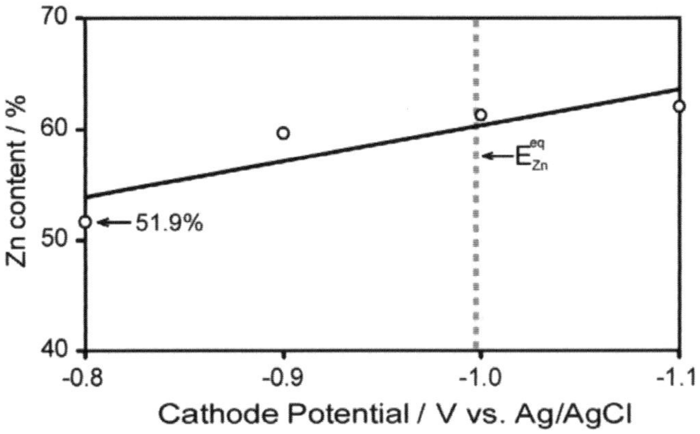

Figure 3. Effect of cathode potential on Zn content in electrodeposits obtained from aqueous solutions containing Zn^{2+} and $TeCit^{+}$ ions.

Figure 4 shows effect of cathode potential on the time-dependence of cathodic current during electrodeposition of ZnTe nanowires. To determine the growth rate of nanowires, the filling time of a nanochannel 6000 nm in length was estimated by monitoring the time-dependence of deposition current at each cathode potential as shown in **Fig.4**. When the nanochannels are filled and the nanowires reach the membrane surface, the current will increase drastically due to the formation of hemispherical caps. The growth rates were estimated by dividing channel length by channel-filling time. At –0.7 V, the filling time is around 3000 s and the deposition rate is estimated to be about 2 nm s^{-1}, while the filling time is close to 2000 s at –0.8 V and the deposition rate is estimated to be around 3 nm s^{-1}.

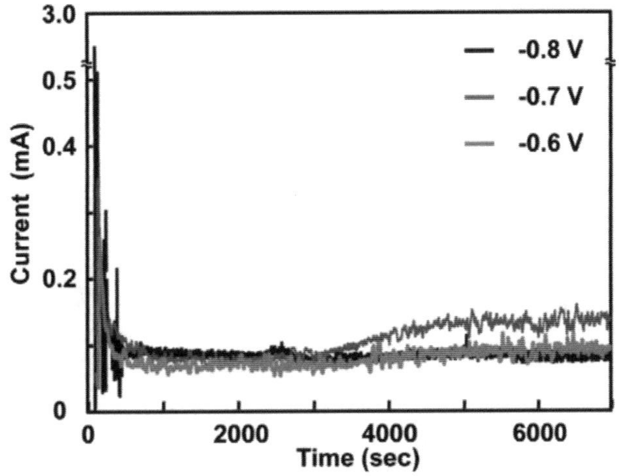

Figure 4. Effect of cathode potential on the time-dependence of cathodic current during electrodeposition of ZnTe nanowires.

SEM images of electrodeposited ZnTe nanowires separated from polycarbonate membrane filters with pore diameter 160nm are shown in **Fig.5**. The cylindrical shape was precisely transferred from the nanochannels to the nanowires and the aspect ratio reached up to ca. 40.

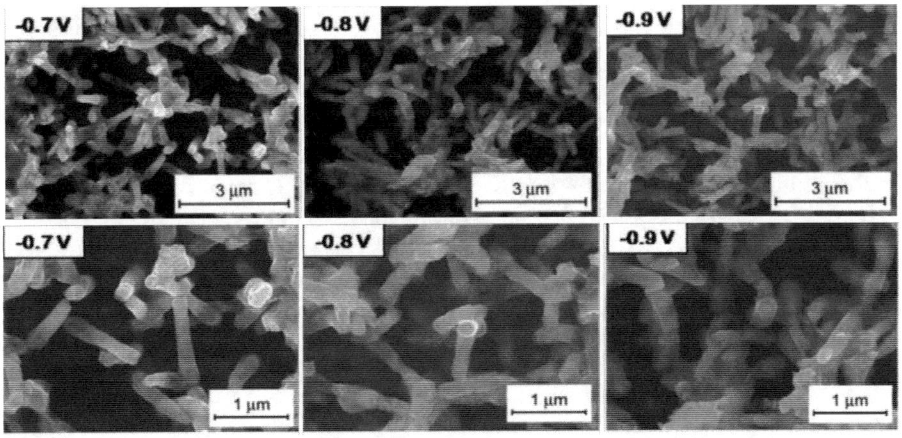

Figure 5. SEM images of ZnTe nanowires electrodeposited at -0.7 V, -0.8 V and -0.9 V. Nanowires were separated from polycarbonate membrane filters with pore diameter 160nm.

3-2. Crystallization process and optical properties of electrodeposited ZnTe

XRD patterns for as-deposited ZnTe films and annealed ZnTe films are shown in **Fig.6**. As-deposited ZnTe was composed from an amorphous phase, while the deposit was crystallized with annealing at 653 K for 5 hours and that contained binary phases of ZnTe and Te. On the other hand, the deposit with annealing at 683 K for 5 hours contained a single phase of ZnTe. A. B. Kashyout *et al.* reported that ZnTe electrodeposition from an aqueous solution containing $ZnCl_2$, TeO_2 and HCl [22]. They found that the electrodeposits annealed at 623 K composed from Te single phase, while the deposits annealed at 648 K and 673 K consist of ZnTe single phase. Melting points of Te and ZnTe are 718 K and 1563 K. Therefore, higher annealing temperature would be necessary to obtain ZnTe single phase.

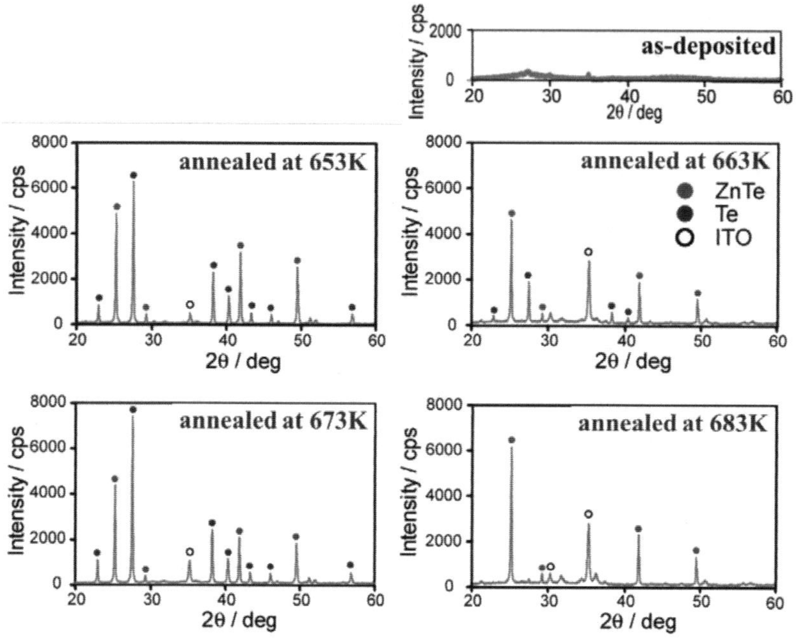

Figure 6. XRD patterns for as-deposited ZnTe, annealed at 653 K, 663 K, 673 K and 683 K.

UV-VIS absorption spectra and $(ah\nu)^2$ vs. $h\nu$ plot for determining the band gap energy of electrodeposited ZnTe are shown in **Fig.7**. In the wave length range less than around 550nm, optical absorption behavior was observed in the deposit with a single phase of ZnTe. Therefore, the band gap energy of the deposit with ZnTe single phase was estimated to be around 2.13 eV. M. Neumann-Spallart *et al.* reported that ZnTe electrodeposition on TCO (F-doped SnO_2 covered glass) from an aqueous solution containing $ZnSO_4$, TeO_2 and K_2SO_4 [23]. They estimated that the band gap energy of electrodeposited ZnTe are around 2.15 eV which is quite close to our experimental data.

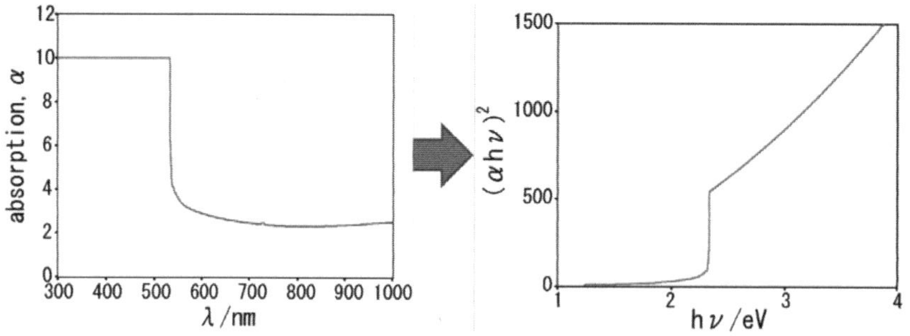

Figure 7. UV-VIS absorption spectra and $(ahn)^2$ vs. hn plot for determining the band gap energy of electrodeposited ZnTe.

4. Conclusions

Optimum cathode potential for electrodeposition of ZnTe was determined to be -0.8 V and the growth rate of ZnTe nanowires was around 3 nm•sec^{-1}. As-deposited ZnTe was composed from an amorphous phase, while the deposit was crystallized with annealing and band gap energy of the deposit with ZnTe single phase showed 2.13 eV.

Acknowledgement

This work was supported in part by Japan Society for the Promotion of Science (Grant-in-aid for Scientific Research C : No. 21560748), The President's Discretionary Fund of Nagasaki University, JFE 21st Century Foundation and Kyushu Industrial Technology Center.

References

[1] Gu J, Tonomura K, Yoshikawa N, Sakaguchi T 1973 *J. Appl. Phys.* **44** 4692
[2] Winnewisser C, Jepsen P U, Schall M, Schiyja V, Helm H 1997 *Appl. Phys. Lett.* **70** 3069
[3] Pal U, Saha S, Chaudhuri A K, Rao V V, Banerjee H D 1989 *J. Phys. D: Appl. Phys.* **22** 965
[4] Bellakhder H, Outzourhit A, Ameziane E L 2001 *Thin Solid Films* **382** 30
[5] Wolf K, Stanzl H, Naumov A, Wagner H P, Kuhn W, Kahn B, Gebhardt W 1994 *J. Cryst. Growth* **138** 412
[6] Bulent M. Basol, Vijay K. Kapur 1988 *Thin Solid Films* **165** 237
[7] Bozzini B, Baker M A, Cavallotti P L, Cerri E, Lenardi C 2000 *Thin Solid Films* **361-362** 388
[8] Bozzini B, Lenardi C, Lovergine N 2000 *Mater. Chem. Phys.* **66** 219
[9] Bouroushian M, Kosanovic T, Karoussos D, Spyrellis N 2009 *Electrochem. Acta* **54** 2522
[10] Chaure N B, Nair J P, Jayakrishnan R, Ganesan V, Pandey R K 1998 *Thin Solid Films* **324** 78
[11] Lin M C, Chen P Y, Sun I W 2001 *J. Electrochem. Soc.* **148** C653
[12] Heo P, Ichino R, Okido M 2006 *Electrochem. Acta* **51** 6325
[13] Islam A B M O, Chaure N B, Wellings J, Tolan G, Dharmadasa I M 2009 *Mater. Character.* **60** 160
[14] Ohgai T, Gravier L, Hoffer X, Ansermet J P 2005 *J. Appl. Electrochem.* **35** 479
[15] Kim D J, Seol J K, Lee M R, Hyung J H, Kim G S, Ohgai T, Lee S K 2012 *Appl. Phys. Lett.* **100** 163703
[16] Ohgai T, Enculescu I, Zet C, Westerberg L, Hjort K, Spohr R, Neumann R 2006 *J. Appl. Electrochem.* **36** 1157
[17] Ohgai T Hjort K, Spohr R, Neumann R 2008 *J. Appl. Electrochem.* **38** 713
[18] Spohr R, Zet C, Bernd Fischer B E, Kiesewetter H, Apel P, Gunko I, Ohgai T, Westerberg L 2010 *Nucl. Inst. Meth. Phys. Res.* **B268** 676
[19] Gandhi T, Raja K S, Misra M 2009 *Thin Solid Films* **517** 4527
[20] Liu Y, Zhang X, Liu R, Yang R, Liu C, Cai Q 2011 *J. Solid State Chem.* **184** 684
[21] Ishizaki T, Saito N, Takai O, Asakura S, Goto K, Fuwa A 2005 *Electrochem. Acta* **50** 3509
[22] Kashyout A B, Arico A S, Antonucci P L, Mohamed F A, Antonucci V 1997 *Mater. Chem. Phys.* **51** 130
[23] Neumann-Spallart M, Königstein C 1995 *Thin Solid Films* **265** 33

Morphology Control of Zn-SiO_2 Composite Films Electrodeposited from Aqueous Solution Containing Quaternary Ammonium Cations

T. Ohgai[1], K. Ogushi[2] and K. Takao[3]

[1]Division of Chemistry and Materials Science, Nagasaki University, 1-14 Bunkyo-machi, Nagasaki 852-8521, JAPAN
[2]Graduate School of Science & Technology, Nagasaki University, 1-14 Bunkyo-machi, Nagasaki 852-8521, JAPAN
[3]Technical Division, Faculty of Engineering, Nagasaki University, 1-14 Bunkyo-machi, Nagasaki 852-8521, JAPAN

E-mail: ohgai@nagasaki-u.ac.jp

Abstract. Zn-SiO_2 alloys were electrodeposited from acidic aqueous solution containing cationic surfactants such as diallyl-dimethyl-ammonium chloride, trimethyl-tetradecyl-ammonium chloride, trimethyl-stearyl-ammonium chloride and dimethyl-distearyl-ammonium chloride. Zn-SiO_2 alloy thin films were obtained at the wide current density range from 5 A/dm^2 to 100 A/dm^2. SiO_2 content in deposits was ca. 10 % at the maximum using the solution containing a surfactant with molecular weight range of 200 to 300.

1. Introduction

Electrodeposition of Zn alloy films on steel sheets has been investigated to achieve an excellent corrosion resistance rather than conventional hot-dip galvanized steel sheets [1-3]. Recently, much attention has also been paid to a Zn alloy films containing metal oxide particles electrodeposited on a steel sheet because of its excellent corrosion resistance properties and cost performance [4-6]. The co-deposition of Zn and metal oxide particles can be achieved using acidic aqueous solution containing cationic surfactants, which can adsorb on metal oxide particles. K. Kondo *et al.* reported that Zn-SiO_2 composite can be electrodeposited from the aqueous solution containing Sn^{2+} ions [7]. They found that SiO_2 particles precipitated along the laterally growing macro-steps of Zn hexagonal crystals and randomly dispersed on the terrace of Zn hexagonal crystals. T. J. Tuaweri *et al.* reported that the electrodeposition behavior of Zn-SiO_2 composite from the aqueous solution containing N, N-dimethyl-dodecyl-amine [8]. They obtained Zn-33wt.%SiO_2 composite using SiO_2 particles with the diameter of 2 μm. M. Azizi *et al.* reported that the co-deposition process of Zn-SiO_2 composite from the aqueous solution containing gelatin [9]. They found that Zn-20mass%SiO_2 composite can be electrodeposited using the solution containing 50 g/L SiO_2 particles. However, the role of surfactants on the co-deposition process of Zn and SiO_2 is still in an unsettled issue.

In this paper, effect of surfactants molecular weight on the morphology of electrodeposited Zn-SiO_2 composite films was investigated using quaternary ammonium cations such as diallyl-dimethyl-ammonium chloride, trimethyl-tetradecyl-ammonium chloride, trimethyl-stearyl-ammonium chloride and dimethyl-distearyl-ammonium chloride.

2. Experimental Procedures

Electrolytic solutions were synthesized from the following chemicals ($ZnSO_4{\cdot}7H_2O$: 300 g/L, Na_2SO_4: 50 g/L, H_2SO_4: 25 g/L, SiO_2 particles: 50 g/L and cationic surfactants: 0.01 mol/L). Average size of SiO_2 particles is ca. 0.5 μm and that contains 6.0 at.% Ca^{2+} ions. Molecular structures of cationic surfactants are shown in **Fig.1** (A: diallyl-dimethyl-ammonium chloride, B: trimethyl-tetradecyl-ammonium chloride, C: trimethyl-stearyl-ammonium chloride, D: dimethyl-distearyl-ammonium chloride). The solution temperature was kept to 323K. A steel sheet and a gold wire were used as a cathode and an anode, while a Ag/AgCl electrode was used as a reference electrode. Optimum current density for electrodeposition of Zn-SiO_2 alloys was determined by a cathodic polarization curve measured over the wide range of cathode potential. Zn-SiO_2 alloys were galvanostatically electrodeposited over the wide range of current density from 5 A/dm^2 to 100 A/dm^2. Crystal phase, structure and chemical composition of electrodeposited Zn-SiO_2 alloys was investigated by using X-ray diffraction (XRD), scanning electron microscope (SEM) and energy dispersive X-ray spectroscopy (EDX).

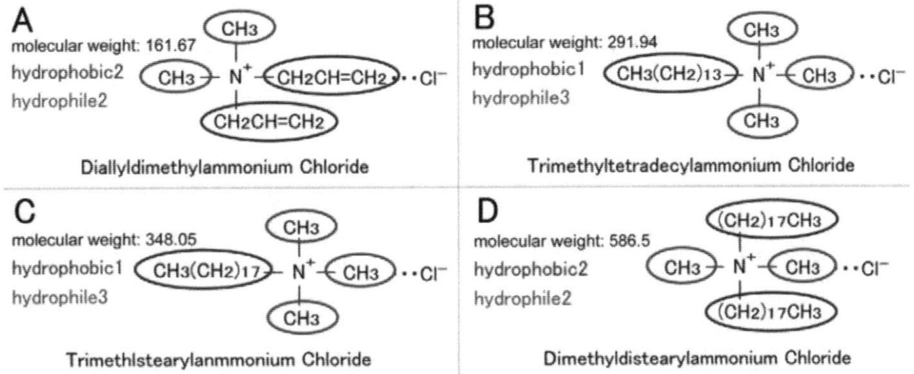

Figure 1. Molecular structures of cationic surfactants (A: diallyl-dimethyl-ammonium chloride, B: trimethyl-tetradecyl-ammonium chloride, C: trimethyl-stearyl-ammonium chloride, D: dimethyl-distearyl-ammonium chloride).

3. Results and Discussion

3-1. Effect of current density on the electrodeposition of Zn-SiO_2

Cathodic polarization curves for Zn-SiO_2 alloy deposition from acidic aqueous solutions with and without cationic surfactant-A are shown in **Fig.2**. At the potential of ca. -0.5 V, H^+ ions began to be reduced. With increasing the current density up to 1 A/dm^2, the potential polarized to less-noble direction in the solution containing cationic surfactant-A. In this potential range, the pH in the vicinity of the cathode will increase up to around 6 due to the diffusion limit of H^+ ions, then $Zn(OH)_2$ forms in the vicinity of cathode [10-12]. Subsequently the cathode potential reaches to the equilibrium potential of Zn (-0.96 Vvs.Ag/AgCl), cationic SiO_2 particles will adsorb on cathode surface. Electrodeposition of Zn-SiO_2 will proceed due to the following reaction.

$$Zn^{2+} + 2OH^- \rightarrow Zn(OH)_2$$
$$Zn(OH)_2 + SiO_2 + 2e^- \rightarrow Zn\text{-}SiO_2 + 2OH^-$$

Furthermore, with increase in the current density more than 100 A/dm^2, the cathode potential polarized significantly to less-noble region rather than -3.0 V due to the diffusion limit of Zn^{2+} ions. Consequently, optimum current density range for Zn-SiO_2 alloy deposition was determined to be the range from 5 A/dm^2 to 100 A/dm^2.

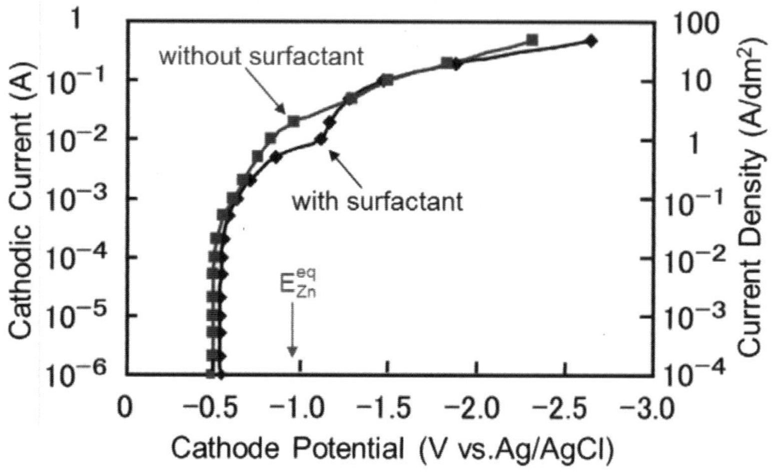

Figure 2. Cathodic polarization curve for electrodeposition of Zn-SiO_2 from acidic aqueous solutions with and without cationic surfactant-A.

SEM images of electrodeposited Zn-SiO_2 thin films obtained from an acidic aqueous solution without cationic surfactants under the different current density (5, 20, 100 A/dm^2) are shown in **Fig.3**. As shown in this figures, electrodeposits are mainly composed from Zn small crystals with the shape of hexagonal plates. With increase in current density, the size of Zn crystals also increased up to around 5 μm in the diameter.

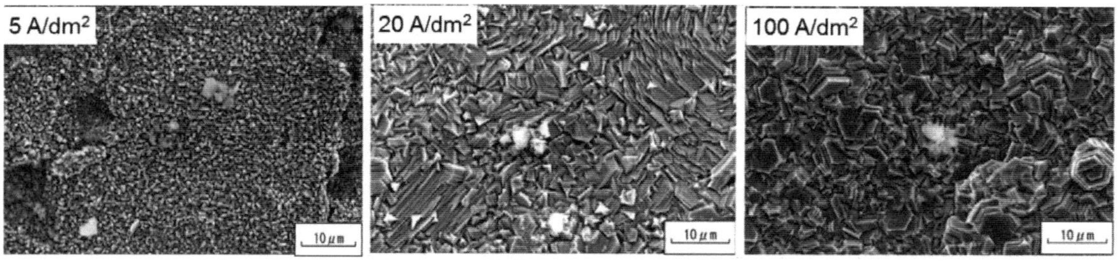

Figure 3. SEM images of electrodeposited Zn-SiO_2 thin films obtained from an acidic aqueous solution without cationic surfactants under the different current density (5, 20, 100 A/dm^2).

XRD patterns of electrodeposited Zn-SiO_2 thin films obtained from an acidic aqueous solution without cationic surfactants under the different current density (5, 20, 100 A/dm^2) are shown in **Fig.4**. It is well known that the electrodeposited crystal growth behavior depends on the deposition overpotential. Usually, with decrease in the deposition overpotential, the electrodeposited crystal growth rate will decrease. If the electrodeposited crystal growth rate is small, the electrodeposits are composed from preferentially oriented crystal plane with most close-packing in atomic arrangement. While, the deposition overpotential will decrease in decreasing current density. Therefore, this electrodeposited crystal growth behavior results in the following phenomenon. The electrodeposits obtained at the current density of 5 A/dm^2 are composed from Zn crystals with the preferentially oriented crystal plane in (002), which is most close-packed in atomic arrangement. On the other hand, the electrodeposits obtained at the current density of 100 A/dm^2 are composed from Zn crystals with the preferentially oriented crystal plane in (101), which is not close-packed.

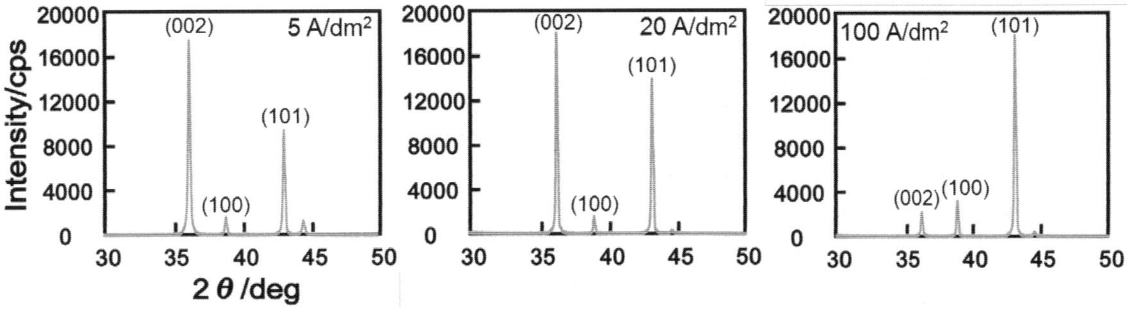

Figure 4. XRD patterns of electrodeposited Zn-SiO_2 thin films obtained from an acidic aqueous solution without cationic surfactants under the different current density (5, 20, 100 A/dm^2).

3-2. Effect of surfactant molecular weight on the electrodeposition of Zn-SiO_2

Figure 5 shows effect of surfactant molecular weight on SiO_2 content in deposits obtained from acidic aqueous solution with and without cationic surfactants under the current density of 100 A/dm^2. The molecular weights of each surfactant are as followings; surfactant-A: 161.67, surfactant-B: 291.94, surfactant-C: 348.05 and surfactant-D: 586.5. SiO_2 contents in deposits obtained from the solution containing each surfactant are as followings; surfactant-A: 6.6 %, surfactant-B: 9.6 %, surfactant-C: 3.8 % and surfactant-D: 1.0 %. It is well-known that adsorption behavior of surfactants depends on their molecular weight [3]. Usually, with increase in molecular weight, the adsorption capacity will increase. Therefore, this adsorption behavior results in decreasing SiO_2 content in deposits obtained from the solution containing the surfactants with molecular weight lower than 100. On the other hand, with increase in molecular weight, the solubility in aqueous solution will decrease [3]. Therefore, this solubility behavior results in decreasing SiO_2 content in deposits obtained from the solution containing the surfactants with molecular weight higher than 400. As shown in **Fig.5**, SiO_2 content in deposits will be ca. 10 % at the maximum using the solution containing a surfactant with molecular weight range of 200 to 300.

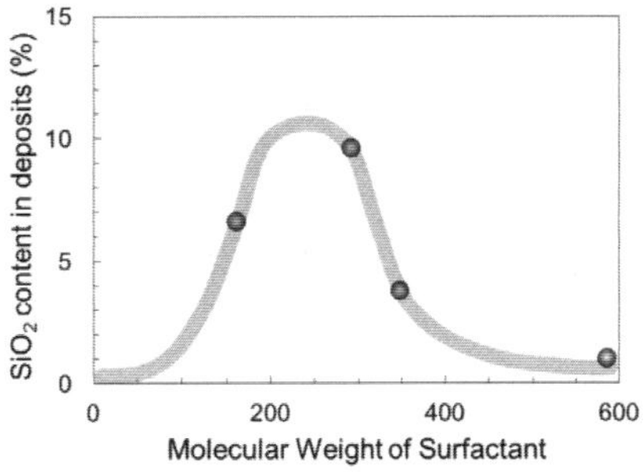

Figure 5. Effect of surfactant molecular weight on SiO_2 content in deposits obtained from acidic aqueous solution with and without cationic surfactants under the current density of 100 A/dm^2. The molecular weights of each surfactant are as followings; surfactant-A: 161.67, surfactant-B: 291.94, surfactant-C: 348.05, surfactant-D: 586.5.

SEM images of electrodeposited Zn-SiO_2 thin films obtained from acidic aqueous solution with and without cationic surfactants under the current density of 100 A/dm^2 are shown in **Fig.6**. As shown in these figures, morphology of the electrodeposits strongly depends on the molecular weight of surfactant. Electrodeposits obtained from the solution without and with surfactant D are composed from Zn crystals with the shape of hexagonal plates. On the other hand, electrodeposits obtained from the solution containing surfactant A, B and C are composed from Zn crystals with the shape of small thin leaves.

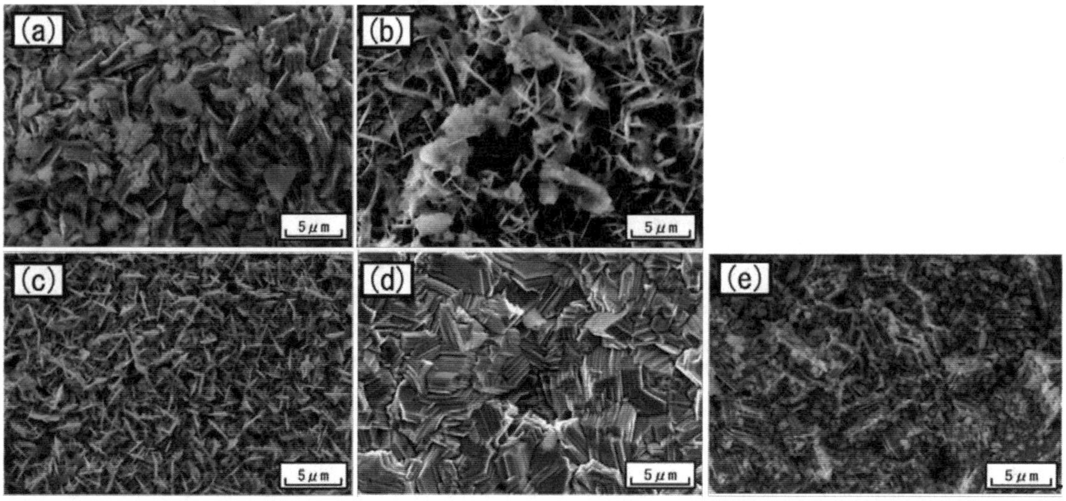

Figure 6. SEM images of electrodeposited Zn-SiO_2 thin films obtained from acidic aqueous solution with and without cationic surfactants under the current density of 100 A/dm^2. (a): surfactant-A, (b): surfactant-B, (c): surfactant-C, (d): surfactant-D, (e): without surfactant.

XRD patterns of electrodeposited Zn-SiO_2 thin films obtained from acidic aqueous solution with and without cationic surfactants under the current density of 100 A/dm^2 are shown in **Fig.7**. Electrodeposits obtained from the solution containing each surfactant are composed from Zn crystals with the preferentially oriented crystal plane in (101), which is not close-packed. Usually, with increase in the deposition overpotential, the electrodeposited crystal growth rate will also increase. Therefore, independently in the molecular weight of surfactants, non-close-packed (101) was preferentially oriented in the electrodeposits obtained at the current density of 100 A/dm^2, which results in large deposition overpotential.

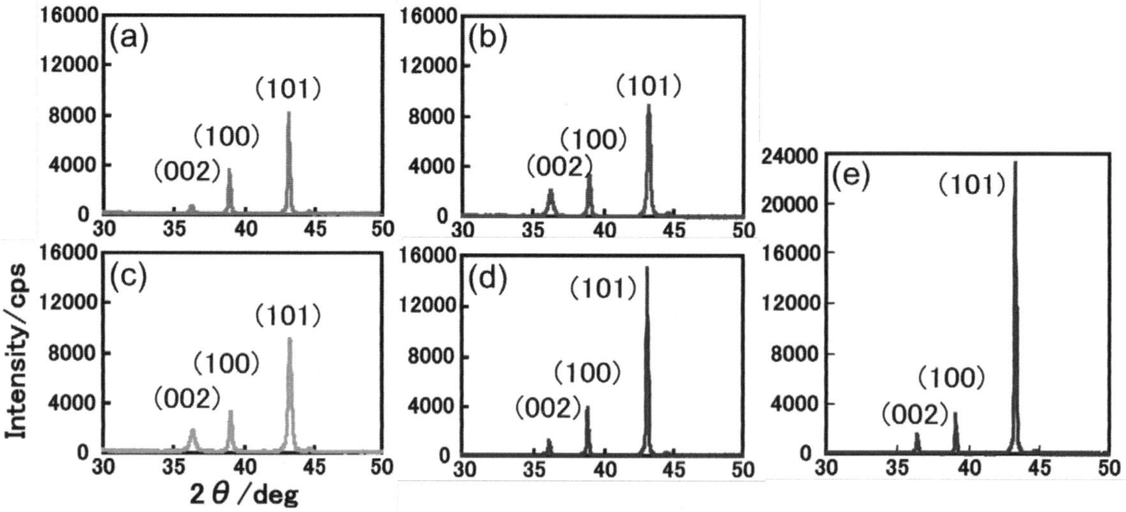

Figure 7. XRD patterns of electrodeposited Zn-SiO_2 thin films obtained from acidic aqueous solution with and without cationic surfactants under the current density of 100 A/dm^2. (a): surfactant-A, (b): surfactant-B, (c): surfactant-C, (d): surfactant-D, (e): without surfactant.

4. Conclusions

Zn-SiO_2 alloy thin films were obtained at the wide current density range from 5 A/dm^2 to 100 A/dm^2. Electrodeposits obtained at current density of 5 A/dm^2 are composed from Zn crystals with preferential orientation in (002), while electrodeposits obtained at the current density of 100 A/dm^2 are composed from Zn crystals with preferential orientation in (101). SiO_2 content in deposits was ca. 10 % at the maximum using the solution containing a surfactant with molecular weight range of 200 to 300.

References

[1] Giridhar J, Ooij W J 1992 *Surf. Coat. Tech.* **52** 17
[2] Hashimoto S, Abe M 1994 *Corrosion Science* **36** 2125
[3] Tulio P C, Rodrigues S E B, Carlos I A 2007 *Surf. Coat. Tech.* **202** 91
[4] Miranda F J F, Barcia O E, Diaz S L, Mattos O R, Wiart R 1996 *Electrochim. Acta* **41** 1041
[5] Eliaz N, Venkatakrishna K, Hegde A C 2010 *Surf. Coat. Tech.* **205** 1969
[6] Khan T R, Erbe A, Auinger M, Marlow F and Rohwerder M 2011 *Sci. Technol. Adv. Mater.* **12** 055005
[7] Kondo K, Ohgishi A and Tanaka Z 2000 *J. Electrochem. Soc.* **147** 2611
[8] Tuaweri T J, Wilcox G D 2006 *Surf. Coat. Tech.* **200** 5921
[9] M. Azizi, W. Schneider, W. Plieth 2005 *J. Solid State Electrochem.* **9** 429
[10] Ohgai T, Fukushima H, Akiyama T and Hirai Y 1997 *Metallurgical Review of MMIJ* **14** 24
[11] Akiyama T, Kobayashi S, Ki J, Ohgai T, Fukushima H 2000 *J. Appl. Electrochem.* **30** 817
[12] Nakano H, Ohgai T, Fukushima H, Akiyama T and Kammel R 2001 *METALL* **55** 676

Investigation of Inhomogeneity of TiO_2 Thin Films Using Spectroscopic Ellipsometry

M Horprathum[1,a], J Kaewkhao[2,3,b], P Eiamchai[1], P Chindaudom[1] and P Limsuwan[3,4]

[1] Optical Thin-Film Laboratory, National Electronics and Computer Technology Center, Pathumthani 12120, Thailand
[2] Center of Excellence in Glass Technology and Material Science (CEGM), Nakhon Pathom Rajabhat University, Nakhon Pathom 73000, Thailand
[3] Thailand Center of Excellence in Physics, CHE, Ministry of Education, Bangkok 10400, Thailand
[4] Department of Physics, Faculty of Science, King Mongkut's University of Technology Thonburi, Bangkok, 10140, Thailand

E-mail: [a]mati.horprathum@nectec.or.th, [b]mink110@hotmail.com

Abstract. Titanium dioxide thin films have been deposited on silicon wafers (100) substrates by pulse DC reactive magnetron sputtering. The spectroscopic ellipsometry is the method use to determine the degree of inhomogeneity of titanium dioxide thin films. The effect of operating pressure on the micro-structural and optical properties of inhomogenity Titanium dioxide thin films were characterized by grazing-incidence X-ray diffraction, spectroscopic ellipsometry, field emission scanning electron microscope and high resolution transmission electron microscopy. The optical properties of the films were examined by variable-angle spectroscopic ellipsometer. Several spectroscopic ellipsometry models, categorized by physical models, were proposed based on the 'simpler better' rule and curve-fits, which were generated and compared to experimental data using regression analysis. It has been found that the triple-layer physical model with the Cody-Lorentz dispersion model offered the most convincing result. Titanium dioxide thin film was found inhomogeneous and a more detailed analysis of high resolution transmission electron microscopy and field emission scanning electron microscope are discussed.

1. Introduction

The titanium dioxide (TiO_2) exists in three different crystalline forms: anatase (tetragonal), rutile (tetragonal) and brookite (orthorhombic), rutile being the most stable of the three, and the formation of its phase depends on starting material, calcinations temperature and deposition method. There are many techniques were used to compare the optical properties of TiO_2 thin films. In case of classical reactive evaporation, the large dispersion of the values for the refractive index and the extinction coefficient is observed, caused by small changes in the process conditions. The influence on the optical properties of the deposition parameters has been studies by several authors in the case of evaporation [1], of DC reactive magnetron sputtering [2], etc. They usually focus at the effects of the deposition parameters, O_2 pressure, deposition rate, and substrate temperature [3-7] which influence the packing fraction, the crystallinity and the optical properties of the films. Only few groups are interested in the influence of the operating pressure on films structural and optical properties [4, 7-8]. This work focuses on the spectroscopic ellipsometry (SE) characterization and analysed of sputtering-derived films. The properties in crystallinity, homogeneity degree and understanding of the inhomogeneity are interested.

2. Experiments

The TiO_2 thin films were prepared by pulse DC reactive magnetron sputtering in both high purity argon (99.999%) and oxygen (99.999%) ultra-high vacuum system (AJA International, Inc. ATC 2000-F), using a Titanium (99.995%; KJ. Lesker) target of 2 inch-diameter. The distance between the substrate holder to target was 90 mm, and the incline angle of the gun was 40° with respect to the plane of substrate holder. The base pressure of the deposition chamber was about 1 x 10^{-7} Torr. The operating pressure was selected as a variable parameter. The gas flows of argon and oxygen were set at 10 and 20 sccm, respectively. The discharge was generated with a 400 W pulsed DC power. The silicon wafers (100) substrate were prepared by ultrasonic washer with isopropanol and acetone successively, and dried in nitrogen atmosphere, before loaded into the deposition chamber. Prior to deposition, the substrate was cleaned by Ar plasma ion for 5 minutes to remove surface contamination of the substrate.

Crystal structures of the films were studied using Rigaku TTRAX III grazing-incidence X-ray diffraction (GIXRD). The Cu-Kα radiation was operated at 50 kV, 300 mA with scanning speed of 2θ per minute at 2θ step of 0.02θ. The measurements were conducted from 20-60θ incident angles. The morphologies of the TiO_2 thin films were investigated using high resolution transmission electron microscopy (HRTEM) experiment performed on a JEM-2010 electron microscope (JEOL) with an acceleration voltage of 200 kV. The optical parameters, i.e. the index of refraction, the extinction coefficient and the film thickness, were calculated from ellipsometric measurements using a variable-angle spectroscopic ellipsometer (VASE, J.A. Woolam). The VASE measurements were done at 70° incident angles, in the photon energy range of 0.75-6 eV, with autoretarder activated.

3. Results and Discussion

3.1. Grazing-incidence X-ray diffraction

It is well known that operating pressure is the main parameters for the formation of different TiO_2 crystal structures. The structural of sputtered 120 nm thick TiO_2 thin films strongly depend on operating pressure (Figure 1). TiO_2 thin films deposited at high operating pressure (OP:5-7 mTorr) exhibit pure anatase phase (Fig 2(c). For TiO_2 thin films prepared at low operating pressure of 3 mTorr composed of a mixed anatase-rutile phase. The higher losses of particles energy in collisions causes decrement on the crystallinity of the rutile phase, that it could form at high temperature and activation energy. From the previous study of M. Yamagishi et al. [9], the rutile phase on unheated film at low operating pressure due to the higher energy of the particles impinging on the growing film surface

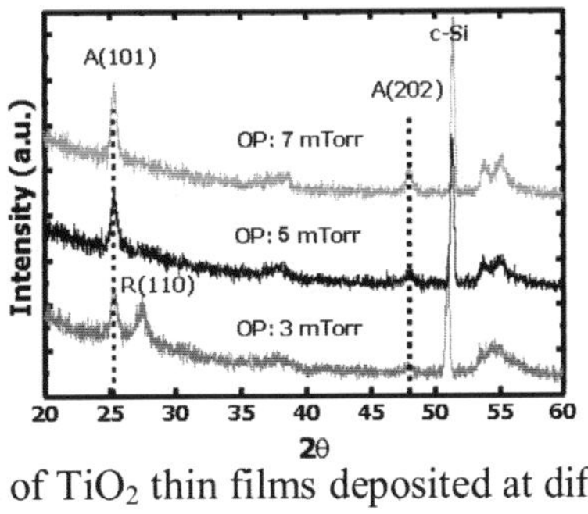

Figure 1 GIXRD pattern of TiO_2 thin films deposited at different operating pressure

3.2 Spectroscopic Ellipsometry (SE) Analysis

3.2.1 SE Modelling for Pure Anatase TiO_2 Thin Films

In this experiment, we use three physical models in analyzing the ellipsometric data of pure anatase TiO_2 thin films prepared at 5 and 7 mTorr. Three fitting models, namely, single-layer model (containing a native oxide and homogeneous layer of TiO_2 thin film (Figure 2(a) top), double-layer model (containing a native oxide, TiO_2 void free layer and TiO_2 with void layer (Figure 2(a) middle) and triple-layer model (containing a native oxide, TiO_2 void free layer, TiO_2 with void layer and surface roughness (Srough) (Figure 2(a) bottom). An interfacial layer native Si oxide which assumed thickness of 2 nm was included because the Si substrate not etched by HF before film deposition. The optical properties of TiO_2 were modeled by a single Cody-Lorentz Oscillator [5, 10-11]. Bruggeman effective medium approximation (BEMA) is used to determine of optical constant of the composition. The proposed physical model corresponding to each film sample was generated based on appropriate generalized oscillator for the optical dispersion, and eventually curve-fitted with the experimental data. The goal is, by a regression algorithm for the ellipsometry equation, to minimize the mean square error function:

$$MSE = \sqrt{\frac{1}{2N-M}\sum_{i=1}^{N}[(\frac{\Psi_i^{\mathrm{mod}}-\Psi_i^{\mathrm{exp}}}{\sigma_{\Psi,i}^{\mathrm{exp}}})^2+(\frac{\Delta_i^{\mathrm{mod}}-\Delta_i^{\mathrm{exp}}}{\sigma_{\Delta,i}^{\mathrm{exp}}})^2]} \quad (1)$$

N is the number of measured Ψ and Δ pairs, M is the total number of variable parameters and σ is the standard deviations. The 'mod' means the theoretical calculation, and the 'exp' means the experimentally measured data.

The experimental SE data are plotted in circle symbol, and the solid line is the regression fitting result. In the first attempt, the model of a single homogeneous layer of TiO_2 on c-Si substrate was tested. Figure 2(b) shown agreement between the SE data and the fitted data of all TiO_2 thin films. It was found that this single-layer and double-layer models were not given a good fit to measured Ψ and Δ spectra. Obviously, the triple-layer physical model proposed the best curve-fits (low MSE value), confirmed that the TiO_2 thin films is far from being homogeneous. The results from these three different models are shown in Table 1. The SE model can be confirmed by cross-section HR-TEM micrograph shown in next section (Figure 3(a)).

3.2.2 SE Modelling for Mixed Phase TiO_2 Thin Film

Next is the case of mixture phase (anatase-rutile) TiO_2 thin film deposited at 3 mTorr operating pressure on silicon wafer substrate. The best fit of physical model of mixture TiO_2 thin film show different structure from previous experiment. These films can be modeled as composed of the native oxide layer, TiO_2 with void layer, TiO_2 void free layer, and surface roughness (Srough). Figure 2(c) shows the agreement between the experimentally observed Ψ and Δ spectra. Notice that the order off void free and the voided layer is different from the films prepared at higher operating pressure. The reason of the phenomenon will be discussed in the next section.

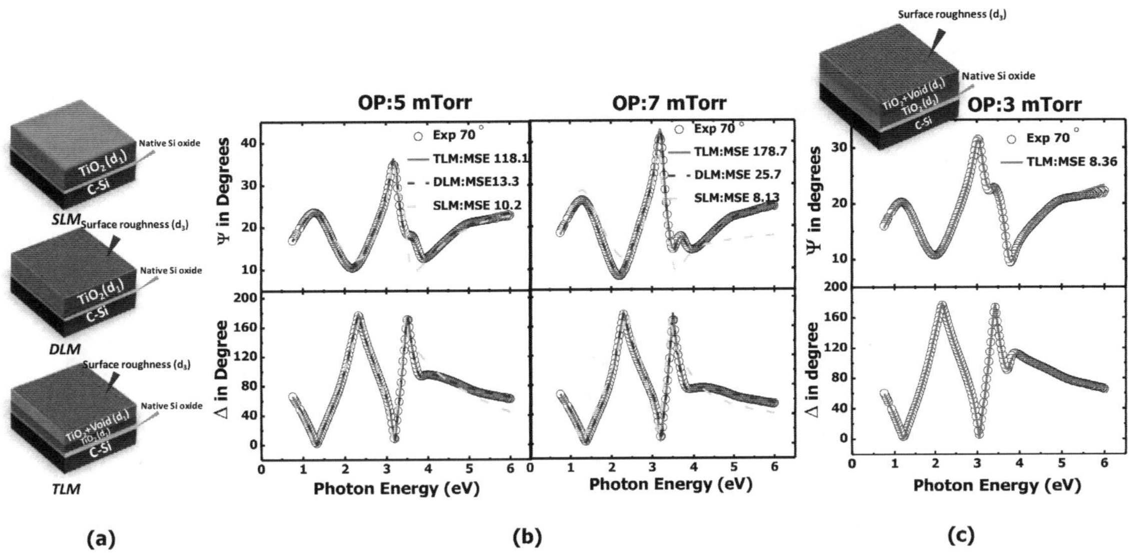

Figure 2 (a) Best-fit single layer model (SLM), double layer model (DLM) and triple layer model (TLM) of TiO_2 thin film on silicon substrate. (b) Measured and calculated ellipsometric Ψ and Δ spectra for the TiO_2 thin films deposited at operating pressure 5 and 7 mTorr on silicon wafer substrate of Angle of incident 70°. (c) Measured and calculated ellipsometric Ψ and Δ spectra for the TiO_2 thin films deposited at operating pressure 3 mTorr on silicon wafer substrate of Angle of incident 70° (inset figure is best-fit model of mixture TiO_2 thin film on silicon substrate)

Table 1 Summary results of the spectroscopic ellisometry characterization from inhomogeneous pure anatase TiO_2 thin films.

Operating Pressure (mTorr)	Dense Layer (nm)	TiO2 with void layer (nm)		Srough layer (nm)		MSE
	d1	d2	%Void	d3	%Void	
5						
SLM	114.2 ± 1.3	-	-	-	-	118.1
DLM	102.2 ± 0.3	-	-	11.8 ± 0.2	31.9 ± 0.5	13.3
TLM	32.7 ± 4.0	71.8 ± 4.1	2.9 ± 0.4	10.6 ± 0.2	35.6 ± 0.9	10.2
7						
SLM	120.2 ± 2.5	-	-	-	-	178.7
DLM	98.7 ± 0.5	-	-	19.0 ± 0.2	36.3 ± 0.4	25.7
TLM	54.6 ± 1.1	46.7 ± 1.2	5.2 ± 0.1	18.0 ± 0.1	37.9 ± 0.2	8.13

Table 2 Summary results of the spectroscopic ellisometry characterization from inhomogeneous anatase-rutile TiO_2 thin films.

Operating Pressure (mTorr)	TiO_2 with void layer (nm)		Dense Layer (nm)	Srough layer (nm)		MSE
	d_1	%Void	d_2	d_3	%Void	
3	23.4 ± 3	7.6 ± 1	82.4 ± 3	11.2 ± 0.2	24.4 ± 0.6	8.36

3.3 Depth Profiles of Inhomogeneous TiO_2 Thin Films

These observed inhomogeneities in the growth direction depend on the plasma conditions. Figure 3(a), shows the depth profiles of the refractive index at 550 nm wavelength for the films deposited on silicon substrate. All films show triples layer with different index properties, due to the difference in films crystallinities. Films deposited at high operating pressure (5 and 7 mTorr) show higher refractive index at the bottom layer than top layer. Films deposited at 3 mTorr show lower refractive index at the bottom layer than top layer. In the case of 3 mTorr, the crystallization at the top layers is increased due to higher energy from substrate temperature which results from deposition condition. At low operating pressure, the high energy gas particles are able to reach the substrate without inter-collision due to the longer mean free path. Higher concentration of the ionized or excited species resulting from the higher electron temperature in the plasma and the higher energy of the particles impinging on the growing film surface leads to the formation of mixture phase and high refractive index at low operating pressure. As an example, HRTEM image (OP 5 mTorr) clearly shows the film microstructure matching with the layers and their thickness in the SE model (Figure 3(a)). From depth profile ellipometric analysis, a dense structure at the bottom of the film was indentified to have higher refractive index from the triple layer model. The middle part of the film that shows a columnar structure has a slightly lower refractive index compared to the dense region at the bottom. On top of the films, a very thin rough layer which its index was calculated from BEMA.

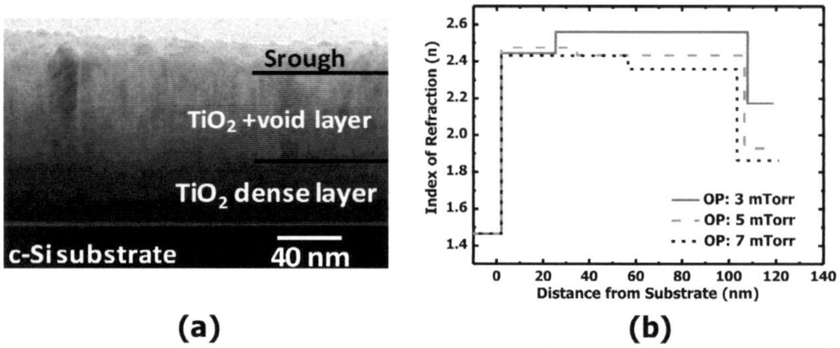

Figure 3 (a) Cross section HRTEM of TiO_2 thin film deposited at 5 mTorr (b) Depth profiles of the refractive indices for the sample (OP 3-7 mTorr) at 550 nm wavelength

3.4 Analysis of Inhomogeneous TiO_2 Films by Scanning Electron Microscopy

Figure 4 shows the surface morphologies and the cross-sectional structures of three samples. The film prepared at lower operating pressure (OP 3 mTorr) has a more compact structure, the film prepared at higher operating pressure (OP 7 mTorr) has a more porous structure, and the films prepared at intermediate pressure (OP 5 mTorr) has a mixed compact and porous structure. That is, the film structure changes gradually from compact into porous as the operating pressure increase. We can explain these qualitatively properties by the Thornton structure zone model [11] with HR-TEM.

4. Conclusion

We characterized the inhomogeneous TiO_2 thin films by SE. The results were compared with field emission scanning electron microscope (FE-SEM) and high resolution transmission electron microscopy (HRTEM). Three physical models were used in analyzing the ellipsometric data. First, the single-layer model contained only the homogeneous TiO_2 layer. Second, the double-layer model has an additional surface roughness on top of the homogeneous layer. Third, the triple-layer model proposed the homogeneous TiO_2 layer, TiO_2 with void layer and a surface roughness. From the study, the triple-layer model suggested the best fits. In addition, it is confirmed that the TiO_2 thin films is far from being homogeneous, when compared with the results from the cross-sectional FE-SEM and HRTEM micrographs

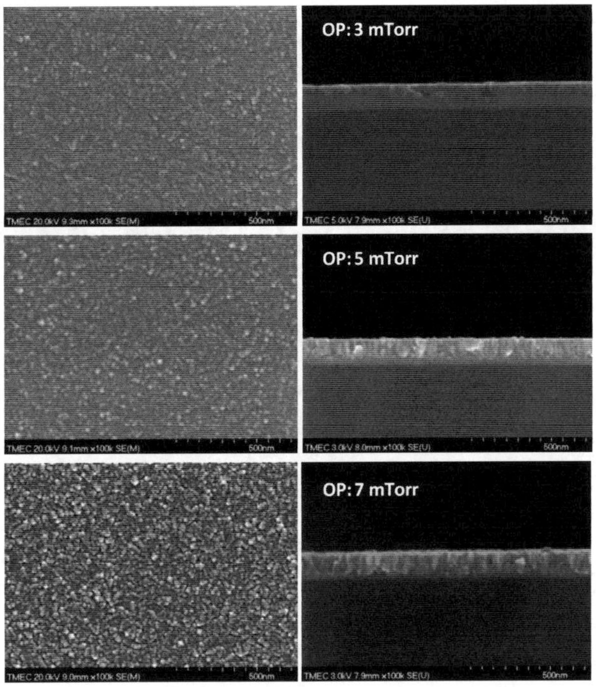

Figure 4 FE-SEM micrograph of TiO_2 thin film eposited at different operating pressure: top view and cross-section.

Acknowledgments

This work was technically and financially supported by the National Electronics and Computer Technology Center (NECTEC), a member of National Science and Technology Development Agency (NSTDA). This work was partially supported by Thailand Center of Excellence in Physics (ThEP) and King Mongkut's University of Technology Thonburi under The National Research University Project.

Reference

[1] Kim S Y 1996 *Appl. Opt.* **35** 6703
[2] Kim S H and Hwangbo C K 2005 *Thin Solid Films* **475** 155
[3] Šícha J, Musil J, Meissner M, Čerstvý R 2008 *Appl. Surf. Sci.* **254** 3793
[4] Musil J, Heřman D, Šícha J 2006 *J. Vac. Sci. Technol. A* **24** 521
[5] Horprathum M, Chindaudom P, Limnonthakul P, Eiamchai P, Nuntawong N, Patthanasettakul V, Pokaipisit A, Limsuwan P 2010 *J. Appl, Phys.* **108** 013522
[6] Li-Joan M, Santos M P 1993 *Thin Solid films* **226** 22
[7] Zhang Y, Ma X, Chen P, Yang D 2007 *J. Crys. Grow.* **300** 551
[8] Glöß D, Franch P, Zywitzki O, Modes T, Klinkenberg S, Gottfried C 2005 *Surf. Coat. Technol.* **200** 967
[9] Yamagishi M, Kuriki S, Song PK, shigesato Y 2003 *Thin Solid Films* **442** 227
[10] Eiamchai P, Chindaudom P, Pokaipisit A and Limsuwan P 2008 *Cur. Appl. Phys.* **9** 707
[11] Horprathum M, Chindaudom P and Limsuwan P 2007 *Chin. Phys. Lett.* **24** 1505
[12] Thonton J A 1978 *J. Vac. Sci. Technol.* **15** 171

Surface Segregation Behavior of B, Ga, Sb, and As Dopant Atoms on Ge(100) and Ge(111) Examined with a First-principles Method

F Iijima[1], K Sawano[1], J Ushio[2], T Maruizumi[1], and Y Shiraki[1]

[1]Advanced Research Laboratories, Tokyo City University, 8-15-1 Todoroki, Setagaya-ku, Tokyo 158-0082, Japan
[2]Central Research Laboratory, Hitachi Ltd., Akanuma 2520, Hikigun, Saitama 350-0395, Japan

E-mail:tmarui@tcu.ac.jp

Abstract. To understand surface segregation behaviour of B, Ga, As, and Sb dopant atoms on Ge (100) and Ge (111) surfaces, the potential energies of these dopant atoms in the five top layers of the surfaces were evaluated by first-principles molecular orbital calculations of model clusters. The obtained potential energy curves of these dopant atoms for Ge (100) closely resemble those for Si (100) surface. On the other hand, Ge (111) surface exhibits different potential profiles as compared with Ge (100) surface. The thermal activation energy between the adsorbed state and sub-surface state was also evaluated for each dopant atom. These energy data determining the driving force and dynamical property of dopant segregation help us to understand the difference in dopant surface segregation phenomena on Ge(100) and Ge(111) surfaces from an atomistic point of view.

1. Introduction

Molecular beam epitaxy (MBE) is an established crystal growth method for forming sharp doping profile and abrupt heterointerfaces. However, many elements have been shown to segregate to the epitaxial surface during MBE growth process. It is known that B, Ga, Sb, and Ge show different behaviors in surface segregation during-Si MBE [1-3]. We previously examined these segregation behaviors on Si (100) surface with calculations using a first-principles method and showed that the trend of surface segregation is primarily prospected by seeing the bond energy difference between the dopant-Si and Si-Si bonds [4]. Aggressive downscaling of high-performance metal-oxide semiconductor field effect transistors (MOSFET) is bringing about the revival of Ge as a promising candidate material for high mobility complementary MOS (CMOS). One of the crucial problems associated with the fabrication of aggressively scaled Ge channel CMOS is the formation of n-type ultrashallow junctions with ion implantation techniques [5-6]. It was recently reported that the formation of ultrashallow ohmic contacts for n-type Ge was made possible by Sb δ-doping followed by Ge homoepitaxial growth [7]. It was also shown that the segregation effect of Sb on Ge (111) surface is lower than that on Ge (100) surface. Unfortunately, there are very few experimental data with regard to the dopant surface segregation phenomena on Ge surface. Judging from the importance of Ge channel as a promising electronic material and δ-doping process as an appropriate formation technique for ultrashallow junctions/contacts, a first-principles study to reveal atomistic mechanism of the dopant surface segregation on Ge surfaces is indispensable for design and improvement of the δ-doping process with Ge homoepitaxial growth. Hence, we investigate the surface segregation behaviors of B, Ga, As, and Sb on Ge (100) and Ge (111) surfaces with a first-principles method and compare them with the previously reported segregation behavior on Si (100) surface.

2. Calculation details

The behavior of a dopant atom incorporated into semiconductor crystal can be explained by a two-state model with an appropriate potential for a dopant atom [8]. A schematic energy diagram is shown in Figure 1. The jumping rate for surface segregation is determined by the potential barrier, so-called activation energy *E*a, and the driving force of surface segregation is the energy difference between the surface (adsorbed) and subsurface (incorporated) states, ΔG_{seg}. We evaluated the potential energies of B, Ga, As, and Sb dopant atoms in the five top layers of Ge (100) and Ge (111) surfaces and also the activation energy of each dopant atom between the adsorbed state and the incorporated state at the first subsurface by using a first-principles molecular orbital program with model clusters in this study. For the calculation, we used a $Ge_{17}H_{25}$ cluster and a $Ge_{16}H_{29}$ cluster as a Ge (100) surface model and a Ge (111) surface model, respectively (see Figure 2). No surface reconstruction was explicitly considered as a first approximation for both surfaces in this study because there remains no reconstructed structures in the MBE grown epitaxial layer. The Ge (100) surface model has the same surface configuration as modeled in our previous paper for Si (100) calculations and the bridge site was taken as the adsorption site for Ge (100) surface due to its largest adsorption energy for B dopant on Si (100) surface [4]. For the Ge (111) surface model, the adsorption site was assumed to be a T4 site with 3-fold symmetry located on the top of a Ge atom in the second layer. The T4 site on Si (111) surface was reported to give the largest adsorption energies for group IIIA metals among various adsorption sites [9].

The potential energy for each dopant atom was obtained taking into account the structural relaxation of the Ge lattice by optimizing the both positions of the dopant atom and the nearest Ge atoms around the adsorbed or incorporated dopant atom. And all the hydrogen atoms terminating the cluster models were fixed throughout the structural optimization. All calculations were performed using the first-principles molecular orbital program package "Gaussian 03" [10]. QST3 keyword option for structural optimization routine of Gaussian 03 program was thoroughly employed in exploration for the transition states in this study. Density functional calculations were carried out using B3LYP hybrid functional and the DGDZVP basis set (double-ζ valence plus polarization) [11,12].

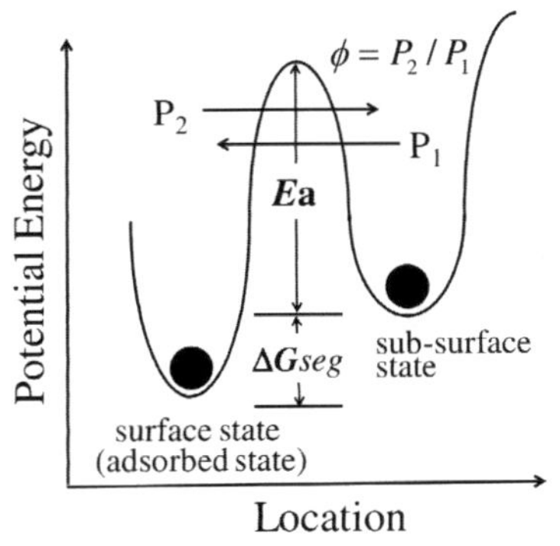

Figure 1. A two-state model to describe dopant exchange process between surface and sub-surface states.

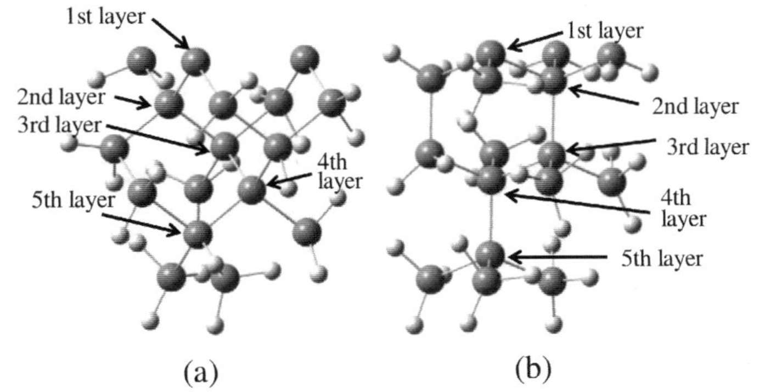

Figure 2. Cluster models used to calculate ΔG_{seg} and *E*a: (a) Ge (100) surface model ($Ge_{17}H_{25}$) and (b) Ge (111) surface model ($Ge_{16}H_{29}$). The actual locations of the dopant atoms are shown by the arrows.

3. Results and discussion

The obtained adsorption energies for B, Ga, As, and Sb dopant atoms on the Ge (100) and Ge (111) surfaces are listed in Table 1. The calculated adsorption energies on Si (100) and Si (111) surfaces are also tabulated for comparison [4, 9]. B atom has the largest adsorption energies for both Ge (100) and Ge (111) surfaces due to its largest bonding energy to Ge atom. Sb has the second largest adsorption energies. The same tendency in the adsorption energy is observed for the dopant adsorption on Si (100) surface.

Table 1. Calculated adsorption energies and bond energies (in eV) of B, Ga, As, and Sb atoms on the Ge and Si surfaces.

Surface	B	Ga	As	Sb
Ge(100)	4.26	2.56	2.99	3.20
Ge(111)	3.50	2.92	3.08	3.42
Si (100)[a)]	7.35	4.56	----	5.28
Si (111)[b)]	4.83	2.53	----	----
Dopant-Ge [c)]	3.49	2.80	2.54	2.34

a) , b): reported values in [4] and [9], respectively.
c) Calculated for the bonds in H_2B-GeH_3, $H_2Ga-GeH_3$, $H_2As-GeH_3$, and $H_2Sb-GeH_3$.

The adsorption structures of these dopant atoms on the Ge (100) surface were very similar to those reported for Si (100) surface [4]. The adsorption structures at T4 site on Ge (111) surface are shown in Figure 3. All the dopant atoms except B atom make four bonds with three Ge atoms on the top surface and one Ge atom in the second layer as shown in Figure 3. The average of bond lengths between the dopant atom and the Ge atoms are 2.10 Å for B, 2.78 Å for Ga, 2.89 Å for As, and 2.99 Å for Sb. The average of the bond lengths for Ga, As, and Sb atoms are longer by 0.4 Å than the sum of atomic covalent radii of dopant and Ge atoms while the averaged B-Ge bond length is only longer by 0.07 Å than the sum of atomic covalent radii. And it is clearly seen in Figure 3 that the adsorption site of B atom is very close to the top surface layer. From these results it can be expected that B atom will distort Ge lattice structure near Ge top surface much more than other dopant atoms due to its small covalent radius and strong preference for trivalent coordination.

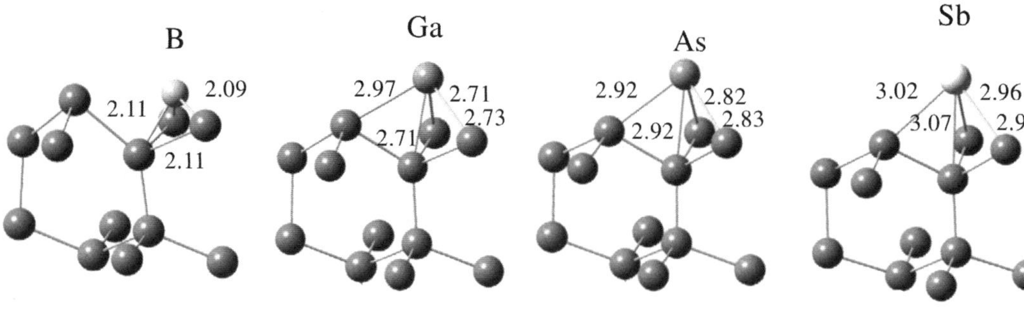

Figure 3. The adsorption structures of B, Ga, As, and Sb dopant atoms at T4 site on Ge (111) surface. The number denotes bond length (in Å) between a dopant atom and a Ge atom. The deeper layer than the fifth layer is omitted and the terminating hydrogen atoms are also omitted for clarity.

The calculated total energies of the clusters as functions of the locations of the dopant atoms – that is, the potential energies for the dopant atoms – are shown in Figure 4. The location of a dopant atom is denoted by the layer number of the layer where the dopant atom is located. As can be seen in

Figure 4, B and As atoms have more energetically stable incorporated states than the adsorbed states in the subsurface layers of Ge (100). On the other hand, all the energies of the incorporated states of Ga and Sb atoms are higher than those of the adsorbed states. The incorporated-state energies for both Ga and Sb atoms increase monotonically as they go deeper from the adsorption site to the fifth Ge layer. These potential energy curves for B, Ga, and Sb atoms on Ge (100) surface are very close to those obtained for Si (100) surface [4]. This simply implies that surface segregation behavior on Ge(100) surface for these four dopant atoms is expected to be similar to that on Si (100) surface ; B and As atoms will not segregate to the Ge (100) surface during MBE process, but Ga and Sb will do. Although the potential energy for Ge (111) surface changes quite differently as compared with the potential energy curve obtained for Ge (100) for all the four dopant atoms, both B and As dopant atoms have more stable states than the adsorbed states in the incorporated layers, the second and fourth layers for B atom and second layer for As atom, respectively.

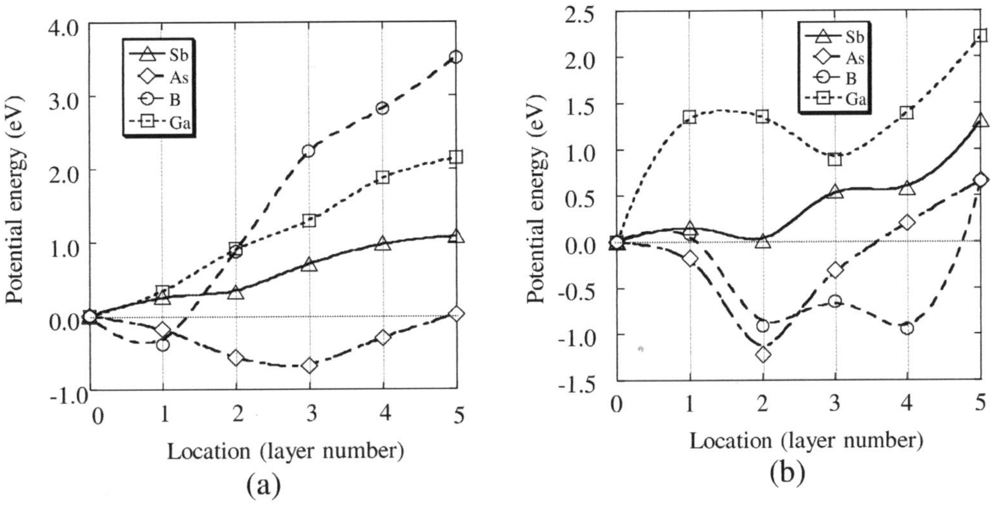

Figure 4. Total energies of the Ge surface cluster model with dopant atoms of B, Ga, As, and Sb in the five surface layers of the Ge surfaces: (a) Ge (100) and (b) Ge(111). The adsorbed state energy is taken as zero. The actual locations of the dopant atoms are depicted by the arrows in Figure 2.

To compare more intensively the surface segregation behavior on Ge (100) with that on Ge (111) surfaces and to draw informative conclusions, it is necessary for us to examine both segregation energy ΔG_{seg} and thermal activation energy Ea. The two-state model tells that a surface segregation process is controlled by a segregation energy for the case of growth at high temperatures and at relatively slow growth-rates while it is controlled by a thermal activation energy for the case of growth at lower temperatures or faster growth rates. Here, we summarize the calculated segregation energies and thermal activation energies for B, Ga, As, and Sb dopant atoms on Ge (100) and Ge (111) surfaces in Table 2.

Table 2. ΔG_{seg} and Ea for Ge (100) and Ge(111) surfaces (in eV).

Surface	ΔG_{seg} / Ea	B	Ga	As	Sb
Ge(100)	ΔG_{seg}	-0.39	0.33	-0.18	0.26
	Ea	0.98	0.96	0.56	1.20
Ge(111)	ΔG_{seg}	0.09	1.36	-0.17	0.16
	Ea	1.17	1.54	2.28	3.00

Considering the magnitude of the activation energies listed in Table 2, we can expect less surface segregation effects on Ge (111) surface than Ge (100) surface for δ-doping process of all the four dopant atoms followed by the homoepitaxial growth at lower temperatures or faster growth rates. It is also expected that B and As dopant atoms will not segregate to Ge surfaces due to the existence of their stable positions in the subsurface layers while Ga and Sb dopant atoms will segregate to the surface. As for the surface segregation of Sb dopant atom on Ge (100) surface, it was estimated that the segregation energy was around 0.4eV and the thermal activation energy was around 1.4 eV based on the temperature dependence of the segregation length in the temperature range from 100 °C to 500 °C [13]. Our computational results agree well with these values. It was also stated in the report that the shift of the temperature dependence curve of the segregation length toward lower temperature region is well explained by the decrease of the thermal activation energy rather than by the increase of the segregation energy within the framework of the two-state model. The lower thermal activation energy calculated for Ge (100) than that for Ge (111) corresponds directly to the experimentally observed temperature dependence curve for Ge (100) in the lower temperature region as compared with that for Ge (111) [7]. Unfortunately there are very few experimental data on surface segregation behaviors of other dopant atoms on Ge (100) and Ge (111) surfaces, we cannot compare our computational results directly with the experimental data. We are hoping that our first-principles calculation results will accelerate the development of δ-doping process as an appropriate formation technique for ultrashallow junctions/contacts.

4. Summary

The surface segregation process for B, Ga, As, and Sb dopant atoms on Ge (100) and Ge (111) surfaces were investigated with a theoretical approach using accurate first-principles method. Segregation energy, ΔG_{seg} ,and activation energy, Ea, were quantitatively evaluated with appropriate density functional and basis functions. Based on the computational results, we can expect less surface segregation effects on Ge (111) surface than Ge (100) surface for δ-doping process of all the four dopant atoms followed by the homoepitaxial growth at lower temperatures or faster growth rates. The expected behavior of Sb surface segregation on Ge (100) and Ge (111) surfaces is in agreement with experimentally observed behavior.

References

[1] Nakagawa K , Miyao M , and Shiraki Y 1989 *Thin Solid Films* **183** 315
[2] Jorke H 1988 *Surf. Sci.* **193** 569
[3] Nakagawa K , Miyao M 1991 *J. Appl. Phys.* **69** 3058
[4] Ushio J, Nakagawa K , Miyao M , Maruizumi T 1998 *Phys. Rev.* **B 58** 3932
[5] Chui C O , Kulig L , Moran J , Tsai W, and Saraswat K C 2005 *Appl. Phys.Lett.* **87** 091909
[6] Satta A *et al* 2006 *Appl. Phys. Lett.* **88** 162118
[7] Sawano K et al 2010 *Appl.Phys. Lett.* **97** 162108
[8] Harris J J *et al* 1984 *Appl. Phys.* **A 33** 87
[9] Tzeli D, Petsalakis I D, and Theodorakopoulos G 2009 *J. Phys. Chem.* **113** 13924
[10] Frisch M J *et al.* Gaussian 03, Revision E.01, 2004 Gaussian, Inc., Wallingford CT.
[11] Becke A D 1993 *J. Chem. Phys.* **98** 5648
[12] Godbout N , Salahub D , Andzelm J , and Winner E 1992 *Can. J. Chem.* **70** 560
[13] Nakagawa K, Sugii N, Yamaguchi S, and Miyao M 1999 *J. Crystal Growth* **201/202** 560

Acknowledgements

This work was partly supported by a project for strategic advancement of research infrastructure for private universities, 2009–2013 and by a Grant-in-Aid for Scientific Research (C) (Grant No. 20560025) from MEXT, Japan.

Measurement of Energy Band Structure of MgO, MgSrO and MgCaO Thin Film by their Secondary Electron Emission Coefficient due to Auger Neutralization

K A Lee, B K Min, Y S Byeon, J H Choi, R J Jung, H S Uhm and E H Choi[1]

Department of Electrophysics, Kwangwoon University,
Kwangwoon-gil 26, Nowon-gu, Seoul, Korea 139-701

E-mail : ehchoi@kw.ac.kr

Abstract. The energy band structure of MgO, MgSrO and MgCaO thin films has been investigated based on the Auger neutralization theory. Profiles of secondary electron emission from MgO, MgSrO and MgCaO thin films have been measured by gamma focused-ion-beam (γ-FIB) system. We have measured the energy band structure, $f_e(\alpha)$, function in their respective valence band from the secondary electron emission characteristics by γ-FIB system with He ion beam whose energy is less than 200 eV in the surface of MgO, MgSrO and MgCaO thin films. For this work, we have employed a Fast Fourier Transform (FFT) and its Inverse-Fast Fourier Transform of the secondary electron current profile emitted from the thin films due to the Auger neutralization mechanism for the determination of the energy band structure $f_e(\alpha)$.

1. Introduction

Nowadays the trend of display industry today is going to the three-dimensional (3D) image system, where plasma display panel (AC-PDP) and liquid crystal display (LCD) are the major players in the world display markets. It is known that the AC-PDP has the better response time and the color impression than the LCD panel, because the visible color light is noticeably comfortable reducing eyestrains. In order to lower the firing voltage in AC-PDP, a high emission coefficient (γ) of the secondary electrons from the protective layers like CaSrO film [1-2] is needed. However, this CaSrO is extremely sensitive to H_2O and CO_2, forming hydroxides and carbonates of Ca and Sr. It is therefore necessary to avoid exposure of deposited CaSrO protective layers to atmosphere by processing the PDP front glasses in inert gases or vacuum [3]. Hence we propose the MgSrO and MgCaO thin film for the protective layers for AC-PDP, in which small amounts of strontium Sr and calcium Ca are included in the MgO pellet, repsectively, in this research. Therefore it is essential and necessary to evaluate the material properties of MgSrO and MgCaO whether these materials could be used for protective layer instead of conventional MgO fim and also can be used for the alterrnative material for CaSrO in AC-PDP.

In this experiment, we have compared characteristics of MgO, MgSrO and MgCaO thin films by X-ray diffraction pattern (XRD) and secondary electron emission coefficient (γ). The measurement of electron energy band structure of MgO single crystal is one of the essential factors to improve the efficiency in AC-PDP, in which the work-function ($\varphi_{(w)}$), energy band width χ, central energy ε_0 of the valence band and any other important information of MgO thin film are also accordingly analyzed and obtained. The work function $\varphi_{(w)} = E_{i0}/2$ of specified MgO thin film has been experimentally obtained from the direct

[1] To whom any correspondence should be addressed

measurement of ion-induced secondary electron emission co-efficient γ versus the various ionization energies E_i for slow ions and its extrapolation of least-squared-fitting for determinations of E_{i0}, which is the ionization energy satisfying the zero emission of ion-induced secondary electrons. During the measurement of the work function $\varphi_{(w)}$ of MgO thin film from the ion-induced secondary electron emission co-efficient γ, the slow ions of He^+, Ne^+, Ar^+, N_2^+, and Xe^+, whose energy is less than 100 eV have been used throughout the experiment [4]. We have investigated the respective electron energy band structure of MgO, MgSrO and MgCaO thin films, using the results of secondary electron emission characteristics (γ) based on the Auger neutralization by gamma focused-ion-beam (γ-FIB) system.

2. EXPERIMENTAL CONFIGURATIONS

Figure 1 shows a schematic of the γ-focused ion beam (FIB) system consisted of the thermal electron source, the ionization region of ions, the electrostatic single Einzel lens that focused the ion beam, and the collector and copper pad for the measurement of the secondary electron emissions from the surface of the thin films.

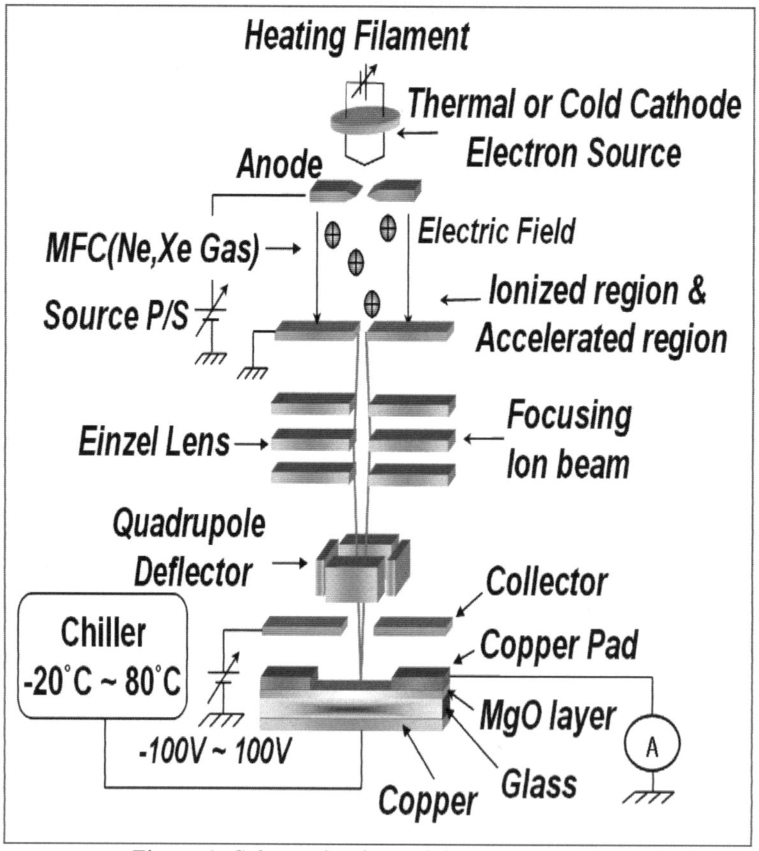

Figure 1. Schematic view of the γ-FIB system.

The thin films of MgO, MgSrO and MgCaO have been deposited by electron beam evaporation at 300℃ and evaporation speed of 20Å/sec, where their thicknesses are about 6000~8000Å in this experiment. We have employed the He ion whose ionization energy E_i is 24.58eV with low energy below 200 eV in this experiment. The He ion approaches the surface of film material, as shown in the figure 1. It is noted that the electric field is polarized toward the collector from the grounded copper pad when the collector potential is negative biased. The secondary electrons emitted from the surface by this slow ion then come back, and only the ion current (I_i) coming to the surface is then measured. On the other hand, the positively biased collector makes the electric field toward the grounded copper pad. Due to this electric field, the secondary electrons emitted from the surface of thin film by the ion beam bombardment move up toward the collector, registering the current (I_t) in the ampere meter shown in the figure 1, in which the emitted secondary electron and ion beam currents (I_i) are included together. The secondary electron emission γ is obtained from $\gamma = (I_t-I_i)/I_i$ [5].

Figure 2 shows the schematic diagram for the Auger neutralization. When an ion approaches on the surface of the respective MgO, MgSrO and MgCaO thin film, an electron in the valence band is combined to the approaching ion by quantum mechanical tunneling effect for their neutralization and this ion becomes to be dropped to the ground state. During the neutralization process, the energy for tunneling electron for neutralization has been transferred to the nearby electron. When the transferred energy to the nearby electron is higher than the work-function, then the secondary electrons could be emitted from the surface at the vacuum. We define the characteristic energy spread of electrons in the valence band as χ and energy level located in the center of the valence band as ε_0, which is shown in figure 2. The symbol α_{min} for an electron in figure 2 is the work function of respective MgO, MgSrO and MgCaO thin film and we defined as $\alpha_{min}= \varepsilon_0-(\chi/2)$, $\alpha_{max}= \varepsilon_0+(\chi/2)$, and also β for nearby electron is similar to that for α. If minimum kinetic energy E_k^{mim} from the electrons located at α_{max} and β_{max} is almost close to zero, we can set $E_k^{mim}=E_i-(\alpha_{max}+\beta_{max})=0$ and $\varepsilon_0= (E_i- \chi)/2$ [4].

The Auger transform $T(E_k)$ defined as

$$T(E_k) = \frac{1}{\chi^2}\int_0^{\infty}\int_0^{\infty} f_e(\beta)\delta(E_k - E_i + \alpha + \beta) f_e(\alpha) d\alpha d\beta \qquad (1)$$

For convenience in the subsequent analysis, the dimensionless energy variables x and x' are defined as $x = (\alpha - \varepsilon_0)/\chi$ and $x' = (\beta - \varepsilon_0)/\chi$, where χ is the characteristic energy spread of electrons in the valence band. Then, Eq. (1) is equivalently expressed as $x' = y - x$, where the energy variable y is defined by

$$y = \frac{E_i - 2\varepsilon_0 - E_k}{\chi} \qquad (3)$$

The state density function f_e in the valence band is a function of the electron energy α in the band. The Auger transform $T(E_k)$ defined by Eq. (1) could be expressed by

$$T(E_K) = \frac{1}{\chi^2}\int_{-\infty}^{+\infty}\int_{-\infty}^{+\infty} f_e(\beta)\delta(x' - y + x) f_e(\alpha) d\beta d\alpha \ . \qquad (3)$$

Now the $T(E_k)$ function can be expressed in terms of variables y and x as

$$T(y) = \int_{-\infty}^{\infty} f_e(x) f_e(y-x) dx. \quad (4)$$

For a specified density $N(E_k)$ of final states, the Auger self-convolution $T(y)$ in Eq. (4) can be experimentally determined from $T(y) = N_0(E_k)/N(E_k)$ by measuring the distribution in energy, $N_0(E_k)$, of electrons escaping from the thin film. It is noted that the Auger self-convolution for ultra-thin film may have the same profile as the distribution of electrons in energy because of 2-dimensional structure. The Auger self-convolution $T(y)$ in Eq. (4) is a typical convolution of the state density function $f_e(x)$ in the valence band. It is convenient in the subsequent analysis to define the Fourier transform by

$$F(k) = \frac{1}{\sqrt{2\pi}} \int_{-\infty}^{+\infty} f(t) \exp(ikt) dt, \quad (5)$$

Taking the Fourier transform of both sides in Eq. (4), it is straightforward to show that the Fourier transform $S(k)$ of the Auger self-convolution $T(y)$ in Eq. (4) relates to the Fourier transform $F(k)$ of the normalized density function f_e by

$$S(k) = \sqrt{2\pi} \left[F(k) \right]^2 . \quad (6)$$

Finally, the state density function f_e in the valence band could be obtained by using the inverse-FFT (IFFT) of $F(k)$, i.e.,

$$f_e(x) = \frac{1}{\sqrt{2\pi}} \int_{-\infty}^{\infty} F(k) e^{-ikx} dk \quad (7)$$

Hence we will have the information of the density-of state function of $f_e(x)$ in the valence band of thin film surfaces of MgO, MgSrO and MgCaO [4].

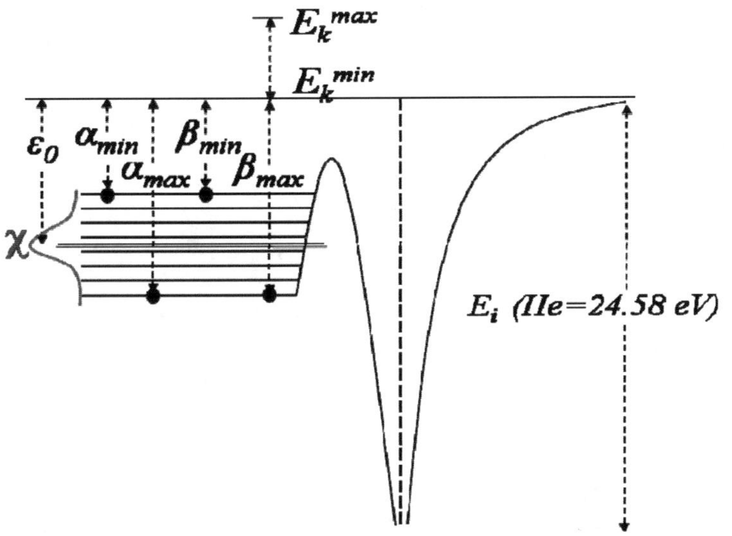

Figure 2. Schematic diagram illustrating the Auger neutralization of an ion at the surface

3. RESULTS AND DISCUSSIONS

The thin films of MgO, MgSrO and MgCaO have been deposited by electron beam evaporation under the 10^{-5} Torr working pressure and the gas temperature 300 ℃. Figure 3 shows the top and side images of MgO, MgSrO and MgCaO thin films at the upper and lower sides, respectively, by scanning electron microscopy (SEM). The surface shapes of MgO and MgCaO are shown to be triangualr with columnar structure of (111) crystal orientation. On the other hands, the surface image of MgSrO has shown to be smoother than others, and shown to be without any crystal orientation. Thicknesses of MgO, MgSrO and MgCaO thin films are about 780nm, 640nm and 740nm, respectively. It is noted that the MgO, MgSrO, and MgCaO films have been deposited by pure MgO pellet, MgSrO pellet doped with 15% strontium, MgCaO pellet doped with 10% calcium, repsectively, in this experiment.

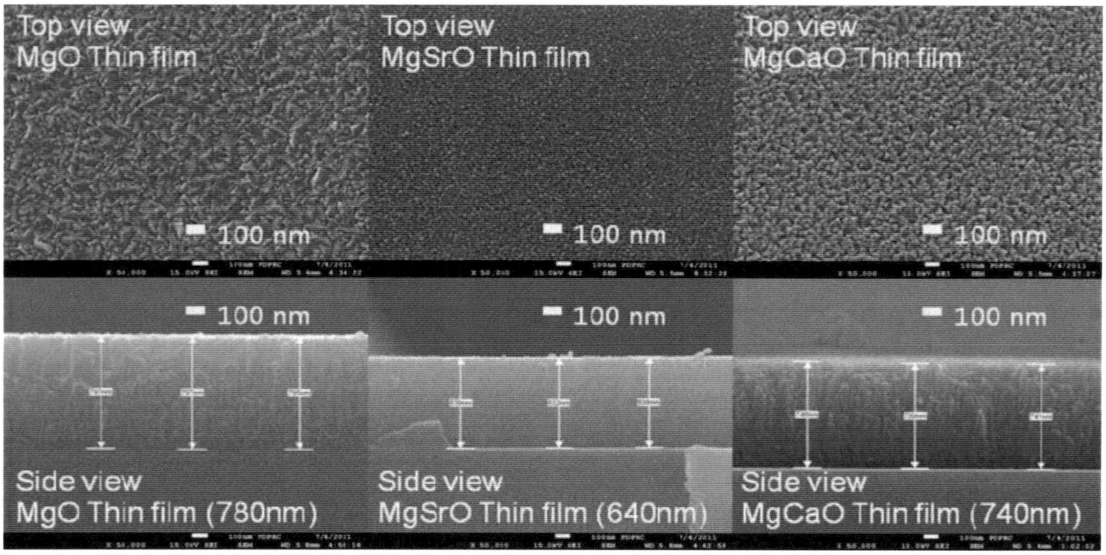

Figure 3. The top (upper) and side images (lower) of MgO, MgSrO and MgCaO thin films by scanning electron microscopy (SEM)

Figure 4 shows the X-ray diffraction pattern of the MgO, MgSrO and MgCaO thin films, in which MgO and MgCaO shows the poly crystallinities of (111), (200) and (220). Also it is noted that there is no crystal orientation for the MgSrO thin film. It has been found that the MgCaO thin film has an obvious peak of (111) orientation, by which the secondary electron emission coefficient (γ) might be generated higher than others.

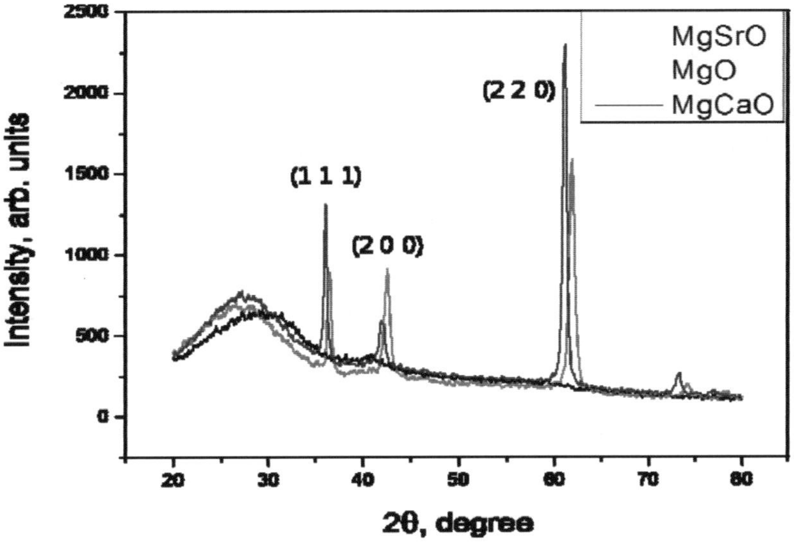

Figure 4. X-ray diffraction pattern of MgO, MgSrO and MgCaO thin films deposited by E-beam evaporation

Figure 5 shows the secondary electron emission coefficients (γ) from MgO, MgSrO and MgCaO thin films, which have been measured by gamma FIB system, versus the He ion energy ranged from 130 eV to 200 eV.

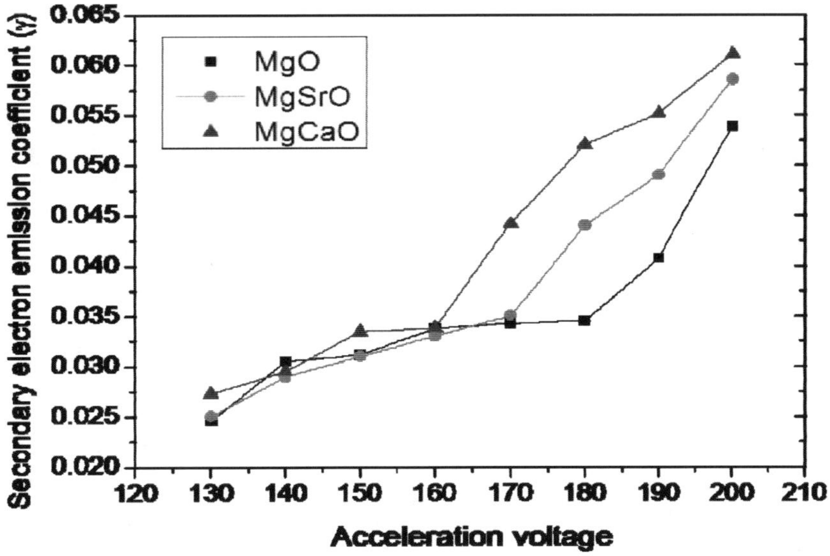

Figure 5. Secondary electron emission coefficient (γ) of the MgO, MgSrO and MgCaO thin films.

It is shown that the secondary electron emission coefficient (γ) for MgCaO thin film has the highest value over He ion energy of 160 eV since it has poly crystalline orientation of (111) and (220), as shown in figure 4. It is noted that the secondary electron emission coefficients (γ) for MgO, MgSrO, and MgCaO thin films are 0.054, 0.059, and 0.061, repectively, under He ion energy of 200 eV. When a He^+ ion approaches on the MgO, MgSrO, and MgCaO thin films, the secondary electron is released to outside of these thin film surfaces by Auger neutralization theory and the pico-ammeter measures this released electron. We have measured the current signals with respect to the collector voltage in the γ-FIB system, represented by the solid square-dot in figures 6(a), 7(a) and 8(a) for the MgO, MgSrO, and MgCaO thin films. These current signals have been captured by using a pico-ammeter connected to the grounded copper pad in figure 1, where the collector voltage was swept from -8 V to +15 V, with resolution of 0.1 V.

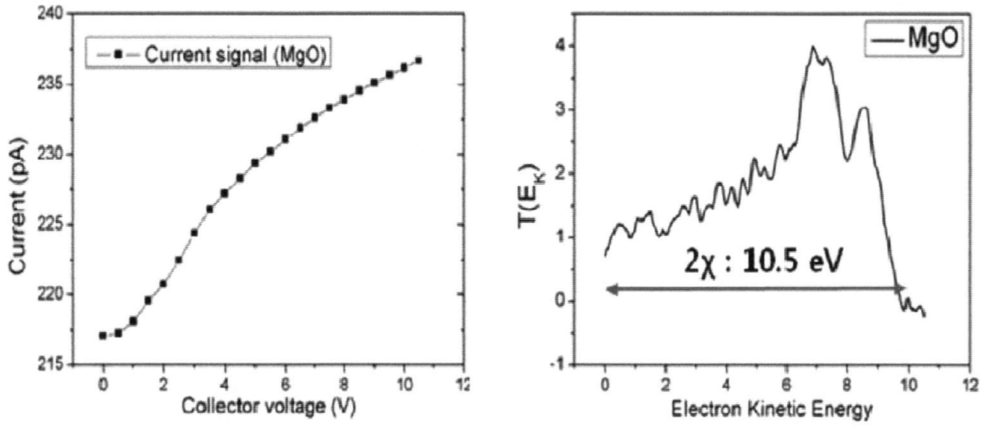

(a) Secondary electron current vs. collector voltage **(b) Plots of Auger transform $T(E_k)$ vs. E_k**

Figure 6. Secondary electron emission current versus collector voltage (a) and plot of Auger transform $T(E_k)$ in terms of the secondary electron kinetic energy E_k emitted from the MgO thin film

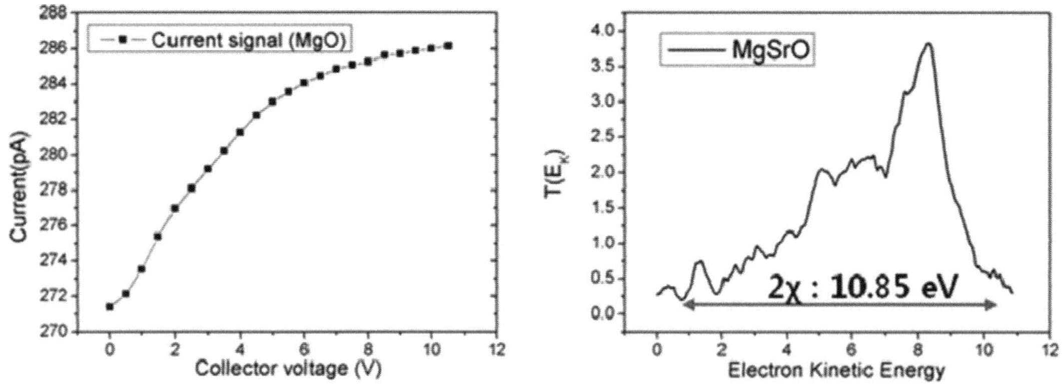

(a) Secondary electron current vs. collector voltage **(b) Plots of Auger transform $T(E_k)$ vs. E_k**

Figure 7. Secondary electron emission current versus collector voltage (a) and plot of Auger transform $T(E_k)$ in terms of the secondary electron kinetic energy E_k emitted from MgSrO thin film

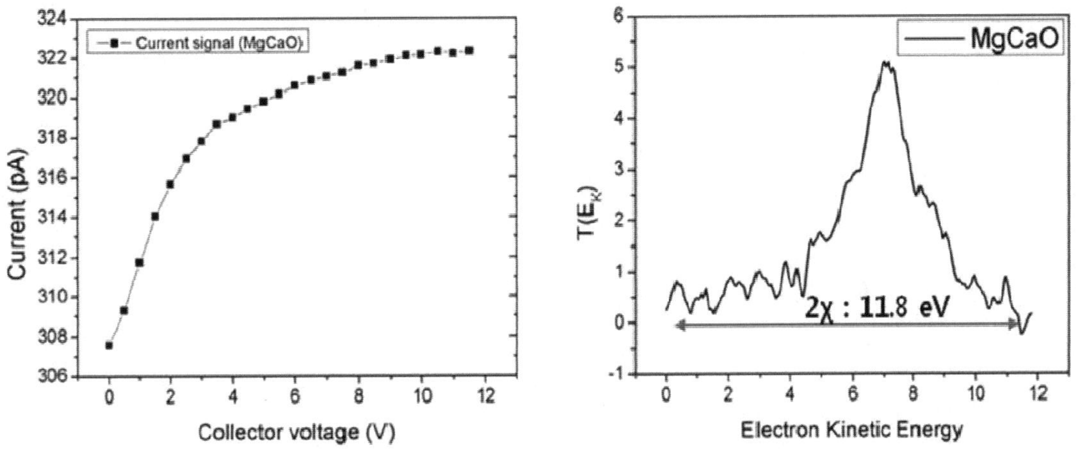

(a) Secondary electron current vs. collector voltage **(b) Plots of Auger transform $T(E_k)$ vs. E_k**

Figure 8. Secondary electron emission current versus collector voltage (a) and plot of Auger transform $T(E_k)$ in terms of the secondary electron kinetic energy E_k emitted from MgCaO thin film.

The energy distribution for the emitted secondary electron from the film surfaces could be represented by differentiating the current signal with respect to the collector voltage. It is noted that the emitted electrons has maximum kinetic energy at the low collector potential corresponding to the current starting–up, but it has a minimum at the collector potential around 11V correponding to current saturation point, as shown in figures 6(a), 7(a) and 8(a). Rescaling this electron distribution in terms of the electron kinetic energy, we can experimentally obtain the Auger transform $T(E_k)$ defined in Eq. (1). The Auger transforms of $T(E_k)$ have been plotted for the thin films of MgO, MgSrO, and MgCaO in figures 6(b), 7(b) and 8(b), respectively, in terms of the their electron kinetic energy, in which the characteristic energy spread χ of electrons in the valence band could be measured since the energy spread for the secondary electrons are denoted by 2χ and the center of the valence band ε_0 could be estimated by $\varepsilon_0 = (E_i - \chi)/2$. Making use of the Auger transform for $T(E_k)$, which is the energy distribution profiles for the emitted secondary electrons, as shown in figures 6(b), 7(b) and 8(b), and carrying out the Fourier transform and inverse transform numerically as mentioned in Eqs. (6) and (7), the electron energy band structures, i.e., density-of-state function $f_e(\alpha)$, of the thin films can be obtained by Eq. (7).

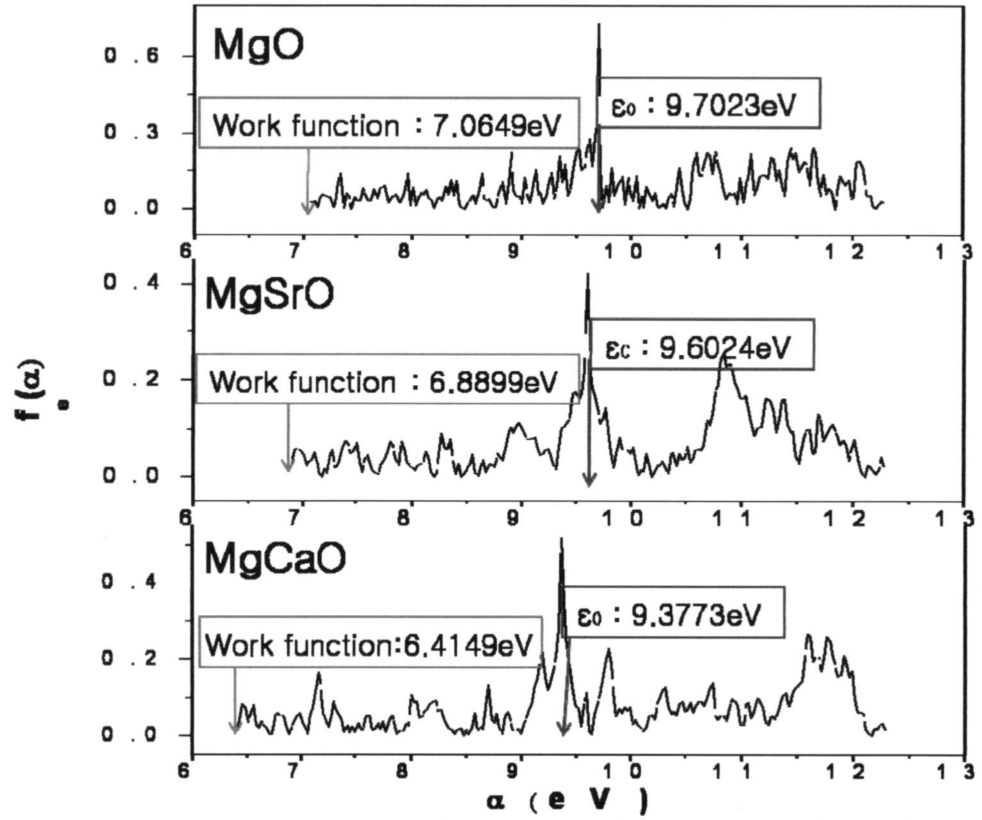

Figure 9. Electron energy band structures of the MgO, MgSrO and MgCaO thin films derived from the secondary electron emission current by γ- FIB system

Figure 9 shows the electron energy band structure, i.e., the density of state function for the electrons in valence band in MgO, MgSrO, and MgCaO thin films, where the work function φ_w and the center of the valence band ε_0 in the valence band are represented, respectively. It is noted that the work functions φ_w are found to be about 7.1 eV, 6.9 eV, 6.4 eV for MgO, MgSrO, and MgCaO thin films, repectively. Also the central energies ε_0 of valence band are found to be about 9.7 eV, 9.6 eV, and 9.4 eV for MgO, MgSrO, and MgCaO thin films, repectively. It is also noted in this experiment that the kinetic energy spread, which is given by 2χ, of the emitted secondary electrons are 10.5 eV, 10.9 eV, and 11.8 eV for MgO, MgSrO, and MgCaO thin films, repectively. From these data, the MgCaO thin film could be used for the best protective layer for the AC-PDP in comparison with other MgO and MgSrO films since the work function φ_w and the central energy ε_0 of the valence band are minimum values for the highest value of the secondary electron emission coefficient.

4. CONCLUSIONS

The electron energy band structure $f_e(\alpha)$ of the MgO, MgSrO and MgCaO thin films has been investigated, respectively, under the Auger neutralization theory by measument of the secondary electron emission current emitted by low energy He ion beam whose energy less than 200 eV with the gamma focused ion beam (γ-FIB) system. Making use of the Auger transform for $T(E_k)$, which is the energy distribution profiles for the emitted secondary electrons, and its inverse Fourier transform yield the state density function $f_e(\alpha)$ of MgO, MgSrO, MgCaO in the valence band. MgCaO is found to have the highest secondary electron emission coefficient (γ) for He ion energy in the range of 130 eV and 200 eV. It is also noted that MgCaO film has the lowest central energy ε_0 = 9.4 eV and the work function φ_w = 6.4 eV. Therefore, our experimental observation indicates that MgCaO thin film looks like to be the most efficient material for emission of the secondary electron, which will in turn eventually reduce the firing voltage and provide a high brightness in plasma display panel.

ACKNOWLEDGEMENT

This research was supported by the MKE (The Ministry of Knowledge Economy), Korea under the ITRC (Information Technology Research Center) Support Program supervised by the NIPA (National IT Industry Promotion Agency)" (NIPA-2009-C1090-0902-0018) and also supported by the SRC program (20100029418). This work also partially supported by Kwangwoon University 2012.

5. REREFENCES

[1] Motoyama Y, Kurauchi T, 2006 SID 06 Digest, 1384
[2] Matulevich Y T et al, 2007 IMID 07 Digest, 213
[3] Matulevich Y T et al, 2009 IMID 09 Digest, 181-183
[4] Uhm H S, Choi E H, and Cho G S, 2009 Appl. Phys. Lett. **94**, 0031501
[5] Lim J Y, Oh J S, Ko B D, Cho J W, Kang S O, Cho G S, Uhm H S, Choi E H, J. Appl. Phys. 2003 **94**, 764
[6] Choi J H, Son C G, Hong Y J, Park B J, Uhm H S and Choi E H, ADVANCES IN NATURAL SCIENCES : Nanosci. Nanotechnol. 2010 1, 045014, 5

Structured nanocarbon on various metal foils by microwave plasma enhanced chemical vapor deposition

G Rius and M Yoshimura
Toyota Technological Institute, 2-12-1 Hisakata, Tempaku, 468-8511, Nagoya, Japan

E-mail: rius@toyota-ti.ac.jp, yoshi@toyota-ti.ac.jp

Abstract. We present a versatile process for the engineering of nanostructures made of crystalline carbon on metal foils. The single step process by microwave plasma-enhance chemical vapor deposition is demonstrated for various substrate materials, such as Ni or Cu. Either carbon nanotubes (CNT) or carbon nanowalls (CNW) are obtained under same growth conditions and without the need of additional catalyst. The use of spacer and insulator implies a certain control over the kind of allotropes that are obtained. High density and large surface area are morphological characteristics of the thus obtained C products. The possibility of application on many metals, and in the alloy composition, on as-delivered commercially available foils indicates that this strategy can be adapted to a bunch of specific applications, while the production of C nanostructures is of remarkable simplicity.

1. Introduction

The potential applications of nanostructured materials can play a major role for the ultimate performance of devices. A particular attention is being paid to the nanosized crystalline C materials. Beyond their use as building blocks for fundamental investigations, they show promise for many applications. There is an intense interest to exploit their electro-optical, mechanical, chemical properties resulting from some specific morphological and structural characteristics. Functional devices made with nanostructured C, such as batteries, supercapacitors, field emission devices, etc. show promise to fulfill the present requirements of charge storage and energy conversion, specially, in terms of efficiency. In other words, they are expected to have capabilities, for instance, to supply both enough energy and power densities for practical operations. Many demonstrators for the proof of concept can be found in the literature, as examples of their advantageous performance. In 2007, Kaempgen et al. [1] demonstrated the supercapacitor performance of CNT electrodes thanks to the creation of superior electrical double layer capacitance. In this case, network of CNTs were deposited onto the metal contacts from solution. More recently, ac line filtering by graphene sheets based on the double layer has also been presented. [2] In this case, the production of C material is based on a growth process similar to the one shown in the present paper. Other strategies for advanced capabilities include nanoparticle-carbon compound materials able to catalyze reactions, such as in water splitting or fuel cells.

Limitations on nanosized C material synthesis include the high control over the crystallinity, purity, and so on. Understanding the process conditions and mechanisms that lead to the different morphologies is essential for their vast consolidation as functional materials. Yet, for transferring such devices from the lab to the market, technological developments for batch fabrication and inexpensive and compatible processing methods are needed. In this communication, we present our findings for the direct synthesis of carbon nanotubes (CNT) [3] and carbon nanowalls (CNW) [4] on metal substrates. This method allows obtaining high density and large surface area C material in a single process and upon conductive supports, which is ideal for charge storage devices. It is produced by microwave plasma enhanced chemical vapor deposition (MPECVD) on as-purchased metal foils.

2. Experimental

We employ a commercial apparatus, CVD-CN-100 from Ulvac Japan Ltd., similar to the setup used in [4]. The synthesis conditions are common for all the different substrates presented herein. Whole process is realized at a constant pressure of 1.5 Torr and assisted by microwave field of 2.45 GHz and 500 W. In particular, nanocarbon material synthesis consists of two steps. First, surface conditioning by H_2 flow during 3 minutes and applying an electrode bias of 100 V; followed by 10 minutes C deposition. The C precursor is CH_4 mixed with H_2 in a 1:1 ratio (20:20 sscm) and 200 V bias for the generation of plasma. The preliminary stage under H_2 flux has a double function; it cleans and prepares the metal for C deposition, but also preheats the chamber and sample. There are no sample stage or chamber tube heaters, so temperature increase is driven by microwave absorption and resulting energy transfer in the plasma environment. Our group has been studying the conditions of synthesis for the CNWs as a function of feedstock gases. [5] Based on that investigation, we test and analyse here the possibility of diversifying the substrates for specific applications (the metal foils) and their effect on the C nanostructures deposition.

As for the substrates, we use commercially available metal foils of high purity by Nilaco (Ni, Cu, constantan, Mo, Ti, etc.). For convenience, we center the presentation of the experimental results and discussion only to metal foils containing Ni and Cu. The choice of Ni and Cu is rooted on the fact that they are the most commonly used catalysts for graphene deposition. Specifically, we show the products obtained on Cu foil of 500 μm thickness, Ni foil 300 μm and Ni-Cu alloy containing 45 % Ni and 55% Cu (namely constantan) of a thickness of 500 μm. SiO_2 on Si chips are also used, either as deposition sites or for isolating the metal foils from the sample stage (underneath the metal foil and onto the Mo sample stage, lower electrode). As insulator and spacer, alumina crucible is employed as well. The necessity and function of the electrical and spatial isolation are presented in the description of results and discussion for the experimental products. Typical sample size is smaller than 1 cm^2.

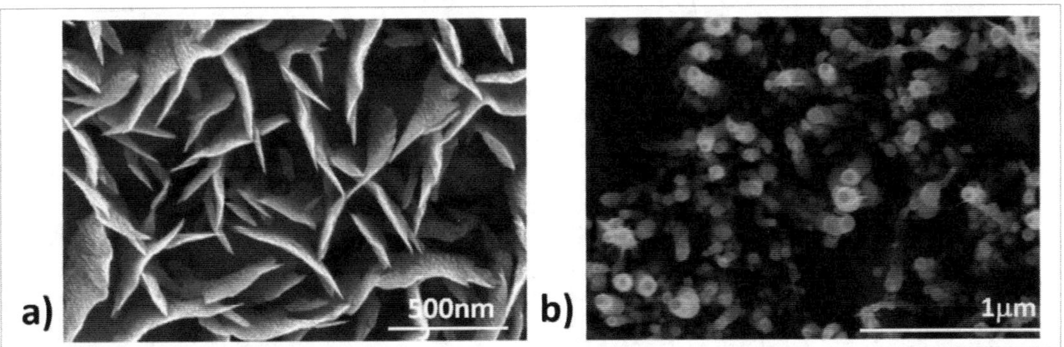

Figure 1. Top view SEM images of nanostructured C deposited on the metal foil using 100 nm thick SiO_2 on Si as the stage-metal separator. a) CNWs are obtained on the Cu foil. b) For Ni foil, CNTs are grown without the addition of catalyst particles to promote the CNT nucleation sites.

Characterization is here based on scanning electron microscopy (SEM) and Raman spectroscopy for the observation of deposited C morphology, arrangement, density, etc. and the structural identification of nanocarbon products, respectively. Optical observation is applied to the determination of extension and uniformity of C deposition coverage. SEM instrument is a field emission Hitachi

S4700, operated at 20 kV, and Raman spectroscopy is done with an inVia Reflex from Renishaw, using a 532 nm excitation laser, ~0.3 mW, 10 s accumulation times.

3. Results and discussion

Two major parameters have been investigated as deterministic conditions for C nanostructuring on various metal foils by MPECVD, i) the substrate material, [6] and ii) the effect of local distribution or configuration of electric field during C deposition. [7] Depending on substrate material, changes of C solubility in metal, absorption of microwave field (leading to local temperature increase), and creation of particles to promote CNT formation, deposition sites for CNWs, etc. are reported to determine C deposition. [8] The influence of field and plasma distribution can also be found in the literature. [9, 10] We present the potential and strength of our approach as follows.

Figure 1 shows the results of C deposition onto samples of Cu and Ni foils. Different from other supports that will be introduced later on (see Figure 2), metal foil pieces were placed on top of a 100 nm SiO_2 on Si chip during MPECVD process. This way, we isolate physically the metal foil from the stage to have a reference with bare SiO_2 substrate as C deposition site [5] to evaluate the metal role and to prevent stage contamination. On Cu foil, packed randomly-oriented CNWs are formed (Figure 1, a)). The vertically arranged graphene flakes are less than 2 μm in height, and ~ 200 nm to 1 μm across. On the other hand, vertically aligned multiwalled CNTs (MWCNTs) of diameters ranging from ~ 15 to 60 nm are produced onto the Ni foil (Figure 1, b)). It is worth remarking this point. CNTs are produced without using specific or additional catalytic particles and that CNT formation does not occur for Cu foil under same processing conditions. These two facts suggest that the tubular growth is possible in the case of Ni due to the solubility of C. The CNT formation may proceed by C nucleation and diffusion [6] in the protrusions formed on the Ni surface due to plasma bombardment and heating, probably, stretching metal from the foil to form catalytic particle.

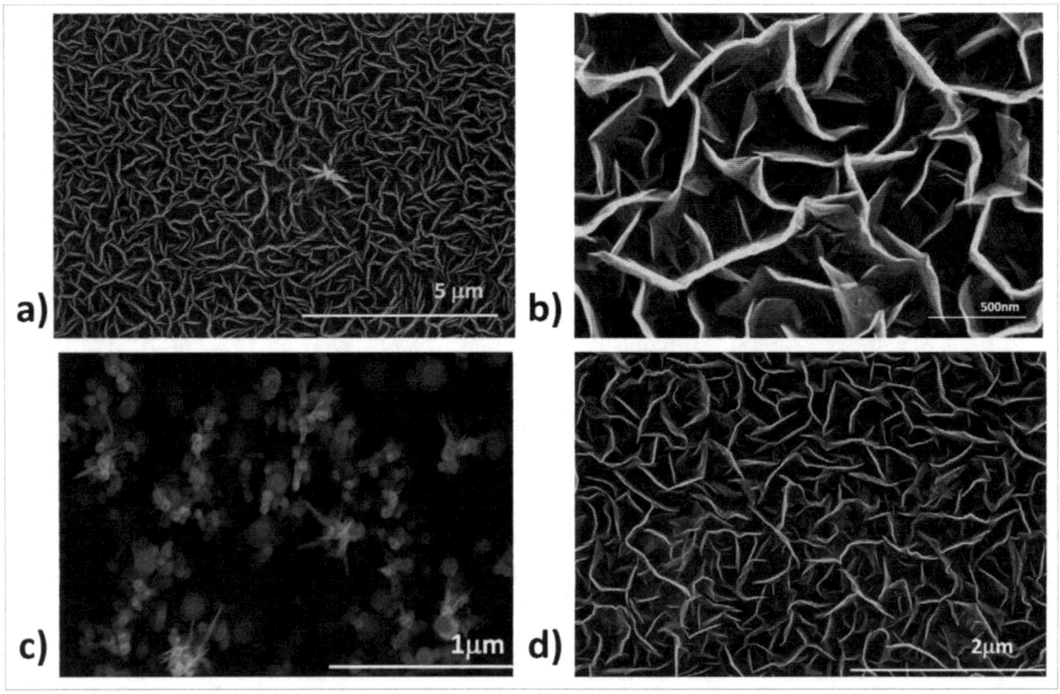

Figure 2. CNWs deposited on Cu using a) 500 nm SiO_2 on Si and b) alumina plate, underneath the metal foil. c) CNTs and d) CNWs grown on Ni using 500 nm SiO_2 on Si and alumina plate, respectively, underneath the metal foil.

The deposition of CNWs in a Cu foil presents a higher density as compared to the bare SiO_2 substrate (data not shown [11]). This can be attributed to the metal through an increase of catalytic activity and, probably, effective temperature. The low solubility of C in Cu promotes the C species to accumulate on the surface and prevents the CNT formation, but the state of the substrate after H_2 flow conditioning can affect as well. Atomic force microcopy imaging [11] demonstrates that pre-growth H_2 treatment smoothes the Cu foil at the micron scale, but causes tiny protuberances in the nanoscale range at the same time. These Cu nanograins on the same Cu foil perhaps additionally promote the vertical arrangement of the multilayer graphene sheets when assuming the Volmer-Weber growth mechanism [2] consisting in horizontal C nucleation islands that become perpendicularly arranged to the substrate by mutual impingement during growth.

The fact that C deposition does not occur on Si samples placed in contact with the Mo electrode (sample stage), but it does in SiO_2 motivated the analysis of spacer effect. The results are exemplified and summarized in Figure 2. In the case of Cu foil pieces, we obtain always similar results using SiO_2 on Si of 100 nm (Figure 1, a)) and 500 nm SiO_2 thickness (Figure 2, a)) or the alumina crucible (Figure 2, b)), a high density of CNWs. However, for Ni foils, lower amount of CNTs are produced if spacer of 500 nm SiO_2 on Si is used (Figure 2, c)) (to be compared to Figure 1, b) for 100 nm SiO_2 on Si) and CNWs are obtained when using the alumina crucible (Figure 2, d)). We understand this phenomenon as a demonstration of the importance of the plasma distribution and related electric field for the conditions that lead to C deposition and structuring. [9, 10]

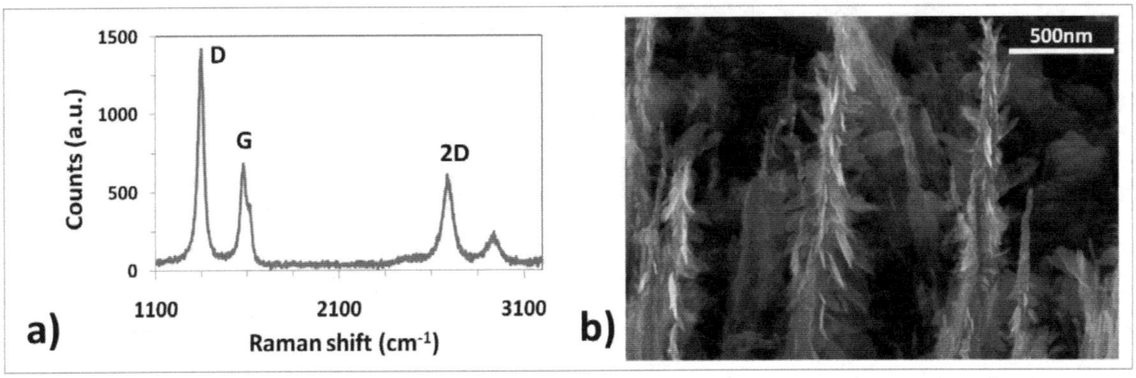

Figure 3. a) Raman spectrum of CNTs deposited on a Ni foil. b) Tilted SEM image of CNTs* grown on constantan foil.

An example of structural C nanomaterial characterization is shown in Figure 3, a), in particular, for the CNTs deposited on Ni foil. Raman spectra present the typical D-band, G-band and 2D-band Raman signatures of graphene-like material at ~1350 cm^{-1}, ~1590 cm^{-1}, and ~2695 cm^{-1}, respectively (similar would obtained for CNWs). The ratio and FWHM of G to 2D peaks confirm the multilayered structure of both CNWs and CNWs. [12] The high D peak accounts as well for the multilayered CNTs and the defects and imperfections of CNWs, high density of edges [13]. In addition, a small shoulder in the G-band is observed at 1620 cm^{-1} and is assigned to the D'-band, which also relates to disorder. [14] The common Raman features of CNTs and CNWs limit further analysis of materials, but could quantify and discern the presence of graphitic C or solid C fibres, if abundant. In this sense, Raman spectroscopy may not be informative of hybrid structures such as the all-C flake-decorated CNTs, named as CNTs*, shown in Figure 3, b). These kinds of materials are extremely attractive due to their intrinsic larger surface area or field emission properties. [15] Actually, the fact that micro-Raman characterization averages data from microns wide area does not allow to obtain individual CNT or CNW information in order to formulate more precise structural statements, such as the number of layers or inner structure of CNTs. Transmission electron microscopy would be ideal, especially to observe the resulting CNT catalyst particle size and shape. However, this analysis goes beyond the scope of present communication and no pristine representative data would be easily obtained.

The CNTs* can be easily obtained when using Constantan foil pieces (Figure 4), likewise processed as on Ni or Cu foils. Again, the use of SiO_2 on Si spacer makes possible the C structuring in the form of CNTs (Figure 4, a) and b)), while CNWs are deposited if using the alumina crucible as separator (Figure 4, c)). More uniform diameter of the CNTs (Figure 4, a)) is obtained, which indicates that certain particle aggregation is prevented by metal alloying. In the basis of Ni versus Cu results, where CNT were only grown upon Ni and lower density (Figure 4, a) in opposition to Figure 1, b)), the use of Ni-Cu alloy reinforces the idea of the need of C solubility inside the metal to form CNTs in present MPECVD conditions. Furthermore, the use of Constantan combined with 500 nm SiO_2 on Si spacer, in Figure 4, b), generates amount of CNTs*, that is the flake decorated together with protruded CNTs, which may be related to the presence of Cu in the form of particles. Other alloys, such as so-called Advance (Ni-Cu-Mn), lead to similar structuring mechanism and stressed the material selectivity, and consequential, density dependence as well. The possibilities of directly tuning the C nanostructuring, using combination of many materials and adjusting of synthesis process are of application-oriented significance.

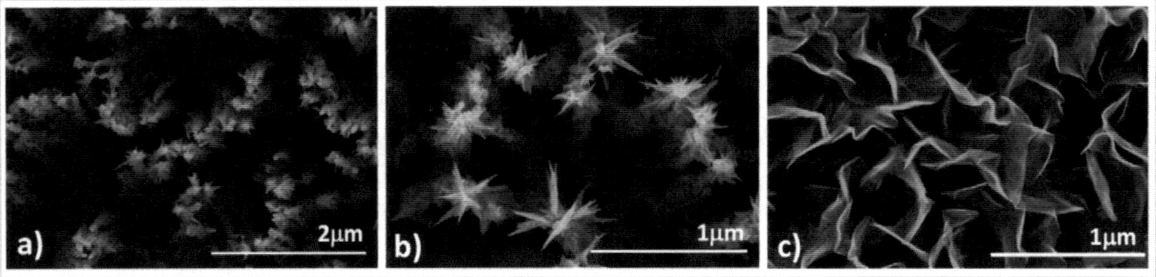

Figure 4. Nanostructured C materials produced on Ni-Cu alloy as a function of the mediator between metal sample and sample stage, a) CNTs, by 100 nm SiO_2 on Si, b) CNTs*, by 500 nm SiO_2 on Si, and c) CNWs, by alumina crucible.

Final remarks related to the present development are included to highlight the importance of the spacer. In addition to what has already been demonstrated, we observe a drastic change in nanocarbon catalysis and its arrangement if spacer is not used. For instance upon Cu foils, whereas electric isolation with respect to sample stage promotes the vertical arrangement of C nanostructures (the CNWs), metal foil making contact with Mo electrode leads to the layered deposition of some amount of low graphitized carbon, horizontally covering the Cu piece. This behavior suggests that physical and electrical contacts are deterministic and intrinsically related for the plasma conditions used in this synthesis processing conditions. The same MPECVD system is being used in another of our technological approaches to integrate CNTs into atomic force microscopy probes. In this application, we use the vertically aligned multiwalled CNTs to functionalize only the vicinity of the tip apex of commercial Si probes. [16] In this application, it is also primordial to introduce some sort of discontinuity that compensates the tip morphology effect. We use a homemade sample holder with membranes to screen the electric fields and ion species during growth, in such a way that their deceleration permits the CNT growth. A dedicated study to analyze this effect in base of CNW growth in Cu foils has been realized and will be soon released for publication. [11]

Spacer / Support	SiO_2 100nm	SiO_2 500nm	Alumina plate
SiO_2	CNWs	CNWs	CNWs
Copper	CNWs	CNWs	CNWs
Nickel	CNTs	CNTs *	CNWs
Nickel/Copper	CNTs	CNTs *	CNWs

Table 1. Summary of carbon allotropes obtained for each support and configuration.

4. Conclusions

In summary, we have presented a versatile process for the engineering of nanostructures made of crystalline carbon on metal foils. The single step process has been applied to various materials to obtain CNTs, CNTs* and CNWs under same growth conditions and without the need of additional catalyst particles. The use of insulator and spacer implies a certain control over the kind of allotropes that are obtained. High density and large surface area are morphological characteristics of the products.

Present strategy to tune the morphology of deposited C nanostructures by processing configuration and elements is an alternative procedure to the C nanoengineering by feedstock gases and use of nanoparticles. The properties of CNWs on Cu foils as electrochemical double layer capacitor are being tested and show an excellent performance for supercapacitor application. The possibility to be applied to other metals and on as-delivered commercial samples indicates that this strategy could be adapted to many specific device requirements. For example, it could be exploited for a bunch of materials (compound contents, e.g. alloys) or to adjust C nanostructures density or inorganic functionalization, at the same time as the production is of remarkable simplicity and controllability.

Acknowledgments

This study was financed in part by a grant of Support Program for Forming Strategic Research Infrastructure for Private Universities from the Ministry of Education, Culture, Sport, Science, and Technology, Japan.

References

[1] Kaempgen M, Ma J, Gruner G, Wee G and Mhaisalkar S G 2007 *Appl. Phys. Lett.* **90** 264104
[2] Miller J R, Outlaw R A and Holloway B C 2010 *Science* **329** 1637
[3] Iijima S 1991 *Nature* **354** 6348
[4] Wu Y, Qiao P, Chong T and Shen Z 2002 *Adv. Mater.* **14** 1 64
[5] Suzuki S, Chatterjee A, Cheng C-L and Yoshimura M 2011 *Jpn. J. Appl. Phys.* **50** 01AF08
[6] Okai M, Muneyoshi T, Yaguchi T and Sasaki S 2000 *Appl. Phys. Lett.* **77** 3468
[7] Wu Y and Yang B 2002 *Nano Lett.* **2** 4 355
[8] Wu Y, Yang B, Zong B, Sun H, Shen Z and Feng Y 2004 *J. of Materials Chemistry* **14** 469
[9] AuBuchon J F, Chen L-H, Gapin A I and Jin S 2006 *Chem. Vap. Deposition* **12** 370
[10] Hassouni K, Silva F and Gicquel A 2010 *J. Phys. D: Appl. Phys* **43** 153001
[11] More details about these results will be introduced in future communications and presented at the International Symposium on Surface Science, ISSS-6, Tokyo, Japan, 11-15th Dec 2011
[12] Dresselhaus M S, Dresselhaus G, Saito R and Jorio A 2005 *Physics Reports* **409** 47

[13] Ni Z H, Fan H M, Feng Y P, Shen Z X, Yang B J and Wu Y H 2006 *J. Chem. Phys.* **124** 204703
[14] Pimenta M A, Dresselhaus G, Dresselhaus M S, Cançado L G, Jorio A and Saito R 2007 *Phys. Chem. Chem. Phys.* **9** 1276
[15] Darbari S, Abdi Y, Mohjerrzadeh S and Asl Soleimani E 2010 *Carbon* **48** 2493
[16] Thomas I T, Rius G, Matsuoka Y and Yoshimura M 2010 *J. Vac. Sci. Technol. B* **28** 1148

Correlation between the electrical and structural properties of aluminium-doped ZnO thin films obtained by direct current magnetron sputtering

L. Avril[1], Ph. Guaino[2], F. Maseri[2], K. Muthukaruppasamy[2], J.-J. Pireaux[1]
[1]Centre de Recherche en Physique de la Matière et du Rayonnement, LISE, FUNDP, 61 rue de Bruxelles,
B-5000 Namur, Belgium
[2]CRM Group-AC & CS, 57 Boulevard de Colonster,
B-4000 Liège, Belgium

E-mail: ludo.avril@laposte.net

Abstract. High quality Aluminium-doped Zinc Oxide (AZO) films have been obtained by suitably controlling the magnetron sputtering parameters and the substrate temperature. The X-ray diffraction studies showed that a transition of orientation from (002) plane to (103) plane, versus substrate temperature. The surface morphology characterized by scanning electron microscopy and atomic force microscopy exhibited a dense and compact structure at higher temperature. For 200 nm thick AZO films deposited at temperature 530°C, using a ZnO target with an Al_2O_3 content of 3 wt%, the lowest electrical resistivity is 6.8×10^{-4} Ω.cm and transmittance is over 85% in the visible spectral region. The conductivity improvement of AZO films was closely related to the crystallanity characterized by the (103) orientation and the densely packed structure.

1. Introduction

For the next generation flat panel displays, solar cells, organic light emitting diodes and other optoelectronic devices, transparent conducting oxide (TCO) films have a wide range of applications as transparent electrodes [1-2]. Aluminium-doped zinc oxide (AZO) has attracted much attention because of its good electrical properties with high optical transparency comparable to indium tin oxide (ITO) [3-4]. Moreover, this oxide is abundant, non-toxic, easily fabricated at industrial scale and cheap, compared to ITO, which are the reasons that stimulating the research on In-free electrodes. AZO have been deposited by many methods including sputtering [5], pulsed laser deposition [6] and evaporation [7]. Among the deposition techniques, magnetron sputtering technique has been widely used because of its high deposition rate and high stability for large area films. The most of results in literature agrees with the preferred growth orientation for AZO thin films, which is along the (002) direction at low and high deposition temperature [8-10]. We rather focus on the improvement in the electrical properties when the growth orientation changes from (002) to (103) plane with increasing substrate temperature. We will explain the relationship between the (103) orientation and the electrical properties.

In this paper, AZO films were deposited by DC magnetron sputtering with different substrate temperatures. Structural, electrical and optical properties of AZO thin films were analysed in order to optimise its TCO properties for its application on $Cu(InGa)Se_2$ based solar cells fabricated on metal coil in a continue line process.

2. Experimental Procedures

(AZO) films have been prepared on soda-lime glass substrate by DC magnetron sputtering technique. A target with a mass proportion ZnO and Al_2O_3 (97:03), (99.95% purity), a size of 76.2 mm in diameter and 6 mm in thickness has been employed as source material. Initially, the chamber was evacuated to a pressure of 5×10^{-6} Torr by a turbo-molecular pump and Ar gas was used as a sputtering gas. The substrates were degreased in acetone, rinsed in de-ionized water followed by rinsing in ethanol and then dried under N_2 gas flow, before introducing into the sputtering chamber. The target to substrate distance is 170 mm and the sputter power is kept at 120 W for all depositions. The AZO target surface was pre-sputtered for 5 minutes in order to remove contaminations before deposition. Substrate temperature (T_s) was varied from 25 to 530°C and working pressure has been kept at of 3 mTorr during sputtering process. The thickness of all AZO films was around 200 nm, as-confirmed by a Veeco Dektak surface profilometer.

The structural properties of AZO films were characterized by a grazing angle X-ray diffraction (XRD, Bruker AXS D8 Advance diffractometer) with Cu K_α radiation (λ = 0.15406 nm) at fixed incidence angle of 2°. The surface morphology of the films was observed by scanning electron microscopy (SEM) (JEOL 7500F Company). The topography and surface roughness of the samples were studied by atomic force microscopy (AFM) with a commercial microscopy (Nanoscope IIIa, Digital Instruments) operating in tapping mode (TM), using a standard silicon cantilevers with spring constant ~ 42 $N.m^{-1}$ and 369 kHz driving frequency. All images were recorded in ambient air at a scan frequency of 1-1.5 Hz. The background slope due to piezoelectric non-linearity was corrected using first or second orders polynomial functions. The electrical resistivity was measured by the four probe method and the optical transmittance of AZO films was measured by a Perkin-Elmer UV-visible spectrophotometer in the range 250 – 2000 nm.

3. Results

3.1. Structural properties

The crystalline structure and orientation of the AZO thin films have been examined by X-ray diffraction (Figure 1). No Al_2O_3 phase has been evidenced in the XRD patterns, which implies that Al atoms are substituted in the ZnO hexagonal lattice or segregated to the nanocrystalline region in grain boundary [11]. The XRD patterns show that the AZO films deposited at room temperature are described by a strong peak at 2θ = 34.43° and a weaker peak at 2θ = 62.88°. These reflections were identified respectively as (002) and (103) plane of hexagonal wurzite ZnO [12], in which a preferred orientation along the c-axis is observed. However, when the substrate temperature rises, a clear crystallographic orientation change of AZO films is observed. For high deposition temperature, the intensity of (103) diffraction peak increases while intensity of (002) peak decreased. A transition of the growth mode from (002) vertical to (103) lateral growth is clearly observed due to the crystalline evolution during the deposition at high substrate temperature. Although such orientation change is rarely seen in the literature compared to (002) growth, Huang et al. [13] have also observed the (103) orientation change but with different deposition condition such as working pressure, oxygen content and film thickness.

Table 1. Resistivity, ρ, average transmittance in visible range, T_{aver}, optical band gap E_g, and crystallite size for sputtered AZO films.

T_s (°C)	ρ (Ω.cm)	T_{aver} (%)	E_g (eV)	Crystallite size (nm)
25	1.7×10^{1}	93	3.25	11.8
375	1.1×10^{-2}	86	3.39	18.8
455	1.7×10^{-3}	86	3.48	17.9
530	6.8×10^{-4}	85	3.55	17.2

The crystallite size may be estimated from the full-width at half maximum (FWHM) of the high intensity peak (002 or 103 orientation) by Scherrer equation [14]:

$$D = \frac{0.9\lambda}{B\cos\theta} \qquad (1)$$

where λ, θ and B are the X-ray wavelength, Bragg diffraction angle and FWHM (θ in radian) of the (002) reflections, respectively. For sputtered AZO films, this gives an estimated grain size in the range 16-20 nm (Table 1).

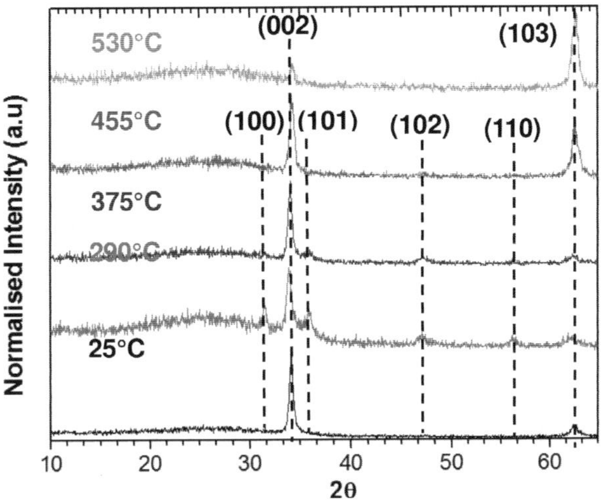

Figure 1. XRD patterns of ZnO:Al films deposited at different substrate temperatures

The AZO surface morphology characterized by SEM and AFM are shown in Figures 2(a) and 2(b). The grain size increases with the substrate temperature. At room temperature, the grains are uniform with repetition and exhibit a distinct grain boundary structure. At 375°C, the films show a textured surface and consist of tapered crystallites separated by voids. Surface morphology and the grain size are strongly influenced by the deposition temperature.

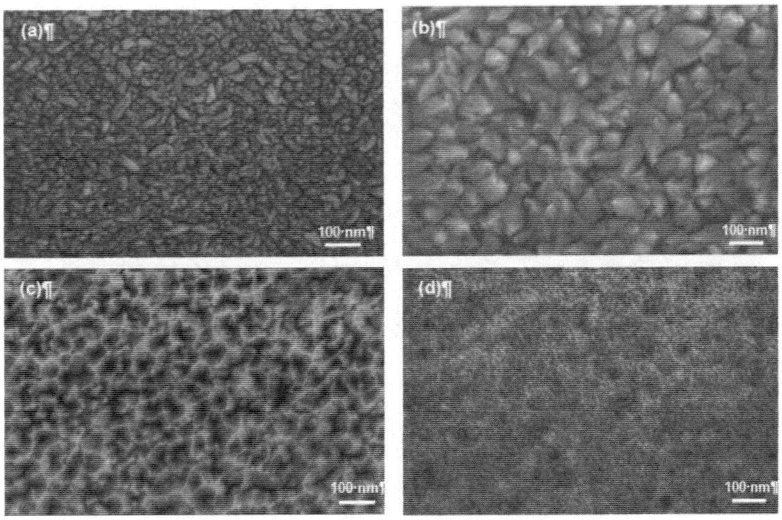

Figure 2. SEM micrographs of AZO films deposited by DC magnetron sputtering at different substrate temperatures: (a) 25°C, (b) 375°C, (c) 455°C and (d) 530°C.

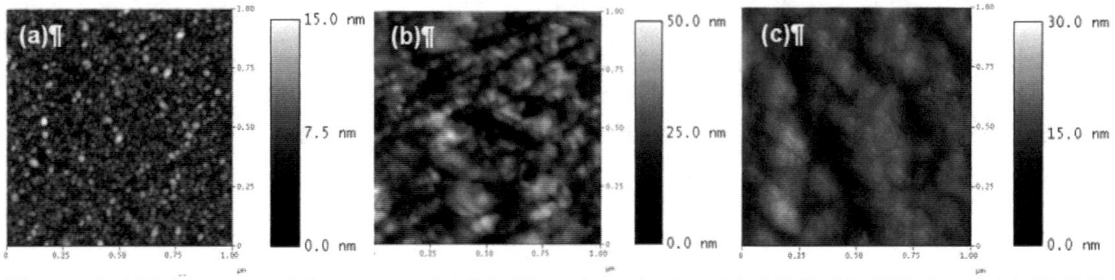

Figure 3. AFM images of the sputtered AZO films deposited at (a) 25°C, (b) 375°C and (c) 530°C.

When the substrate temperature increases from 375°C, grain coalescence is observed as shown in Figures 2(c) and 2(d). Corresponding situation in XRD above this temperature is change in preferential orientation and above this temperature the grain size also does not grow (Table 1) [15]. Due to this grain coalescence, an aggregate structure is formed and the films consist of a dense array of grains without discernible boundaries are observed [16-17]. Hence the higher substrate temperature may give more tightly packed grains with an orientation change in our study.
AFM images of the AZO films confirm the SEM results. Figures 3(b) and 3(c) show the grain coalescence when compared to the well distinct grains (Figure 3(a)). With increasing deposition temperature, a decrease in surface roughness (Figures 3) is observed and this implies the textured surface. The root-mean-square (RMS) surface roughness is RMS = 7.6 nm for T_s = 375°C and RMS = 4 nm for T_s = 530°C., indicating a smooth surface and low roughness of AZO films.

3.2. Electrical properties

As shown in Figure 4, the resistivity of AZO films decreases sharply from 1.7×10^{1} to 6.8×10^{-4} Ω.cm when the deposition temperature ranges from 25 to 530°C. That corresponds to the typical range one can find in the literature [18-19]. The lower value of electrical resistivity occurs with orientation change from (002) to (103) in the XRD patterns, while high resistivity is attributed to the (002) diffraction peak.

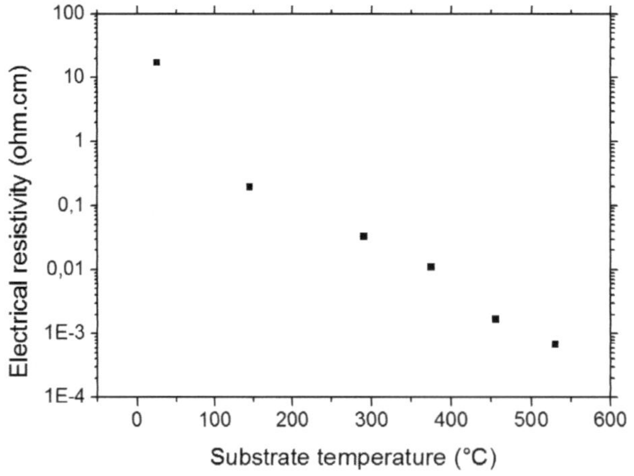

Figure 4. Resistivity of AZO films as a function of substrate temperature.

3.3. Optical properties

Figure 5(a) shows the optical transmittance of 200 nm AZO films prepared at different substrate temperatures, constant working pressure, $P_{working}$ = 3 mTorr and discharge power, P_{DC} = 120 W. The average optical transmittance in the visible range is more 85% for all samples [20-21].
As the deposition temperature increases, the optical absorption edge blue shifts monotonically. This trend is well known and explained by the increase in carrier concentration stated by the Burstein-Moss

effect [22-23]. The decrease in transmittance in the long wavelength region is due the reflection by the collective oscillation of conduction band electrons [24]. Hence this is attributed to the increasing carrier concentration with substrate temperature. For the direct optical gap semiconductor, α and the optical band gap, E_g, are related by [25]:

$$\alpha = \left(h\nu - E_g\right)^{1/2} \tag{2}$$

where h is Planck's constant, and ν is the frequency of the incident photon. Figure 5(b) shows the plot of α^2 vs hν. ZnO is the direct band gap semiconductor, so the x-axis intercept of the linear extrapolation of the curve plotted between α^2 and hν will directly give the band gap ($\alpha^2 = 0$). From Figure 5(b), E_{g0} for the pure ZnO film is E_{g0} ~ 3.3 eV and the E_g for samples prepared at room temperature is E_g ~ 3,25 eV. When the substrate temperature increases, the E_g rises until 3,55 eV at T_s = 530°C. These results are consistent with the results obtained from literature [26-27].

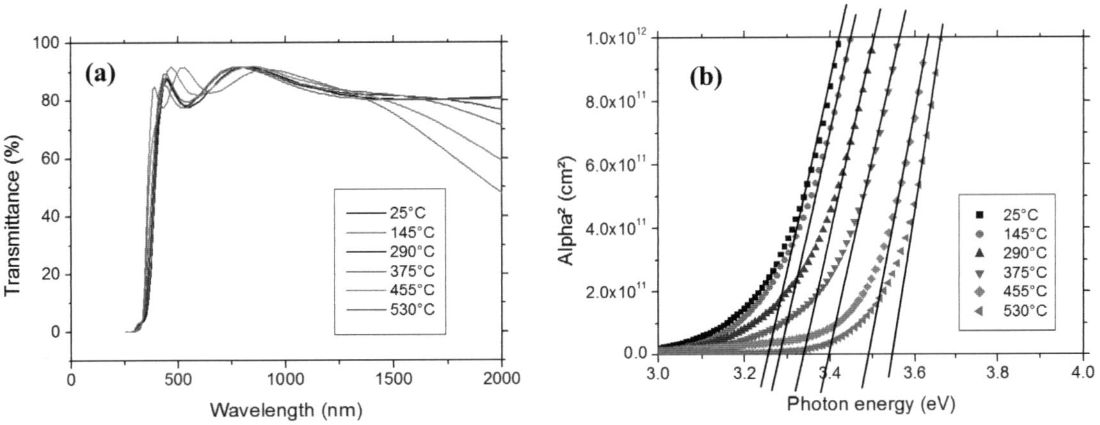

Figure 5. (a) Optical transmittance of AZO thin films prepared at different substrate temperatures and (b) square of the absorption coefficient as a function of photon energy.

4. Discussion

When the substrate temperature is over 375°C, the morphology of AZO films is an aggregate structure due to grain coalescence [16-17]. The higher substrate temperature gives more tightly packed grains. The electrical resistivity is improved, due to the grain coalescence structure and lowest resistivity of 7×10^{-4} Ω.cm is obtained. This is mainly attributed to the decreasing of number of scattering centres and trapping centres in the grain boundaries in which the high dense and packed grains cause an increase of carrier mobility and electron concentration [28].

The observed tightly packed grains growth is associated with the strongly marked preferred orientation change from (002) reflection to (103) reflection which is rarely observed. In our study, orientation change is attributed to the rise in substrate temperature and this accompany with grain coalescence. In many works, the minimum resistivity has been related to the presence of the strong (002) reflection in the XRD patterns. Very few have reported this orientation change. For example Gao et al. [29] reported this orientation change while growing films by pulsed DC sputtering. Even though the explanation is not given, there is a close relation between the resistivity of the film and its crystallographic orientation change. Huang et al. [13] rather attribute low resistivity to the better crystallanity and the larger grain size, even though the same orientation change was observed. But he observes this orientation change, when the working pressure rises up from 0.2 to 0.93 P and no remarkable variation of carrier concentration was observed in his study. Further the film deposited at 0.4 Pa presents lowest sheet resistance and rise up again while (103) peak still grows with working pressure.

In this work, we rather correlate the improvement of the electrical resistivity, not only with the formation of aggregate structure, but also with film growth along with (103) direction. The widening of the band gap is related to the increase of carrier concentration and corresponds to the low electrical resistivity state.

5. Conclusions

Aluminium-doped zinc oxide thin films have been prepared on sodalime glass substrate by DC magnetron sputtering with the ZnO target containing 3 wt% Al_2O_3, at different substrate temperatures. The crystallographic structure of as-prepared AZO films is hexagonal wurzite, in which the films are oriented along the c-axis with a dominant peak (002) in the X-ray diffraction pattern at room substrate temperature, whereas at higher deposition temperature the strongest intensity is associated to the (103) reflection. Moreover, the electrical and optical properties are enhanced at higher deposition temperature. The lowest resistivity of 6.8×10^{-4} Ω.cm and transmittance over 85% in the visible spectral region are obtained at T_s = 530°C. The SEM and AFM results have shown that the small grains are distributed uniformly at low temperature and grain coalescence occurs at T_s > 455°C. The optical energy band gap for AZO films is found to increase with the deposition temperature due to higher carrier concentration.

The highest intensity of (103) diffraction peak leads to the lowest resistivity state, associated to the grain coalescence structure and the high optical band gap. However further studies are going on to bring out exact correlation between the orientation change and lower resistivity.

6. References

[1] Xu D, Deng Z, Xu Y, Xiao J, Liang C and Pei Z 2005 *Physics Letters A* 346 148
[2] Exarhos G J and Zhou X D 2007 *Thin Solid Films* 515 7025
[3] Kim H, Horvitz J S and Chrisey D B 2002 *Thin Solid Films* 420-421 539
[4] Chen M, Pei Z L and Sun C 2001 *Mater. Sci. Eng.* B 85 212
[5] Fang G, Li D and Yao B L 2003 *Vacuum* 68 363
[6] Suzuki A, Nakamura M and Michihata R 2008 *Thin Solid Films* 517 1478
[7] Ma J, Ji F, Ma H L and Li S Y 1996 *Thin Solid Films* 279 213
[8] Li W, Sun Y, Wang Y and Cai H 2007 *Solar Energy Materials & Solar Cells* 91 659
[9] Cebulla R, Wendt R and Ellmer K 1998 *J. Appl. Phys.* 83 1087
[10] Seebera W T, Abou-Helala M O, Bartha S 1999 *Mater. Sci. Semicond. Processing* 2 45
[11] Kar J P, Kim S, Shin B and Park K I 2010 *Solid-State Electronics* 54 1447
[12] Guillen C and Herrero J 2010 *Vacuum* 84 924
[13] Huang C H, Cheng H L and Chang W E 2011 *J. Electrochem. Society* 158 H510
[14] George J and Menon C S 2000 *Surf. Coat. Technol.* 132 45
[15] Tanusevski A and Georgieva V 2010 *Appl. Surf. Sci.* 256 5060
[16] Song D, Aberle A G and Xia J 2002 *Appl. Surf. Sci.* 195 291
[17] Fang Z B, Yan Z J, Tan Y S, Liu X Q and Wang Y Y 2005 *Appl. Surf. Sci.* 241 303
[18] Lee J , Lee D, Lim D and Yang K 2007 *Thin Solid Films* 515 6094
[19] Yang W, Liu Z, Peng D L and Wu Z 2009 *Appl. Surf. Sci.* 255 5669
[20] Oh B Y, Jeong M C, Lee W and Myoung J M 2005 *J. Crystal Growth* 274 453
[21] Lin S S, Huang J L and Sajgalik P 2005 *Surf. Coat. Technol.* 190 39
[22] Sernelius B E, Berggren K F, Jin Z C and Hamberg I 1998 *Physical Review* B 37 244
[23] Munoz M and Pollak F H 2001 *Physical Review* B 63 1
[24] Ginley D S 2010 *Handbook of Transparent conductors* ed Springer p 2
[25] Montero J, Herrero J and Guillen C 2010 *Solar Energy Materials & Solar Cells* 94 612
[26] Wang F H, Chang H P, Tseng C C and Huang C C 2011 *Current Applied Physics* 11 S12
[27] Kim K H, Park K C and Ma D Y 1997 *J. Appl. Phys.* 81 7764
[28] Song D, Widenborg P and Chin W 2002 *Solar Energy Materials & Solar Cells* 73 1
[29] Gao X Y, Lin Q G and Feng H L 2009 *Thin Solid Films* 517 4684

Determination of Energy Band Alignment in Ultrathin Hf-based Oxide/Pt System

A Ohta[1], H Murakami[1], S Higashi[1], and S Miyazaki[2]
[1]Department of Semiconductor Electronics and Integration Science, Graduate School of Advanced Sciences of Matter, Hiroshima University, Kagamiyama 1-3-1, Higashi-hiroshima, Hiroshima 739-8530, Japan
[2]Graduate School of Engineering, Nagoya University, Furo-cho, Chikusa-ku, Nagoya, Aichi 464-8603, Japan

E-mail: semicon@hiroshima-u.ac.jp

Abstract. Effect of incorporating a third element into HfO_2 on the electronic structures has been studied by high resolution x-ray photoelectron spectroscopy (XPS). Hf-IIIa (La, Y, Gd, and Dy) oxide and Hf-Ti oxide films were deposited on a Pt layer by metal organic chemical vapor deposition (MOCVD) and co-sputtering and followed by post-deposition annealing in O_2 ambience at 500°C. The energy bandgap (Eg) of these Hf-based oxide films was determined by analyzing the energy loss spectra of O 1s photoelectrons in consideration of the overlap with Hf 4s core-line signals. From analyses of the valence band signals and the cut-off energy for photoelectrons, the valence band offset between the Hf based-oxide, and the Pt electrode and the work function value of the Pt layer were evaluated. By combining the oxide bandgap values, the valence band line-ups, and the Pt work function value, the energy band profile of the Hf-based oxide/Pt has been determined.

1. Introduction

The further scaling in metal-oxide-semiconductor field effect transistors (MOSFETs) and dynamic random access memory (DRAM) capacitors requires reducing the equivalent oxide thickness (EOT) with an implementation of dielectrics with higher k values [1]. Hafnium-based oxides have been intensively studied as alternative dielectrics because of their favorable barrier height and relatively good thermal stability [2, 3]. We reported an increase in the energy bandgap (Eg) with Al and Si incorporation into HfO_2 (e.g., Hf-Al oxide (Al/(Hf+Al) = 30%) : Eg = 6.30 eV [4] and Hf-Si oxide (Hf/(Hf+Si) = 43%) : Eg = 6.40 eV [5]). Incorporating a more polar element into the oxide network and a structural phase transformation by doping another element are considered as an effective approach to increasing the k value more [6-9]. However, there are some drawbacks such as bandgap shrinkage and reduction of barrier height reflecting the energy bandgap of incorporated material [10]. The impact of incorporating a third element of the HfO_2 network on the energy band profile has yet to be studied in detail.

In this work, we have studied the effect of incorporating a third element (La, Y, Gd, Dy, and Ti) into HfO_2 on the electron barrier height at the Hf-based oxide/Pt interface using high resolution x-ray photoelectron spectroscopy (XPS).

2. Experimental Procedures

After the formation of a Pt layer on thermally-grown SiO_2/Si by direct-current (DC) sputtering, 3-5 nm-thick HfO_2 and Hf-IIIa oxides (Y, La, Gd, and Dy) were deposited by metal-organic chemical vapor deposition (MOCVD) at 450°C in O_2 ambience at 10 Pa using di-pivaloy-methanato (DPM) precursors. And, ~5 nm-thick Hf-Ti oxide films with different Ti content were deposited by a co-sputtering method. Subsequently, post-deposition annealing was performed at 500°C in dry-O_2 ambience under atmospheric pressure to densify the films so prepared. From XPS core-line analyses, the cation content of the IIIa element in the films after annealing was evaluated to be La/(Hf+La) = ~54 %, Y/(Hf+Y) = ~62 %, Gd/(Hf+Gd) = ~54 %, and Dy/(Hf+Dy) = ~60 %. And, in each Hf-based oxide film, a fairly good compositional uniformity was confirmed from the photoelectron take-off angle dependence of core-line signals.

To evaluate the energy band structure of the samples, O1s energy loss signals, valence band spectra, and the cut-off energy for photoelectrons were measured by utilizing monochromatized Al Kα radiation (hv = 1486.6 eV: S-XPS). To accurately determine the energy bandgaps of the films, energy loss spectra of O1s photoelectrons were measured by hard x-ray photoelectron spectroscopy (H-XPS) using excited synchrotron radiation (hv = 7939 eV) in SPring-8 as well as by S-XPS.

3. Results and Discussion

The Eg values for the thin oxide films can be determined from the threshold energy of the energy loss signals of O 1s photoelectrons [4, 5, 11, 12]. In the case of Hf-based oxides, the superimposing of Hf 4s core-line signals on the energy loss signals of primary O 1s core-line signals makes it difficult to determine the onset of the energy loss signals. To overcome this difficulty, the binding energy position and spectral shape of Hf 4s signals were determined by x-ray excitation energy at about 7939 eV (Synchrotron: SPring-8) and 1486.6 eV (Al Kα) as shown in figure 1. Because the ratio of the photo-ionized cross section of the Hf 4s core line to that of the Hf 4f core line increases significantly with increasing excitation energy [13], the Hf 4s signals excited by the synchrotron radiation with a binding energy of around 538 eV are clearly distinguished in comparison to the case excited by Al Kα radiation. As indicated in figure 1, one can determine that the energy separation between Hf 4s and Hf 4f core lines is about 519.65 eV. By subtracting the Hf 4s component from the measured O 1s photoelectron energy-loss spectrum, the energy separation between Hf 4s and Hf 4f signals was taken into account and the intensity was normalized by Hf 4f signal intensity (figure 2(a)). By applying the same procedure to the measured spectra, the inherent O 1s energy loss spectra were obtained as seen in figures 2(b) and 2(c), and then in each spectrum, the onset of the energy loss spectrum was defined by linearly extrapolating the segment of maximum negative slope to the background level. As a result,

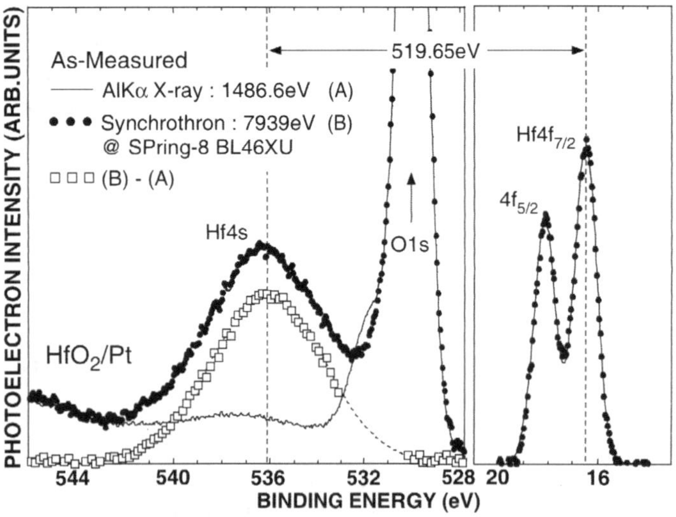

Figure 1. O 1s, Hf 4s, and Hf 4f spectra for HfO_2 after O_2 annealing at 500°C using excited Al Kα (1486.6 eV) and hard X-ray (7939 eV) radiation.

the Eg values of pure HfO_2 and Hf-Y oxide were determined to be 6.10 eV within an accuracy of 0.1 eV. The Eg value is slightly increased by the incorporation of lanthanoids such as La, Gd, and Dy. For Hf-Ti oxides shown in figure 2(c), by incorporating Ti atoms, the Eg value of Hf-Ti oxide was gradually decreased. As the Ti/(Hf+Ti) content increased up to ~74 %, the Eg value decreased by 2.0 eV from the value of pure HfO_2.

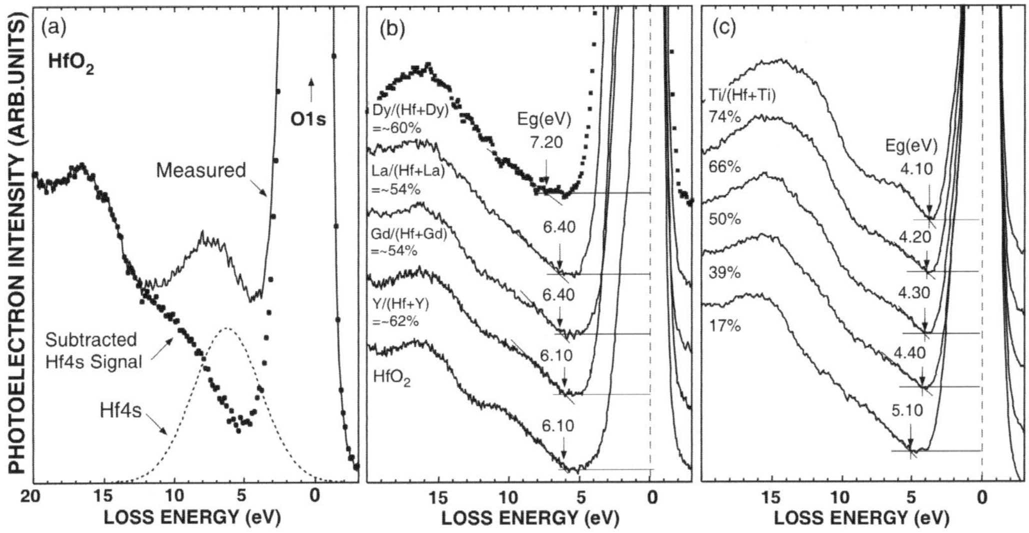

Figure 2. O 1s energy loss spectrum measured for HfO_2 thin films and a demonstration of subtracting Hf4s signals from the measured spectrum (a) and O1s energy loss spectra deconvoluted for Hf-IIIa (La, Y, Gd, and Dy) oxide films (b) and for Hf-Ti oxide films (c). For the spectra of (b) and (c) , Hf4s signals were subtracted completely from the measured spectra as demonstrated in (a). The onset of the energy loss signals for each sample was determined with an accuracy of ±0.05 eV.

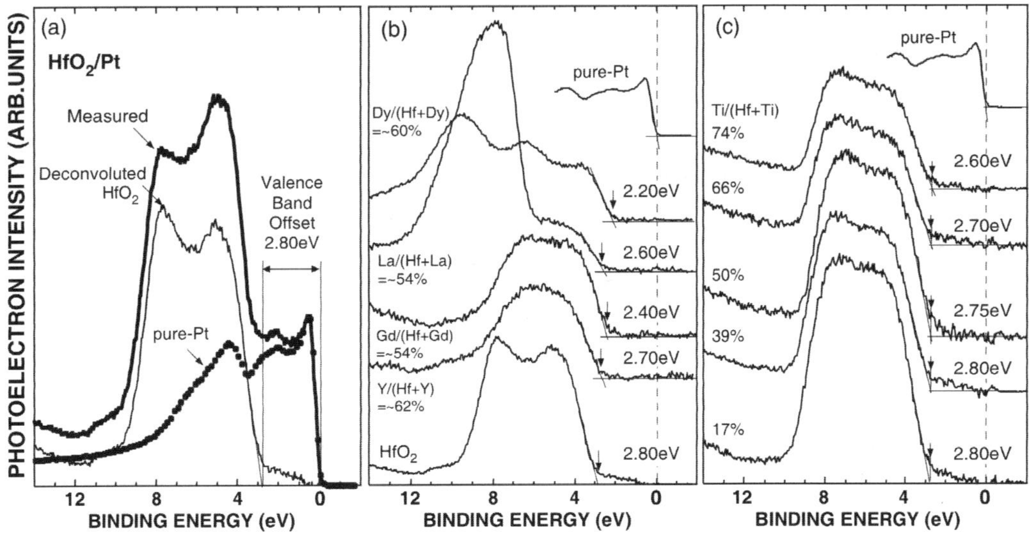

Figure 3. Valence band spectra of HfO_2/Pt structure before and after spectral deconvolution (a) and deconvoluted valence band spectra of Hf-IIIa (La, Y, Gd, and Dy) oxide films (b) and of Hf-Ti oxide films (c) as shown in figure 2.

To evaluate the valence band offset, the valence band spectra for the samples were measured and deconvoluted into mainly two components originated from the Hf based oxide and Pt (figure 3(a)). A reference valence band spectrum of the Pt layer was used in the spectral deconvolution, which was separately measured for the sample before Hf-based oxide deposition. And, in each deconvoluted Hf based oxide signal, taking into account the potential drop in the dielectric layer, the binding energy was normalized by surface C-H bonding units in C 1s spectra. From the energy separation of the tops of the deconvoluted valence band spectra, the valence band offset between Pt and each of the Hf-based oxides was evaluated as shown in figures 3(b) and 3(c). The valence band offset with respect to the Fermi Level of Pt gradually decreases from 2.80 eV for Pure HfO_2 incorporating a rare-earth oxide such as Y, Gd, La, and Dy. In addition, the valence band offset between Hf-Ti oxide and Pt is slightly decreased from 2.80 to 2.60 eV with Ti content.

The work function (WF) of the Pt surface was evaluated from the cut-off energy for photoelectrons (figure 4). Photo-excited electrons passing though materials can suffer inelastic scatterings and lose their energy with some electronic excitations. If the kinetic energy of excited electrons becomes or is below the WF of a metallic material of interest, such low energy electrons can no longer emit to the outside. Thus, the WF value can be determined from the threshold energy for electrons near the lower limit in the kinetic energy scale. To determine precisely the WF value considering the thermal broadening in the metal Fermi surface, curve fitting to the photoelectron yield spectra near the lower limit in the kinetic energy scale was conducted by using a Fowler function [14] given by

$$\varphi(x)=\frac{\pi^2}{6}+\frac{x^2}{2}-e^{-x}+\frac{e^{-2x}}{2^2}-\frac{e^{-3x}}{3^2}+\cdots\cdots(x\geq 0),\ E_F=kT\cdot x \qquad (1)$$

The validity of this method was confirmed from the WF determination of pure metals such as Au and Ag. From a best fit result, the WF value for Pt was determined to be 5.55 eV.

Based on the results of figures 2, 3, and 4, the conduction band (CB) offsets between Pt and each Hf-based oxide, namely the barrier heights for electrons, were estimated and summarized as shown in figures. 5, 6, and 7. The conduction band and valence band offset for the HfO_2/Pt interface were obtained to be 3.30 and 2.80 eV, respectively (figure 5). And, electron barrier heights over 3.3 eV

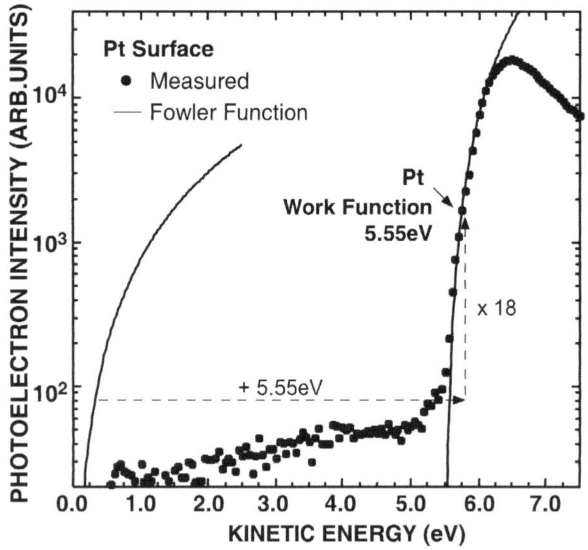

Figure 4. The photoelectron yield spectra from Pt surface near the cut-off in the kinetic energy scale. Curve fitting with a Fowler function was also demonstrated.

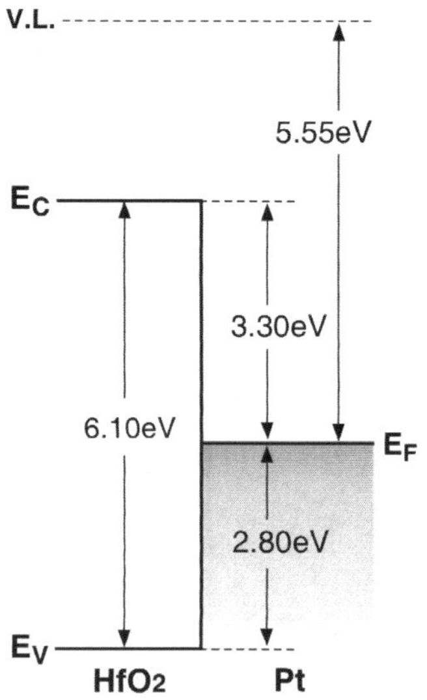

Figure. 5 Energy band profile of HfO_2/Pt structure.

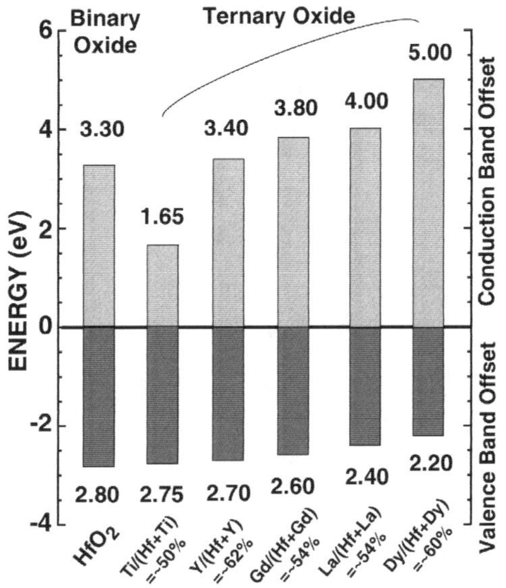

Figure 6. Conduction band and valence band offset at Hf-based oxides/Pt interface.

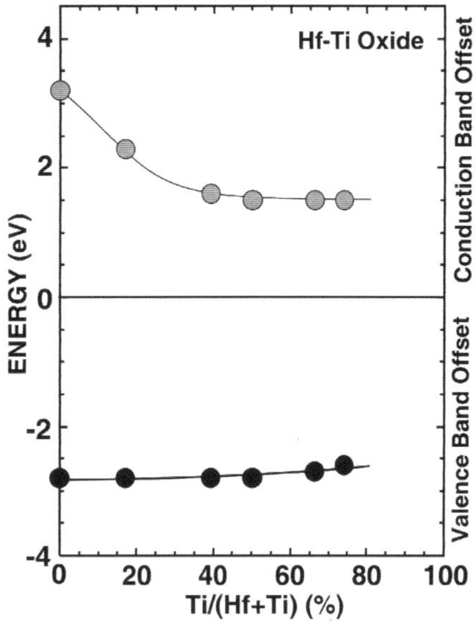

Figure 7. Changes in the conduction band and valence band offsets between Hf-Ti oxide and Pt as functions of the Ti/(Hf+Ti) content.

were obtained at the Pt/Hf-IIIa (Y, La, Gd, and Dy) oxide interfaces (figure 6). Obviously, the electron barrier height decreases with increasing Ti content up to 39 % and then saturates at ~1.5 eV with further increase in the Ti content (figure 7). While the valence band offset slightly decreases with Ti content, significant changes in the conduction band offset by third element addition can be interpreted in terms of the bottom of the oxide conduction band edge changing from unoccupied Hf 5d state to Y 4d, La 5d, Gd 5d, Dy 5p, and Ti 4d states. And, slight changes of valence band offset can be explained by the top of the oxide valence band being derived from non-bonding O 2p states irrespective of the third element addition.

4. Conclusion

The energy band profile of Hf-based oxide/Pt has been determined by combining the oxide bandgap values and the valence band line-ups. The conduction band and valence band offset for HfO_2/Pt interface were obtained to be 3.30 and 2.80 eV, respectively. Hf-IIIa (Y, La, Gd, and Dy) ternary oxides show an electron barrier of 3.4 eV and higher. With an increase in Ti content up to 39%, the conduction band offset is significantly decreased from 3.20 to 1.50 eV. The valence band offset is slightly decreased by incorporating Ti.

5. Acknowledgements

The synchrotron radiation experiments were performed at SPring-8 with the approval of the Japan Synchrotron Radiation Research Institute (JASRI/SPring-8) as Nanotechnology Support Project of the Ministry of Education, Culture, Sports, Science and Technology (Proposal No. 2007A0005, 2011A0026/BL-No.46XU). The authors would like to thank Prof. S. Zaima (Nagoya Univ.), Drs. J-Y. Son, M. Machida, H. Oji (JASRI), Drs. M. Tanioku, M. Horikawa, and A. Ogishima (Elpida.) for their fruitful discussions and assistance with the experiments. A part of this work was supported by the Research Institute for Nano-device and Bio Systems (RNBS), Hiroshima University, Japan.

6. References

[1] For example, "The International Technology Roadmap for Semiconductors (ITRS), 2011"

[2] Wilk G D, Wallace R M and Anthony J M 2000 *J. Appl. Phys.* **87** 484

[3] Koyama K, Kaneko A, Ino T, Koike M, Kamata Y, Iijima R, Kamimuta Y, Takashima A, Suzuki M, Hongo C, Inumiya S, Takayanagi M and Nishiyama A 2002 IEDM Tech Dig., 847

[4] Ohta A, Nakagawa H, Murakami H, Higashi S, Kawahara T, Torii K and Miyazaki S 2004 *Japan. J. Appl. Phys.* **43** 7831

[5] Ohta A, Nakagawa H, Murakami H, Higashi S, Miyazaki S, Inumiya S and Nara Y 2006 *Transactions of the Materials Research Society of Japan* **31** 125

[6] Chen F, Bin X, Hella C, Shi X, Gladfelter W L and Campbell S A 2004 *Microelec. Eng.* **72** 263

[7] Toriumi A, Kita K, Tomida K and Yamamoto Y 2006 *ECS Trans.* **1** 185

[8] Govindarajan S, Böscke T S, Sivasubramani P, Kirsch P D, Lee B H, Tseng H H, Jammy R, Schröder U, Ramanathan S and Gnade B E 2007 *Appl. Phys. Lett.* **91** 062906

[9] Ketsmana I, Losovyj Y B, Sokolov A, Tang J, Wang Z, Natta M L, Brand J I, Dowben P A 2008 *Appl. Surf. Sci.* **254** 4308

[10] Afanas' ev V V, Stesmans A, Chen F, Li M and Campbell S A 2004 *J. Appl. Phys.* **95** 7936

[11] Miyazaki S 2001 *J. Vac. Sci. Technol. B* **19** 2212

[12] Ohta A, Yamaoka M and Miyazaki S 2004 *Microelect. Eng.*, **72** 154

[13] Yeh J Y and Lindau I *Atomic Data and Nuclear Data Table* **32**

[14] Fowler R H 1966 "Statistical Mechanics: The Theory of the Properties of Matter in Equilibrium," 2nd ed. (Cambridge, England: University Press)

Characterization of Ultrathin Ta-oxide Films Formed on Ge(100) by ALD and Layer-by-Layer Methods

K Mishima[1], H Murakami[1], A Ohta[1], S K Sahari[1], T Fujioka[1], S Higashi[1], and S Miyazaki[2]

[1]Grad. School of Advanced Sciences and Matter, Hiroshima Univ., 1-3-1 Kagamiyama, Higashi-Hiroshima, 739-8530, Japan
[2]Grad. School of Eng., Nagoya Univ., Furo-cho, Chikusa-ku, Nagoya, 464-8603, Japan

semicon@hiroshima-u.ac.jp

Abstract. Atomic layer deposition (ALD) and Layer-by-Layer deposition of Ta-oxide films on Ge(100) with using tris (*tert*-butoxy) (*tert*-butylimido) tantalum have been studied systematically. From the analysis of the chemical bonding features of the interface between TaO_x and Ge(100) using x-ray photoelectron spectroscopy (XPS), Ge atom diffusion into the Ta oxide layer and resultant $TaGe_xO_y$ formation during deposition at temperatures higher than 200 °C were confirmed. Also, we have demonstrated that nanometer-thick deposition of Tantalum oxide as an interfacial layer effectively suppresses the formation of GeO_x in the HfO_2 ALD on Ge. By the combination of TaO_x pre-deposition on Ge(100) and subsequent ALD of HfO_2, a capacitance equivalent thickness (CET) of 1.35 nm and relative dielectric constant of 23 were achieved.

1. Introduction

Germanium (Ge) is a promising channel material for next generation metal–insulator–semiconductor field-effect transistors (MISFETs) because of its higher carrier mobility than Si [1-3]. To implement a Ge-MISFET, the formation of high-quality high-k gate dielectric scalable to an equivalent oxide thickness (EOT) as thin as 0.5 nm and even below on Ge is also required [4]. Thus, one of major technological challenges is a control of the interfacial reaction between the high-k dielectric and Ge [5]. It is also important to minimize the interfacial defect state density [6,7]. From viewpoints of suppression of interfacial oxidation and atomic diffusion, interfacial barrier layers such as GeON, SiO_2, and Al_2O_3 have been studied so far [8,9]. The formation of the interfacial barrier layer with a higher dielectric constant is a key for the EOT scaling.

In this study, we focused on tantalum oxide (TaO_x) as an interfacial control layer between hafnium oxide (HfO_2) and Ge substrate, taking into account the facts that Tantalum pentoxide (Ta_2O_5) has a relatively high dielectric constant among various binary oxides [10] and Ta^{5+} ion radius is slightly smaller than Hf^{4+} ion radius.

2. Experimental Procedures

The substrates used in this study were p-type Ge(100) wafers with a resistivity of ~10 Ω·cm. After a pure water rinse, the Ge(100) surface was chemically oxidized in a 15 % H_2O_2 solution at room temperature. Subsequently, the chemical oxide was completely removed by immersion in a 30 % diluted HCl solution. Just after the HCl treatment almost no signals due to O and Cl atoms were detected on the Ge (100) surface with x-ray photoelectron spectroscopy (XPS).

In the TaO_x formation, tris (*tert*-butoxy) (*tert*-butylimido) tantalum (Ta-TTT)[11] was used as a precursor for metal organic chemical vapor deposition (MOCVD) due to its higher reactivity of Ta-TTT with O_3 compared to pentaethoxy tantalum (PET) used conventionally. Tantalum oxide films were deposited on the pre-cleaned Ge(100) surfaces at substrate temperatures ranging from 150 and 400 °C by alternately supplying UV-O_3 and Ta-TTT using a N_2 bubbling method. For comparative study, TaO_x formation on thermally-grown SiO_2 was also carried out. The process sequence for a deposition cycle is schematically illustrated in Figure 1. To avoid excessive oxidation of Ge surface with UV-O_3 in the early stages of TaO_x layer formation, ultrathin TaO_x layers were also deposited at 400 °C in a layer-by-layer fashion in which Ta-TTT was introduced intermittently to promote thermal decomposition of ligands without UV-O_3. Subsequently, HfO_2 was deposited at 280 °C using atomic layer deposition (ALD) with tetrakis (ethylmethylamino) hafnium (TEMA-Hf).

The chemical bonding features of the sample so-prepared was characterized by hard x-ray photoemission spectroscopy (HAXPES) using a synchrotron radiation (7939 eV) as well as high-resolution XPS using a monochromatized Al Kα radiation (1486.6 eV). The dielectric layer thickness was measured using spectroscopic ellipsometry (SE).

3. Results and Discussion

The temperature dependence of TaO_x deposition was first studied to define ALD conditions. For the samples after 40 deposition cycles in the process sequence seen in Figure 1, the TaO_x thickness was evaluated and summarized as a function of substrate temperature. Figure 2 shows the temperature dependence of growth rate per cycle obtained on thermally-grown SiO_2 (~4 nm), where the compositional intermixing at the interface between TaO_x and SiO_2 is negligible. The deposition at temperatures higher than 250 °C features the CVD mode since the oxide thickness increases exponentially with temperature. On the other hand, the deposition in temperature range from 150 to 250 °C shows less temperature dependence, in which thermal decomposition of precursors is markedly suppressed and the deposition under the pulse sequence is likely to be approaching to the ALD mode. From the Arrhenius plot of the growth rate as shown in Figure 3, the growth rate in the high temperature region is characterized with an activation energies E_a of 0.44 eV while in the temperature range below 200 °C the Ea value as small as 0.18 eV is obtained as often reported in the thermal decomposition of metal-organic precursors [12,13]. To get an insight into the thermal decomposition of Ta-TTT, the chemical bonding features of TaO_x films were evaluated using XPS. As shown in Figure 4, N1s signals due to the Ta-N bonds were clearly observable at ~402 eV in peak binding energy for the sample formed at 200 °C. It is suggested that in the temperature range of 150 – 250 °C, thermal decomposition of butyl ligands from the precursor proceeds selectively and consequently Ta-N bonding units are incorporated in the TaO_x network. At temperatures higher than 250 °C, the thermal decomposition of Ta=N double bonds becomes significant and promotes the N removal as a

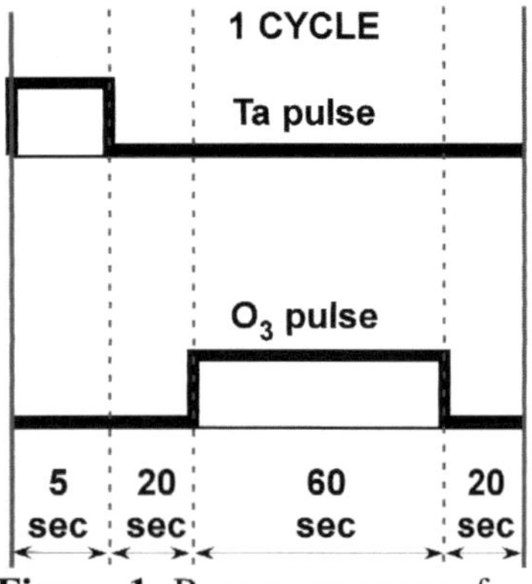

Figure 1. Process sequence for a deposition cycle of Ta oxide (TaO_x).

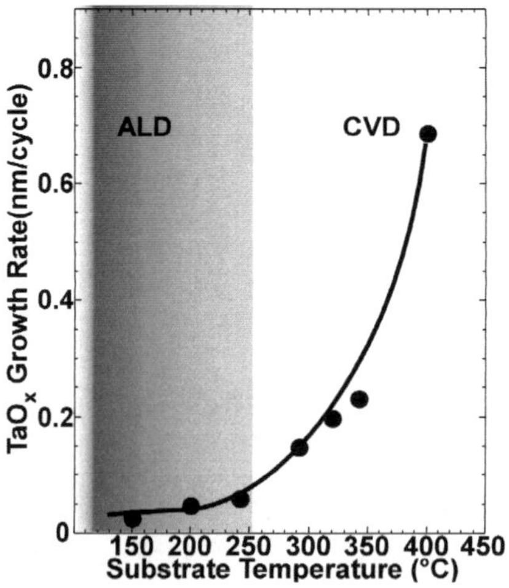

Figure 2. TaO_x thickness deposited as a function of substrate temperature, which was obtained from the SE evaluation of the samples deposited on 4 nm-thick thermally-grown SiO_2/Si(100).

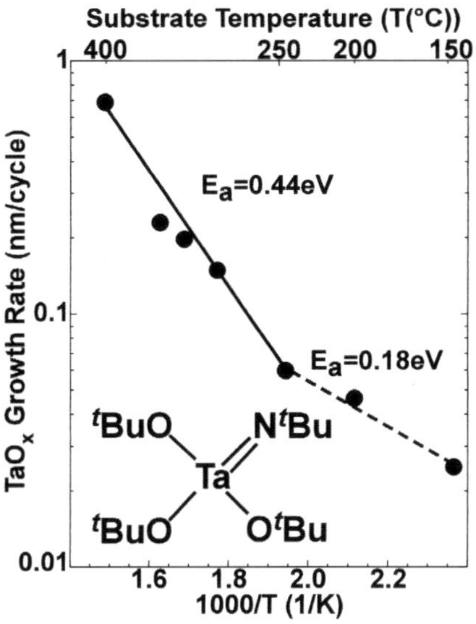

Figure 3. An Arrhenius plot of the temperature-dependent growth rate of TaO_x.

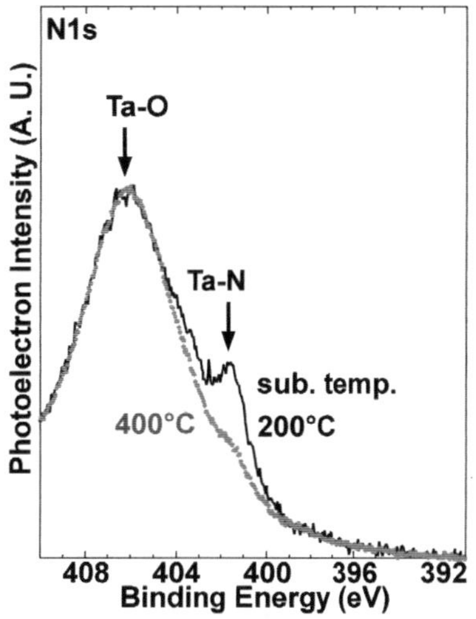

Figure 4. N1s and $Ta4p_{3/2}$ spectra for the samples prepared at 200 °C and 400 °C on SiO_2 in the process sequence represented in Figure 1.

result of the formation of volatile molecules such as NO_x. As the next step, we examined the interfacial reactions in TaO_x layer formation on Ge(100) with the gas pulse sequence shown in Figure 1. As shown in Figure 5, for both the samples prepared at 200 and 400 °C having almost the same thickness, the signals due to the Ge-O bonding units were observed at ~33 eV with no clear difference in the signal intensity. This result indicates the interfacial GeO_x layer formation presumably due to UV-O_3 supply during deposition cycles, especially in the early stages of the TaO_x formation. We also evaluated Ge2p and O1s spectra for TaO_x prepared at 400 °C by only Ta-TTT pulses without subsequent UV-O_3 supply to confirm enhanced Ge oxidation due to UV-O_3 supply (Figure 6). Obviously, the Ge2p signals due to GeO_x components for the case without UV-O_3 at 400 °C are much lower than those for the case with UV-O_3 at 200 °C. The decrease in GeO_x components for the case without UV-O_3 was also confirmed from the reduced O1s signals in the higher binding energy side in comparison to the case with UV-O_3. Base on this result, it is likely that short pulses of Ta-TTT without UV-O_3 introduction at a temperature as high as 400 °C is quite effective to suppress the GeO_x formation in the TaO_x formation on Ge. To gain a better understanding in the chemical structure after the TaO_x formation on Ge(100), $Ge3d_{5/2}$ signals and $GeL_2M_{45}M_{45}$ Auger electron signals taken after TaO_x formation were evaluated, as shown in Figure 7. The chemical shift reflecting chemical bonding features can be evaluated using the Auger parameter independent of positive charging during XPS measurement [14]. The Auger parameter is calculated using the following equation:

Auger parameter = Ge $3d_{5/2}$ (B.E.) + GeLMM (K.E.)

First we confirmed that the Auger parameter of non-oxidized Ge signals after TaO_x formation on Ge(100) showed the same value (1205.6 eV) as obtained for thermally-grown GeO_2/Ge(100). As for the oxide components, the Auger parameter for TaO_x/Ge(100) is 1202.2 eV, which is slightly larger than that for GeO_2/Ge (1201.6 eV). The result implies the Ge atom diffusion into TaO_x layer. After 7 nm-thick HfO_2 deposition on Ge (100) and pre-deposited TaO_x/Ge(100), chemical bonding features were also characterized using HAXPES which enable us to probes the buried interface. As indicated in Figure 8, the formation of GeO_x during HfO_2 deposition is suppressed with pre-deposition of TaO_x thicker than 1 nm. Notice that, the Ge2p spectrum for the case of pre-deposition of 5 nm-thick TaO_x

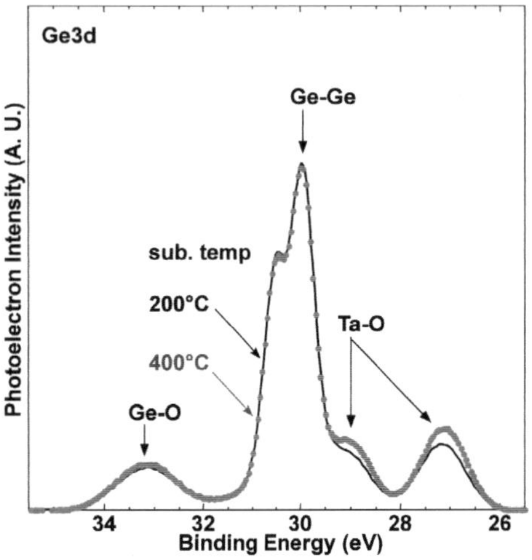

Figure 5. Ge3d and Ta4f spectra for the samples prepared at 200 °C and 400 °C on Ge(100) in the process sequence represented in Figure 1. Intensity normalization and binding energy calibration were made by using $Ge3d_{5/2}{}^{0+}$ signals.

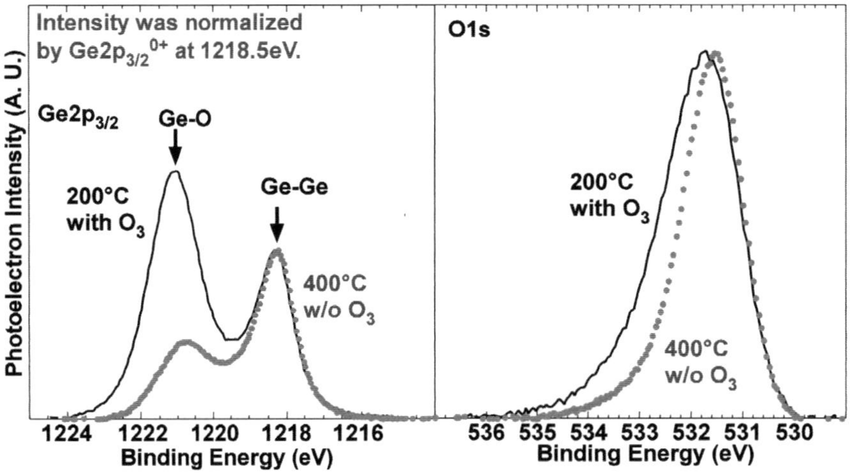

Figure 6. (a) $Ge2p_{3/2}$ and (b) O1s spectra for the samples prepared with UV-O_3 at 200 °C and without O_3 at 400 °C.

shows very light difference from the spectrum taken just after HCl cleaning and indicates little amount of oxide components

Based on these results, we fabricated Pt gate MIS capacitors with ~1.5 nm-thick TaO_x pre-deposition on p-Ge(100) followed by ~6.5 nm HfO_2 deposition and post-deposition anneal in N_2 ambient at 300 °C, and measured capacitance-voltage (C-V) characteristics at room temperature without any post metallization anneal (Figure 9). The capacitance equivalent thickness (CET) of 1.35 nm and dielectric constant of 23 were obtained although, for the sample prepared without TaO_x pre-deposition, C-V curves were hardly measured due to high leakage current. However, C-V hysteresis due to charge injection of ~100 mV and flat band voltage (V_{fb}) shift due to fixed charge as high as -1.08 V were measured. Leakage current density (J_g) value was 40 mA/cm^2 at V_g of V_{fb}-1 V(= -0.9 V). Reduction of residual carbons and defects in HfO_2 and/or at HfO_2/TaO_x interface is needed with optimization of the deposition and anneal conditions as well as post-metallization anneal.

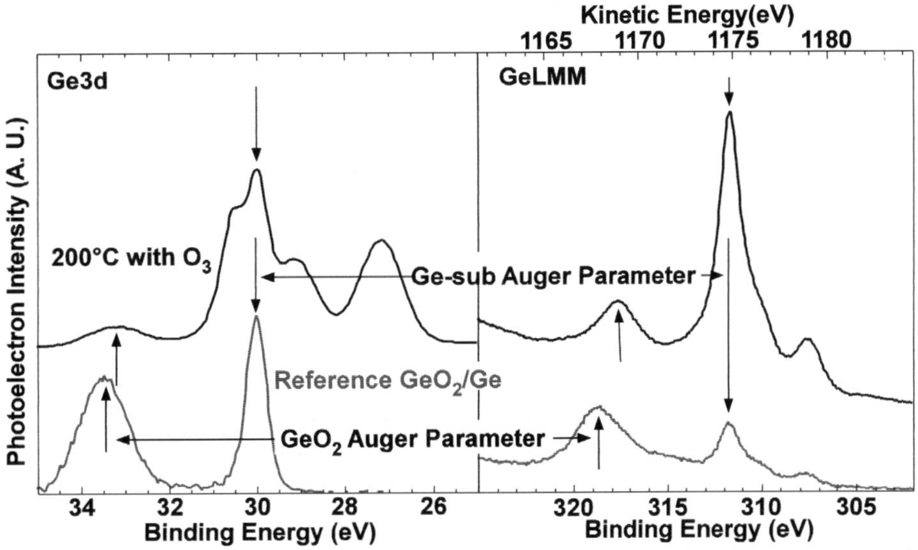

Figure 7. Ge3d$_{5/2}$ core-line and GeL$_2$M$_{45}$M$_{45}$ Auger spectra for TaO_x/Ge. The spectrum taken for a thermally –grown GeO_2 on Ge(100) is also shown as a reference.

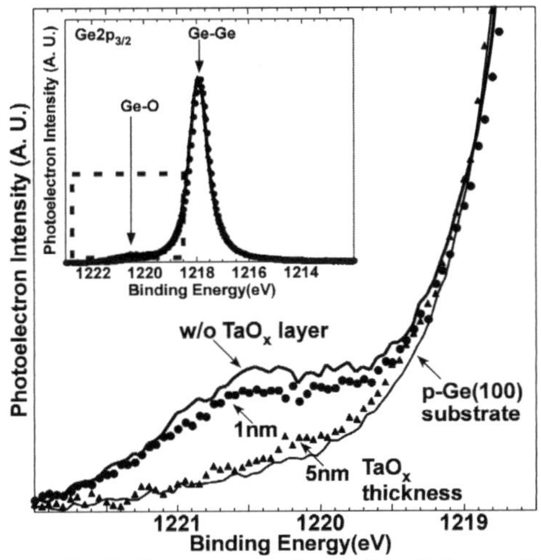

Figure 8. Ge2p spectra taken after 6.5 nm-thick HfO_2 deposition on Ge(100) and on 1 nm-thick and 5 nm-thick TaO_x pre-deposited on Ge(100). All the samples were annealed at 300 °C for 10 min in N_2 ambience. The spectrum taken for the wet-chemically-cleaned Ge substrate is also shown as a reference.

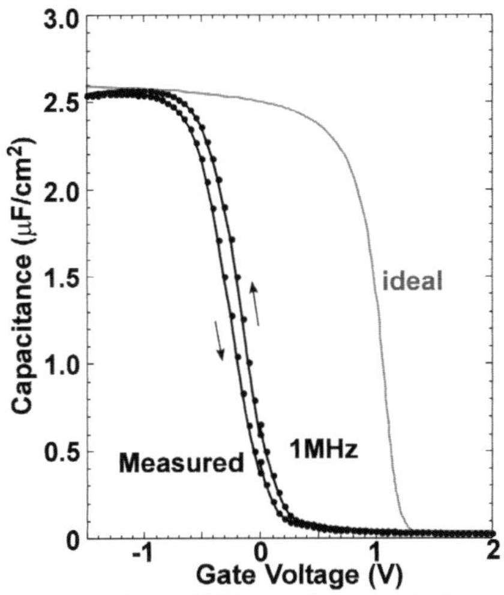

Figure 9. C-V characteristics for Pt/HfO_2(~6.5 nm)/TaO_x(~1.5 nm)/Ge(100). Prior to Pt gate formation the sample was annealed at 300 °C for 10 min in N_2 ambience but no thermal anneal was preformed after Pt gate formation. An ideal C-V curve is also shown as a reference.

4. Conclusions

We demonstrated TaO_x deposition on Ge(100) wafers using Ta-TTT. The Ge diffusion during the TaO_x deposition results in the formation of $TaGe_xO_y$, which is applicable to interfacial control for high-k dielectric stack on Ge. In fact, with pre-deposition of 1.5 nm-thick TaO_x on Ge(100) and subsequent deposition of ~6.5 nm-thick HfO_2, Pt-gate MIS capacitors with a CET of 1.35 nm and a relative dielectric constant of 23 was obtained.

5. Acknowledgements
A part of this work was supported by Research Institute for Nano-device and Bio Systems (RNBS), Hiroshima University, Japan. The Ta-TTT precursor was provided by TOSOH Corporation. We would like to thank J. –Y. Son, H. Oji, and Y. T. Cui (JASRI) for their help with the XPS measurements at SPring-8 as the member of research proposal No. 2011A0026/ BL-No.46XU.

References
[1] Sze S M and Ng K K 2007 *Physics of Semiconductor Devices* (New York, Wiley 3rd ed) p. 786.
[2] Chui C O, Kim H, McIntyre P C, and Saraswat K C 2003 *IEDM Tech.* Dig., p. 437.
[3] Shang H, Okom-Schmidt H, Chan K K, Copel M, Ott J A, Kozlowski P M, Steen S E, Cordes S A, Wong H-S P, Jones E C, and Haensch W E 2002 *IEDM Tech.* Dig. P. 441.
[4] 2009 *International Technology Roadmap for Semiconductor.*
[5] Hyoungsub M, and McIntyre C P 2004 *appl. Phys. Lett.* 85 14.
[6] Murakami H, Fujioka T, Ohta A, Bando T, Higashi S, and Miyazaki S 2010 *ECS Trans.* 33 253.
[7] Ohta A, Fujioka T, Murakami H, Higashi S, and Miyazaki S 2011 Japan J. *Appl. Phys.* 50 to be published.
[8] Houssa M, Jaeger B De, Delabie A, Elshocht S V, Afanas'ev V V, Autran J L, Stesmans A, Meuris M, Heyns M M 2005 *of Non-Crystalline Solids,* 351 1902.
[9] Kato R, Kyogoku S, Sakashita M, Kondo H, and Zaima S 2009 Japan *J. Appl. Phys.*, 48 05DA04.
[10] Robertson J 2004 *Eur. Phys. J. Appl. Phys.* 28 265.
[11] Yotsuya T, Chiba H, Furukawa T, Yamamoto T, Inaba T, Tada K, Suzuki T, Fujimoto K, Funakubo H, Yamakawa T, Oshima N 2008 ECS Trans. 16 243.
[12] Elam J W, Schuisky M, Ferguson J D, George S M 2003 *Thin Solid Film.*
[13] Norton T E, Amato W C 2001 *Chem. Mater 13* 4655-4660.
[14] Wagner C D, Gale L H, Raymond R H 1979 *Anal. Chem.* 51 466.

Kinetics of thermally oxidation of Ge(100) surface

S K Sahari[1], A Ohta[1], M Matsui[1], K Mishima[1], H Murakami[1], S Higashi[1], and S Miyazaki[2]

[1]Graduate School of Advanced Sciences of Matter, Hiroshima University, Higashihiroshima, Hiroshima, 739-8530 Japan
[2]Graduate School of Engineering, Nagoya University, Nagoya, Aichi, 464-8603 Japan

[1] semicon@hiroshima-u.ac.jp, [2] miyazaki@nuee.nagoya-u.ac.jp

Abstract. Thermal oxidation of a Ge(100) surface was investigated by using spectroscopic ellipsometry (SE) and x-ray photoelectron spectroscopy (XPS). Ge oxide was grown in the temperature range of 375 to 550°C in dry-O_2 ambience at atmospheric pressure. Although the Ge-oxide growth rate shows a linear relationship in a log-log plot at a fixed temperature, and the slope indicates an enhancement of GeO desorption at oxidation temperatures over 490ºC. The GeO desorption was also confirmed from the XPS analysis of the Si surface which was oxidized simultaneously with the Ge(100) surface. Thus, the Ge thermal oxidation at atmospheric pressure cannot be explained simply by the Deal-Grove model, in which the contribution of thermal desorption of Ge monoxide must be taken into account.

1. Introduction

As Ge has higher carrier mobilities than those of Si [1-3], it is a promising channel material for metal-oxide-semiconductor (MOS) field effect transistors (FETs). Controlling the MOS interface and minimizing the generation of defect state densities in the gate dielectric and at the dielectric/Ge interface are the most critical issues for the development of Ge-channel MOSFETs [4, 5]. The detailed mechanism of Ge oxidation and the nature of the interfaces between Ge and GeO_2 are still research matters, which are presumably related to the fact that GeO_2 is chemically and thermally unstable in contrast to SiO_2 [6, 7]. In particular, the suppression of Ge oxide desorption is a key to obtaining a high FET performance. So far, the GeO desorption has been intensively studied [8-12]. Thermal desorption spectroscopy (TDS) measurements of thermally grown GeO_2 films on Ge(100) clarify that the film desorbs mainly in the form of GeO and the peak temperature depends on the GeO_2 film thickness [10]. In addition, the GeO desorption is thought to be enhanced by the reaction at the interface between GeO_2 and Ge substrate [11, 12]. Since there are few published reports of the thermal oxidation of Ge surface with taking into account of both Ge oxidation and GeO desorption, the understanding of the oxidation mechanism is very limited.

In our preliminary report, chemical bonding features of thermally-grown Ge oxide/Ge(100) were investigated by using high-resolution x-ray photoelectron spectroscopy (XPS). It is found that Ge sub-oxides signals due to Ge^{2+} and Ge^{3+} components slightly increased with increasing thermal oxidation temperatures from 375 to 490°C, which indicates that oxygen deficient region in Ge-O network increased with oxidation temperatures. In addition, the amount of these sub-oxides decreases by 60% after being annealed at 375°C in dry oxygen ambience at atmospheric pressure [13]. In this work, we have studied the Ge oxide growth on Ge(100) surface at different oxidation temperatures in dry oxygen ambience at atmospheric pressure by spectroscopic ellipsometry (SE) and XPS.

2. Experimental Procedures

A p-type Ge(100) with a resistivity of ~10 Ω·cm was used. The Ge wafers were first dipped in deionized pure water to dissolve the native oxide and followed by immersing them in 15% H_2O_2 to re-oxidize the surface intentionally. Then the wafers were rinsed with deionized pure water and subsequently dipped in 30% HCl solution to remove the surface oxide layer. Finally, the Ge wafers were briefly dipped in deionized pure water to minimize residual Cl atoms. After the wet-chemically cleaning, a Ge oxide layer was grown at the temperature ranging from 375 to 550°C in a quartz furnace with flowing dry-O_2 at a rate of 4 L/min under atmospheric pressure. The thicknesses of thermally-grown Ge oxide were evaluated with an accuracy of ±0.1 nm by SE with wavelengths ranging between 192 and 1000 nm. The chemical bonding features of thermally grown Ge oxides so-prepared were characterized by XPS under monochromatized AlKα (hν = 1486.6 eV) radiation.

3. Results and Discussion

The Ge oxide growth on Ge(100) by thermal oxidation for 10 min in the temperature ranging from 375 to 550°C is shown in figure 1. At oxidation temperatures below 490°C, the Ge oxide thickness is almost exponentially increased with temperature. Further increase in oxidation temperature significantly enhances the growth rate of Ge oxide, implying that the oxygen diffusion to the Ge surface through Ge oxide becomes significant. To get a better understanding of Ge thermal oxidation, the Ge oxide growth with different oxidation temperatures was summarized as a function of oxidation time (Figure 2). As indicated in figure 2, we have extracted the slope *n* at each oxidation temperature from linear extrapolation in the log-log plot, and found that the *n* value of the 375°C oxidation is close to that of the native oxide growth (0.12) [14], where in an air-exposure time within 10 min at 24°C, native Ge oxide with a thickness as low as 0.6 nm is formed. The result suggests that the oxidation at 375°C and below is rate-limited by surface reaction. With increasing oxidation temperature from 490 to 550°C, the *n* value decreases slightly from 0.38 to 0.36, which implies evolving thermal desorption of Ge oxide.

To investigate the GeO desorption during the thermal oxidation at temperatures over 500°C, both Ge(100) and Si(100) were oxidized at the same time as schematically illustrated in figure 3. In this experiment, wet-chemically cleaned p-type Si(100) wafer with a resistivity of 8-12 Ω·cm was used. The $Si2p_{3/2}$ and Ge3d spectra taken for the Si(100) surface after oxidation at 505 and 550°C for 10 min are shown in figure 4. With an increase of oxidation temperature from 505 to 550°C, not only an

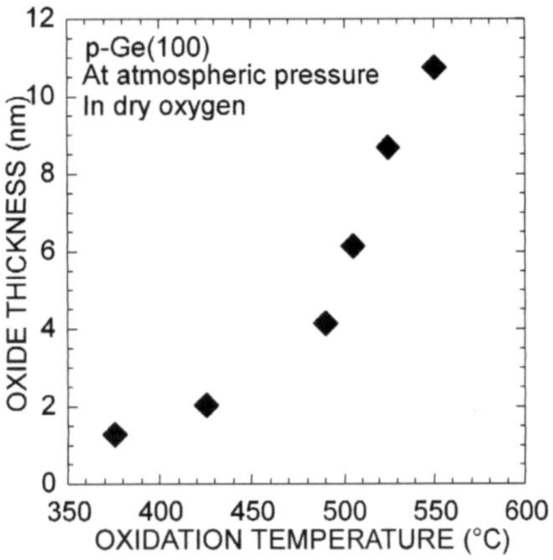

Figure 1. Thermal oxidation of Ge(100) for 10 min in dry-O_2 ambience as a function of oxidation temperature.

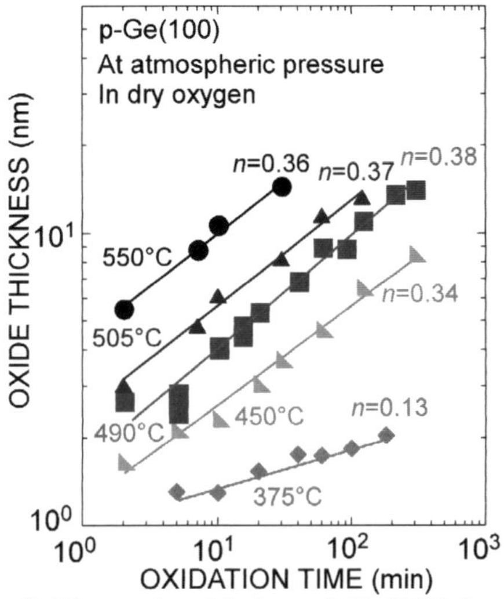

Figure 2. Thermal oxidation of Ge(100) in dry-O_2 ambience at different temperatures as a function of oxidation time.

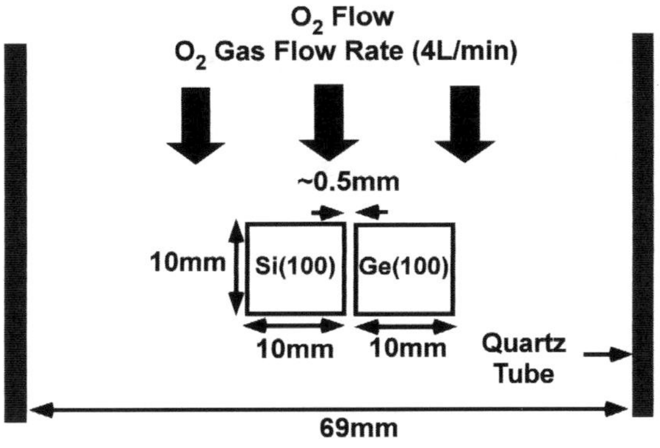

Figure 3. Experimental setup to confirm GeO desorption during Ge thermal oxidation.

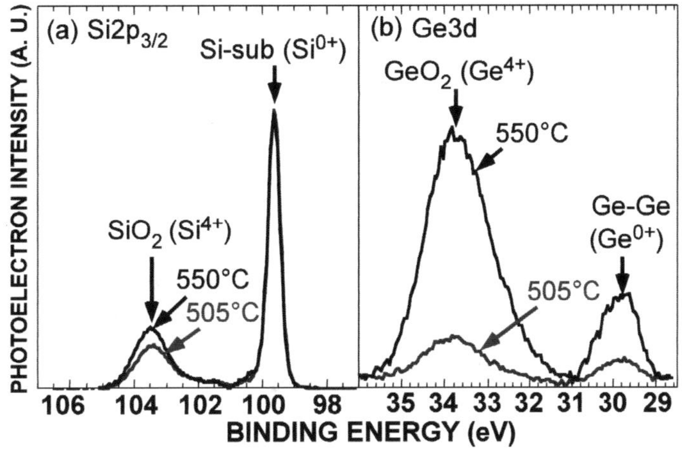

Figure 4. (a) Si $2p_{3/2}$ and (b) Ge 3d spectra taken for Si(100) surfaces after oxidation at 505 and 550°C in the experimental setup as shown in figure 3. The photoelectron take-off angle was set at 30°. The photoelectron intensity in each set of the spectra was normalized by Si^{0+} signals.

increase in chemically-shifted Si $2p_{3/2}$ signals due to the Si oxide components but also a significant increase of Ge 3d signals was clearly observed. The result of figure 4 indicates that the GeO desorption occurs even at 505°C oxidation and is enhanced with increasing oxidation temperatures, which is consistent with the results of SE analysis (Figure 2). Notice that Ge3d signals observed on oxidized Si(100) surfaces consist mainly of metallic Ge (Ge^{0+}) and Ge oxide (Ge^{4+}) components. It is likely that, in the deposition of Ge-rich oxide from the thermal decomposition of Ge monoxide thermally-desorbed from oxidizing Ge substrate, reduction of Ge oxide layer and/or phase separation into two thermally stable states (Ge^{0+} and Ge^{4+}) proceed. The detailed chemical bonding features of this metallic Ge component on Si surface were evaluated from the Auger parameter without influence of the positive charge-up during the XPS measurement as shown in figure 5 [15]. The Auger parameter was calculated by using the following equation.

$$\text{Auger Parameter} = \text{Ge3d}_{5/2}\,(\text{B.E}) + \text{Ge}\ L_3M_{45}M_{45}\,(\text{K.E}) \qquad (1)$$

As a result, the Auger parameter of a metallic Ge component observed on the oxidizing Si surface was obtained to be 1174.6 eV, which is higher than the reference value of Ge-Ge bonding units (1173.8 eV). This energy shift can be interpreted as a formation of Si-Ge bonding units. Also, the formation

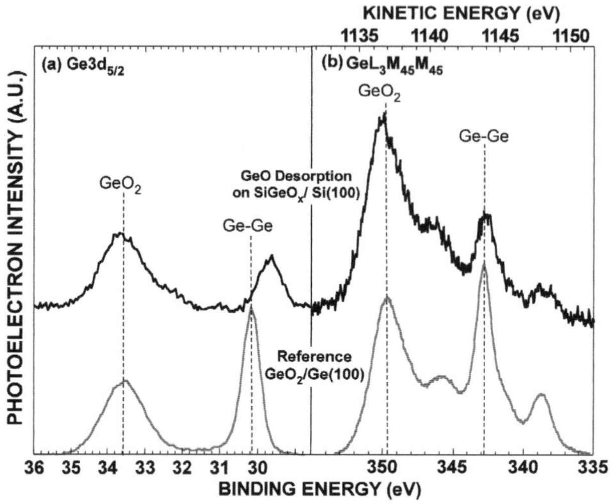

Figure 5. (a) Ge $3d_{5/2}$ and (b) Ge $L_3M_{45}M_{45}$ Auger spectra taken for Si(100) surface oxidized at 550°C shown in Figure 4. The spectra from thermally-grown GeO_2/Ge(100) is also shown as a reference. The photoelectron take-off angle was set at 90°. The photoelectron intensity was normalized by GeO_2 signals.

of Si-Ge bonding units in the oxide layer was confirmed from photoelectron take-off angle dependence of XPS core-line signals.

Next, to get an insight into the mechanism of Ge thermal oxidation, we made a trial estimation by applying the general oxidation equation proposed by Deal and Grove [16] to experimentally-obtained results. In the Deal-Grove model which is well-known to explain Si thermal oxidation well, the growth of oxide thickness of x_0 with oxidation time of t is given by

$$x_0^2 + Ax_0 = B(t+\tau) \qquad (2)$$

$$x_0 = B(t+\tau)/x_0 - A \qquad (3)$$

where the quantity τ corresponds to a shift in the time coordinate which corrects for the presence of the initial oxide layer. Furthermore, the coefficients of A and B are the positive value in the Si oxidation case. In figure 6, the obtained Ge oxide thickness shown in figure 2 is replotted as a function of $(t+\tau)/x_0$ based on the eq. (3). In each oxidation temperature, the τ value was estimated by using the same method as described in Ref. 16. The oxidation at 375°C only shows a linear relationship but has a negative value of intercept of A, which is different from the situation of the Si oxidation. For the higher oxidation temperature from 490 to 550°C, a logarithmic relationship was obtained. Obviously, the Deal and Grove model cannot be applied to the Ge thermal oxidation at atmospheric pressure in all temperature range. Thus, the contribution of GeO desorption must be considered in thermal oxidation of Ge surface and particularly for oxidation over 400°C at atmospheric pressure.

To gain a better understanding of thermal desorption process, the time-dependent model as described in Ref. 17 was applied, in which nonsteady-state oxidation in the early stages is taken into account. Based on the model, the Ge oxide growth rate, dx_0/dt, was evaluated as a function of oxide thickness in the consideration of the thickness dependent GeO-desorption [18]. Figure 7 shows the results obtained at 490°C, where the oxidation rate was calculated from $(x_0-x_i)/t$, where x_i is the initial thickness of oxide at t=0 min. The oxidation rate increases at with an increase in oxidation thickness until it reaches its maximum at ~5 nm in thickness and then decreases with increasing oxide thickness. In the case of Si thermal oxidation at atmospheric pressure, oxidation proceeds with no desorption and the oxidation rate in the early stages show an exponential decay with progressive oxidation [17], which is theoretically explained the oxidation reaction rate at the interface is controlled by the

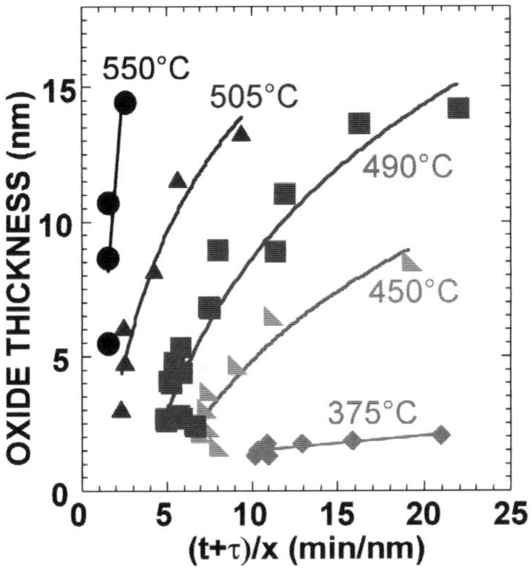

Figure 6. Evaluation of rate constants for oxidation of Ge(100) in dry-O_2. Values of (t+τ)/x for 375 and 450°C are the re-scaling value by dividing the actual values by 6 and 2, respectively.

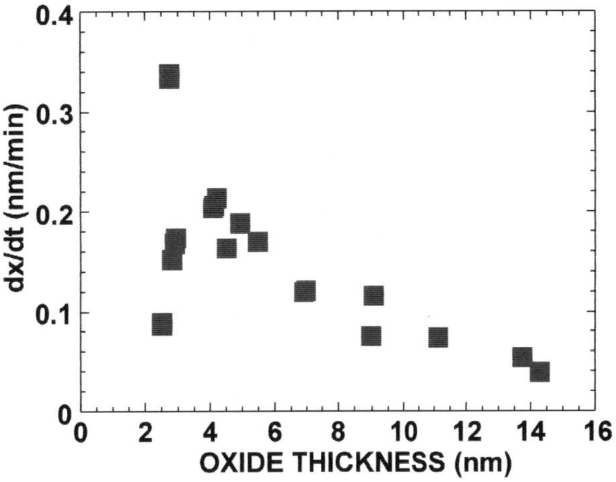

Figure 7. The oxide growth rate as a function of oxide thickness for 490°C oxidation in dry-O_2.

behavior of the emitted Si atoms. [19]. Taking into account of this fact, the observed increase in the Ge oxide growth rate in the early stages indicates that the GeO desorption proceeds simultaneously during thermal oxidation at atmospheric pressure.

4. Conclusion

The thermal oxidation of the Ge(100) surface was investigated. The slope of Ge oxide growth at 375°C with oxidation time shows the same value as native oxide growth and this slope gradually decreases with increasing temperatures over 490°C. The rate constant for Ge oxidation at 375°C is a linear relationship while oxidation at temperatures over 490°C shows a logarithmic relationship. The GeO desorption during thermal oxidation was confirmed from the XPS analysis of Si substrate placed next to the Ge during oxidation over 505°C. Also, the GeO desorption involves in thickness dependent oxidation rate for Ge thermal oxidation at temperatures over 490°C.

Acknowledgement

A part of this work was supported by the Research Institute for Nano-device and Bio Systems (RNBS), Hiroshima University, Japan.

References

[1] Sze S M and Ng K K 2007 *Physics of Semiconductor Devices* (New York, Wiley 3rd ed) p.786.
[2] Chui C O, Kim H, McIntyre P C and Saraswat K C 2003 *IEDM Tech. Dig.* p.437.
[3] Shang H, Okorn-Schmidt H, Chan K K, Copel M, Ott J A, Kozlowski P M, Steen S E, Cordes S A, Wong H-S P, Jones E C and Haensch W E 2002 *IEDM Tech. Dig.* p.441.
[4] Murakami H, Fujioka T, Ohta A, Bando T, Higashi S and Miyazaki S 2010 *ECS Trans.* **33** 253
[5] Ohta A, Fujioka T, Murakami H, Higashi H and Miyazaki S 2011 *Japan. J. Appl. Phys.***50**10PE01
[6] Murakami H, Miura M, Ohta A, Yogauchi R, Higashi S and Miyazaki S 2008 *Ext. Abst. of 4th Int. SiGe Technol. and Device Meeting* p.165
[7] Prabhakaran K, Maeda F, Watanabe Y and Ogino F 2000 *Appl. Phys. Lett.***76** 2244.
[8] Nagashio K, Lee C H, Nishimura T, Kita K and Toriumi A 2009 *Mater. Res. Soc. Symp. Proc.* 1155-C06-02

[9] Wang S K, Kita K, Nishimura T, Nagashio K and Toriumi A 2011 *Japan J. Appl. Phys.* **50** 04DA01
[10] Toriumi A, Wang S K, Lee C H, Yoshida M, Kita K, Nishimura T and Nagashio K 2010 *ECS Trans* **28(2)** 171
[11] Kita K, Suzuki S, Nomura H, Takahashi T, Nishimura T and Toriumi A 2008 *Japan J. Appl. Phys.* **47** 2349.
[12] Kita K, Lee C H, Nishimura T, Nagashio K, and Toriumi A 2009 *ECS Trans.* **19** 101.
[13] Sahari S K, Ohta A, Matsui M, Murakami H, Higashi S, Miyazaki S 2011 *Ext. Abst. of 2011 SSDM* p. 14
[14] Sahari S K, Murakami H, Fujioka T, Bando T, Ohta A, Makihara K, Higashi S, Miyazaki S 2011 *Japan J. Appl. Phys.* **50** 04DA12
[15] Wagner C D, Gale L H and Raymond R H 1979 *Anal. Chem.* **51** 466.
[16] Deal B E and Grove A S 1965 *J. Appl. Phys.* **36(12)** 3770
[17] Massoud H Z, Plummer J D and Irene E A 1985 *J. Electrochem. Soc. Solid-State Science and Technology* **132 (11)** 2685
[18] Wang S K, Kita K, Lee C H, Tabata T, Nishimura T, Nagashio K and Toriumi A 2010 *J. Appl Phys.* **108** 054104
[19] Kageshima H, Shiraishi K and Uematsu M 1999 *Japan. J. Appl. Phys.* **38** L971

Physical and electrical properties of ultra-thin nickel silicide Schottky diodes on Si (100)

Y Tamura[1], R Yoshihara[1], K Kakushima[2], H Nohira[3], O Nakatsuka[4], P Ahmet[1], Y Kataoka[2], A Nishiyama[2], N Sugii[2], K Tsutsui[2], K Natori[1], T Hattori[1], and H Iwai[1]

[1]Frontier Research Center, Tokyo Institute of Technology, Yokohama 226-8502, Japan
[2] Interdisciplinary Graduate School of Science and Engineering, Tokyo Institute of Technology, Yokohama 226-8502, Japan
[3]Tokyo City University, Tokyo 1-28-1, Japan
[4]Nagoya University, Nagoya 463-8603, Japan

E-mail: tamura.y.al@m.titech.ac.jp

Abstract. The physical and electrical properties of Ni silicides, reactively formed by a thin Ni layer of 3 nm, have been investigated. The existence of $NiSi_2$ phase has been confirmed at low temperature annealing by x-ray photoelectron spectroscopy. The silicides have shown flat surfaces up to an annealing temperature of 800 °C and a stable sheet resistance can be achieved. The Schottky barrier heights extracted from diode characteristics have shown stable values against annealing temperature owing to the stability of the film with an ideality factor nearly to unit.

1.Introduction

Ni silicides have been widely investigated for contact materials in microelectronic devices [1]. One of the issues of Ni silicides is that the formed NiSi, a low-resistive phase in Ni silicides, starts to agglomerate at an annealing temperature of 600 °C [2]. As the agglomeration is the results of minimizing the gain boundary between NiSi grains and the interface energy of NiSi and Si substrate, the temperature for agglomeration decreases when the thickness of NiSi is reduced [3]. Recently, Ni films with thicknesses less than 4 nm were found to be resistant to agglomeration up to an annealing temperature of 850 °C [4]. The difference in the silicide films is the formation of $NiSi_2$ phase even at an annealing temperature of 300 °C, which is commonly formed at an annealing temperature of 800 °C [5]. Therefore, the suppression of the agglomeration can be understood by the difference in the energy of $NiSi_2$ and NiSi phases. The suppression of agglomeration is also advantageous to obtain a flat interface between silicides and Si substrates. The interface of silicide and Si substrates tends to form pyramid shapes with Si(111) plane, which results not only the increase in the junction leakage, but also variability due to different Shocttky barrier height (ϕ_{Bn}) between Si(100) and (111) planes [6]. Therefore, another advantage of agglomeration-resistant Ni silicides is the suppression of forming interfaces with different crystallographic orientations which reduces the ϕ_{Bn} variation. However, there is little report on the electrical properties of Schottky diodes. This paper confirms the phase and the surface morphology changes on annealing temperature. And then the electrical characteristics of diodes are measured to extract the ϕ_{Bn}.

2. Device fabrication

The wafers used in this study were *n*-type Si(100) with a doping density of 3×10^{15} cm^{-3}. Ni thin films with thicknesses of 3.0 and 5.5 nm were deposited by RF magnetron sputtering system after chemical cleaning including hydrogen termination of the substrates. The wafers were annealed in nitrogen gas at annealing temperature ranging from 200 to 800 °C for 1 minute. The sheet resistance (ρ_{sh}) of the films was obtained by four terminal measurements and the chemical composition was measured by hard x-ray photoelectron spectroscopy at SPring-8 BL46XU [7]. The surface morphology of the films was measured by atomic force microscope (AFM). For Schottky diode fabrication, a SiO_2 isolated Si wafers were used. An Al film was formed as a back contact for diode fabrication. The ϕ_{Bn} and ideality factor (*n*-factor) of the diodes were carefully extracted to exclude the leakage current at diode periphery.

3. Composition and morphology of Ni silicide films

Figure 1 shows ρ_{sh} of the film on annealing temperature. For the sample with 5.5-nm-thick Ni layer, ρ_{sh} showed a large decrease over 300 °C, which is attributed to the formation of NiSi phase in the silicide. When the annealing temperature is over 500 °C, the ρ_{sh} showed a large increase due to agglomeration of the silicides. On the other hand, for the sample with Ni thickness of 3 nm, a gradual reduction in the ρ_{sh} was observed from 250 °C, and the value became stable at annealing temperatures from 400 to 800 °C, which is in good agreement with previous reports. Considering the resistivity of typical bulk $NiSi_2$ (34~50 Ωcm) [8], the phase of the silicide film can be considered as $NiSi_2$.

Ni $2p_{3/2}$ spectra of the samples annealed at various temperatures are shown in figure 2. The binding energy at the peak intensity was found to shift to higher energy from pure Ni, which was obtained by as-deposited sample. At an annealing temperature of 250 °C, the spectrum indicates the main composition is Ni-rich phase, which is in good agreement with the ρ_{sh} as high resistivity is reported for Ni-rich phase. The sample with 5.5-nm-thick-Ni showed a single peak at 500 °C annealing, which corresponds to the NiSi phase at 853.72 eV [9]. For 3-nm-thick-Ni samples, while increasing the temperature, two peak intensities, one at 854.19 eV with large intensity and the other Ni-rich phase with small intensity, were observed at 500 °C. The intensity at 854.19 eV further increased when the sample was annealed at 800 °C, indicating that the residual Ni-rich phase was converted to $NiSi_2$ phase by annealing [10].

The surface morphology images of the samples on annealing are shown in figure 3. The surface roughness increases with annealing with 5.5-nm-thick Ni samples and agglomeration was observed when the sample was annealed at 600 °C. On the other hand, 3-nm-thick Ni samples showed little difference up to 800 °C. The morphology change can be understood from the ρ_{sh} change on annealing temperature.

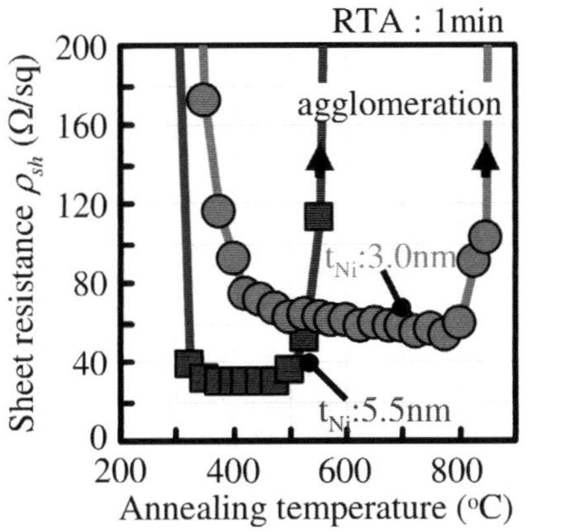

Figure 1. ρ_{sh} of silicides with 3.0 and 5.5 nm thick Ni layers on annealing temperature (t_{Ni} is Ni thickness).

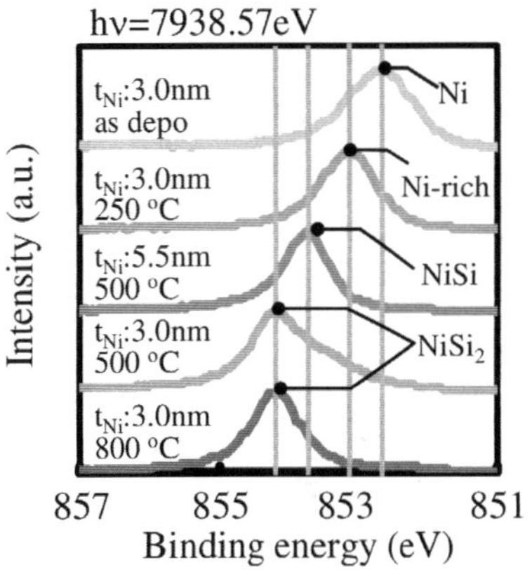

Figure 2. Ni $2p_{3/2}$ spectra of the silicide films.

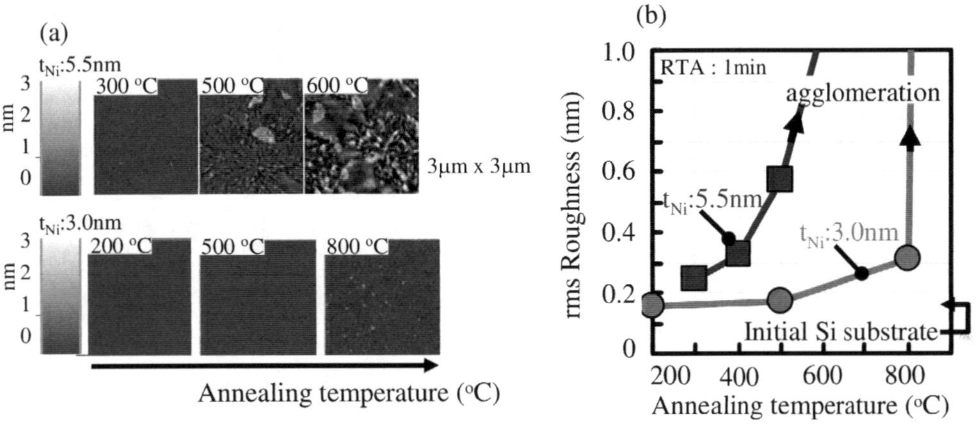

Figure 3. (a) Atomic force microscope (AFM) images the samples annealed at different temperatures. (b) Summary of the surface roughness on annealing temperature.

4. Electrical characteristics of Ni silicide diodes

Figure 4 shows diode characteristics of the fabricated Ni silicide Schottky diodes on various annealing conditions. While increasing the annealing temperature from 300 to 500 °C, the reverse currents with 3-nm-thick Ni diodes were found to be increased, indicating the reduction in the ϕ_{Bn}. At annealing temperatures from 500 to 700 °C, the ϕ_{Bn} showed a stable value of 0.54±0.02 eV, owing to the smooth surface of silicides and uniform formation of $NiSi_2$ phase. However, at 800 °C, the diode showed large increase in the reverse current from theoretical value and large reduction in the forward current. As the diodes were formed on a SiO_2 isolated Si wafer, different reactions at the periphery of the electrodes might be the reason for the change in the diode characteristics. On other hand, the reverse currents continued to increase with annealing temperature for 5.5-nm-thick Ni diodes, even with surface agglomerated samples. This fact might be due to interface roughening to modify the local ϕ_{Bn} [11], so that it could be concluded that the control of the ϕ_{Bn} was difficult with thick Ni layers. The *n*-factors of the diodes were less than 1.1 for all the diodes.

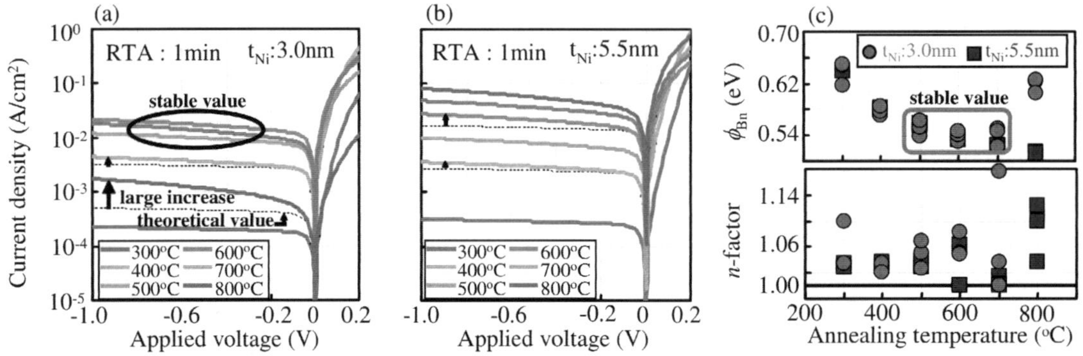

Figure 4. Diode characteristics of silicide Schottky diodes formed from (a) 3.0-nm and (b) 5.5-nm thick Ni layers (c) ϕ_{Bn} and n-factor of the diodes on annealing temperature.

5. Conclusions

Ni silicides, reactively formed by 3-nm-thick Ni, have shown a stable sheet resistance, flat morphology and stable Shottky barrier heights against annealing temperature up to 700 °C compared to those formed with thicker Ni film. These properties are owing to the formation of stable $NiSi_2$ phase at a temperature as low as 500 °C.

Acknowledgements

This project was supported by NEDO. The authors would like to thank Park Systems Japan Inc. for AFM observation.

References

[1] Iwai H, Ohguro T, Ohmi S 2002, Microelectron. Eng. 60 157-169
[2] Tsutsui K, Xiang R, Nagahiro K, Shiozawa T, Ahmet P, Okuno Y, Matsumoto M, Kubota M, Kakushima K, Iwai H 2008 Microelectron. Eng. 85 315-319
[3] Nolan T P and Sinclair R, Beyers R 1992 J. Appl. Phys. 71 720-724
[4] Keyser K De, Bockstael C Van, Meirhaeghe R L Van, Detavernier C, Verleysen E, Bender H, Vandervorst W, Jordan-Sweet J and Lavoie C 2010 Appl. Phys. Lett. 96 173503
[5] Lavoie C, Detavernier C, Cabral Jr C, d'Heurle F M, Kellock A J, Jordan-Sweet J, Harper J M E 2006 Microelectron. Eng. 83 2042-2054
[6] Tung R T and Gibson J M 1985 J. Vac. Sci. Technol. A. 3 987-991
[7] Kobayashi K and Yabashi M, Tanaka Y, Tokushima T and Shin S, Tamasaku K, Miwa D and Ishikawa T, Nohira H and Hattori T, Sugita Y, Nakatsuka O, Sakai A and Zaima S 2003 Appl. Phys. Lett. 83 1005-1007
[8] Chen L J 2004 *Silicide Technology for Integrated Circuits* (London: The Institute of Electrical Engineers)
[9] Grunthaner P J and Grunthaner F J, Madhukar A 1982, J. Vac. Sci. Technol. 20 680-683
[10] Rey S, Hommet J, Schmerber G, Normand F Le 2000, J. Crystal Growth 216 225-234
[11] Nakatsuka O, Suzuki A, Akimoto S, Sakai A, Ogawa M and Zaima S 2008 Jpn. J. Appl. Phys. 47 2402-2406

Embedding of Copper into 30-nm-wide trench by a high-magnetic-field magnetron sputtering method with oxygen added sputtering gas

Masatoshi Itoh , and Shigeru Saito
Department of Electrical Engineering, Tokyo University of Science
1-3 Kagurazaka, Shinjuku-ku, Tokyo 162-8601, Japan

E-mail: labsaito@rs.kagu.tus.ac.jp (M. Itoh)
saito@ee.kagu.tus.ac.jp (S. Saito)

Abstract. We successfully embedded Cu in 30-nm-wide trenches with an aspect ratio of 10 using a high-magnetic-field, medium vacuum magnetron sputtering method having high deposition speeds. We used a $Ru/TaN/SiO_2/Si$ substrate featuring a Ru barrier layer and a sputtering gas formed by adding O_2 at 5×10^{-5} Pa to 1-at.% N_2-Ar. The trenches could be completely filled with Cu in a sputtering time of 5 min. This result is discussed on the basis of capillary theory focusing on the wettability of Cu with the Ru barrier layer.

1. Introduction

Copper (Cu) wiring used in ultra-large-scale-integration (ULSI) devices has been patterned by methods such as electroplating and sputtering. Electroplating is currently the mainstream process for Cu embedding at wiring widths of 100 nm and narrower. Next-generation ULSIs, however, will call for even finer Cu wiring widths of 32 or 22 nm, which means that preparation of the Cu seed layer required for electroplating will be difficult as will Cu embedding by electroplating.

The sputtering method has, however, been used for Cu embedding at wiring widths greater than 100 nm. At present, it is being used for preparing the Cu seed layer and barrier layer in electroplating, but it is not being used for Cu embedding at wiring widths of 100 nm and narrower. The sputtering method, however, has features that cannot be overlooked. For example, it has been used since the time of aluminum (Al) wiring and is consequently a mature technology, and it is expected that existing sputtering equipment used for Al wiring could be appropriated for use in Cu wiring. These strong points underscore a need for developing new sputtering technology that can support Cu embedding for wiring widths of 100 nm and narrower.

We have been researching the formation of Cu wiring for ULSIs using high-vacuum planar magnetron sputtering. In this research, we found that Cu embedding characteristics can be improved by employing optimal embedding conditions in which O_2 is added to 1-at.% N_2-Ar sputtering gas[1] and by using a ruthenium (Ru) barrier layer with which Cu has excellent wettability. We have also been successful in achieving complete Cu embedding in a $Ru/TaN/SiO_2/Si$ substrate for a trench width of 30 nm and an aspect ratio (AR) = 10 by high-vacuum planar magnetron sputtering [2]. The Cu deposition speed by this sputtering method, however, is slow, as about one hour of sputtering is needed to complete Cu embedding.

Increasing the speed of Cu deposition would provide a manufacturing advantage since it would shorten the time taken up by the Cu wiring process. With this in mind, we set out to achieve high deposition speeds and shorten sputtering time by embedding Cu using medium vacuum magnetron sputtering in a high magnetic field. We used two types of Ru/TaN/SiO_2/Si substrates in our experiments: one of trench width 80 nm (AR=5) and the other of trench width 30 nm (AR=10). In short, we performed Cu-embedding experiments by high-magnetic-field, medium vacuum magnetron sputtering under the optimal Cu-embedding conditions used for high-vacuum planar magnetron sputtering. In this way, we were able to embed Cu in these two types of trenches in a sputtering time of only 5 min.

2. Experiment

This experiment used a SiO_2/Si substrate with a trench width of 100 nm and AR = 4.5 on which a 20-nm-thick TaN barrier layer was formed. On this structure, a Ru barrier layer was also formed by high-vacuum planar magnetron sputtering equipment using an oil diffusion pump. The Ru sputtering time was varied to obtain two types of substrates: a Ru(10 nm)/TaN(20 nm)/SiO_2/Si substrate with a trench width of 80 nm (AR=5) and a Ru(35 nm)/TaN(20 nm)/SiO_2/Si substrate with a trench width of 30 nm (AR=10). The Ru barrier layer was formed under the following conditions: an arrival vacuum level of $P=1\times10^{-5}$ Pa, a target – substrate distance of d = 200 mm, an Ar sputtering gas pressure of $P_S=3.0\times10^{-2}$ Pa, a sputtering voltage of V_S = 6 kV, a sputtering current of I_B = 7.5 mA, a substrate temperature of Ts = 200°C, sputtering gas consisting of pure Ar, and a target composed of Ru.

After being momentarily drawn into the surrounding atmosphere, the fabricated substrate was transported to ultra-high-vacuum equipment having a cryopump. This equipment was used to perform Cu embedding by the high-magnetic-field, medium vacuum magnetron sputtering method under the following conditions: $P = 1\times10^{-8}$ Pa, d = 200 mm, P_S = several Pa, V_S = 200 V, I_B = 1.65 A, Ts = 250 – 400°C, sputtering gas: 1-at.% N_2-Ar, added gas: O_2, added O_2 partial pressure: $P_{O2} = 10^{-3} - 10^{-6}$ Pa, and target: Cu.

The 1-at.% N_2-Ar sputtering gas was provided by a cylinder containing a mixture of these gases, and variable leak valves were used to supply the sputtering gas and added gas to the sputtering equipment. The Cu sputtering time was varied according to the objective in question.

The state of Cu embedding was observed by scanning electron microscope (SEM) after cutting away samples of the Cu-sputtered substrate with nippers. Resistivity (ρ) was calculated based on the Cu film thickness and the sheet resistance measured by the four probe method.

3. Results and Discussion

The embedding of Cu using the high-magnetic-field, medium vacuum magnetron sputtering method was performed using 1-at.% N_2-Ar sputtering gas to which O_2 at 5×10^{-5} Pa was added. The substrate temperature was 350°C. These conditions are the same as the Cu-embedding conditions found to be optimal for the high-vacuum planar magnetron sputtering method [2]. However, when using the high -magnetic-field, medium vacuum magnetron sputtering method, Cu deposition speed under these conditions was 80 nm/min, which is more than 20 times that of the high-vacuum planar magnetron sputtering method.

Results of embedding Cu in 80-nm-wide (AR = 5), trenches with Ru(10 nm)/TaN(20 nm) barrier layer are shown in Figure 1. The Cu sputtering time was 5 min. For comparison purposes, the case of using pure Ar sputtering gas is also shown. An SEM image of the substrate cross section before Cu embedding is shown in Figure 1(a), and the result of Cu embedding under optimal conditions is shown in Figure 1(b). As can be seen, the trenches are completely filled under optimal conditions. The result of Cu embedding using pure Ar sputtering gas is shown in Figure 1(c). In this case, some of the trenches are incompletely filled. These results demonstrate that optimal Cu embedding conditions for the high-vacuum planar magnetron sputtering method are also effective for the high-magnetic-field, medium vacuum magnetron sputtering method.

On the basis of these results, we next attempted to embed Cu in 30-nm-wide (AR = 10), trenches with Ru(35 nm)/TaN(20 nm) barrier layer under optimal embedding conditions. Results are shown in

Figure 1. Embedding of Cu in 80-nm-wide, AR = 5 trench with Ru(10 nm)/TaN(20 nm) barrier layer

(a) Before embedding of Cu, (b) after embedding of Cu under optimal Cu-embedding conditions, and (c) after embedding of Cu using pure Ar sputtering gas

Cu sputtering time: 5 min; substrate temperature: 350°C

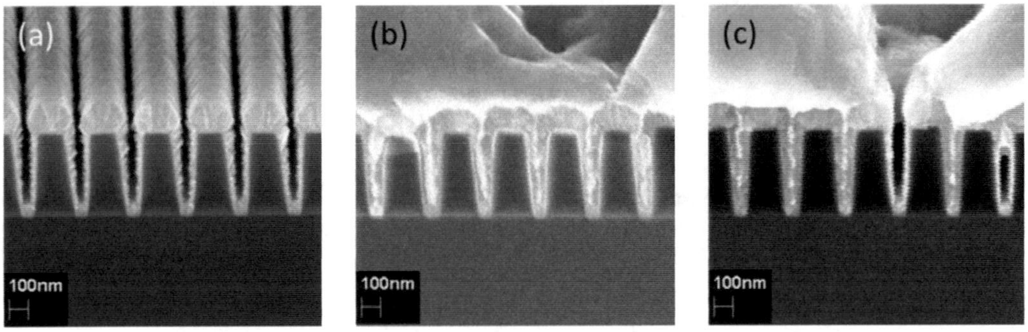

Figure 2. Embedding of Cu in 30-nm-wide, AR = 10 trench with Ru(35 nm)/TaN(20 nm) barrier layer

(a) Before embedding of Cu, (b) after embedding of Cu under optimal Cu-embedding conditions, and (c) after embedding of Cu using pure Ar sputtering gas

Cu sputtering time: 5 min; substrate temperature: 350°C

Figure 2. The Cu sputtering time here was 5 min, and the result of using pure Ar sputtering gas is also shown for comparison purposes. An SEM image of the substrate cross section before Cu embedding is shown in Figure 2(a). The result of Cu embedding under optimal conditions is shown in Figure 2(b), where it can be seen that the trenches are completely filled. The case of Cu embedding using pure Ar sputtering gas is shown in Figure 2(c). In this case, some of the trenches are incompletely filled, as was the case with the 80-nm-wide trenches.

We discuss the reason for these results on the basis of capillary theory focusing on the wettability of Cu with the Ru barrier layer [3, 4].

According to capillary theory, the critical free-energy barrier for nucleation ΔG^* is given by the following equation:

$$\Delta G^* \propto \frac{(2-3\cos\theta+\cos^3\theta)}{4}, \qquad (1)$$

Here, θ denotes the contact angle. Now, ΔG^* can be used to express nuclei density N^* as in the following equation:

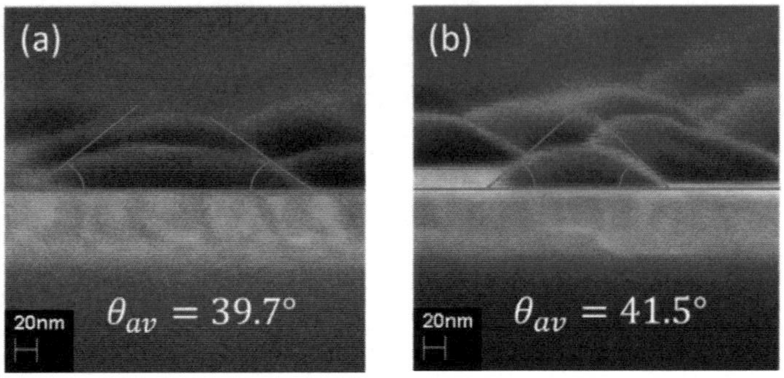

Figure 3. SEM image of substrate cross section
(a) optimal Cu-embedding conditions, (b) pure Ar sputtering gas
Cu sputtering time: 2 sec; substrate temperature: 350°C

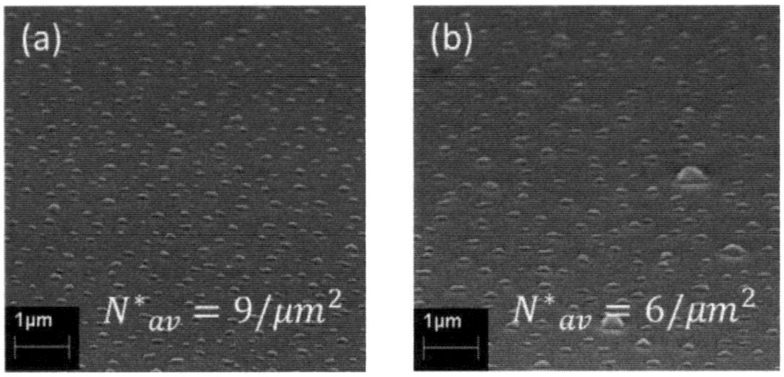

Figure 4. SEM image of substrate surface
(a) optimal Cu-embedding conditions, (b) pure Ar sputtering gas
Cu sputtering time: 2 sec; substrate temperature: 350°C

$$N^* = n_s \exp(-\Delta G^*/kT), \qquad (2)$$

Here, n_s denotes nucleation-site density.

The results of observing the wettability of Cu with the Ru barrier layer are shown in Figure 3. The Cu sputtering time was 2 sec, and the case of using pure Ar sputtering gas is shown for comparison purposes. When performing Cu embedding under optimal conditions as shown in Figure 3(a), the average contact angle θ_{av} was found to be 39.7°, but when performing Cu embedding with pure Ar sputtering gas (Figure 3(b)), it was found to be 41.5°. In other words, the average contact angle when embedding Cu under optimal conditions is about 2° less compared to that when using pure Ar sputtering gas. The standard deviation of the average nuclei angle under optimal Cu-embedding conditions was 4.7°, and 6.3° for using pure Ar sputtering gas. Next, the results of observing the nuclei density of Cu on the substrate surface is shown in Figure 4. These SEM images were used to determine average Cu nuclei density per 1 μm^2. Under optimal Cu-embedding conditions, average Cu nuclei density was $9/\mu m^2$ as shown in Figure 4(a), and when using pure Ar sputtering gas, average Cu nuclei density was $6/\mu m^2$ as shown in Figure 4(b). In short, average Cu nuclei density under optimal Cu-embedding conditions was found to be about 1.5 times that when using only pure Ar sputtering gas. From Equations (1) and (2), this increase in nuclei density N^* means a decrease in the critical free-energy barrier for nucleation ΔG^*, which supports the decrease observed in the contact angle. This result shows that Cu wettability improves under optimal Cu-embedding conditions.

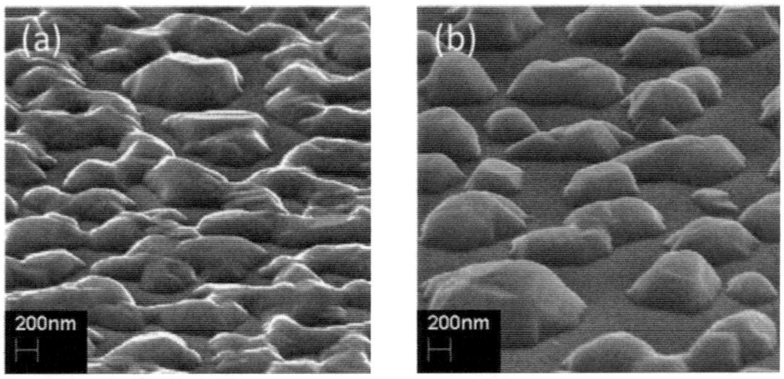

Figure 5. SEM image of substrate surface

(a) optimal Cu-embedding conditions, (b) pure Ar sputtering gas

Cu sputtering time: 1 min; substrate temperature: 350°C

Finally, the Cu surface shape when extending the Cu sputtering time is shown in Figure 5. Sputtering time was 1 min, and the case of using pure Ar sputtering gas is also shown for comparison purposes. Under optimal Cu-embedding conditions, agglomeration of Cu on the Ru barrier layer is suppressed and a Cu film is beginning to form as shown in Figure 5(a). However, when using pure Ar sputtering gas, Cu agglomerates on the Ru barrier layer and undergoes crystal growth as shown in Figure 5(b). Under optimal Cu-embedding conditions, the surface of Cu sputtered on the Ru barrier layer enters a flowing-like state and Cu wettability improves. This suggests that improvement in Cu wettability is associated with improvement in Cu-embedding characteristics.

The resistivity (ρ) of embedded Cu was 2.0 μΩ · cm under optimal Cu-embedding conditions. The surface of Cu sputtered on the Ru barrier layer under optimal Cu-embedding conditions was smooth. From this and data obtained from previous measurements by secondary ion mass spectrometry (SIMS), we estimated that the density of O_2 in the Cu film was less than 2% [5].

4. Conclusion

We demonstrated that Cu embedding can be performed in a relatively short time using a high -magnetic-field, medium vacuum magnetron sputtering method having high deposition speeds. This method required the use of optimal Cu embedding conditions as used in high-vacuum planar magnetron sputtering. Under these conditions, Cu deposition speed and Cu resistivity (ρ) were 80 nm/min and 2.0 μΩ · cm, respectively. We were able to completely embed Cu in 30-nm-wide, AR = 10 trenches in a sputtering time of 5 min. The improvement in Cu wettability at this time was observed from a drop in contact angle θ, increase in Cu nuclei density, and shape of the Cu surface. These results suggest that improvement in Cu wettability is associated with improvement in Cu embedding characteristics.

References

[1] Itoh M, Urano T, Uhara Y, Hayashi H, Manba Y, Taniseki A, Jyan H, Nishikawa E, and Saito S 2009 *Appl. Surf. Sci.* **256** 1240
[2] Itoh M, Ito H, Tsuru Y, Uhara Y, and Saito S 2011 *Proc. 11th Int. Symp. on Sputtering & Plasma Processes, 6-8 July 2011*, Kyoto, Japan, pp 524-527
[3] Milton O 1992 *The Materials Science of Thin Films* (Academic Press) p 40, P 198
[4] Kim H, and Shimogaki Y 2007 *J. Electrochem. Soc.* **154** G13
[5] Itoh M, Urano T, Uhara Y, Koh Y, and Saito S 2009 *Proc. 10th Int. Symp. on Sputtering & Plasma Processes, 8-10 July 2009*, Kanazawa, Japan, pp 483-485

Suppressive Effect of Interface Reaction and Water Absorption by Al Incorporation into Pr-oxide Film

W Takeuchi[1,*], K Furuta[1], K Kato[1,2], M Sakashita[1], H Kondo[1], O Nakatsuka[1], and S Zaima[1]

[1]Department of Crystalline Materials Science, Graduate school of Engineering, Nagoya University, Furo-cho, Chikusa-ku, Nagoya 464-8603, Japan

[2] Research Fellow of Japan Society for the Promotion Science

*E-mail: wtakeuti@alice.xtal.nagoya-u.ac.jp

Abstract. We have investigated the effects of Al incorporation into a Pr-oxide/Si gate stack formed by atomic layer deposition. The $PrAlO_x$ (PAO) layers show an amorphous structure by the incorporation of Al into the Pr-oxide. The PAO sample with 10%-Al shows good C-V characteristics without a hump, and the interface state density (D_{it}) is as low as 5×10^{10} cm^{-2} eV^{-1}. The amount of Si in the Pr-oxide film decreases by the Al incorporation. We deduce that Al incorporation into a Pr-oxide layer inhibits reaction at the interface of the PAO/Si substrate. We found that Al incorporation into the Pr-oxide is effective against the inhibition of moisture incorporation into the oxide film. The $Pr(OH)_3$ component, estimated by x-ray photoelectron spectroscopy, increases near the surface of the Pr-oxide and PAO films with 10%-Al after 300 days, while the SiO_x component does not change. The $Pr(OH)_3$ component decreases with increasing Al incorporation. The D_{it} of the PAO/Si sample with 10%-Al increases after 300 days. After post metallization annealing of the sample in N_2, D_{it} decreases with increasing annealing temperature.

1. Introduction

The thickness of a silicon dioxide (SiO_2) layer for a gate insulator has decreased to the nanometer scale with shrinking of complementary metal-oxide-semiconductor (CMOS) transistors, and its leakage current increases significantly. Recently, Hf-oxide has been shown to be one of the most promising candidates for a high dielectric (high-k) gate insulator to reduce leakage current with increasing physical thickness. However, gate insulator materials having a higher dielectric constant than Hf-oxide (higher-k materials) are required for future CMOS devices. Rare-earth metal oxides such as praseodymium (Pr) - and lanthanum (La)-oxides are promising because their bulk dielectric constants are as high as 30 [1]. Moreover, it is expected that Pr-oxide has a large electron effective mass since the conduction band structures of Pr-oxide are very flat. The leakage current that can be attributed to the presence of such large electron masses in the Pr-oxide is very low (J_g: 5×10^{-9} A/cm^2, Vg=1 V, EOT: 1.4 nm) [2].

Pr-oxide films have been grown by molecular beam epitaxy (MBE) [1], chemical vapor deposition (CVD) [3], and atomic layer deposition (ALD) [4]. Recently, we successfully deposited Pr-oxide films by CVD and ALD using a $Pr(EtCp)_3$ precursor by controlling the film thickness and its fluctuation [5-7]. However, an interfacial (IF) SiO_x layer often forms between the high-k materials and Si substrate, when high-k materials are deposited directly on Si [8-10]. The dielectric constant of the IF layer is usually low, which reduces the effective dielectric constant of the gate stack structure. Furthermore, the dielectric constant of the high-k oxide film decreases due to diffusion of Si into the high-k materials. It is thus important to prevent the formation of the IF layer. In addition, rare-earth metal oxides are highly hygroscopic in general. It is known that Pr-oxide tends to react with water vapor to form praseodymium hydroxides ($Pr(OH)_3$) [4,11].

The diffusion coefficient of oxygen in an Al oxide thin film is known to be smaller than those in the majority of oxides. There are some reports on the insertion of ultrathin Al_2O_3 film into the HfO_2/Si interface [12] and Al incorporation into La-oxide films [13-14]. The effects of the insertion of Al_2O_3 film include suppression of interfacial reaction, raising the thermodynamical stability of a high-k material on silicon, and modulation of the energy band gap and flatband voltage. Therefore, we focused on the incorporation of Al oxide into a Pr oxide layer in order to prevent formation of the IF layer and suppression of highly hygroscopic. Also, it is known that $PrAlO_x$ (PAO) has a high dielectric constant (ε_r=18–25) [15-16]. The PAO film can be deposited by CVD and ALD. However, the affect of the IF layer and hygroscopic properties on the electrical properties of the PAO film have not been clarified yet.

In this study, we examined the formation of PAO films by ALD using $Pr(EtCp)_3$, trimethylaluminum (TMA), and H_2O precursors. We also investigated the crystallinity and electrical properties of PAO gate insulators.

2. Experiments

N-type Si(001) wafer was used as the substrate. Prior to the oxide film deposition, the Si substrate was cleaned with the conventional RCA solutions. Then, the native oxide on the substrate was removed and a hydrogen (H)-terminated surface was formed with a diluted HF-solution. We deposited PAO films on the hydrogen terminated n-type Si substrates by ALD using a SUNALE™ R-150B reactor (Picosun). Pulsed supply of $Pr(EtCp)_3$ and TMA was repeated alternately in an ALD chamber. The compositions of Pr and Al (Al/(Al+Pr)) were controlled by changing the duration of their subcycle values (x,y) in the $[TMA/H_2O]_x/[Pr(EtCp)_3/H_2O]_y$ ALD sequence. The subcycle values (x,y) used were (0,54), (1,8), (1,1), and (100,0). The composition ratio Al/(Al+Pr) was estimated from each deposition rate of TMA/H_2O and $Pr(EtCp)_3/H_2O$. Samples with Al/(Al+Pr) composition of 0%, 10%, 60%, and 100% were prepared. The ALD process was carried out at a substrate temperature of 300°C. After the ALD, top and back Al electrodes were formed by vacuum evaporation. Crystalline structures were determined using transmission electron microscopy (TEM). X-ray photoelectron spectroscopy (XPS) was carried out to analyze the chemical bonding states and elemental composition. Capacitance-voltage (C-V), deep level transient spectroscopy (DLTS), and current-voltage (I-V) measurements were performed for the Al/PAO/Si MOS capacitors.

3. Results and Discussion

Figures 1(a)-(f) show TEM images and transmission electron diffraction (TED) patterns of PAO films on Si(001) substrates with an Al composition of 0% [Fig. 1(a), Fig. 1(d)], 10% [Fig. 1(b), Fig. 1(e)], and 60% [Fig. 1(c), Fig. 1(f)]. The thickness of PAO films with Al composition of 0%, 10%, and 60% were estimated to be 8.4, 6.1, and 5.9 nm, respectively, from the TEM observations. In the sample without Al, many diffraction spots appear as shown in Fig. 1(d). This indicates the formation of polycrystalline cubic Pr_2O_3. On the other hand, only diffraction patterns related to the Si substrate are observed for the samples with Al/(Al+Pr) of 10% and 60%. This indicates that the structure of the oxide is amorphous. These results mean that the crystalline structure of the oxide film changed to an amorphous structure by Al incorporation during deposition.

Figure 2 shows the C-V characteristics of Al/PAO/Si MOS capacitors for various Al ratios at RT with a measurement frequency of 1 MHz. The PAO films were deposited with ALD at 300°C. All C-V

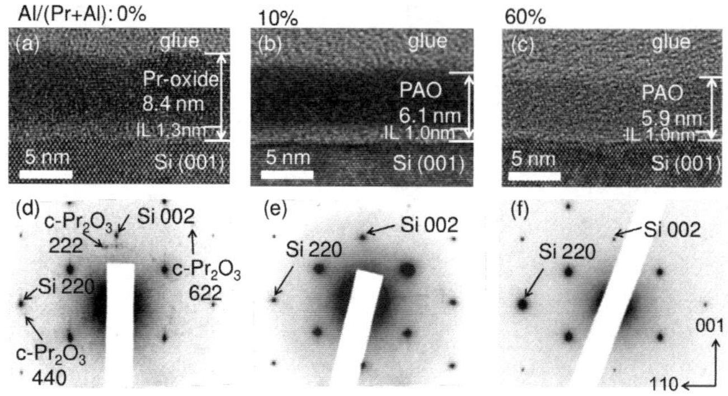

Figure 1. TEM images and TED patterns of PAO films with an Al composition of 0% [Fig. 1(a), (d)], 10% [Fig. 1(b), (e)], and 60% [Fig. 1(c), (f)] deposited on a Si(001) substrate.

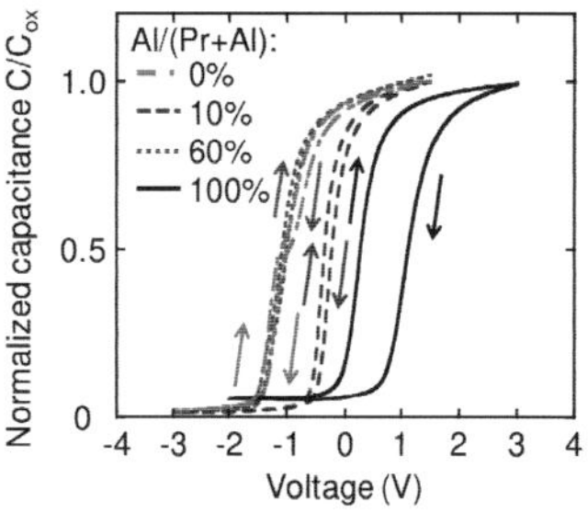

Figure 2. C-V characteristics of Al/PAO/Si MOS capacitor with various Al compositions.

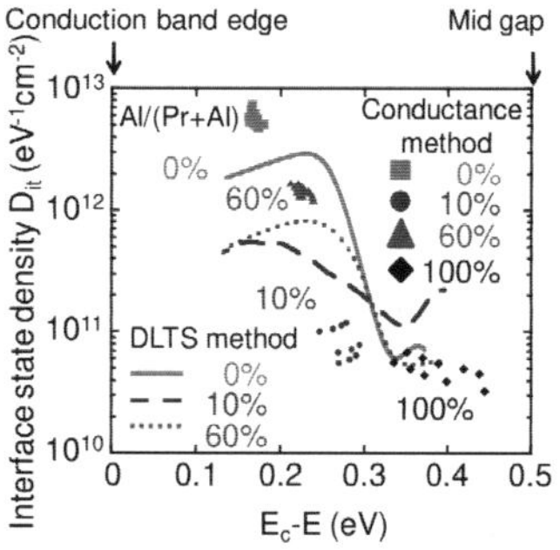

Figure 3. Interface state density (D_{it}) of Al/PAO/Si MOS capacitor with various Al compositions measured by the conductance method and the DLTS method.

curves were measured from the inversion to accumulation conditions, and then these were continuously returned to the accumulation condition. In the C-V curves of all samples, a hysteresis loop was observed. The hysteresis loop direction was always clockwise (the direction is shown by an arrow). This indicates carrier injection into traps near the PAO/Si interface from the Si substrate. Also, a hump in the C-V characteristics can be observed in the sample without Al, which indicates a large interface state density (D_{it}). In contrast, in the PAO samples, no such hump is observed in the C-V curves. Figure 3 shows the D_{it} values estimated using the conductance method and DLTS as a function of the energy from the conduction band edge. The energy distribution of the interface state density shows a general U-shape, as shown in the DLTS results. The interface state densities of the PAO samples were smaller than that of the Pr-oxide sample without Al. In particular, a D_{it} value of 5×10^{10} cm^{-2} eV^{-1} was achieved for the sample with Al/(Al+Pr) of 10% by the conductance method. This indicates that the interface state density can be decreased by the incorporation of Al into the Pr oxide/Si structure.

Figures 4(a)-(c) show Si2p photoemission spectra from the Pr oxide sample with various Al ratios Al/(Pr+Al) of (a) 0%, (b) 10%, and (c) 60%. Samples were sputtered with Ar ions at an acceleration energy of 3 keV. The photoelectron spectra for various sputtering times of the samples are shown in the figures. In the case of Al/(Pr+Al) of 0%, a peak is observed at 102.5 eV for a sputtering time of 1000 s to 4000 s. This indicates that Pr silicate ($PrSiO_x$) and SiO_x are formed [5]. In contrast, in the case of sputtering time above 5000 s, this peak is not observed. On the other hand, this peak is also observed in the PAO samples.

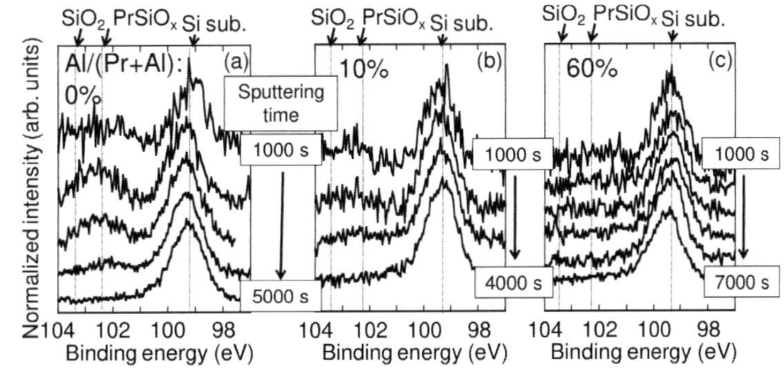

Figure 4. Si2p photoemission spectra of the Pr oxide sample with various Al ratios (Al/Pr+Al): (a) 0%, (b) 10%, and (c) 60% for various sputtering times.

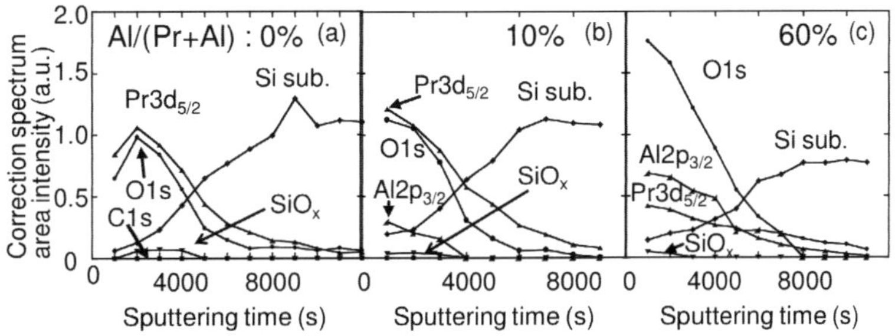

Figure 5. The depth profiles of the area intensity of XPS spectra for each element of PAO samples with the ratio of Al to Pr of (a) 0%, (b) 10%, and (c) 60%, respectively.

$Pr3d_{5/2}$, O1s, Si2p, and Al2p core spectra were obtained by XPS. Figures 5(a)-(c) show depth profiles of the area intensity of each element core spectra for PAO samples with Al/(Pr+Al) of (a) 0%, (b) 10%, and (c) 60%, respectively. The existence of the SiO_x component is observed through the films in all samples. This indicates that $PrSiO_x$ is formed. In other words, Si diffuses to the surface region during the ALD process. On the other hand, SiO_x composition for Al/(Pr+Al) of 0%, 10%, and 60% samples was 5.4%, 1.8%, and 1.6%, respectively. The amount of Si in the PAO film decreases with increasing amount of incorporated Al. Thus, we deduce that Al incorporation into the Pr-oxide film inhibits interfacial reaction at the PAO/Si interface. The presence of Al is observed through the films in all samples. This shows that the Al atoms were mixed into the Pr-oxide films during the ALD process. In fact, the Al composition in the samples with the designed Al/(Pr+Al) composition of 10% and 60% was estimated to be 22% and 67%, respectively. We deduce that the difference between designed and practical values of the Al composition in the oxide is due to the difference in the surface condition of the Pr-oxide and Al_2O_3. The small value of the interface state density can be attributed to prevention of the formation of SiO_x and $PrSiO_x$ at the interface by the Al incorporation. The Al incorporation into the Pr-oxides thus effectively improves the electrical properties of the Pr oxide MOS capacitors.

Next, we investigated the effect of Al on the hygroscopic characteristics of the Pr oxide film. We kept the PAO/Si samples in a vacuum desiccator for 300 days. We then performed XPS and C-V measurements. Figures 6(a)-(f) show O1s photoemission spectra for the Pr oxide sample with various Al ratios Al/ (Pr+Al) of 0%, 10%, and 60%. We measured XPS spectra for the samples before and after Ar sputtering for 1000 s. These spectra were normalized by the area intensities of the Pr-oxide at the $Pr3d_{5/2}$ core spectra for each sample. After 300 days without sputtering, the intensity of the peak around an energy of 532 eV increased in the samples with Al/(Pr+Al) of 0% and 10%. After sputtering, the peak intensity dramatically decreased. On the other hand, in the sample with Al/(Pr+Al) of 60%, the spectrum hardly changed after 300 days regardless of the surface sputtering. These results indicate that this component exists near the surface. Also, the peak intensity of SiO_x in the Si2p photoemission spectra hardly changed after 300 days in all samples (*not shown*). Thus, we consider that this peak at

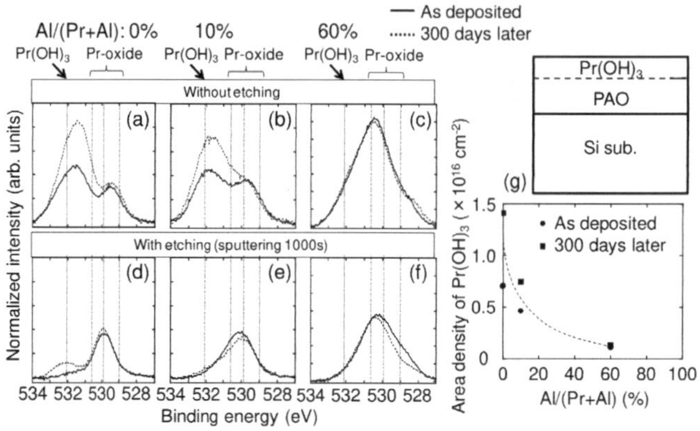

Figure 6. Comparison of hygroscopic properties in the as-deposited sample and the sample after 300 days. The O1s photoemission spectra for the Pr oxide sample with various Al ratios Al/(Pr+Al): (a), (d) 0%, (b), (e) 10%, and (c), (f) 60% without and with sputtering, respectively. (g) Area density of $Pr(OH)_3$ as a function of Al/(Pr+Al).

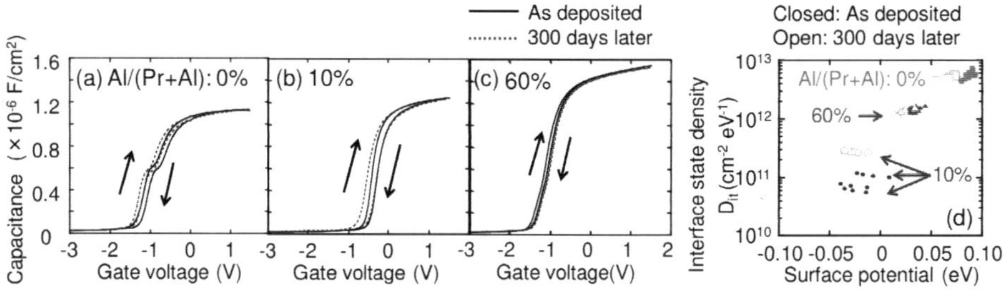

Figure 7. (a)-(c) C-V characteristics and (d) D_{it} of the Al/PAO/Si MOS capacitor with various Al compositions in the as-deposited sample and the sample after 300 days.

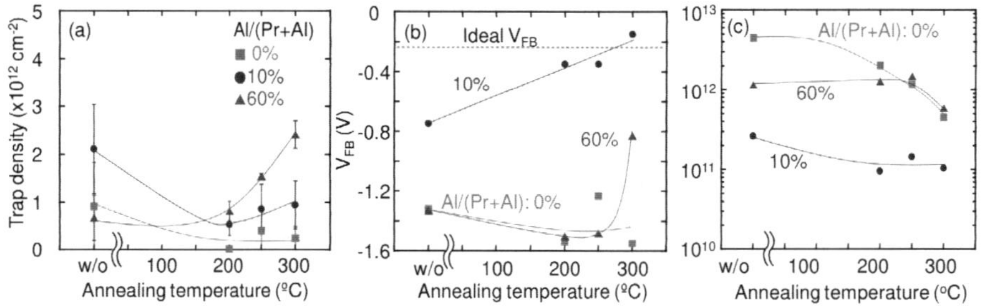

Figure 8. (a) Trap density, (b) flat band voltage (V_{FB}), and (c) minimum interface state density (D_{it}) for the Al/PAO/Si MOS capacitors with various Al compositions at several annealing temperatures in the sample after 300 days.

an energy of 532 eV in the O1s spectra is related to $Pr(OH)_3$ due to the hygroscopic properties of the Pr oxide. The area density of the $Pr(OH)_3$ peak decreased with increasing Al component in the Pr-oxide, as shown in Fig. 6(g). These results mean that hygroscopicity of the Pr-oxide film can be reduced by Al incorporation.

We also investigated the time degradation of electrical properties. Figures 7(a)-(c) show C-V characteristics of the Pr oxide sample with various Al ratios Al/(Pr+Al) of 0%, 10%, and 60%, respectively. The flat band voltage, estimated from the C-V characteristics of the sample without Al, slightly changed to the negative voltage direction after 300 days compared to the as-deposited samples as shown in Fig. 7(a). In the C-V curve of Al/(Pr+Al): 10% sample, the hysteresis width increased after 300 days, as shown in Fig. 7(b). On the other hand, the C-V curve of the sample with Al/(Pr+Al): 60% hardly changed even after 300 days.

Figure 7(d) shows the interface state density value as a function of the surface potential energy for samples with the various Al ratios as-deposited and after 300 days. In the sample with Al/(Pr+Al): 10%, the D_{it} value increased after 300 days. Meanwhile, the D_{it} values for the samples with Al/(Pr+Al): 0 and 60% hardly increased even after 300 days. The variation in the D_{it} value was too small in the sample without Al since D_{it} is higher than for the sample with Al. Thus, we might not have observed a variation in D_{it} after 300 days. In the low Al composition samples, the C-V characteristics changed. Thus, we deduce the degradation of electrical properties for the low Al composition samples by moisture.

Next, the samples after 300 days were annealed at 200-300°C in a N_2 ambient atmosphere to investigate the recovery effect of annealing for the time degradation sample. Figures 8(a) and (b) show the trap density near the interface and the flat band voltage (V_{FB}) estimated from the C-V characteristics as a function of the annealing temperature. In the samples with Al/(Pr+Al) of 0% and 10%, the trap density near the interface decreased with an increase in the annealing temperature. Meanwhile, the Al/(Pr+Al): 0% and 10% sample of the trap density near the interface increased with the annealing temperature. V_{FB} hardly changed with annealing in the sample without Al while V_{FB} of the PAO samples changed toward the ideal V_{FB} value. The minimum value of D_{it} at each temperature is shown in Fig. 8(d), which decreased with increasing annealing temperature.

4. Conclusions

We investigated the effects of Al incorporation into the Pr-oxide films for suppression of the formation of an IL layer as well as the hygroscopic properties. The amount of Si diffusing from the substrate into the Pr-oxide decreased with increasing Al composition. The interfacial reaction was effectively suppressed by the Al incorporation into the Pr-oxide. Good C-V characteristics without humps and a low interface state density as low as 5×10^{10} cm^{-2} eV^{-1} were obtained in the PAO sample with the 10%-Al component. The $Pr(OH)_3$ component increased in the Pr-oxide with a low Al component after 300 days, while that in the PAO sample with the 60%-Al component did not change. The hygroscopic properties could be suppressed by Al incorporation into Pr-oxide that effectively improved the interfacial properties of the Pr-oxide/Si structure.

Acknowledgments

This work was partly supported by a Grant-in-Aid for Specially Promoted Research (No. 22000011) from the Ministry of Education, Culture, Sports, Science and Technology of Japan. The authors acknowledge Prof. Yutaka Tokuda of Aichi Institute of Technology for his help in establishing the DLTS measurement system.

References

[1] Osten H J, Bugiel E and Fissel A 2003 *Solid-State Electron.* **47** 2161
[2] Osten H J, Dabrowski J, Müssig H J, Fissel A, and Zavodinsky V 2004 *Predictive simulation of semiconductor processing: status and challenges* (Berlin: Springer) p 259
[3] Nigra R L, Toro R G, Malandrino G, Raineri V and Fragalá I L 2003 *Adv. Mater.* **15** 1071
[4] Kukli K, Ritala M, Pilvi T, Sajavaara T, Leskelä M, Jones A C, Aspinall H C, Gilmer D C and Tobin P J 2004 *Chem. Mater.* **16** 5162
[5] Kondo H, Sakurai S, Sakashita M, Sakai A, Ogawa M and Zaima S 2010 *Appl. Phys. Lett.* **96** 012105
[6] Kondo H, Matsui H, Furuta K, Sakashita M and Zaima S 2010 *Jpn. J. Appl. Phys.* **49** 04DA14
[7] Adachi M, Sakashita M, Kondo H, Takeuchi W, Nakatsuka O and Zaima S 2011 *Jpn. J. Appl. Phys.* **50** 04DA08
[8] Watahiki T, Tinkham B P, Jenichen B, Shayduk R, Braun W and Ploog K H 2008 *Appl. Surf. Sci.* **255** 758
[9] Müssig H J, Dąbrowski J, Ignatovich K, Liu J P, Zavodinsky V and Osten H J 2002 *Surface Science* **504** 159
[10] Schmeiβer D and Zschech E 2006 *Materials Science in Semiconductor Processing* **9** 934
[11] Mullica D F, Milligan W O and Beall G W 1979 *J. Inorg. Nucl. Chem.* **41** 525
[12] Park M, Koo J, Kim J, Jeon H, Bae C and Krug C 2005 *Appl. Phys. Lett.* **86** 252110
[13] Arimura H, Brown S L, Callegari A, Kellock A, Bruley J, Copel M and Watanabe H 2011*IEEE Electron D'ev. Lett.* **32** 288
[14] Sivasubramani P, Kim M J, Gnade B E, Wallacea R M, Edge L F, Schlom D G, Craft H S and Maria J P 2005 *Appl. Phys. Lett.* **86** 201901
[15] Rouffignac P and Gordon R G 2006 *Chem. Vap. Deposition*, **12** 152
[16] Schroeder T, Lupina G, Sohal R, Lippert G, Wenger C, Seifarth O, Tallarida M and Schmeisser D 2007 *J. Appl. Phys.* **102** 014103

Nb thin films fabricated by a magnetic flux distribution variable-type facing target sputtering system

S. Morohashi, Y. Masumoto, K. Tanaka, K. Usui and Y. Kuzuo

Faculty of Engineering, Yamaguchi University, 2-16-1 Tokiwadai, Ube 755-8611, Japan

E-mail: smoro@yamaguchi-u.ac.jp

Abstract. A magnetic field distribution variable-type facing target sputtering system has been newly devised. The magnetic field distribution between facing targets can be easily changed by moving a magnet without breaking the vacuum. By changing the magnetic field distribution between facing targets, Niobium deposition rate has changed about five times with the same sputtering conditions, such as RF applied electric power and Ar pressure.

1. Introduction

Among thin-film deposition techniques, sputtering is widely used for the fabrication of electronic devices with multilayer film structures, such as displays, solar batteries, and ferromagnetic or superconducting tunneling junctions in which electrons tunnel through a barrier of about 1-nm thickness, [1,2]. In recent years, for solar batteries and image displays in which thinner structures are preferred, there has been increasing demand for multilayer electronic devices to be fabricated on organic films instead of hard and heavy glass substrates, in order to realize lightweight, low-cost, and flexible systems [3]. For the fabrication of tunneling devices and devices with multilayer film structures for film-based electronics and displays, we require a so-called low-temperature sputtering technology that can realize clear interfaces without any damage to the heat-sensitive substrate and the interfaces between layers during film deposition [4][5]. The conventional facing-target sputtering system is characterized by low damage because of the unique cathode structure, in which γ electrons are not incident on the substrate surface [6]. However, when compared with the widely used magnetron sputtering system, it became clear that the challenge was to improve both its (a) compactness and (b) through-put.

2. Facing-target sputtering system with new functions

2.1 Comparison between magnetron sputtering and conventional facing-target sputtering

Figure 1 shows a comparison between magnetron sputtering and conventional facing-target sputtering in terms of the damage to Teflon tape when a Si substrate is wrapped with Teflon tape and a SiO_2 film is deposited at approximately the same deposition rate (13.0 nm/min) [7]. Using magnetron sputtering, damage was observed in the Teflon tape after 7 min and 40 s from the start of sputtering (see Figure

1(c)). In contrast, when using the conventional facing-targets sputtering technique, the shape of the Teflon tape was still retained after 21 min and 30 s (see Figure 1(d)). As shown schematically in Figures 1(a) and (b), in normal magnetron sputtering in which the substrate faces the target, there is a structurally high probability that the electrons (γ electrons), which are released by ion bombardment of the target surface and the recoiled Ar ions, will hit the substrate; the deposited film is then damaged when these high-energy particles are incident on the substrate. In contrast, in the conventional facing-targets sputtering system, as shown schematically in Figure 1(b), the magnet is placed right in the back of the target, so that there are opposite magnetic poles between the facing targets; therefore, the γ electrons and the resulting plasma are confined in the space between the targets, and the deposited film is not damaged because the γ electrons are not incident on the substrate surface [6]. It is thought that the difference in the damage to Teflon, as shown in Figs. 1(c) and (d), reflects this difference in the cathode structure.

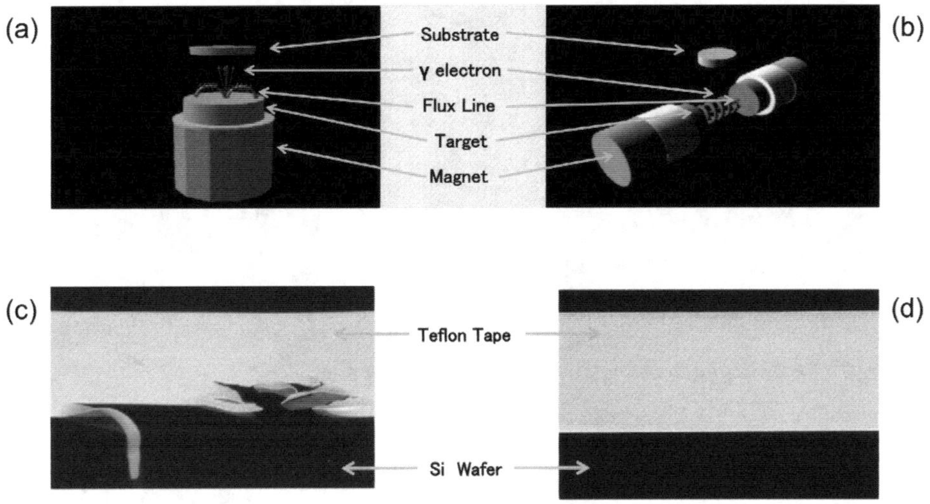

Fig. 1 Schematic illustration of the damage caused by magnetron sputtering and conventional facing-target sputtering. (a) Schematic diagram of magnetron sputtering. (b) Schematic diagram of conventional facing-target sputtering. (c) Surface state of Teflon tape after 100 nm-thick SiO2 film deposition by magnetron sputtering. (d) Surface state of Teflon tape after 300 nm-thick SiO2 film deposition by conventional facing-target sputtering.

2.2 Revolving-type facing-target sputtering system with multi-target

The volume of the cathode in conventional facing-target sputtering does not pose a problem in the case of single-target sputtering for the deposition of a single-layer film. However, in the case of multi-target sputtering for the continuous deposition of multilayer films, the multiple parallel-facing structures requires a larger cathode capacity and a larger vacuum pump; the system therefore has high initial and running costs, that is, it is uneconomical in terms of energy usage and is rather large. The revolving facing-target sputtering system was devised to enhance the compactness while causing little damage, as is characteristic of facing-target sputtering systems [7][8]. Figure 2 shows a schematic diagram of a sputtering system with four targets. The magnets in each revolving facing cathode-box form a closed magnetic circuit, and the magnets of the facing sides of a revolving facing cathode-box have opposite magnetic poles. Four targets are installed on the outer side of each revolving facing cathode-box. Compact sputtering with multi-target is realized by revolving these facing cathode boxes.

.

Figure 3 shows the revolving facing-target sputtering system with six targets fabricated in the Consortium R&D Project for Regional Revitalization from METI in 2005-2006 [9]. The inset shows a schematic diagram of the revolving facing cathode-box with six targets within the chamber. The total cathode capacity of the system is estimated to be less than about one third of that for conventional facing-targets sputtering with six targets, and the system is advantageous from the viewpoint of energy economy and compactness.

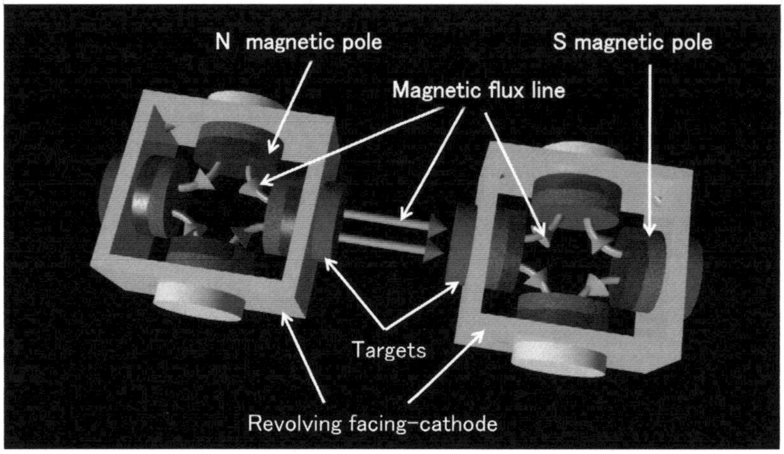

Fig. 2 Schematic diagram of the revolving facing cathode-box in the revolving-type facing-target.

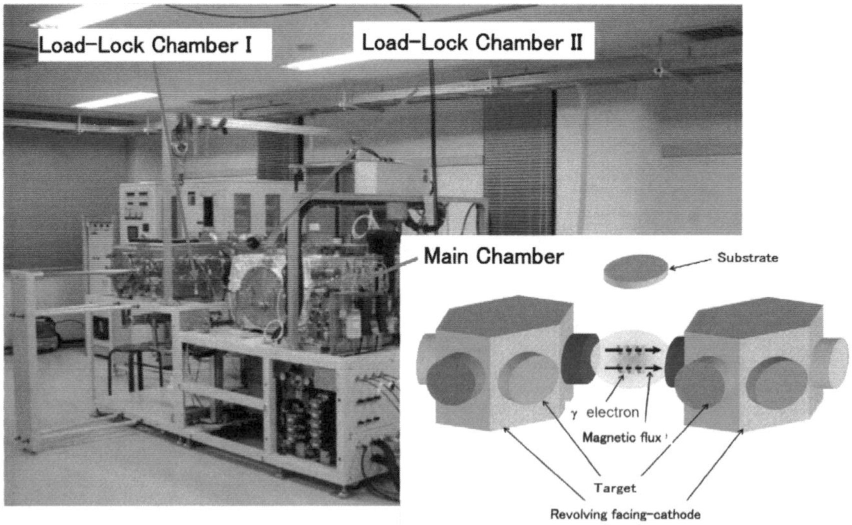

Fig. 3 Revolving facing-target sputtering system with six targets fabricated in the "Development of a new low-temperature sputtering system for organic electroluminescence electrodes and passivation films" (Principal investigator: Shinichi Morohashi, Yamaguchi University), the Consortium R&D Project for Regional Revitalization from METI, 2005-2006.

2.3 Magnetic flux distribution variable-type facing-target sputtering system

Figure 4 shows the magentic flux distribution variable facing-target sputtering system devised to realize high-speed deposition for improved throughput, while satisfying the requirements for low damage [10] [11]. The system can increase the degrees of freedom of sputtering, that is, sputtering can be achieved under different magnetic flux distribution between facing-target, where the magnetic flux distribution is controlled with a moving rod magnet without breaking the vacuum. This idea defies the conventional notion of "one cathode, one sputtering system." Figure 4 shows the cross-section of a facing-cathode, with moving rod magnets that change a magnetic flux distribution between facing-target; the use of a moving yoke [11] or an electromagnet would also be possible.

The outer stationary cylindrical magnets are of the type also found in the conventional facing-target sputtering system. The facing cylindrical magnets are of opposite magnetic poles; in the figure, the cylindrical magnet in the back of the left target is an N-type magnet, and that in the back of the right target is an S-type magnet. The bar magnets in the middle are moving magnets; the cylindrical magnets and the moving magnets have opposite magnetic poles at the back of each target. With such a magnet arrangement, three magnetic flux distributions are created: (1) a magnetic flux distribution between the facing-target due to the facing cylindrical magnets; (2) a magnetic flux distribution between the facing-target due to the facing moving rod magnets, directed oppositely to the field created by the cylindrical magnets; and (3) a magnetic field distribution directed from the N to S magnetic poles, as found in magnetron sputtering, created by the stationary cylindrical magnet and the moving rod magnet at the back of each target. The magnetic flux distributions (1), (2), and (3) between the facing-target can be controlled by the positions of the moving rod magnets without breaking the vacuum. The magnetic flux distribution is shown schematically for the two cases in Figure 4. Figure 4(a) shows the case where the magnetic flux distribution of (1) is strong, and Figure 4(b) shows the case where the magnetic flux distributions of (2) and (3) are added to the distribution of (1). The

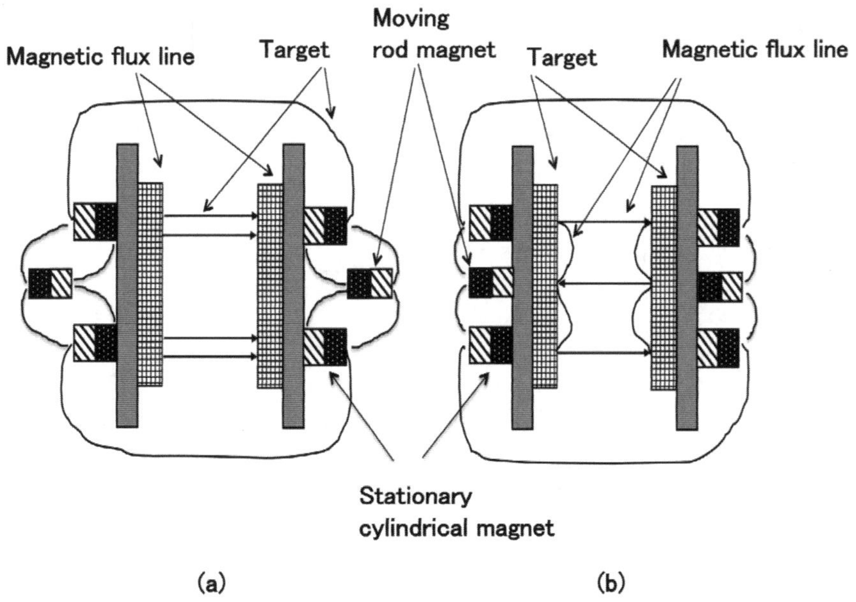

Fig. 4 Schematic diagram of the magnetic flux distribution between facing-cathode in magentic flux distribution variable facing-target sputtering.
(a) Magnetic flux distribution when the moving rod magnets are farthest from the targets, (b) Magnetic flux distribution when the moving rod magnets are nearest to the targets.

magnetic flux distributions of (2) and (3) can be controlled continuously by the positions of the moving rod magnets (2).

Nb, a high melting point metal was used as a target (purity 3N and diameter 90 mmφ), and the Nb deposition rate and the substrate temperature were evaluated as a function of the positions of the moving rod magnets under the conditions of Ar pressure of 0.7 Pa, RF applied power of 1 kW, and with a 10-cm distance from a target center to a substrate. Kapton tape was stuck on the silicon substrate, and a thermo label heater was stuck underneath the substrate; the film thickness and the substrate temperature were then measured simultaneously. The deposition rate was calculated by the ratio of the deposited film thickness over the sputtering time (2 min). The substrate temperature was the temperature measured without the use of cooling water. The moving rod magnet was moved within the 99-mm range of moving distance as measured from the origin, which was defined by the closest position of the moving rod magnet to the back of the target. Figure 5(a) corresponds to the magnetic flux distribution shown in Figure 4(a), where electric discharge plasma was observed similarly to that found in conventional facing-target sputtering. Figure 5(b) corresponds to the magnetic flux distribution shown in Figure 4(b), where the electric discharge plasma with a ring erosion area, as found in magnetron sputtering, was observed right at the front of the facing-target. The behavior depends on the variation of the magnetic flux distributions (1), (2), and (3) between the facing-target through the movement of the moving rod magnets.

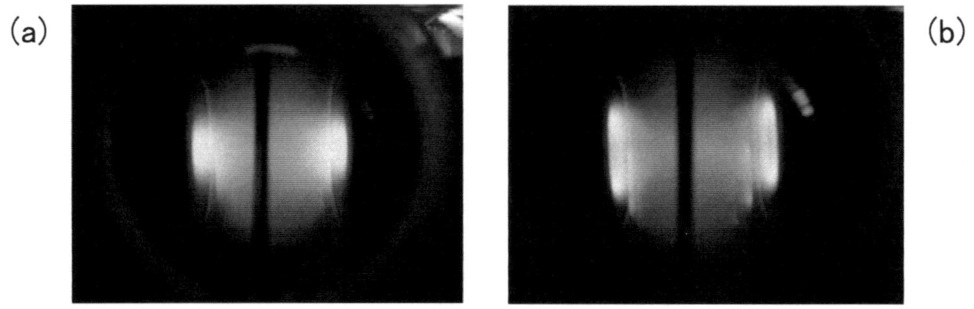

Fig. 5 Discharge plasma during Nb deposition using magentic flux distribution variable facing-target sputtering. (a) when the moving rod magnets are farthest from the targets, (b) when the moving rod magnets are nearest to the targets

The deposition rate changed by a factor of approximately five, from 50.3 nm/min to 11.0 nm/min, and the substrate temperature changed from 110°C to 40°C (see Figure 6) [12] [13]. The deposition rate and substrate temperature were highest at the position in which the magnetic flux distribution (3) was strongest, at the moving distance of zero. The deposition rate and substrate temperature became small in the region where the influence of the magnetic flux distribution (3) became small. It is considered that the variation in the magnetic flux distribution between the facing-target influences the electron density between the plasmas, the ionization density of the Ar gas, and the energy of the sputtering particles, ultimately changing the Nb deposition rate and the substrate temperature.

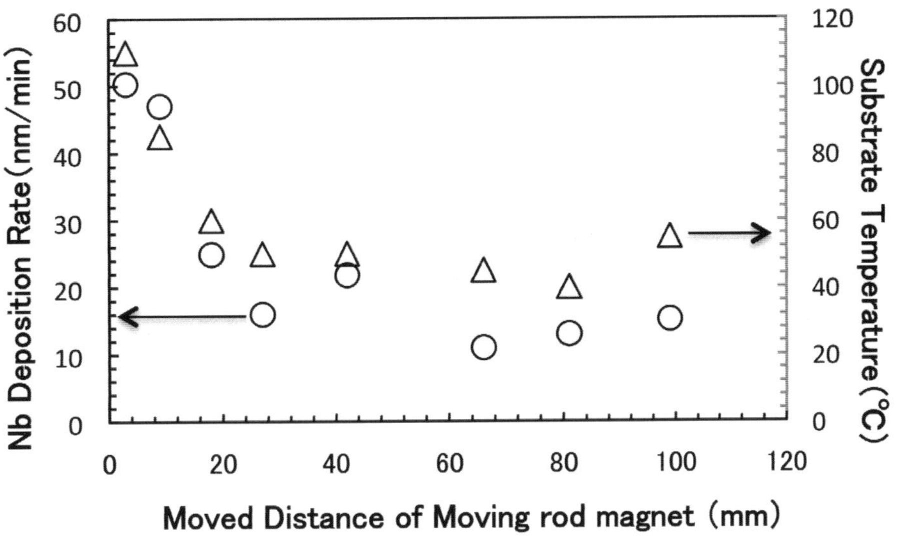

Fig. 6 Deposition rate and substrate temperature as a function of the positions of the moving rod magnets for controlling the magentic flux distribution variable facing-target sputtering process.

3. Conclusion ‡

A magentic flux distribution variable facing-target sputtering system was developed to control the magnetic flux distribution between facing-target by using moving magnets. We could change the deposition rate of a high-melting-point metal (Nb) by a factor of approximately five simply by changing the magnetic flux distribution between the facing-target. In this new facing-target sputtering system using only one cathode, we can expect the following features. (1) The magnetic flux distribution between the facing-target can be changed easily on the air side without breaking the vacuum, and electric discharge plasma between the facing-target is formed accordingly. (2) Low-damage and high-throughput plasma discharge are realized. (3) The new film deposition method provides high functionality and performance: the first layer is deposited in the low-damage plasma discharge, and the remaining layers are deposited in the high-throughput discharge, or the films are deposited in the magnetic flux distribution suited to the material. Furthermore, (4) we can expect applications in the fabrication of film-based electronic devices with multilayer film structures, making the sputtering system which combines magnetic flux distribution variable facing-target sputtering and the previously presented revolving facing-target sputtering.

‡

Acknowledgement

Part of this study was made in Morohashi's laboratory, in the Graduate School of Science and Engineering, Yamaguchi University, as part of the Open Advanced Facilities Initiative for Innovation (Nanotechnology Network) of the Ministry of Education, Culture, Sports, Science and Technology, We deeply thank all members concerned.

References

[1] S. Morohashi, F. Shinoki, A. Shoji, M. Hayakawa and H. Hayakawa: *Appl. Phys. Lett.*, 46, (1985) 1179.

[2] S. Morohashi and S. Hasuo: *J. Appl. Phys.*, 61, (1987) 4835.
[3] F.Kessler, D. Herrmann and M. Powalla: *Thin Solid Films.*, 480 - 481, (2005) 491.
[4] S. Takasawa, S. Ukishima, N. Tani and S. Ishibashi: *ULVAC TECHNICAL JOURNAL*, 64, (2006) 18.
[5] S. Morohashi, M. Ikuta, T. Miyoshi, D. Matsumoto, S. Ariyoshi, M. Ukibe, M. Ohkubo and H. Matsuo: *IEEE Trans. Applied Superconductivity.*, 15 (2005) 98.
[6] S. Ono and M. Naoe: *Suppl. Trans. JIM.*, 29 (1988) 57.
[7] JPN Patent, Pat. No. 3936970 (Inventor: Shinichi Morohashi, Yamaguchi University).
[8] S. Morohashi: *Shinku.*, 113 (2007) 14.
[9] "Development of a new low temperature sputtering system for organic electroluminescence electrodes and passivation films" (Principal investigator: Shinichi Morohashi, Yamaguchi University), Consortium R&D Project for Regional Revitalization (METI) 2005-2006.
[10] Two international patent applications, PCT/JP2008/058621 and PCT/JP2009/058976 (Inventor: Shinichi Morohashi, Yamaguchi University).
[11] S. Morohashi: *J. Vac. Soc. Jpn.*, 128 (2010) 26.
[12] Nikkan Kogyo Shimbun press release, Business & technology page, Nov. 12, 2010.
[13] S. Morohashi, Y. Matsumoto, K. Tanaka, K. Usui and E. Komatsu: *J. Vac. Soc. Jpn.*, 54 (2011)to be published.

Journal of Physics: Conference Series **417** (2013) 012019 doi:10.1088/1742-6596/417/1/012019

All-Wet Metallization Process for Transparent Polyimide Films

S Ikeda[1], Y Kobayashi[1], Y Fujiwara[1], K Akamatsu[2], and H Nawafune[2]

[1] Electronic Materials Research Division, Osaka Municipal Technical Research Institute, 1-6-50 Morinomiya, Joto-ku, Osaka 536-8553, Japan

[2] Department of Nanobiochemistry, Frontiers of Innovative Research in Science and Technology (FIRST), Konan University, 7-1-20 Minatojimaminami, Chuo-ku, Kobe 650-0047, Japan

E-mail: ikeda-s@omtri.or.jp

Abstract. Electrically conductive Ag thin films were successfully formed onto optically transparent polyimide films via all-wet chemical process. The process involves alkaline-induced surface modification of polyimide films, subsequent incorporation of Ag^+ ions through ion exchange reaction followed by reduction of incorporated Ag^+ ions. Initial alkaline treatment time determines thickness (depth) of surface-modified layer and amount of Ag^+ ions loading consequently. Higher Ag^+ ions loading result in forming electrically conductive Ag thin films and its granular interfacial structure. Cu electroplating is able to be conducted on the resulting Ag thin films directly, and adhesive strength between the Cu(/Ag) film and transparent polyimide substrate reached up to 2 kN m^{-1}.

1. Introduction

Optically transparent polyimide films have received much attention due to their thermal stability, flexibility, and chemical durability as an alternative to glass substrates used for various optoelectronic devices including solar cells, electronic papers and displays. For such applications, surface metallization technique is essential to fabricate electrodes and circuits onto film surfaces. Generally, dry process including physical vapor deposition (PVD) and chemical vapor deposition (CVD) is employed in order to fabricate metallic thin films onto polyimide film substrates [1-3]. However, dry process consumes larger energy and higher cost than wet process involving electroless and electroplating is applied for metallizing polyimide surfaces. To carry out successful electroless plating for polymer substrates, proper roughness yielding anchoring effect and consequent preferable adhesion and catalytic activity initiating electroless plating reaction have to be induced to the substrate surface through relatively tedious process consisting of a lot of steps, e.g., pre-cleaning, water rinsing, chemical etching, sensitizing, and activating [4].

On the other hand, direct metallization process has been developed as a facile technique for metallizing polyimide surfaces [5-14]. The direct metallization process for polyimide films mainly consists of alkaline-induced surface modification, metallic ions incorporation through ion exchange reaction, and reduction of incorporated metallic ions. In the previous report [14], Cu plating film fabricated onto the polyimide (Kapton) film through direct metallization process showed 1 kN m^{-1} of

90° peel strength, the value of which is sufficient to microelectronic uses. In this study, direct metallization process for transparent polyimide substrate (Figure 1) has been investigated and the adhesion between resulting Cu plating films and transparent polyimide substrates has been evaluated.

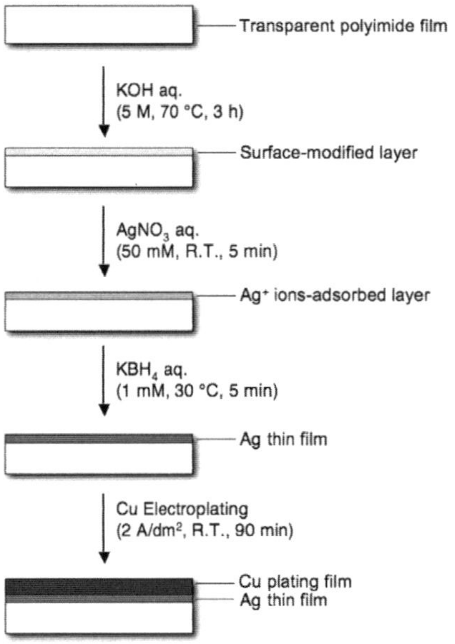

Figure 1. Schematic illustration for direct metallization process for transparent polyimide films.

2. Experimental

Optically transparent polyimide films are commercially received from Mitsubishi Gas Chemical Company, Inc. (Neopulim L-1000, 100 and 200 μm in thickness). All chemicals excepting for Cu electroplating additives were purchased from Nakalai Tesque Company, Inc. and used without any purification. A brightener (Top Lucina) added into Cu electroplating solution was kindly provided by Okuno Chemical Industries Co., Ltd.

Transparent polyimide films were immersed into 5 M KOH aqueous solution at 70 °C for 3 h. After KOH treatment, the films were rinsed with large amount of water and immersed into 50 mM $AgNO_3$ aqueous solution at ambient temperature for 5 min. After immersion into $AgNO_3$ aqueous solution, the films were rinsed with water and immersed into 1 mM KBH_4 aqueous solution at 30 °C for 5 min in order to reduce Ag^+ ions incorporated into the film. After reduction of Ag^+ ions, the films were washed with copious amount of water. Finally, Cu electroplating was conducted on Ag thin film formed on the polyimide film with plating bath composition and conditions expressed in Table 1.

Table 1. Cu electroplating bath composition and conditions.

$CuSO_4 \cdot 5H_2O$	70 g dm^{-3}
H_2SO_4	200 g dm^{-3}
Cl^-	50 mg dm^{-3}
Brightener	5 ml dm^{-3}
Temperature	R.T.
Current density	2.0 A dm^{-2}

FT-IR spectra of the transparent polyimide film surfaces before and after KOH treatment were recorded on a Nicolet 4700 spectrometer with an ATR attachment. Amount of Ag^+ ions incorporated in surface-modified polyimide films were quantified by using an ICP atomic emission spectrometer (Shimadzu ICPS-8100). UV-VIS absorption spectra of the films were collected with an UV-3150 spectrometer (Shimadzu). Sheet resistance of Ag thin films was measured with a low resistivity meter (Loresta EP MCP-T360, Mitsubishi Chemical Analytech) using a 4-pin probe method. Cross-sectional TEM observations were performed with JEM-2100, JEOL operating at 200 kV. Cross-sections of the films for TEM observations were prepared with a microtome (Ultracut UCT, Leica) using a diamond knife. Adhesive strength between Cu plating film and transparent polyimide substrate was evaluated by means of 90° peel test with using a bond tester (Dage 4000, Nordson DAGE) operating at 25 mm/min of peel rate.

3. Results and discussion

3.1. Surface modification of transparent polyimide films

Changes in molecular structure of the polyimide films before and after KOH treatment were characterized with ATR-FTIR spectroscopy and the obtained spectra were shown in Figure 2. The spectrum obtained from a bare transparent polyimide film shows typical absorption peak (i) assigned to imide C=O stretching modes at 1780 and 1710 cm^{-1} (Figure 2A) [15-17]. On the other hand, after KOH treatment for 3 h, absorbance for these absorption bands are significantly decreased and new absorption bands attributed to (ii) amide I band (carbonyl stretching) and (iii) amide II band (coupling C-N stretch and N-H deformation) are appeared at 1680 and 1550 cm^{-1}, respectively (Figure 2B) [15-17]. In addition, absorption band (iv) corresponds to carboxylic acid groups is obviously appeared at 1410 cm^{-1} with neutralizing the KOH-treated film with 1% HCl aqueous solution (Figure 2C) [15-17]. These results indicate that amide bonds and carboxyl groups are successfully formed via hydrolysis reaction of five-membered imide rings with KOH treatment in the present condition. According to the previous investigation, in the case of Kapton film, similar degree of surface modification had been achieved by treating with 5 M KOH aq. at 50 °C for 5 min [8,9]. In order to modify a Neopulim film surface similarly, it is found that a much stronger hydrolysis condition has to be required. These behaviors may be related to intrinsic water absorption properties of 2.9 and 1.6%, for Kapton and Neopulim films, respectively.

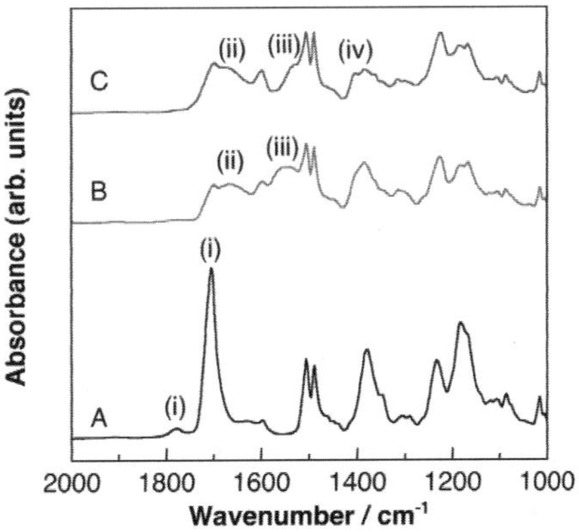

Figure 2. ATR-FTIR spectra obtained from transparent polyimide films (A) before and (B) after KOH treatment for 3 h (C) followed by neutralization with using 1% HCl aqueous solution.

3.2. Incorporation of Ag^+ ions

As reported elsewhere [8,9], Ag^+ ions will be incorporated into the surface-modified layer of transparent polyimide film via ion exchange reaction between K^+ and Ag^+ ions with immersing the surface-modified film into $AgNO_3$ aqueous solution. Figure 3 shows relationship between initial KOH treatment time and amount of incorporated Ag^+ ions in the films after ion exchange step (2nd step in figure 1). Ag^+ ions loading is linearly increased with increasing initial KOH treatment time. From the previous studies [7], it has been found that thickness of surface-modified layer in depth direction is increased in proportional to initial KOH treatment time. Depth profiling revealed that incorporated metallic ions were homogeneously distributed in the surface-modified layers in macroscopic scale [11,14]. In other words, cation exchangeable site, i.e., carboxyl groups, would be homogeneously formed over entire surface-modified layer through initial KOH treatment. Therefore, Ag^+ ion loading can be readily controlled with initial KOH treatment time.

Influence of KOH treatment-induced surface modification and subsequent Ag^+ ions incorporation on optical transparency of the films was investigated with UV-VIS absorption spectra (Figure 4). There is no change in UV-VIS absorption spectra for before and after KOH treatment and following Ag^+ ions incorporation, which indicate those two steps (1st and 2nd step in figure 1) yield no deterioration in optical transparency of the present polyimide films.

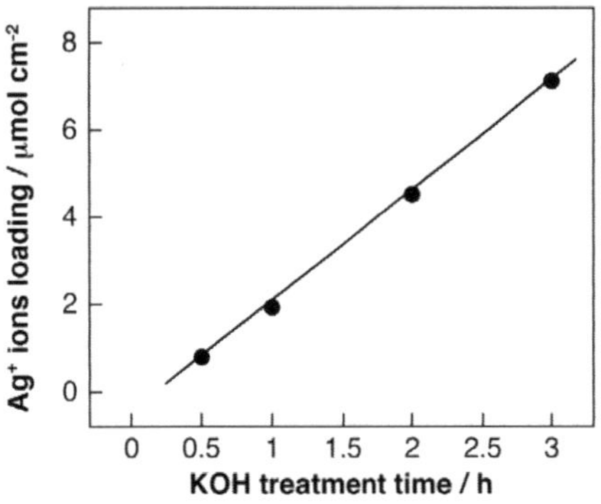

Figure 3. Effect of KOH treatment time on Ag^+ ion loading after ion exchange reaction in $AgNO_3$ aqueous solution.

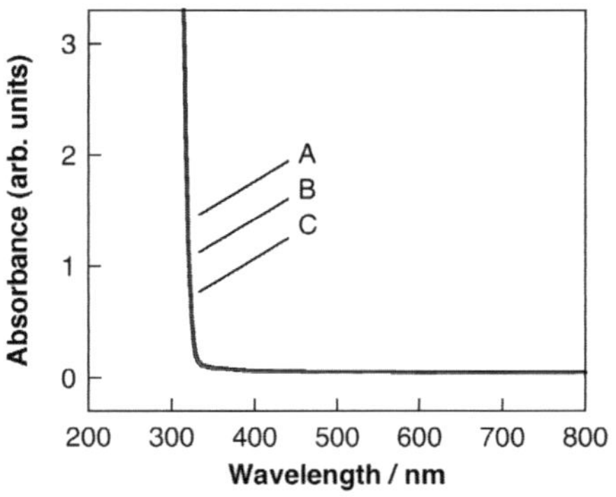

Figure 4. UV-VIS absorption spectra obtained from the films (A) before and (B) after KOH treatment for 3 h and (C) subsequent Ag^+ ions loading.

3.3. Formation of Ag thin films

Film appearance of the films after reduction of incorporated Ag^+ ions, which were fabricated with different initial KOH treatment time, is shown in Figure 5. Initial KOH treatment for 0.5 and 1 h gives rise to formation of the films colored light yellow and brown, respectively, and these colors can be attributed to surface plasmon resonance absorption of Ag nanoparticles. On the other hand, mirror-bright Ag films are successfully formed onto the surface of the transparent polyimide films which were initially KOH-treated for 2 and 3 h. While former two samples (corresponding to figure 5B and C) showed too high resistivity to measure, latter two samples (corresponding to figure 5D and E) showed relatively lower sheet resistance of 10.3 and 3.02 Ω/sq, respectively, the values of which are acceptable to carry out direct electroplating.

Figure 6A–C show cross-sectional TEM images of the films after reduction treatment of incorporated Ag^+ ions and subsequent Cu electroplating. Ag particles of few tens of nanometers in diameter are deposited at the surface of transparent polyimide film and it seems that the particles coagulate with each other (Figure 6A). This film, which is corresponding to the sample in figure 5C, showed almost no electric conductivity. The cross-sections prepared for TEM observation had at least 100 nm in thickness, thus Ag thin film shown in figure 6A would be made of isolated Ag particles, in other words, that would be not continuous film. Initial KOH treatment for 3 h results in formation of thicker Ag film and granular interface consisting of Ag nanoparticles and surface-modified polyimide layer as shown in Figure 6B. After Cu electroplating, uniform Cu film is directly deposited onto the Ag thin film and the specific granular interfacial nanostructure is maintained. Such interfacial microstructure would be an effective adhesion promoter between metallic films and polymer substrates [14].

Adhesive strength between Cu/Ag film and transparent polyimide substrate was evaluated by means of 90° peel test. Typical peel strength profiles against peeling distance are shown in Figure 6D. Extremely high peel strength (>1.5 kN m^{-1}) is demonstrated for the most part of the films, while there are a few poor adhesive regions. Interestingly, the peel strength value over 2 kN m^{-1} is much higher than the value (1 kN m^{-1}) of Cu plating films on Kapton films which were fabricated in the same way reported in the reference [14]. The detailed mechanism for the difference in peel strength between Neopulim samples and Kapton samples is under the investigation. One of the probable reasons is assigned to difference in intrinsic mechanical properties for each type of polyimide film.

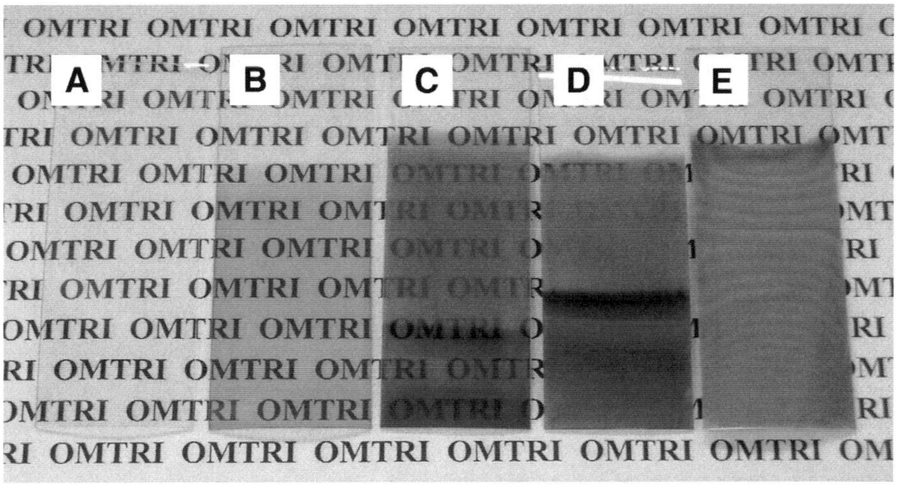

Figure 5. Changes in film appearance (A) before and (B-E) after reduction of incorporated Ag^+ ions prepared with initial KOH treatment for (B) 0.5 h, (C) 1 h, (D) 2 h, and (E) 3 h.

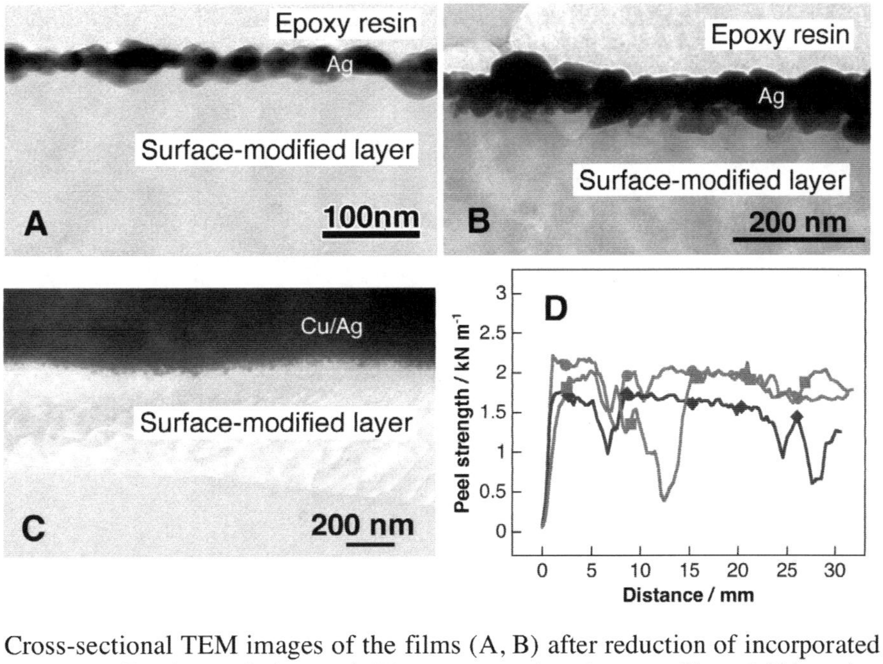

Figure 6. Cross-sectional TEM images of the films (A, B) after reduction of incorporated Ag^+ ions and (C) subsequent Cu electroplating and (D) representative three profiles of 90° peel strength of Cu plating film corresponding to the sample (C). Initial KOH treatment time; (A) 1 h, (B–D) 3 h.

4. Conclusion

We have successfully demonstrated that mirror-bright Ag thin films are readily deposited onto transparent polyimide films via direct metallization process with relatively stronger surface modification condition. Cu electroplating can be performed on electrically conductive Ag thin films having specific interfacial microstructure. Extremely high adhesive strength between Cu films and transparent polyimide substrates is suitable for optoelectronic applications.

References

[1] Kien M, Strunskus T, Peter R and Faupel F 1998 *Adv. Mater.* **10** 1357
[2] Murdey R and Stuckless J T 2003 *J. Am. Chem. Soc.* **125** 3995
[3] Denis Y W Y, and Spaepen F 2004 *J. Appl. Phys.* **95** 2991
[4] Mallory G O and Hajdu J B 1990 *Electroless Plating: Fundamentals and Applications* (New York: Noyes Publications)
[5] Ikeda S, Akamatsu K and Nawafune H 2001 *J. Mater. Chem.* **11** 2919
[6] Akamatsu K, Nakahashi K, Ikeda S and Nawafune H 2003 *Eur. Phys. J.* D **24** 377
[7] Akamatsu K, Ikeda S, Nawafune H and Deki S 2003 *Chem. Mater.* **15** 2488
[8] Akamatsu K, Ikeda S and Nawafune H 2003 *Langmuir* **19** 10366
[9] Ikeda S, Akamatsu K and Nawafune H 2003 *Transactions of the Materials Research Society of Japan* **28** 1267
[10] Akamatsu K, Ikeda S, Nawafune H and Yanagimoto H 2004 *J. Am. Chem. Soc.* **126** 10822
[11] Ikeda S, Akamatsu K, Nawafune H, Nishino T and Deki S 2004 *J. Phys. Chem.* B **108** 15599
[12] Akamatsu K, Shinkai H, Ikeda S, Adachi S, Nawafune H and Tomita S 2005 *J. Am. Chem. Soc.* **127** 7980
[13] Tomita S, Akamatsu K, Shinkai H, Ikeda S, Nawafune H, Mitsumata C, Kashiwagi T and Hagiwara M 2005 *Phys. Rev.* B **71** 180414(R)1
[14] Ikeda S, Akamatsu K, Nawafune H and Yanagimoto H 2007 *Adv. Funct. Mater.* **17** 889
[15] Snyder R W, Thomson B, Bartges B, Czerniawski D and Painter P C 1989 *Macromolecules* **22** 4166
[16] Lee K-W, Kowalczyk S P and Shaw J M 1990 *Macromolecules* **23** 2097
[17] Lee K-W, Kowalczyk S P and Shaw J M 1991 *Langmuir* **7** 2450

Synthesis of Polystyrene-based Nanocomposite Thin Films with Domain Structure Consisting of Au Nanoparticles

T Tsuruoka, S Kumazaki, I Osaka, H Nawafune and K Akamatsu

Department of Nanobiochemistry, Frontiers of Innovative Research in Science and Technology (FIRST), Konan University, 7-1-20 Minatojimaminami, Chuo-ku, Kobe 650-0047, Japan

E-mail: akamatsu@center.konan-u.ac.jp

Abstract. We have presented a very simple approach, the incorporation of metal complex into polymer matrix followed by thermal decomposition of complex, to the fabrication of metal nanoparticles/polymer composite thin films and have determined their domain structure. The approach involves the polymerization of monomer with metal complex into polymerized matrix, and annealing in vacuum condition. Here, the key is to use a gold (I) complex with phenyl groups that have strong interaction between metal complexes. This method allows us to synthesize Au nanoparticles dispersed in polystyrene-based polymer thin films with Au domain structures.

1. Introduction

Inorganic nanoparticles are emerging as very attractive nanomaterials for advanced applications in catalysis, transistors, and sensors.[1-3] In particular, metal nanoparticles, such as gold, silver, and copper, have been of great interest as nanomaterials for optics, because of their surface plasmon resonances (SPR).[4-6] Despite progress in synthesis of metal nanoparticles, controlling the SPR, which depends on size and shape of nanoparticles, distance between nanoparticles, remains challenging. Whereas a variety of methods have been developed to prepare the metal nanoparticles and are now accessible for controlling the size and shape of nanoparticles, controlling the distance between nanoparticles is now being actively explored. [7-12] Therefore, plasmonic engineering of nanoparticles is currently faced with the challenge of developing arrangement technology of metal nanoparticles for feature distance between nanoparticles.

Recently, great efforts have been focused on fabricating composite nanostructures consisting of inorganic nanoparticles and organic polymers.[13-15] These include efforts to improve their physical and chemical properties, and to introduce additional functionalities because such nanocomposite can be useful in the field of optics and electronics. On the other hand, they have been of great interest because they provide a good model system for studying experimentally interaction with inorganic and organic materials.[16,17] Thus, preparation of nanocomposite containing nanoparticles has become a significant research field. Various methods have been developed to prepare nanocomposites, including incorporation of inorganic nanoparticles into polymer matrices[18-23] and vacuum deposition on polymer substrate.[24] In these works, however, distribution of nanoparticles in polymer matrices relies on interaction between organic ligands or surface energy of nanoparticles, which makes it difficult to control the interparticle distance in polymers. To resolve this problem, we have attempted *in-situ* synthesis of metal nanoparticles into polymer matrices by hydrogen reduction using ion-doped polymers.[25-27] This synthesis was demonstrated to be a highly efficient approach for formation of metal nanoparticles uniformly dispersed in matrices because the doped-ions were homogeneously distributed in the polymers. Although the preparation of nanocomposite containing Ag, Cu, Ni, and Pt nanoparticles with a diameter of 2-10 nm can be achieved by using polymers bearing ion-exchange sites such as carboxylic groups, the synthetic approach for synthesis of Au nanoparticles using ion-doped polymer is still rare.

Here, we demonstrate a novel ion-doped route to the fabrication of nanocomposite thin films made of polystyrene-based polymer as a matrix embedded with Au nanoparticles formed through thermal decomposition of gold (I) complex. Incorporation of complex is achieved by using $AuCl(PPh)_3$ [Chlorotriphenyl- phosphine gold (I)]. The present process allows one to fabricate the nanocomposite thin films with unique domain structure consisting of Au nanoparticles.

2. Method

Preparation of Gold(I) Complex: Gold (I) complex, $Au(OCOCH_3)(PPh)_3$ complex, was prepared by ligand exchange reaction. $Ag(OCOCH_3)$ crystals were added into a chloroform solution of $AuCl(PPh)_3$ at room temperature. After reaction for 1 h, the resulting solution was filtered through a 0.2 μm membrane filter to separate white precipitates.

Preparation of Au Nanoparticles/PS Composite: A chloroform solution of styrene, divinylbenzene, and 2, 2'-azobisisobutyronitrile was heated at 60 °C for 12 h under Ar atmosphere to form cross-linked polystyrene (PS)/chloroform solution. Au (I) complex/PS thin films were prepared by a spin-coating using the mixture of chloroform containing cross-linked PS and $Au(OCOCH_3)(PPh)_3$ complexes. The concentration of Au (I) complex was changed from 0.5 to 2.0 M. The thin films were annealed at 120-200 °C for 30 min in vacuum.

Measurements: Fourier transform infrared (FT-IR) measurements was performed to confirm the changes in chemical structure of the Au(I) complex by ligand exchange reaction. The FT-IR spectra were obtained using FT/IR 670plus (Jasco) and were recorded at a resolution of 4 cm^{-1} and 100 accumulations. X-ray powder diffraction (XRD) data were collected on a Rigaku RINT-2200 Right System (Ultima IV) diffractometer with CuKα radiation. The microstructure of the composite films, i.e., distribution in the films, mean size, and size distribution of Au nanoparticles was confirmed by transmission electron microscopy (TEM, JEM-2000, JEOL) operating at 120 kV. The Optical properties of the obtained films were characterized by a UV/VIS spectrophotometer (V-550, Jasco).

3. Results and Discussion

$AuCl(PPh)_3$ complex was soluble to chloroform, but the saturation solubility of complex against chloroform is low. To improve the solubility of the complex, we conducted the ligand exchange reaction with chloride anion to acetate anion using $Ag(OCOCH_3)$ crystals. Figure 1 schematically describes the procedure for the preparation of $Au(OCOCH_3)(PPh)_3$ complex. By mixing chloroform solution of $AuCl(PPh)_3$ complex and $Ag(OCOCH_3)$ crystal, the white precipitate was seen in the solution. In order to confirm the chemical structure of complex, FT-IR spectroscopy was used for the Au(I) complex before and after ligand exchange reaction. As shown in Figure 2, the FT-IR spectrum

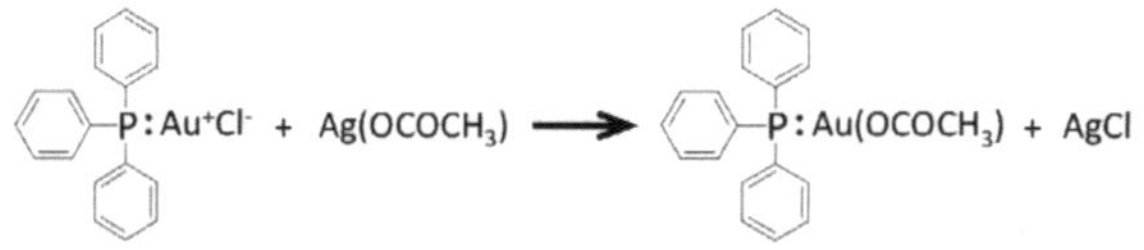

Figure 1. Schematic illustration of ligand exchange reaction for Au(I) complex.

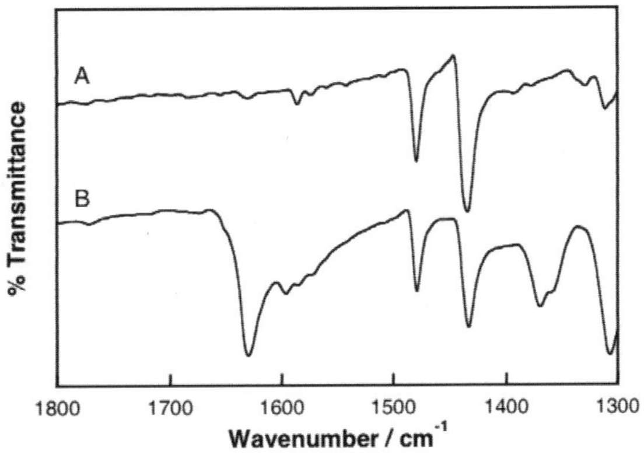

Figure 2. Transmittance FT-IR spectra of (A) $AuCl(PPh)_3$ and (B) $Au(OCOCH_3)(PPh)_3$ complexes.

measured for $AuCl(PPh)_3$ complex showed two bands at 1480 and 1435 cm^{-1} arising from benzene ring. After ligand exchange reaction, two typical bands at 1630 and 1370 cm^{-1} arising from acetate ligand are newly observed, and there are no distinct changes in benzene ring bands. This result indicates that the ligand exchange reaction of Au(I) complexes can be achieved by mixing $AuCl(PPh)_3$ complexes with $Ag(OCOCH_3)$ crystals. For further study of the ligand exchange process, the precipitate obtained by ligand exchange reaction was characterized by XRD measurement. The precipitates exhibited five well-resolved diffraction peaks in the 20-70° (2q) that can be indexed as AgCl with cubic structure. This is expected because silver (I) cation can be readily reacted with chloride anion, leading to the ligand exchange with chloride anion of gold (I) complex and acetate anion of silver acetate.

Au nanoparticles/PS composite thin films were prepared by the spin-coating of mixture solution containing cross-linked PS and $Au(OCOCH_3)(PPh)_3$ complex followed by annealing in vacuum condition. In Figure 3, TEM images of the thin films after annealing show that there is indeed a unique domain structure, and spherical Au nanoparticles with a particle diameter ranging from 10 to 15 nm were formed in PS matrix. Interestingly, according to high-resolution TEM images, the domain structure were found to be constructed by Au nanoparticles with nearly closed-packing assemblies. Recently, there are many reports for preparation of inorganic nanoparticles/polymer thin film composites by mixing with as-prepared nanoparticles and polymer matrix.[18-23] These studies suggest that the microstructure of composite films is dependent on the rate of solvent evaporation and interaction between nanoparticle ligand and polymer matrix.[23] In our present study, during spin-coating for preparation of thin films, the rate of chloroform evaporation may be too fast to form uniformly dispersed Au(I) complexes in PS matrix. In addition, the complex ligands and matrix exhibit insufficient affinity, leading that the phase separation is induced by minimizing the interface surface energies.

The effect of precursor concentration in the mixture solution on the size of nanoparticles and domain was investigated by TEM analysis. In Figure 3, the domain size consisting of Au nanoparticles increased with increasing the concentration of $Au(OCOCH_3)(PPh)_3$ complex in the PS precursor films, although the size of Au nanoparticles was almost constant at 10-15 nm. This effect may be explained in terms of interaction between Au surface and protective molecules. The resulting nanoparticles are stabilized by triphenylphosphine molecules, which show strong affinity against Au nanoparticles because phosphine molecules are often used as particle stabilizer.[28-30] The strong affinity inhibits the growth of Au nanoparticles, resulting in increasing the domain size of nanoparticles in PS matrix.

The changes in domain structure of composite films are clearly related to the changes in optical properties of Au nanoparticles. Before annealing in vacuum condition, the PS thin film containing

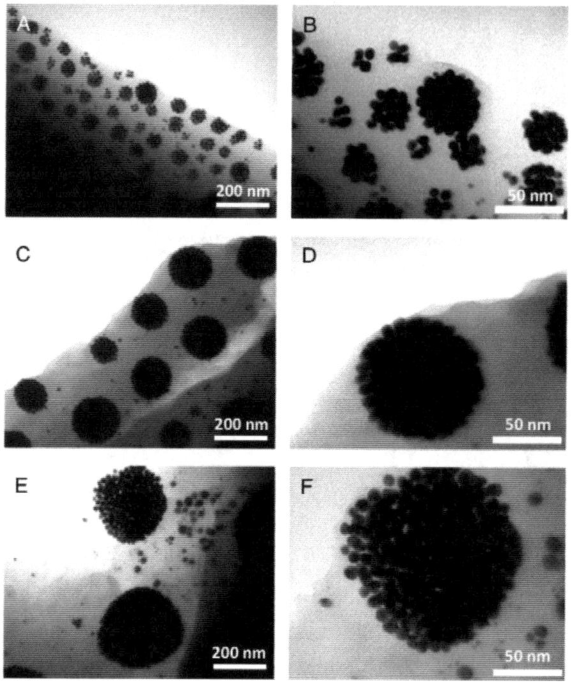

Figure 3. TEM images of PS composite films including Au nanoparticles obtained by using different casting solution containing (A,B) 0.5, (C,D) 1.0, and (E,F) 1.5 M of Au complex.

$Au(OCOCH_3)(PPh)_3$ complex is a transparent and colorless. The color of nanocomposite films after heat treatment changed to red and this change is caused by SPR of Au nanoparticles. The tuning of plasmonic properties of nanocomposites can be readily controlled by the amount of doped $Au(OCOCH_3)(PPh)_3$ complex and annealing temperature. A small amount of complex (the concentration of complex for chloroform solution: 0.5M) led to the formation of nanocomposite thin films with pale red. Doubling the concentration of complex for the spin-coating solution was found to produce nanocomposite thin films with beautiful color of wine red. More concentrated solution above 1.5 M was used to evaluate the effect of amount of $Au(OCOCH_3)(PPh)_3$ complex on SPR properties of nanoparticles. The color of films gradually changed from purple to indigo blue by increasing the concentration to 2.0 M, which indicate that a coupling of SPR signals of Au nanoparticles strongly appeared because the domain size increased with increasing the concentration of complex (as shown in Figure 3).[27]

To clarify the effect of complex concentration on optical properties of nanocomposites, we conducted the UV-vis measurement of the resulting samples. Figure 4 shows UV-vis spectra of the nanocomposite films prepared by different concentration of complex. These spectra indicate that the peak wavelength of SPR depends on the concentration of complex and shift from 550 to 569 nm as the concentration of complex increases from 0.5 to 1.5 M. On the basis of TEM observation, the observed shift of SPR peaks may be attributed to the increasing the size of domain structure including Au nanoparticles. In addition, note that increase in the annealing temperature resulted in increasing in the intensity of SPR although the absorption peaks were almost constant. After annealing at 120 °C, Au nanoparticles with a diameter of ca. 5 nm were observed and the domain of nanoparticles were not dense. This result suggests that the doped-Au ions were not completely reduced below 120 °C. On the other hand, the annealing above 150 °C resulted in the formation of domain structure consisting of

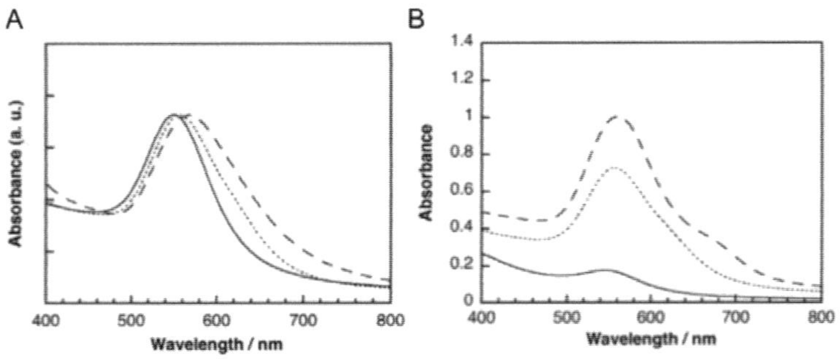

Figure 4. (A) UV-vis spectra of PS composite films including Au nanoparticles obtained by using different casting solution containing 0.5 (solid), 1.0 (dot), and 1.5 M (dashed) of Au complex. (B) UV-vis spectra of PC composite films after annealing at 120 (solid), 150 (dot), and 200 °C (dashed).

Figure 5. TEM images of PC composite films after annealing at (A) 120, (B) 150, and (C) 200 °C.

closely packed Au nanoparticles with a diameter of ca. 10 nm (Figure 5). These experiments provide evidence for SPR properties of Au nanoparticles in polymer thin films depending on the size and packing density of nanoparticles.

4. Conclusion

We have reported that Au nanoparticles/PS composite thin films with an unique domain structure can be prepared by using Au(I) complexes. The domain structure of nanocomposites consists of Au nanoparticles assemblies. The optical properties of the composite films can be readily controlled by changing the concentration of gold (I) complex and annealing temperature, which may be caused by the coupling between nanoparticles and increasing the size of nanoparticles. The present synthetic approach allows one to control the SPR properties of Au nanoparticles, the ability of which is expected to be useful for the fabrication of optical devices.

References

[1] Alivisatos A P 1996 *J. Phys. Chem.* **100** 13226
[2] Kleemann W, Petracic O, Binek C, Kakazei G N, Pogorelov Y G, Sousa J B, Cardoso S and Freitas P P 2001 *Phys. Rev. B* **63** 134423

[3] Ebels U, Duvail J L, Wigen P E, Piraux L, Buda L D and Ounadjela K 2001 *Phys. Rev. B* **64** 144421
[4] Mirkin C A, Letsinger R L, Mucic R C and Storhoff J J 1996 *Nature* **382** 607
[5] Chang S S, Shih C W, Chen C D, Lai W C and Wang C R C 1999 *Langmuir* **15** 701
[6] Wang H, Tam F, Grady N K and Halas N J 2005 *J. Phys. Chem. B* **109** 18218
[7] Turkevich J, Stevenson P C and Hillier J 1951 *Discuss. Faraday Soc.* **11** 55
[8] Brust M, Walker M, Bethell D, Schiffrin D J and Whyman R 1994 *J. Chem. Soc., Chem. Commun.* 801
[9] Pastoriza-Santos I and Liz-Marzán L M 2002 *Langmuir* **18** 2888
[10] Sun Y G and Xia Y 2002 *Science* **298** 2176
[11] Sun Y and Xia Y 2003 *Adv. Mater.* **15** 695
[12] Jin R, Cao Y C, Hao E, Metraux G S, Schatz G C and Mirkin C A 2003 *Nature* **425** 487
[13] Laurent C, Mauri D, Kay E and Parkin S S P 1989 *J. Appl. Phys.* **65** 2017
[14] Heilmann A, Werner J, Schwarzenberg D, Henkel S, Grosse P and Theiss W 1995 *Thin Solid Films* **270** 103
[15] Salz D, Lamber R, Wark M, Baalmann A and Jaeger N 1999 *Phys. Chem. Chem. Phys.* **1** 4447
[16] Akamatsu K, Shinkai H, Ikeda S, Adachi S, Nawafune H and Tomita S 2005 *J. Am. Chem. Soc.* **127** 7980
[17] Tomita S, Jonsson P E, Akamatsu K, Nawafune H and Takayama H, 2007 *Phys. Rev. B* **76** 174432
[18] Tessler N, Medvedev V, Kazes M, Kan S and Banin U 2002 *Science* **295** 1506
[19] Coe S, Woo W K, Bawendi M, Bulovic V 2002 *Nature* **420** 800
[20] Lin X M, Jaeger H M, Sorensen C M and Klabunde K J 2001 *J. Phys. Chem. B* **105** 3353
[21] Rabani E, Reichman D R, Geissler P L and Brus L E 2003 *Nature* **426** 271
[22] Tanaka H 2000 *J. Phys. Condens. Matter.* **21** R207
[23] Coe-Sullivan S, Steckel J S, Woo W K, Bawendi M G and Bulovic V 2005 *Adv. Funct. Mater.* **15** 1117
[24] Akamatsu K, Takei S, Mizuhata M, Kajinami A, Deki S, Takeoka S, Fujii M, Hayashi S and Yamamoto K 2000 *Thin Solid Films* **359** 55
[25] Ikeda S, Akamatsu K, Nawafune H, Nishino T and Deki S 2004 *J. Phys. Chem. B.* **108** 15599
[26] Akamatsu K, Adachi S, Tsuruoka T, Ikeda S, Tomita S and Nawafune H 2008 *Chem. Mater.* **20** 3042
[27] Akamatsu K, Shimada M, Tsuruoka T, Nawafune H, Fujii S and Nakamura Y 2010 *Langmuir* **19** 1254
[28] Weare W W, Reed S M, Warner M G and Hutchison J E 2000 *J. Am. Chem. Soc.* **122** 12890
[29] Prasad B L V, Stoeva S I, Sorensen C M and Klabunde K J 2003 *Chem. Mater.* **15** 935
[30] Balasubramanian R, Guo R, Mills A J and Murray R W 2005 *J. Am. Chem. Soc.* **127** 8126

Preparation and Optical Properties of Spherical Inverse Opals by Liquid Phase Deposition Using Spherical Colloidal Crystals

Y Aoi and T Tominaga

Department of Materials Chemistry, Faculty of Science and Technology, Ryukoku University, Oe, Seta, Otsu 520-2194, Japan

aoi@rins.ryukoku.ac.jp

Abstract. Titanium dioxide (TiO_2) inverse opals in spherical shape were prepared by liquid phase deposition (LPD) using spherical colloidal crystals as templates. Spherical colloidal crystals were produced by ink-jet drying technique. Aqueous emulsion droplets that contain polystyrene latex particles were ejected into air and dried. Closely packed colloidal crystals with spherical shape were obtained. The obtained spherical colloidal crystals were used as templates for the LPD. The templates were dispersed in the deposition solution of the LPD, *i.e.* a mixed solution of ammonium hexafluorotitanate and boric acid and reacted for 4 h at 30 °C. After the LPD process, the interstitial spaces of the spherical colloidal crystals were completely filled with titanium oxide. Subsequent heat treatment resulted in removal of templates and spherical titanium dioxide inverse opals. The spherical shape of the template was retained. SEM observations indicated that the periodic ordered voids were surrounded by titanium dioxide. The optical reflectance spectra indicated that the optical properties of the spherical titanium dioxide inverse opals were due to Bragg diffractions from the ordered structure. Filling in the voids of the inverse opals with different solvents caused remarkable changes in the reflectance peak.

1. Introduction

The fabrication of three-dimensional (3D) ordered macroporous materials with pore sizes in the submicrometer range has been of considerable interest [1]. Such materials have great potential in a variety of areas, including photonics [2-4], catalysis [5] and separation media [6,7]. Of particular interest are dielectric 3D periodic structures. These 3D structure lead to a periodic modulation of the refractive index generating a photonic stop band that does not allow the propagation of light with a definite wavelength [8,9]. They exhibit remarkable optical properties and have great potential in a wide range of scientific and technical applications such as optical filters [10,11], optical switches [12], wave-guides [13], optical fibers [14], and miniaturized spectrometers [15]. A widely used approach to making 3D ordered macroporous materials is to fill the interstitial spaces of a colloidal template (synthetic opal) with the solid and subsequently to selectively remove the template materials [1]. A colloidal crystal template is usually prepared by self-assembly of mono-dispersed spherical particles in an opal structure. Infiltration methods include the sol-gel method, electrodeposition, suspensions of nanoparticles, and chemical vapor deposition (CVD). The subsequent removal of the template yields an inverse opal material with a close-packed arrangement of voids.

One of the important issues in the design of colloidal crystals and inverse opals are the control of their uniformity in size and shape [16]. Several research groups have reported some methods to fabricate colloidal crystals and inverse opals with spherical shape [16,17].

We have used liquid phase deposition (LPD) with a colloidal template to fabricate TiO_2 inverse opals [18]. The LPD method is a wet process to deposit metal oxide thin film. Metal oxide or hydroxide thin films form by means of a ligand-exchange (hydrolysis) equilibrium reaction of metal-fluoro complex species $MF_x^{(x-2n)-}$ and a consumption reaction with boric acid as the F^- scavenger. This method enables the homogeneous metal oxide or hydroxide thin films to be formed on various substrates by simply immersing them in the reaction solution.

Here, we report a simple method to fabrication of spherical titanium dioxide inverse opals. The colloidal crystals with spherical shape were fabricated by the evaporation-induced crystallization of polystylene (PS) latex suspension droplets that were generated by ink-jet technique. Next, the colloidal crystals were used as templates for the LPD. The interstitial spaces of the colloidal crystal were completely filled with titanium dioxide by the LPD process. Finally, PS latex particles were removed by calcination, which left macropores at their sites. The obtained TiO_2 inverse opals with spherical shape were in the range of few tens of micrometers in size. The ordered pore diameters were in the range of a few hundreds of nanometers and could be adjusted by using different size of PS latex particles. These TiO_2 inverse opals are of practical significance in wide areas of applications such as photonic crystals, photocatalysts and separation media.

2. Experimental

2.1. Preparation of PS colloidal crystals with spherical shape by ink-jet drying

The carboxylated PS latex colloidal particles were purchased from Sakuranomiya Chemical Co. Ltd. The average diameters of the particles was 255 nm. 13.6 wt% of PS latex suspension were ejected by ink-jet equipment (Cluster Technology, Pulse Injecter PIJ-40, nozzle diameter 40 μm) into 80 °C heated air. Spherical PS colloidal crystals were generated by evaporation-induced self-assembly of polystyrene spheres.

2.2. Preparation of TiO_2 inverse opals by the LPD

The parent solutions used for the LPD were aqueous solutions of $(NH_4)_2TiF_6$ and H_3BO_3. $(NH_4)_2TiF_6$ (Kishida Chemical Co. Ltd.) and H_3BO_3 (Nacalai Tesque Inc.) were dissolved in distilled water at a concentration of 0.5 mol dm^{-3}. These solutions were mixed to give a concentration of 0.1 mol dm^{-3} for $(NH_4)_2TiF_6$ and 0.2 mol dm^{-3} for H_3BO_3, and used for the deposition solution for the LPD method. The PS spherical colloidal crystals obtained were soaked in the deposition solution of the LPD, and kept for 4 h at 30 °C. The colloidal crystals were then separated from the solution, washed with distilled water, and dried at ambient temperature. The PS latex particles were removed by calcination in air. The temperature was increased at a rate of 0.5 °C min^{-1} to 400 °C and stabilized at 400 °C for 1 h before being decreased to ambient temperature.

2.3. Chracterization

The morphology and microstructure of the synthesized composites and inverse opals were observed by scanning electron microscopy (SEM, JEOL JSM-6301F) and transmission electron microscopy (TEM, JEOL JEM-2000EX). Prior to the SEM observation, the samples were sputter coated with Au. The X-ray diffraction measurements were performed on a Rigaku RINT-2500HFV using Cu Kα irradiation. The optical properties of samples were investigated by recording their transmission and reflection spectra using a Shimadzu UV-3600 spectrophotometer.

3. Results and Discussion

The PS particles inside the suspension droplets self-assembled into spherical colloidal crystals spontaneously when the aqueous solvent was removed from the droplets by evaporation. The capillary

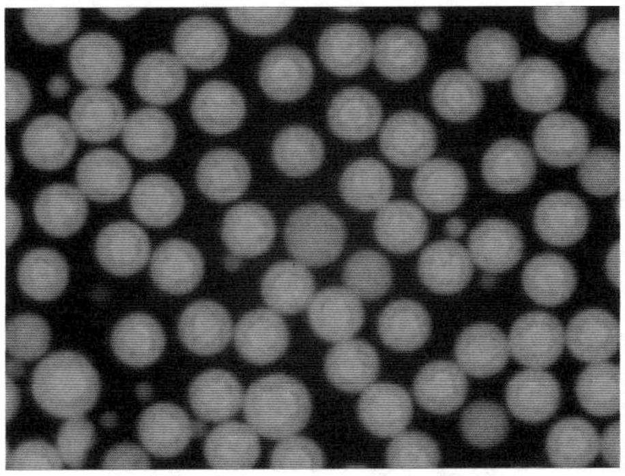

Figure 1. Optical micrograph of spherical colloidal crystals formed by 255 nm PS latex particles.

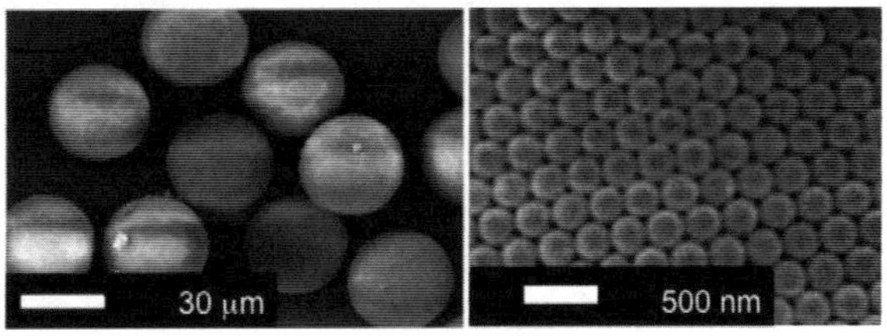

Figure 2. SEM images of spherical colloidal crystals formed by 255 nm PS latex particles.

force close-packes the PS particles into their final configuration, and van der Waals' forces subsequently cement the PS particles together. Figure 1 shows the optical micrograph of the spherical colloidal crystals. The colors arise from light diffraction (opalescence) from the colloidal arrays. This is clear evidence of long-range ordering of the PS particles as shown in the SEM images of figure 2 for the internal arrangement of a colloidal crystal of PS particles. Figure 2 shows SEM images of the spherical colloidal crystals. The spherical colloidal crystals were ~30 μm in diameter and consist of closely packed PS particles in a face-centered cubic arrangement due to a relatively large capillary force at the air-water interface.

The optical reflectance spectrum of the spherical colloidal crystals is shown in figure 3. The reflection peak appears around 600 nm. Because the spherical surface of the colloidal crystal is the (111) plane of the fcc lattice, the angle between the incident beam and the surface is not the same over the surface of the colloidal crystal, resulting in a rather broad reflection band. The reflection peak is consistent with the estimates from Bragg's law.

Next, the prepared spherical colloidal crystals were used as template materials for the LPD process. The prepared spherical colloidal crystals were soaked in a reaction solution of LPD process, which went through the interstices between PS latex particles by capillary force. Then titanium oxide deposited by means of hydrolysis reaction of $(NH_4)_2TiF_6$ on the template surface. After reaction for 4 h, the interstitial spaces of template were completely filled with titanium oxide. In the LPD process, titanium dioxide is deposited via heterogeneous nucleation. In this work, we have used carboxylated polystyrene particles as template. In this case, titanium oxide is preferentially deposited onto hydrophilic polystyrene surface and consequently deposition of free titanium oxide hardly takes place. The as-deposited sample was amorphous in X-ray diffraction, but Raman spectroscopy indicated that the infiltrated sample was anatase TiO_2. After a calcination at 400 °C for 1 h, the PS latex particles as template were removed, leaving ordered spherical air voids at their sites in the matrix of TiO_2. SEM images of the calcined sample are shown in figure 4. As shown in figure 4, spherical inverse opals were successfully produced. The void size was reduced to 214 nm due to the shrinkage during the calcination. Nevertheless, spherical shapes of colloidal crystals were kept, as shown from a low-magnification image of figure 4. No cracks were observed for the prepared spherical inverse opals, although some cracks due to shrinkage during the calcination were observed for inverse opals prepared by the LPD method onto the glass substrate [18].

X-ray diffraction measurement of the calcined sample indicating that the sample becomes crystalline by the calcination, and the observed peaks are corresponded to anatase TiO_2.

The optical reflectance spectrum of the TiO_2 spherical inverse opals is shown in figure 5 (a). The reflection peak appears around 430 nm, consistent with Bragg's law. Filling in the voids of inverse opals with solvents caused a change in the average refractive index of the material. It also caused a change of position of the reflection peak [19]. To investigate the solvent-filling effects, we chose a group of solvents that spanned a range of refractive indices. The solvents were: air (n = 1.00), 1-penthanol (n = 1.42), and 1,2-dibromobuthane (n = 1.54). Figure 5 shows the optical reflectance spectra of TiO_2 spherical inverse opals filled with solvents. The absorption peaks assigned to the stop band shifted to higher wavelengths for higher refractive index solvents.

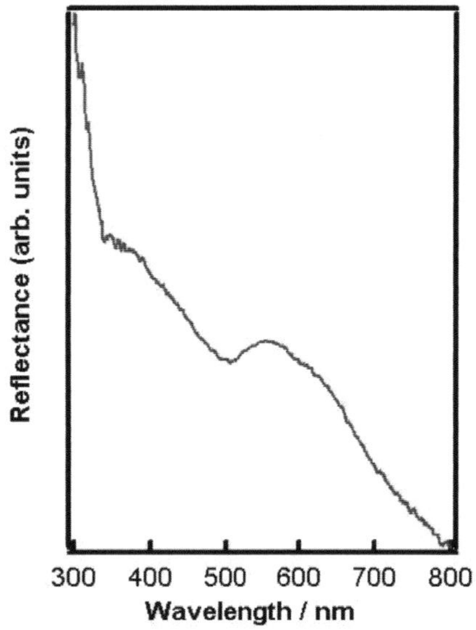

Figure 3. Reflectance spectrum of spherical colloidal crystals formed by 255 nm PS latex particles.

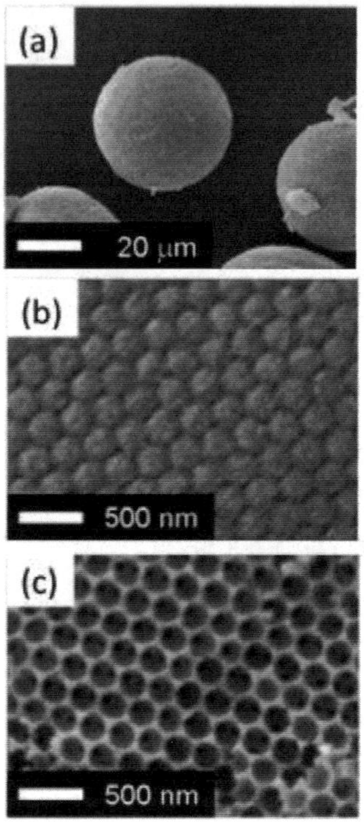

Figure 4. SEM images of the spherical inverse opals. (a) Low-magnification image. (b) Surface image. (c) Fractional section image of broken inverse opals.

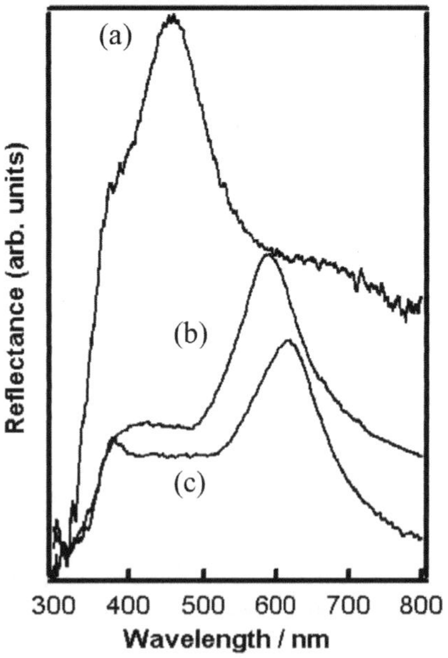

Figure 5. Reflectance spectra of spherical inverse opals filled with various solvents. (a) air, (b) 1-penthanol, and (c) 1,2-dibromobuthane.

4. Conclusions

Spherical PS colloidal crystals and spherical TiO_2 inverse opals were fabricated in this study. Spherical colloidal crystals were produced by ink-jet drying technique. The PS particles inside the suspension droplets self-assembled into spherical colloidal crystals spontaneously when the aqueous solvent was removed from the droplets by evaporation. TiO_2 spherical inverse opals were synthesized by using the LPD method using a spherical colloidal crystal as a template, followed by removing of the PS latex particles by the calcination. SEM observations indicated that spherical inverse opals were successfully produced. The optical reflectance spectra indicated that the optical properties of the spherical TiO_2 inverse opals were due to Bragg diffractions from the ordered structure. Filling in the voids of the inverse opals with different solvents caused remarkable changes in the reflectance peak.

References

[1] Stein A 2001 *Macroporous Mesoporous Mater* **44-45** 227
[2] Zakhidov A, Baughman R, Iqbal Z, Cui C, Khayrullin I, Dantas S O, Marti J and Ralchenko V G 1998 *Science* **282** 897
[3] Schroden R C, Al-Daous M and Stein A 2001 *Chem. Mater.* **13** 2945
[4] Thijssen M S, Sprik R, Wijnhoven J E G J, Megens M, Narayanan T, Lagendijk A and Vos W L 1999 *Phy. Rev. Lett.* **83** 2730
[5] Tanev P T, Chibwe M and Pinnavaia T J 1994 *Nature* **368** 321
[6] Lewandowski K, Murer P, Svec F and Frechet J M J 1998 *Anal. Chem.* **70** 1629
[7] Akolekar D B, Hind A R and Bhargava S K 1998 *J. Colloid Interface Sci.* **199** 92
[8] Yablonovitch E 1998 *Phys. Rev. Lett.* **58** 2059
[9] John S 1987 *Phys. Rev. Lett.* **58** 2486
[10] Noda S, Chutinan A and Imada M 2000 *Nature* **407** 608
[11] Leonard S W, van Driel H M, Schilling J and Wehrspohn R B 2002 *Phys. Rev. B* **66** 161102
[12] Pan G, Kesavamoorthy R and Asher S A 1997 *Phys. Rev. Lett.* **78** 3860
[13] Lin S Y, Chow E, Hietala V, Villeneuve P R and Joannopoulos J D 1998 *Science* **282** 274
[14] Knight J C, Birks T A, Russell P S J and Atkin D M 1998 *Science* **282** 1476
[15] Kim S H, Park H S, Choi J H, Shim J W and Yang S M 2010 *Adv. Mater.* **22** 946
[16] Yi G, Moon J H and Yang S M 2001 *Chem. Mater.* **13** 2613
[17] Li H, Wang H, Chen A, Meng B and Li X 2005 *J. Mater. Chem.* **15** 2551
[18] Aoi Y, Kobayashi S, Kamijo E and Deki S 2005 *J. Mater. Sci.* **40** 5561
[19] Blanford C F, Schroden R C, Al-Daous M and Stein A 2001 *Adv. Mater.* **13** 26

Effects of pulse voltage on the formation of nanoporous Ti oxides by dealloying amorphous TiCu alloy

Shengli Zhu[a, b, c], Lele Pi[a],Guoqiang Xie[b], Zhenduo Cui[a,c], Xianjin Yang[a,c], Akihisa Inoue[b]

a School of Materials Science and Engineering, Tianjin University, Tianjin, 300072, China

b Institute for Materials Research, Tohoku University, Sendai, 980-8577, Japan

c Tianjin Key Laboratory of Composite and Functional Materials, Tianjin, 300072, China

Corresponding author:, email: slzhu@imr.tohoku.ac.jp (Dr. Shengli Zhu)

Abstract. In present study, we prepared nanoporous Ti oxides by potentiostat and pulse dealloying TiCu amorphous alloy in HNO_3 solution. The influence of applied voltage on the morphology was investigated by SEM. The mechanism of dealloying was also discussed. A multilayer nanoporous structure was fabricated using pulse dealloying. The ideal nanoporous structure with mean pore diameter of about 50-100 nm and mean pore wall thickness of about 50 nm was obtained using pulse dealloying with initial voltage of 1.0 V and pulse size of 0.4 V. The static currents for pulse dealloying are higher than those for potentiostat dealloying under the same initial potentials. The nanoporous surface is mainly composed of TiO, Ti_2O_3 and TiO_2.

1. Introduction

Titanium dioxide (TiO_2) is one of the important semiconductor materials which is widely used as an efficient photo-catalyst, because it is chemically and biologically inert, photo-catalytically stable, commercially available and inexpensive, and environmentally friendly [1-3]. The valence band (VB) of TiO_2 (versus natural hydrogen electrode) lies at deep levels and photo-generated holes have strong enough oxidizing power to create OH radicals, which decompose most organic pollutants [4]. Position of its conduction band (CB) is also high enough to reduce oxygen molecules adsorbed on the surface [5]. An important requirement for high TiO_2 electro catalytic efficiency is a large surface area which enhances electron transfer. Nanostructured TiO_2 materials, with a typical dimension less than 100 nm, have recently emerged. Such materials include spheroidal nanocrystallite and nanoparticles together with forming porous film [6-9], nanotubes[10,11], nanosheets[12], and nanofibers[13]. Recently, nanoporous metallic materials have received increasing attention due to large surface-to-volume ratio, light weight, excellent electrical/thermal conductivity and quantum size effect [14]. They can be widely used in catalysis, biofiltration, surface enhanced Raman scattering, surface plasmon resonance, sensor, drug delivery, etc.

Dealloying, as also called the selective corrosion, is a process that the active elements in the original alloy system are dissolved in the electrolyte solution, but the inert elements reserve in the alloy and re-assemble into the porous structure. Due to the strict requirements for the starting alloys of dealloying, only limited crystalline alloy systems have been demonstrated to form uniform nanoporous structures, such as Ag-Au [15], Zn-Au [16], Al-(Au, Ag, Pd, Cu) [17], Cu-Au [18], Si-Pt [19], Cu-Pt [20], Mn-Cu [21], Au-Pd [22], etc. In comparison with crystalline alloys, amorphous alloys are considered as good starting alloy for dealloying due to their monolithic phase with a homogeneous composition and structure down to subnanoscale [23, 24]. There are several attempts to synthesize nanoporous metals using amorphous alloy as the starting alloy [25, 26, 27]. In our previous work, nanoporous structure was synthesized on the surface of Ti-Cu amorphous alloy in nitric acid solution [28].However, it is difficult to fabricate multilayered nanoporous structure on Ti and Ti alloys using traditional electrochemical etching because of their self-passive properties. In present study, we investigated the effect of pulse voltage on the formation of nanoporous Ti oxides by dealloying the amorphous TiCu alloy.

2. Experiments

The $Ti_{30}Cu_{70}$ (nominal atom percent) amorphous alloy ribbons with width of 2 mm and thickness of several 10 μm were prepared by melt spinning. The amorphous alloy ribbons were cut into 40-50 mm in length. After cleaned in deionized water by ultrasonic cleaner, the ribbons were subjected to dealloying process in 30 wt% HNO_3 solution at 70 ℃ using a Gamry Reference 600 electrochemical workstation. A three-electrode electrochemical configuration was employed: a $Ti_{30}Cu_{70}$ amorphous alloy ribbon working electrode, a saturated calomel electrode (SCE) as reference electrode and a platinum mesh as counter electrode. The morphology of nanoporous structure was investigated by Hitachi S4800 Scanning Electron Microscope (SEM). The chemical composition of the nanoporous sample and valence state of elements were characterized by X-ray Photoelectron Spectroscopy (XPS).

3. Results and discussion

Fig. 1 shows the microstructure of nanoporous surface prepared by potentiostat dealloying in 30 wt% HNO_3 solution for 10800 s. The insets in the blue square are the low magnification images. Fig.2 shows the corresponding I-t curves. Lots of shallow pits are observed on the surface of sample subjected to etching under 0.5 V voltage (Fig. 1(a)). Nanoporous structure is obtained for the etching under 1.0 V voltage, as shown in Fig. 1(b). When the voltage increases to 1.5 V, the nanoporous structure is ruined (Fig. 1(c). The I-t curves exhibit similar shape: decreasing sharply, and then increasing slowly, finally reach the static current after about 5000 s. The static current of 1.0 V etching is slightly higher than that of 0.5 V etching, and much lower than that of 1.5 V etching. Three events would occur when applying anodic potential on the TiCu alloy in HNO_3 solution: removing the Cu element, forming oxides film and electrochemical polishing (indeed, the electrochemical polishing includes the corrosion of Ti, Cu and their oxides). The HNO_3 solution can dissolve Cu and its oxides, but hardly dissolve the dense Ti oxides at low voltage. When the voltage is lower than 1.0 V, the removing of Cu is favourable. The pores or pits form on the sample surface with Cu element dissolved continuously. With increasing voltage, the drive force of removing Cu increases, hence the diameter and depth of nano pores increase. At the same time, the formation of Ti oxides film is enhanced. So the nanoporous structure is only formed on the surface. When the voltage is 1.5 V, the electrochemical polishing becomes favourable. The sample exhibits a relative plat surface. Only shallow pits form due to the local energy fluctuations in the interface between the sample and the solution. The electrochemical polishing means that large quantities of react products would generate; hence, the static current of 1.5 V etching was much higher than that under 1.0 V.

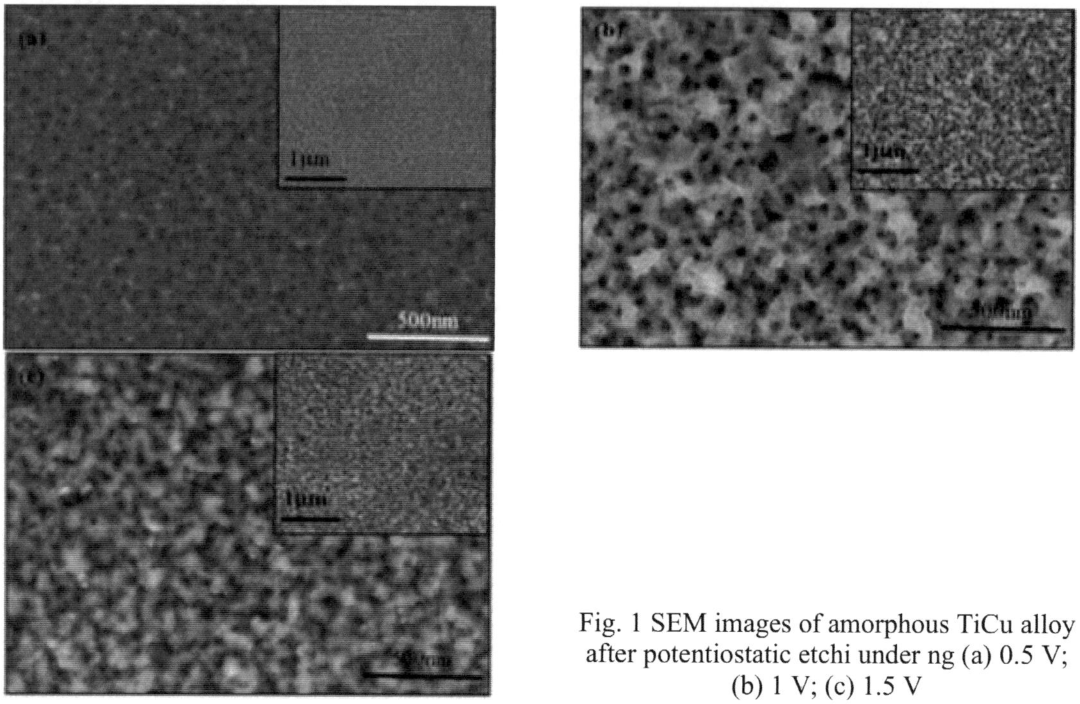

Fig. 1 SEM images of amorphous TiCu alloy after potentiostatic etchi under ng (a) 0.5 V; (b) 1 V; (c) 1.5 V

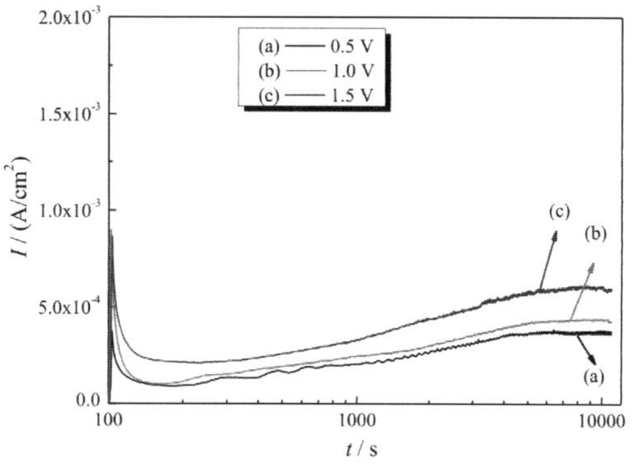

Fig. 2 I-t curves of amorphous TiCu alloy after potentiostat etching under (a) 0.5 V; (b) 1 V; (c) 1.5 V

Fig. 3 shows the microstructure of nanoporous surface prepared by pulse dealloying in 30 wt.% HNO_3 solution for 5400 s. The pulse size is 0.4 V. The insets in the blue square are the low magnification images. Fig. 4 shows the corresponding I-t curves. The Y-axis (I) is mean current density during one pulse interval. The sample subjected to pulse dealloying with 0.5 V initial potential exhibits an inhomogeneous nanoporous surface containing some large holes with diameter of about 100-200 nm and lots of small pits with diameter less than 50 nm, as shown in Fig. 3(a). However, the nano pores exist only on the surface, indicating a single layer nano pores. The pulse dealloying under 1.0 V initial potential results in formation of ideal nanoporous structure with mean pores diameter of about 50-100 nm and mean pore wall thickness of about 50 nm. The nanoporous structure consists of multilayer nano pores. One can observe that inside nano pores exist in the wall or bottom of the surface nano pores, as shown in inset of the Fig. 3(b) (the enhanced image in the red square). The sample subjected to 1.5 V initial potential dealloying exhibits a destructive nanoporous structure (Fig. 3(c)). Low magnification image shows some large pits with diameter of several μm, which consist of destructive nano pores. According to the I-t curves, the static current increases with increasing initial potential. The times to reach static currents for the 0.5, 1.0 and 1.5 V pulse dealloying are about 700 s,

1200 s and 2000 s, respectively. They are shorter than those of potentiostat dealloying, indicating that the pulse dealloying would favour of formation of nano pores compared with potentiostat dealloying. The curve (a) in Fig. 4 appears a strange peak in the starting stage. The reason of the peak formation is not clear and needs to further investigating. Fig. 5 shows the microstructure and corresponding I-t curve prepared by pulse dealloying in 30 wt.% HNO_3 solution for 10800 s. There is no distinct difference in surface morphology between 5400 s etching (Fig. 3 (b)) and 10800 s etching. The I-t curve exhibits similar shape, static current and time to reach static current with Fig. 4 (b). The cross section image shows a multilayer nanoporous structure in the whole ribbon (not shown in the present paper), indicating that the depth of nanoporous structure would be larger than several μm. However, only 500 nm nanoporous layer can form on the TiCu ribbon using potentiostat dealloying [21].

The XPS results show that the as-prepared nanoporous sample mainly composes of Ti, Cu and O, and the contained C (The XPS spectra is not shown here). The proportion of the chemical composition of Ti to Cu is much higher than nominal atomic ratio of $Ti_{30}Cu_{70}$ alloy. The change of proportion of Ti to Cu between initial alloy and prepared nanoporous structure indicates the reduction of Cu due to the dealloying process. The Ti2p spectrum is measured as two doublet peaks and one single peak, which are assigned to Ti^{2+}, Ti^{4+} and Ti^{3+}, respectively. The single peak in the O1s spectrum is corresponding to O^{2-}. The Cu2p spectrum exhibits a doublet peak which is assigned to Cu^{+}. Based on the information above and analysis of Ti, Cu, and O, the composition of nanoporous structure is TiO, TiO_2, Ti_2O_3 and small amount of Cu_2O.

The formation of nanoporous structure from dealloying started with dissolution of active atoms on a flat alloy surface, leaving behind vacancies. Before the next layer attacked, the inert atoms diffuse and start to agglomerate into islands. As a result, the surface is comprised of two distinct kinds of regions: inert clusters and patches of undealloyed material exposed to electrolyte. However, in the TiCu binary alloy system, Ti can be self-oxidated to titanium oxides in the HNO_3 solution, and then develop into dense oxide films. In potentiostat manner, the load voltage of 1 V vs SCE is not enough to break the oxide films consisted of Ti oxides. The potentiodynamic polarization curve of $Ti_{30}Cu_{70}$ was examined in 5.36 mol/L HNO_3 solution (not shown here). According to the polarization curve, the breakdown of the passive film occurs at 1.4 V (vs. SCE). It indicates that the oxide film on the surface is stable under lower applied potential. As a result, further etching would be difficult to carry out. In the case of pulse dealloying, the oxide film is broken at the pulse peaks since the peaks potential reach the critical breakdown potential of TiCu alloy. The further etching would occur. After the breakdown of the oxide film, electrolyte attacks the next layer. As a result, Cu atoms in the next layer dissolve continually and Ti atoms as well as its oxides agglomerated. When the potential returns to the initial potential (1.0 V vs. SCE), the oxide film is self-recovery and the over-corrupt is avoided. Ultimately, this periodic process leads to the formation of multilayered nanoporous structure.

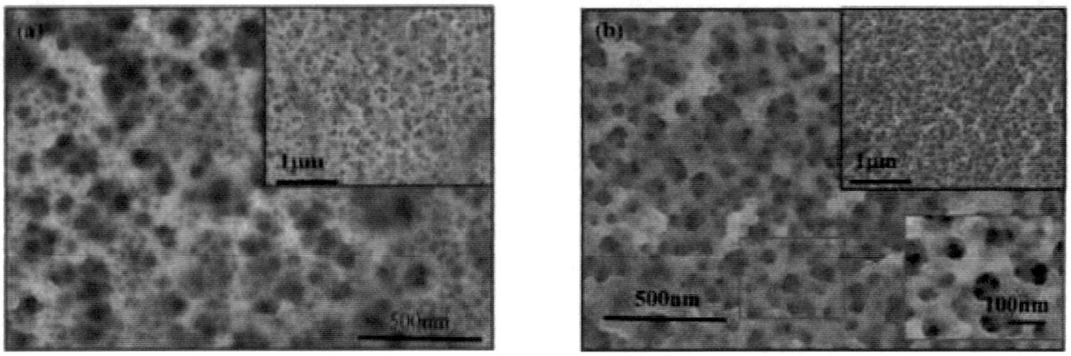

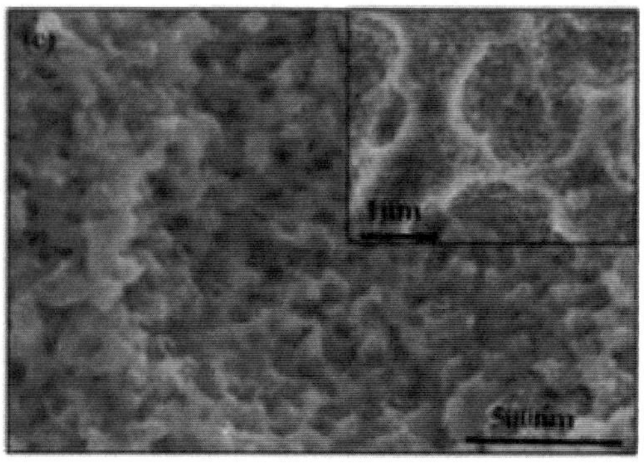

Fig. 3 SEM images of amorphous TiCu alloy after pulse etching under initial voltage of (a) 0.5 V; (b) 1 V; (c) 1.5 V

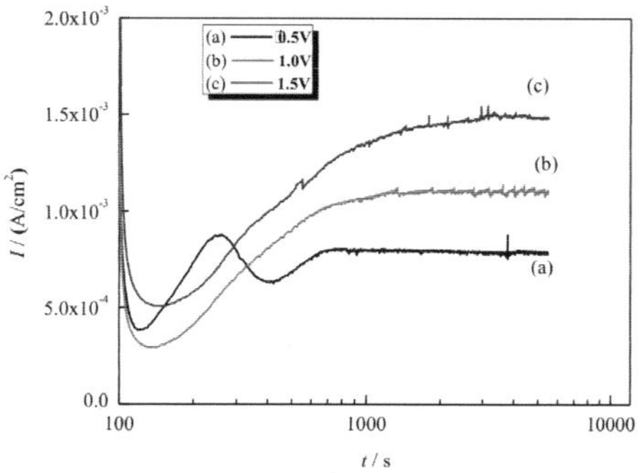

Fig. 4 I-t curves of amorphous TiCu alloy after pulse etching under initial voltage of (a) 0.5 V; (b) 1 V; (c) 1.5 V

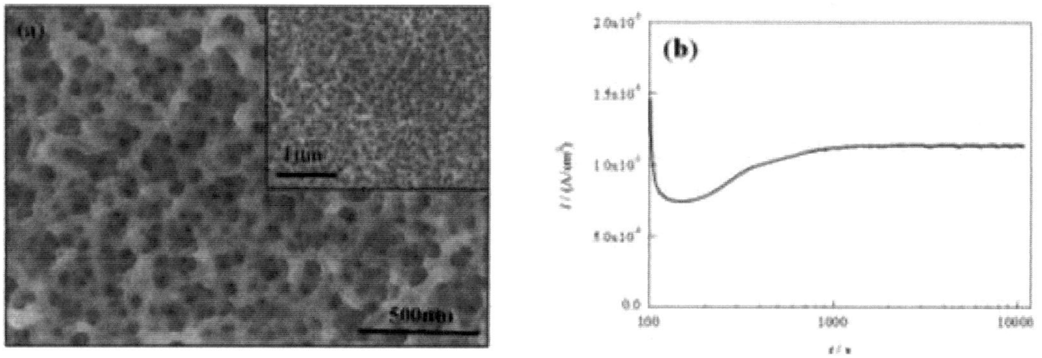

Fig. 5 SEM images and I-t curve of amorphous TiCu alloy after pulse etching 10800 s under initial voltage of 1 V

4. Conclusions

We prepared nanoporous Ti oxides by potentiostat and pulse dealloying TiCu amorphous alloy. Three events would be suggested for potentiostat dealloying TiCu alloy in HNO_3 solution: removing the Cu element, forming oxides film and electrochemical polishing. Multilayer nanoporous structure was fabricated using pulse dealloying. The pulse dealloying under 1.0 V initial potential results in formation of an ideal nanoporous structure with mean pore diameter of about 50-100 nm and mean pore wall thickness of about 50 nm. The static currents for pulse dealloying are higher than those for potentiostat dealloying under the same initial potentials,

Acknowledgments: This work was supported by National Natural Science Foundation of China (50901051) and Key Projects in the Tianjin Science & Technology Pillar Program (09ZCKFGX29100).

Journal of Physics: Conference Series **417** (2013) 012022 doi:10.1088/1742-6596/417/1/012022

References

[1] Grätzel M 2000 Prog. Photovolt. Res. Appl. 8 171
[2] Diebold U 2003 Surf. Sci. Rep. 48 53
[3] Miyauchi M, Nakajima A, Watanabe T and Hashimoto K 2002 Chem. Mater. 14 2812
[4] Bayati M R, Golestani-Fard F and Moshfegh A Z 2010 Mater. Chem. Phy. 120 582
[5] Ambrus Z, Mogyorosi K, Szalai A, Alapi T, Demeter K, Dombi A and Sipos P 2008 Appl. Catal. A 340 153
[6] Priya R, Baiju K V, Shukla S, Biju S, Reddy M L P, Patil K R and Warrier K G K 2009 Catal. Lett. 128137
[7] Golubović A, Šćepanović M, Kremenović A, Aškrabić S, Berec V, Dohčević-Mitrović Z and Popović Z V 2009 J. Sol-Gel Sci. Technol. 49 311
[8] Charbonneau C, Lee K E, Shan G B, Gomez M A, Gauvin R and Demopoulos G P 2010 Electrochem. Solid-State Lett. 13 H257
[9] Baiju K V, Zachariah A, Shukla S, Biju S, Reddy M L P and Warrier K G K 2009 Catal. Lett. 130 130
[10] Liang Y Q, Yang X J, Cui Z D and Zhu S L 2009 J. Mater. Res. 24 3647
[11] Liang Y Q, Cui ZD, Zhu S L and Yang X J 2010 Electrochim. Acta 55 5245
[12] Xiang Q J, Lv K L and Yu J G 2010 Appl. Catal. B: Envir. 96 557
[13] Tang H, Yan F, Tai Q D and Chan H L W 2010 Biosen. Bioelectro. 25 1646
[14] Erlebacher J, Aziz M J, Karma A, Dimitrov N and Sieradzki K 2001 Nature 410 450
[15] Ding Y, Kim Y J and Erlebacher J 2008 Adv. Mater. 16 1897
[16] Fang H J and Sun I W 2005 Adv. Funct. Mater. 15 989
[17] Zhang Q and Zhang Z H 2010 Phys. Chem. Chem. Phys. 12 1453
[18] Renner F U, Gründer Y, Lyman P F andZegenhagen J 2007 Thin Solid Films 515 5574
[19] Thorp J C, Sieradzki K, Tang L, Crozier P A, Misra A, Nastasi M, Mitlin D and Picraux S T 2006 Appl. Phys. Lett. 88 033110
[20] Pugh D V, Dursun A andCorcoran S G 2003 J. Mater. Res. 18 216
[21] Hayes J R, Hodge A M, Biener J, Hamza A V and Sieradzki K 2006 J. Mater. Res. 21 2611
[22] Lang X Y, Guo H, Chen L Y, Kudo A, Yu J S, Zhang W, Inoue A andChen M W 2010 J. Phys. Chem. C 114 2600
[23] Wang W H, Dong C andShek C H 2004 Mater. Sci. Eng. R 44 45
[24] Chen H S and Miller C E 1970 Rev. Sci. Instru. 41 1237
[25] Yu J S, Ding Y, Xu C X, Inoue A, Sakurai T and Chen M W 2008 Chem. Mater. 20 4548
[26] Abe H, Sato K, Nishikawa H, Takemoto T, Fukuhara M and Inoue A 2009 Mater. Trans. 50 1255
[27] Jayaraj J, Park B J, Kim D H, Kim W T andFleury E 2006 Scripta Mater. 55 1063
[28] Zhu S L, He J L, Yang X J, Cui Z D and Pi L L 2011 Electrochem. Commun. 13 250

The Role of Activated Nitrogen Species on Double-folded Screen Nitriding Process

Sanggweon Kim*[1], Jaehoon Lee[1], Nagahiro Saito[2], and Osamu Takai[2]
[1]*Surface Technology & Heat Treatment R&D Department, Korea Institute of Industrial Technology (KITECH)*
7-47, Songdo-Dong, Yeonsu-Gu, Incheon, 406-840, R. O. Korea
[2]*Department of Materials Science and Engineering, Nagoya University, 464-8603, Japan*

Abstract: As clean and energy saving surface hardening technology, plasma nitriding techniques have been evolved with object of higher performance in the last decades. Even though the diffusion of nitrogen inward to steel is occurred at the final step, solid diffusion from surface, energy transition from gas molecule of nitrogen to atomic or an activated state have many different steps depending on the plasma conditions, parameters and the design of each equipment. And this study made comparative on nitrogen sources transfer with conventional DC plasma nitriding and novel nitriding process using plasma diagnosis and metallurgical observation. With different vacuum pressure, gas ratio and new designed electrode (double-folded screen cathode electrode), it showed a different behavior of DC plasma nitriding including the nano-sized nitride on the outer surface of specimen due to nitrogen source of determining plasma species. In this study, plasma species was able to identify with optical emission spectroscopy (OES) studies. From these observations, we could understand better role of ions or neutral nitrogen species, like neutral nitrogen (N), N_2^+ and NH_x radicals in plasma nitriding process with different parameters. And cutting layers of nitride specimen were showed the results due to a different species gas flow ratio or plasma conditions.
Keyword: atomic nitrogen, neutral nitrogen, diffusion rate in steel, hollow cathode discharge (HCD) effect, plasma species in plasma nitriding process, optical emission spectroscopy (OES).

1. Introduction

Generally, plasma nitriding process has an excellent benefit in low deformation, energy saving, eco-friendly, can treat a selected area or parts, product reliability, and it has expectedly merits for economical process with low energy consumption and gases. And plasma nitriding process has been widely accepted in industry uses, for the applications to automobile parts, machinery, dies, and molds to enhance wear and anti-corrosion and fatigue properties.

Although plasma nitriding process has many above-mentioned advantages over conventional gas nitriding, it has still applied to the complicated shape product (especially products with deep hollows) and controlling of thickness of compound layer is relatively difficult in practice. During DC plasma nitriding process, nitrogen molecule is introduced to the product surface where nitrogen molecule is activated via glow discharge area in electric field. And nitrogen molecules should be dissociated to ionic species or several radicals for subsequent reactions and adsorption on steel surface and then atomic reacted nitrogen atoms can diffuse into interstitial sites of steel. At the same time, it has a different diffusion rate and hardness profile in dependence on the plasma condition, energy states, the kinds of activated species and mass transfer difference.

Several models for the plasma nitriding processes have been proposed in the part to explain the mass transfer mechanism in plasma nitriding process, example sputtering model, and role of dominant plasma species etc[1].

Recently, plasma-assisted nitriding technology[1-6], and screen nitriding process and its applications[7-9] that utilizes nitrogen atom with auxiliary electrode has been deeply studied because they are able to produce compound-layer-free nitriding layers on steel alloys and thick nitriding layers on non-ferrous metals.

During the conventional plasma process, nitrogen molecule is introduced to product a surface on which some nitrogen gases are ionized in glow discharge area with high current. Since nitrogen molecules are to dissociate as ions or atoms for subsequent reactions, a high ionizing rate can beneficially affect the properties of surface coatings or nitriding layers by the plasma process.

Therefore, many studies have focused on solving such problems by adopting auxiliary electrodes or post plasma technique in a plasma nitriding system.

The novel nitriding process was conceptionally designed and scale-up to generate a high-density neutral atom including others activated ionic species utilizing trapped electron and ions collisions between double folded-cathodic electrodes with enlarged mean free path under higher vacuum range than general plasma nitriding. Although the novel nitriding process utilized high vacuum, high plasma density was achievable from restricted electrons in the magnetic field between negatively charged screens. Fig. 1 shows and summarizes the simple concept and principles of the novel nitriding process.

In particular, after the novel nitriding process has showed a bright mirror surface without compound layer on the specimen surface.

The novel nitriding process, so called by ATONA process, can describe a brief three step reaction of plasma nitriding procedure from ionic source to diffusion couple in the vacuum chamber.

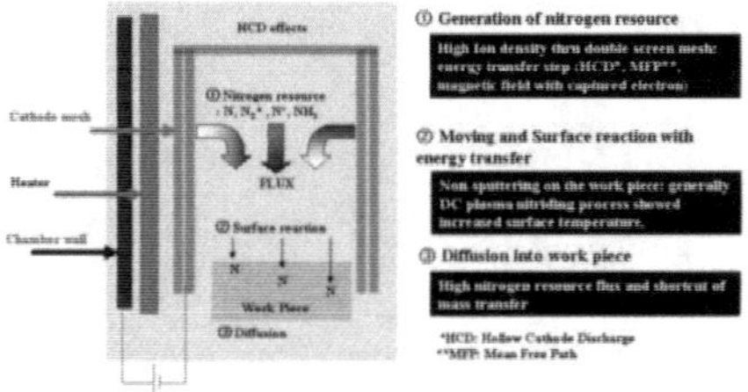

Fig. 1 The concept and a brief of the novel nitriding (ATONA) process.

2. Experimentals procedure

The specimen was a S45C carbon steel. Specimen of S45C was prepared with quenching at 1143K and tempering at 433K and SKD11 mold steel was prepared with quenching at 1303K and tempering at 773K. It was polished with SiC paper followed by fine polishing with 1 μm diamond pastes. Polished S45C carbon steel and SKD11 mold steel specimens were ultrasonically cleaned in acetone.

With the plasma nitriding system, both the novel nitriding and DC plasma nitriding were carried out at the same chamber by applying DC power on the screen and on the jig, for each process, respectively. Base pressures for both the novel nitriding and DC plasma nitriding process were 5×10^{-3} Torr. The temperatures for the novel nitriding and DC plasma nitriding were 500°C, respectively. Before the main process, specimen was ion bombardment with Ar and H_2 gas at a pressure of 5×10^{-2} Torr for 30 minute. During the ion cleaning, a cathode voltage of about -800 V was applied on the jig. After the ion cleaning, N_2 and H_2 gases were backfilled to the nitriding chamber and 10 - 20 KW of power was applied on the screen for the novel nitriding process and on the stage for the DC plasma nitriding process. The working pressure was 0.1 Torr for the novel nitriding and about 2 Torr for DC plasma nitriding process. The ratio of N_2 to H_2 was 1:3 and 3:1 (Fig. 2).

Plasma species was analyzed with optical emission spectroscopy (OES, AVANTE AvaSpec-3648). Microstructure was observed by scanning electron microscope (SEM, HITACH S-4300). The surface hardness and the case depth profiles were determined with a micro-Vickers hardness tester (Future Tech FM-7). An indentation cracking test was carried out with a Rockwell hardness tester (Matsuzawa DXT-3).

3. Results and discussions

Fig. 3 showed results of OES analysis at DC plasma nitriding process with a changing parameters, 1 Torr, 673K, N_2:H_2 = 1:3 about 650V DC power. It has almost same intensity of peak with position of plasma species.

Fig. 4 were showed results of OES analysis of the novel nitriding process with changed parameters, (a) 0.1 Torr, 673K, N_2:H_2 = 1:3, (b) 0.1 Torr, 673K, N_2:H_2 = 3:1, and (c) 1 Torr, 673K, N_2:H_2 = 1:3 12KW DC power on the double-folded screen. In the result of OES analysis of DC plasma nitriding process showed a dominant plasma species with N_2^*, N_2^+.

However, in Fig. 4 (a), Fig. 4(b) and (c) have shown higher intensity of spectra than DC plasma nitriding with some parameters for temperature, gas ratio and pressure.

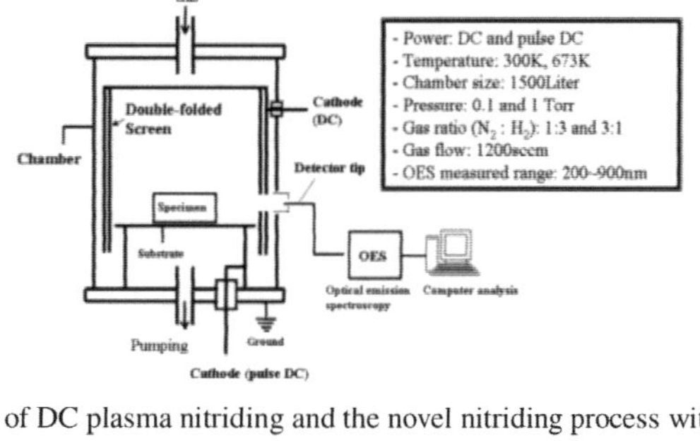

Fig. 2 The parameters of DC plasma nitriding and the novel nitriding process with schematic diagram.

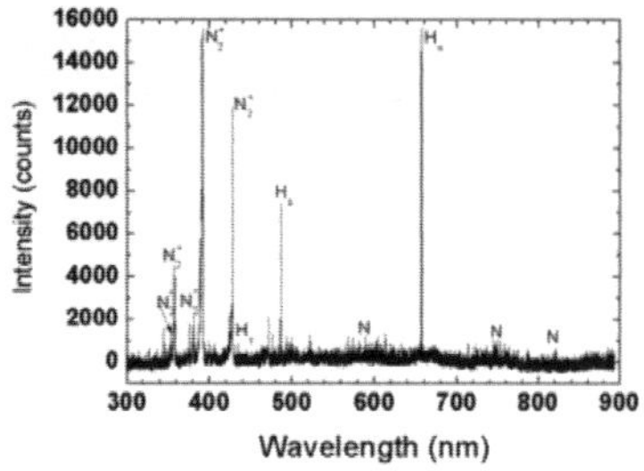

Fig. 3 The results of OES analysis for DC plasma nitriding process at 1 Torr, 673K and gas ratio N_2:H_2 = 1:3.

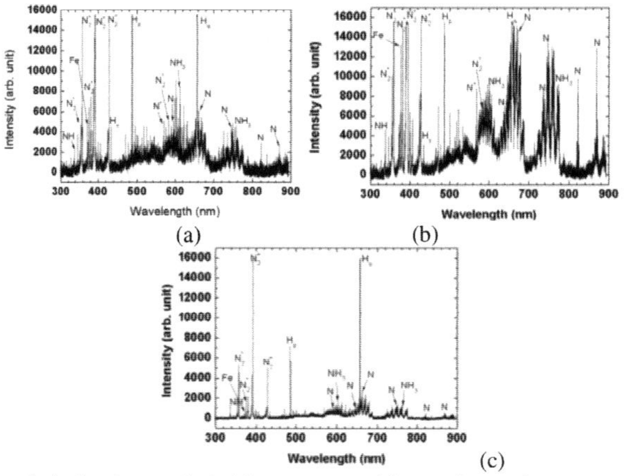

Fig. 4 The results of OES analysis for the novel nitriding process with experimental parameters: (a) 0.1Torr, 673 K, and N_2:H_2 = 1:3, (b) 0.1Torr, 673 K, and N_2:H_2 = 3:1, and (c) 1 Torr, 673 K, and N_2:H_2 = 1:3.

Especially, in the case of the condition of the high nitrogen gas ratio, Fig. 4 (b) showed a peak of high intensity with complicated species, N_2^*, N_2^+, N^+, N, NH_3. It can understand that new concepts and designs for the novel nitriding process, and possibility of high density plasma in higher vacuum chamber and composed auxiliary electrode within double screens. When the N_2 gas flow can make to an increasing plasma species from the crashed N_2 molecules in the double-folded screen, it is summarized of a dominant species in the plasma environment each process in Table 1.

Table 1 Summarize of dominant plasma species of DC plasma nitriding and the novel nitriding process

	DC plasma nitriding	the novel nitriding	
		Low	High
Vacuum	H_2 activity decrease Dominant species : N_2^*, N_2^+	~Torr	10^{-2} ~10^{-1}
		Decreasing H_2 activity	Increasing H_2 activity
Temperature	No affected	300K	673K ~
		H_2 and $N_2^{*(1st, 2nd)}$, N_2^+activity	H_α, H_β and $N_2^{*(1st, 2nd)}$ N_2^+
Gas ratio (10^{-1} Torr, N_2/H_2)	H_2 activity decrease	$N_2^{*(1st, 2nd)}$, N_2^+, NH_x and N^+	Same left but more higher intensity

(a) (b)

Fig. 5 The scanning electron micrographs of S45C carbon steel after 2 hr treatment of (a) the novel nitriding and (b) DC plasma nitriding process.

As shown in the secondary electron image of Fig. 5 (b), 2 hr of the novel nitriding process resulted in the formation of nano-sized grains on the S45C carbon steel. The grain size ranged from 30 to 100 nm. On the other hand, 2 hr of DC plasma nitriding yielded 1 - 2 μm grains on the surface of S45C carbon steel, as shown in Fig. 5 (a). Fig. 6 (a) and 6 (b) shows cross-sectional micrographs of the S45C carbon steel after 2 hr of DC plasma nitriding (a) and the novel nitriding (b) treatments.

Generally, most surface of nitrided-steel has a compound layer composed of γ'-(Fe_4N) and/or ε-($Fe_{2\text{-}3}N$) phases.

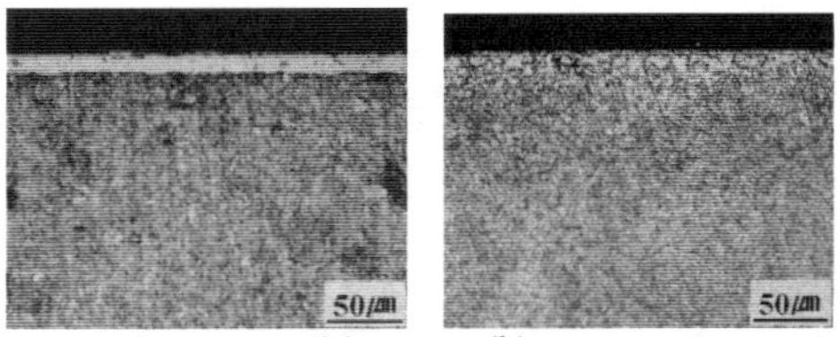

(a) (b)

Fig. 6 Cross-sectional optical microscopy images of S45C carbon steel after 2 hr treatment of (a) DC plasma nitriding and (b) the novel nitriding process.

As shown in Fig. 6 (b), excess thickness of the compound layer does not occur in the novel nitriding treated S45C steel. On the other hand, the thickness of the compound layer of the DC plasma nitriding specimen was about 20 μm during same process time and temperature.

Fig. 7 shows the micro-hardness profiles of DC plasma nitriding and the novel nitriding of S45C steel. At the time, surface hardness of the novel nitrided S45C (black line) steel was very hard indeed about 850 HV. Fig. 8 shows X-ray pattern of the novel nitrided S45C steel. It was composed of γ'-Fe_4N and α-Fe. However it was difficultly distinguished this phase from optical microscopy. Therefore we are supposed to precipitate of γ'-Fe_4N with very small size. And nitrogen content of the nano-sized surface was about 40 atomic percent (Fig. 9). The surface hardness of the novel nitrided S45C carbon steel was about 850 Hv, which was higher than that of DC plasma nitrided S45C carbon steel.

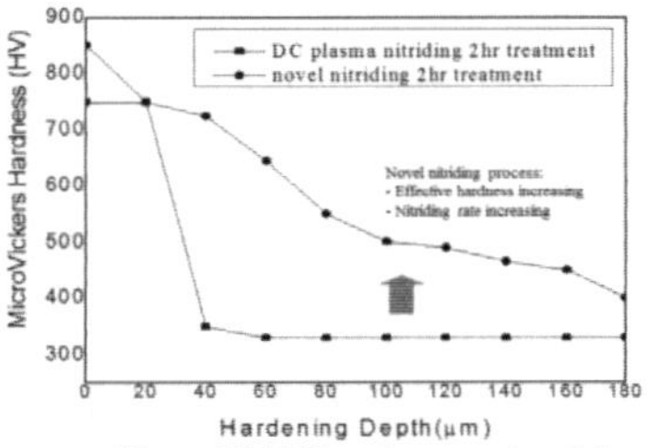

Fig. 7 The competition of micro-hardness profiles of S45C carbon steels of the novel nitrided (square) and DC plasma nitrided (filled circle) process.

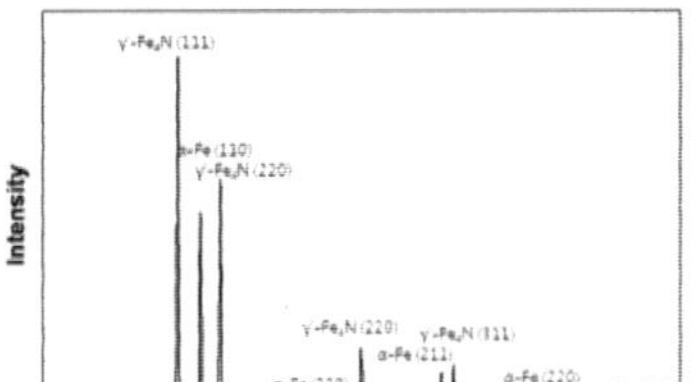

Fig. 8 The X-ray diffraction result of the novel nitrided S45C steel after 2 hr treatment.

To investigate the failure and appearance of the surface of the compound layer, an indentation cracking test was carried out, as shown in Fig. 10. After the indentation cracking test of the DC plasma nitriding specimen Fig. 10 (a), extended cracks were found near the circular indentation. On the other hand, no crack with a mirror surface was found near an indentation in the novel nitriding specimen Fig. 10 (b). These results indicated that mechanical failure did not occur with the novel nitriding specimen because it had a very thin compound layer. Since the compound layer of the novel nitriding specimen was not thick, failure may have not been as severe as I was for the DC plasma nitriding specimen.

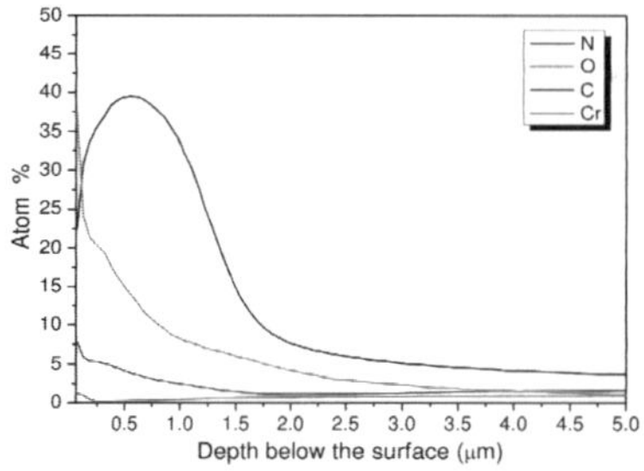

Fig. 9 The result of GDS analysis of the novel nitrided S45C steel.

The novel nitriding process also enhanced the nitrogen diffusion rate of the matrix (Fig. 11). Immediately below the compound layer, a diffusion layer (or diffusion zone) is formed by nitrogen diffusion. The diffused nitrogen atoms are interstitially locating in iron matrix, or forming nitrides with active nitriding elements at grain boundaries, or in grain. This diffusion zone, however, cannot be clearly observed with a microscope.

(a) DC plasma nitrided surface

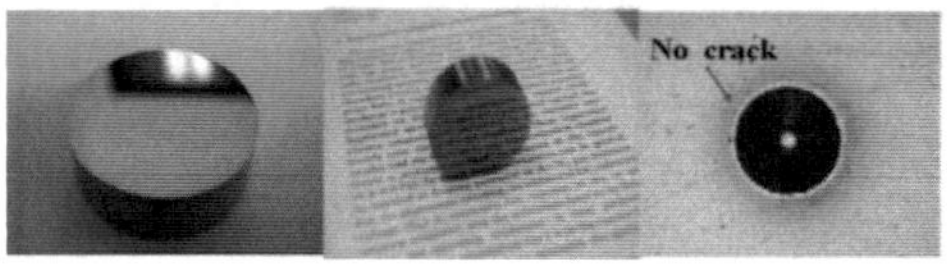

(b) the novel nitrided surface

Fig. 10 Indentation cracking test results of (a) the novel nitrided S45C carbon steel specimen and (b) DC plasma nitrided S45C carbon steel specimen.

This high hardness of the novel nitrided specimen has shown a result from fast nitrogen diffusion to the matrix. Such fast nitrogen diffusion, compared with DC plasma nitriding, may have been due to the surface nano-grains and thin compound layer.

The adsorption of nitrogen may have increased due to the large portion of grain boundaries at the nano-structured nitride layer since chemical affinity was increased by the long grain boundaries.

The novel nitrided iron has composed with a nano-sized nitride (Fig. 11) on surface, α-Fe and precipitation γ'-Fe_4N phase almost (the result of X-ray diffraction in Fig. 8). The diffusivity of γ'-Fe_4N ($D_{N\gamma'} = 3.2 \times 10^{-12}$ cm^2/s) and ε-$Fe_{2\text{-}3}N$ ($D_{N\varepsilon} = 9.8 \times 10^{-11}$ cm^2/s) were slower than α-Fe ($D_{N\alpha} = 1.48 \times 10^{-2}$ cm^2/s) matrix's that at 723K.

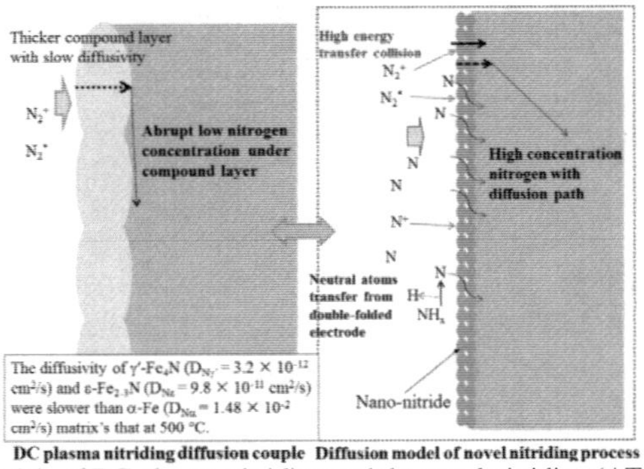

Fig. 11 The diffusion models of DC plasma nitriding and the novel nitriding (ATONA) treated specimen.

Therefore, the amount of adsorbed nitrogen may have been much higher on the novel nitrided specimen than on the DC plasma nitriding treated specimen which caused a high concentration gradient of nitrogen. In our case, the high concentration layer was under 500 nm of the novel nitrided surface, which was at least 4 times lower than that of the DC plasma nitriding treatment. And the nitrogen diffusion rate on the novel nitrided S45C carbon steel was much higher than that on the DC plasma nitriding treated S45C carbon steel.

In particular, plasma nitriding process can make easily the compound layer determined most slowly a diffusion couples via N_2^+, N_2^* at final step of solid diffusion. The compound layer is composed with ε-$Fe_{2\text{-}3}N$ and γ'-Fe_4N phase that has constant chemical concentration with 6.6 ~ 11.2wt. % of nitrogen within diffusion couples. In this case, the novel nitriding process can understand three steps with crashing neutral nitrogen atom and gas diffusion into the vacuum chamber and adsorption on the cold and electric floated specimen surface as following concepts.

At the first step, the generation of neutral nitrogen resource at Fig. 12, nitrogen molecules transferred activated and ionized species extremely by auxiliary electrode between double-folded electrodes with hollow cathode discharge under high vacuum chamber relatively general plasma nitriding process.

Especially, Fig. 12 was showed the captured electron motion in the double-folded screen that it has energetic motion of several effective collisions with higher vacuum state.

In the Fig. 4 (a) and (b) was showed and understood easily for activated species in the novel nitriding process due to nitrogen source of species for DC plasma nitriding, H_α, H_β, H_γ, and including N_2^+, N_2^* with neutral nitrogen and ammonia group in the novel plasma nitriding process [10-13].

And the second step, the motion of activated species has enlarged and moved to specimen surface with mean free path (MFP) via higher vacuum range than general plasma nitriding process.

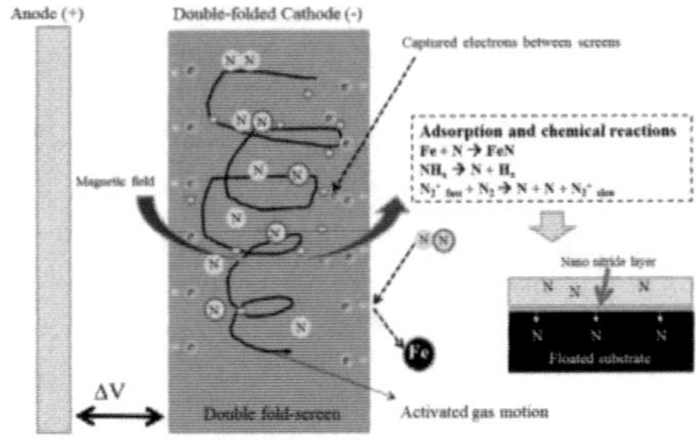

Fig. 12 Behaviors of captured electrons in the double-folded screen electrode and chemical reaction model of plasma species.

At third step, the surface of specimen has homogeneous temperature with an external heating elements, it can sustain a nano-sized nitride without grain growth even though it is composed with the sputtered iron atom from electrode with activated nitrogen species, and after nitriding treatment has a brighten surface. Finally, the enlarged surface area has composed with a high concentration nitrogen atom by nano-sized nitride enhanced high adsorption rate without formation of a compound layer and diffuse into the solid steel surface due to high concentration of neutral nitrogen atom.

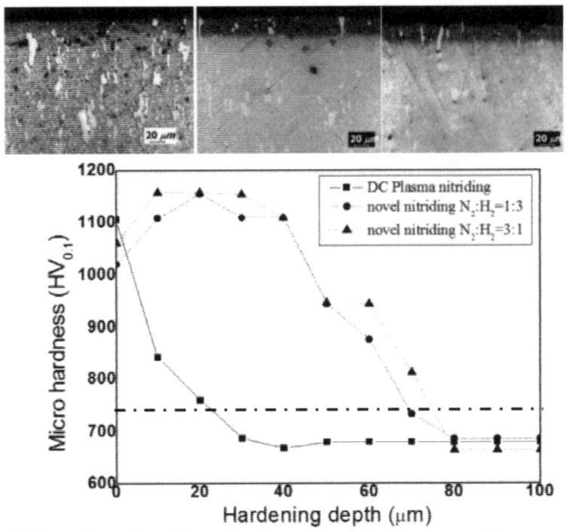

Fig. 13 It has shown a summarizing of analyzed results for the each process with parameter of DC plasma nitriding and the novel nitriding process on SKD 11 mold steel.

4. Conclusions

From the above comparative studies with different pressures and temperature and gas ratio, we could understand different mechanisms of mass transfer depending on the energy and position of glow source. Newly developed the novel nitriding process has the advantage for the treatment of complicated parts with high density plasma using auxiliary electrode in the chamber.

Definitely, the novel nitriding process distinguish with DC plasma nitriding process at dominant reactive species in the reaction chamber by OES analysis that the novel nitriding showed high energized neutral nitrogen atoms, activated nitrogen molecules, N_2^*, N_2^+, and NH_3 but DC plasma nitriding showed dominantly N_2^+ molecule. It can be shown distinguish of the depth profile each nitriding processes after treatment at Fig. 13.

Therefore, the novel nitriding shows a brighten mirror image with nano-sized nitride surface without compound layer even it has shown more deep hardening and high hardness profile than DC plasma nitriding process with same treating time. Exactly, the speed of the novel nitriding process is much faster than that of general plasma nitriding due to fast diffusion at least.

Whatever it is, the novel nitriding system can be used to easily get to the S-phase on austenitic stainless steel, in addition to the interlayer of PVD coatings. Besides this kinds process, as we know, is very useful in surface hardening of pre-hardening mold steel for the plastic forming tools that they have to have super cleaned surface with submicron ordered surface roughness.

References

[1] H. Michel, T. Czerwiec, M. Gantois, D. Ablitzer, and A. Ricard: Surf. Coat. Tech., 1995, **72**, 103.
[2] T. Czerwiec, H. Michel, and E. Bergmann: Surf. Coat. Tech., 1998, **108-109**, 182.
[3] A. Leyland, K. S. Fancey, A. S. James, and A. Matthews: Surf. Coat. Tech., 1990, **41**, 295.
[4] N. Renevier, H. Michel, T. Czerwiec, and P. Collignon: Surf. Coat. Tech., 1998, **98**, 1400.
[5] H.-J. Spies, H. Le Thien, and H. Biermann: HTM11871, Fatigue, 2005, **60** , 1.
[6] J. Musil, J. Vlcek, M. Ruzicka: Vacuum, 2000, **59**, 940.
[7] J. Georges: Heat Treat. Met., 2001, **28**, 33.
[8] C. X. Li, T. Bell and H. Dong: Surf. Eng., 2002, **18**, 174.
[9] Sh. Ahangarani, F. Mahboubi, A.R. Sabour, Vacuum, 2006, **80**, 1032.
[10] L. Petitjean, A. Ricard: J. Phys., 1984, **D17**, 919.
[11] Y. M. Kim, J. U. Kim, J. G. Han: Surf. Coat. Tech., 2002, **151-152**, 227.
[12] Y. M. Kim, J. G. Han: Surf. Coat. Tech., 2003, **171**, 205.
[13] S. G. Kim, S. W. Kim, and P. J. Brand: J. of the Kor. Soc. Heat Treatment, 2010, **23**, 150.

*Corresponding author. Sanggweon Kim, 7-47, Songdo-Dong, Yeonsu-Gu, Incheon, 406-840, Korea.
Tel: +82 32 8500264; Fax: +82 32 8500260,
E-mail: kimsg@kitech.re.kr*

Journal of Physics: Conference Series **417** (2013) 012024 doi:10.1088/1742-6596/417/1/012024

Water on TiO_2 studied by work function change: adsorption in cycles

Nenad Bundaleski*, Ana G. Silva*, Bobbie Jean-Shaw, Orlando Teodoro and Augusto Moutinho

CeFITec, Department of Physics, Faculdade de Ciências e Tecnologia, Universidade Nova de Lisboa, P-2829-516, Caparica, Portugal

Abstract

The nature of water adsorption on $TiO_2(110)$ rutile surface attracts a lot of attention for quite some time. In spite of the considerable experimental and theoretical efforts a lot of details remain unclear. We have been using work function study to follow the adsorption of water on TiO_2 at room temperature, and interpreted the results in terms of fast dissociative adsorption on bridging oxygen vacancies (BOV) and much slower non-dissociative adsorption on Ti_{5f} rows. Additionally, we concluded that water from Ti_{5f} rows efficiently desorbs at room temperature which is not the case for BOV adsorption sites. Here we propose a novel experimental approach which consists of monitoring in real-time the work function change during cycles of water adsorption. Since desorption at BOVs does not take place at room temperature, this method allows us to resolve the adsorption dynamics on the two adsorption sites. The first results changed our understanding of the phenomenon: we show that both, adsorption on BOVs and Ti_{5f} are both very fast. Additionally, slow exponential decay of the work function is observed, which is not directly related to water adsorption. The possible explanation of the third slow contribution could be related to the migration of hydrogen atoms along the bridging oxygen rows.

* Corresponding author: acs@fct.unl.pt and n.bundaleski@fct.unl.pt

Introduction

TiO_2 is one of the most studied metallic oxides due to the possible applications in photo-catalysis, heterogeneous catalysis, as gas sensors, solar cells, bio-compatible materials among other applications. Besides, it is investigated as a model metallic oxide surface. In addition, rutile TiO_2 (110) has attracted particular attention as the most stable TiO_2 surface. The chemical activity of this surface is mainly governed by the bridging oxygen vacancies (BOV), which act as the adsorption sites for different molecules. Water adsorption on TiO_2 (110) is thoroughly investigated using various surface sensitive experimental techniques among which STM particularly shed light to this very complex system [1,2].

Concerning the water adsorption on TiO_2 (110)-(1x1) surface, two sites have been mainly identified: BOVs on which the dissociative adsorption takes place yielding in formation of two hydroxyls with hydrogen atoms facing the vacuum, and the non-dissociative adsorption on titanium rows with two hydrogen atoms facing the vacuum [3]. Knowing that both hydroxyl and water molecules are polar, work function change should be proportional to the dipole concentrations. Work Function Study (WF) has been employed to monitor the water adsorption dynamics on reduced rutile TiO_2 (110) surfaces [4-6]. Recently, we showed that water adsorption on non-reconstructed TiO_2(110) could be also monitored this way in spite of the quite small signal intensity due to the small concentration of BOVs [7, 8].

It is well-known that there will be no desorption from BOVs at room temperature in contrast to Ti rows from which the desorption is efficient [1-3]. Here we use this fact to resolve the two different contributions to the WF change: if the sample is exposed to water exposing/pumping cycles the desorption will take place on both adsorption sites in the first cycle but only on Ti rows in subsiding cycles. This novel approach in WF study will allow us to resolve adsorption dynamics' related to different adsorption sites.

Experimental details

The experiments were performed in the multipurpose surface analysis apparatus described elsewhere [9] with the base pressure in low 10^{-10} mbar region. The water used in the experiments has been purified using several freeze-pump-thaw cycles. Its partial pressure has been followed using an ionization gauge while the purity of the introduced gas has been checked by a residual gas analyzer. WF was monitored as an onset of the secondary electron energy distribution. Secondary electron emission was stimulated by non-monochromatic X-rays emitted from the Mg Kα (1.2536 keV) anode of an X-ray source, routinely used for XPS. In that way we avoid formation of BOVs which can be created even by low energy electrons [5]. The TiO_2 sample was biased to -30 V to improve the detection of electrons with low kinetic energies. The secondary electron energy spectra have been taken in time intervals of about 20 s in fixed analyzer transmission mode with the pass energy of 10 eV. The sample was cleaned through several Ar^+ sputtering (3 keV, at oblique incidence) and annealing cycles, at 773 K for 10 min, until only Ti and O peaks were observed in X-ray photoelectron spectroscopy (XPS) survey spectra. Just before the true experiments the sample is annealed for short time at 773 K in order to suppress the contamination from the residual gas water vapour.

Results and Discussion

In previous WF measurements during the water adsoprtion on non-reconstructed TiO_2 (110) rutile we were observing an initial steep drop of the work function followed by the much slower linear decrease and finally, after turning off the water introduction valve, partial recovery of the work function up to the level close to the elbow defined by the two slopes. The steep drop was considered to be due to the water adsorption on BOVs while the lower slope and the later work function recovery were addressed to adsorption/desorption on/from Ti rows. According to the basic adsorption models [10] slower decrease of the WF signal should be part of an exponential decay. In order to check that we were repeating our measurements with much longer exposing times. The typical result is presented in Figure 1.

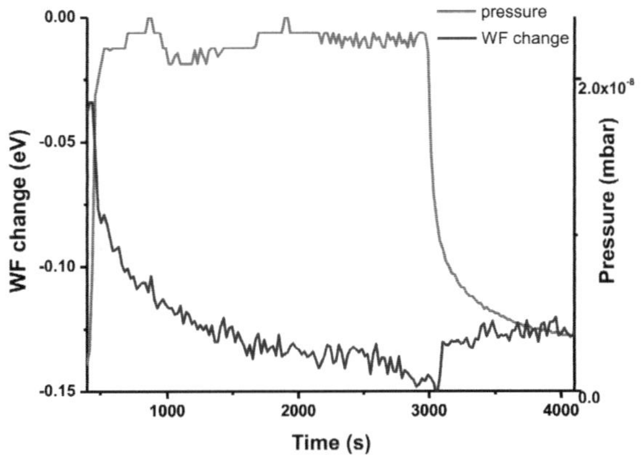

Figure 1. WF change due to the water exposure; water partial pressure is about $2 \cdot 10^{-8}$ mbar.

Although we indeed observe exponential decay for the slower contribution, the recovery of the work function after closing the valve is much lower than expected. Since this result challenges our previous assumption, we decided to perform WF study during the water adsorption in cycles in order two resolve the two contributions corresponding to water adsorption on BOVs and on Ti rows. The result of such an experiment is given in Figure 2. In the frame of the first cycle WF signal should correspond to adsorption on both sites while in other cycles only adsorption on Ti rows should be relevant. The striking result of this experiment is that introduction of water provokes steep drop of the WF signal at the beginning of each cycle. This is clear proof that adsorption dynamics' on BOVs and on Ti rows are both very fast. Therefore, a new question concerning the origin of the slow WF decrease is open.

The slow decrease of the work function is not affected by fast and huge changes of the water partial pressure i.e. it is not directly related to the water adsorption: although the valve is closed and the water partial pressure is low, further decrease of the work function is clearly observed between the first and second exposure cycle. On the other hand, this signal can be observed only after the initial water exposure. Since the water molecules are assumed to be efficiently desorbed from Ti rows at the end of each cycle, the signal is somehow related to water adsorption on BOVs.

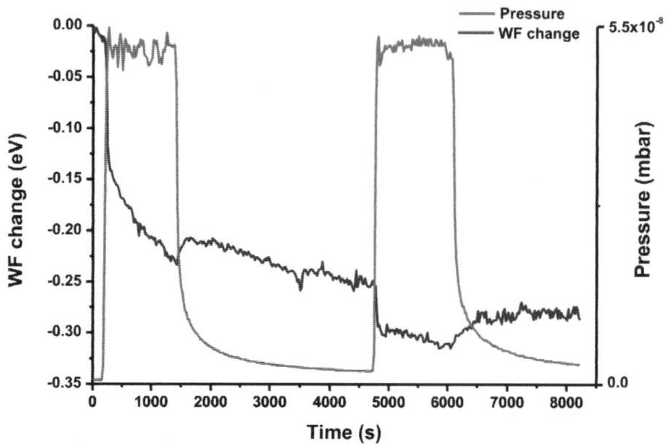

Figure 2. WF change due to the water exposure in two cycles. Steep drop is present after each cycle indicating that adsorption on Ti rows is also fast.

The result of water adsorption on BOVs is creation of hydroxyl pairs i.e. two neighboring dipoles are formed. Due to their Coulomb interaction they are partially depolarized so that the total dipole moment of the pair is smaller that the sum of two hydroxyl dipole moments far from each other. This effect is taken into account in the frame of the Topping model [11] describing the work function change dependence on dipole concentration

$$\Delta WF = -\frac{e}{\varepsilon_0} \cdot \frac{n_{BOV} \cdot 2 \cdot \mu_0}{1 + \dfrac{\alpha}{4\pi\varepsilon_0 \langle r \rangle^3}}, \tag{1}$$

where μ_0 is the moment of a dipole placed far from other dipoles, $\langle r \rangle$ is the average distance between the dipoles, α is the dipole polarizability and n_{BOV} is the concentration of bridging oxygen vacancies.

Migration of hydrogen atoms along bridging oxygen rows, which is equivalent to migration of hydroxyl dipoles, has been observed at room temperature, when the average time between two hops is about 5000 s [12]. Since considerable number of dipole pairs will become separated during the measurement, slow decrease of the work function could indeed be related to the migration of hydrogen atoms along bridging oxygen rows.

In order to test this assumption we model the hydrogen migration using 1D 'random walks' model by Ehrlich [13] according to which the mean distance between two atoms equals

$$\langle r \rangle = \Delta r \cdot \left\{ 1 + \sum_n w(n,t) \cdot n \right\}, \tag{2}$$

where Δr is the distance between the two bridging oxygen atoms, $w(n,t)$ is the probability to migrate to distance $n \cdot \Delta r$ in the time interval t. This probability equals

$$w(n,t) = \exp\left(-\frac{t}{\tau}\right) \cdot I_n\left(\frac{t}{\tau}\right), \tag{3}$$

where τ is averaging time between two hops and I_n is the modified Bessel function of the first kind and order n. Time evolution of the work function change due to hydrogen migration is given in Figure 3. The obtained result is both, qualitatively and quantitatively similar to the experimental ones however, not all parameters used are realistic. Namely, the problem is with the polarizability taken to be 15 $\mathring{A}^3$ which is few times higher than the expected value for the water molecule. Although hydrogen migration has relevance in describing the slow work function change, it is either not considered properly by this simplified approach or it could be that some other effects are dominantly responsible for this signal.

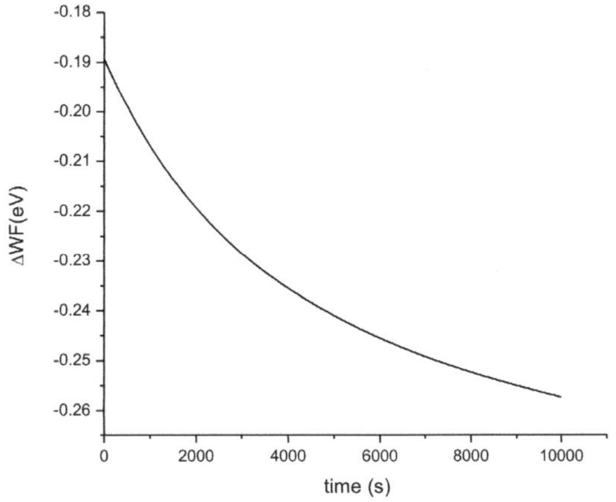

Figure 3. Work function change due to the migration of the hydrogen atoms calculated for the following parameters:

$n_{BOV} = 5 \cdot 10^{13}$ cm^2, $\mu_0 = 0.75$ D, $\tau = 5000$ s, $\alpha = 15$ $\mathring{A}^3$.

Conclusion

WF study is an efficient tool to monitor adsorption of water on TiO_2 in real time. However, the interpretation of experimental results is not straightforward. Usually it is considered that the work function change is directly related to the dipole surface concentration, which was also our starting assumption. By performing water adsorption in cycles we managed to resolve the contributions of two different adsorption sites and identify an additional contribution with slow dynamics which is not directly related to water adsorption. We discussed the possibility of addressing this contribution to migration of hydrogen atoms. It was shown the migration can explain the time evolution of this contribution qualitatively, although there are some difficulties concerning the quantitative description. The observed discrepancy may be due to the inappropriate model, but also due to the presence of other effects which are not yet taken into account.

References

1. Diebold U *Surf. Sci. Report* **48** (2003) 53
2. Pang C L, Lindsay R, Thornton G Chem. Soc. Review **37** (2008) 2328
3. Wendt S et al. Phys. Rev. Letts. **96** (2006) 066107
4. Maksymovych P, Mezhenny S, Yates Jr J T, Chem. Phys. Lett. **382** (2003) 270
5. Valentin C D et al., *J. Am. Chem. Soc.* **127** (2005) 9895
6. Marques H P, Canario A R, Moutinho A M C, Teodoro O M N D Appl. Surf. Sci, **255** (2009) 7389
7. N Bundaleski, A G Silva, U Schröder, A M C Moutinho and O M N D Teodoro, J Phys. Conf. Series, 257 (2010) 012008
8. Ana G. Silva, Nenad Bundaleski, Augusto M Moutinho and Orlando M Teodoro, Appl. Surf. Sci.2011 in press, doi: 10.1016/j.apsusc.2011.05.063
9. O. Teodoro, O. Silva, A. Moutinho, Vacuum 46 (1995) 1205.
10. H. Lüth, Surfaces and Interfaces of Solids, Springer-Verlag, Berlin, (1993)
11. J. Topping, Proc. R. Soc. London, A114 (1927) 67
12. S. C. Li et al, J. Am. Chem. Soc. 130 (2008) 9080
13. G. Ehrlich, J. Chem. Phys. 44 (1966) 1050

Determination of the magnetization profile of Co/Mg periodic multilayers by magneto-optic Kerr effect and X-ray magnetic resonant reflectivity

P Jonnard[1], K Le Guen[1], J-M André[1], R Delaunay[1], N Mahne[2], A Giglia[2], S Nannarone[2,3], A Verna[2], Z-S Wang[4], J-T Zhu[4], S-K Zhou[4]

[1] Laboratoire de Chimie Physique – Matière et Rayonnement, Université Pierre et Marie Curie, CNRS UMR 7614, 11 rue Pierre et Marie Curie, F-75231 Paris Cedex 05, France
[2] Istituto Officina dei Materiali IOM-CNR Laboratorio TASC, SS 14 km 163,5, I-34149 Basovizza, Trieste, Italy
[3] Dipartimento di Ingegneria dei Materiali e dell Ambiente, Universita di Modena e Reggio Emilia, Via Vignolese 905, I-41100 Modena, Italy
[4] Institute of Precision Optical Engineering, Department of Physics, Tongji University, Shanghai 200092, P. R. China

E-mail: philippe.jonnard@upmc.fr

Abstract. The resonant magnetic reflectivity of Co/Mg multilayers around the Co $L_{2,3}$ absorption edge is simulated then measured on a specifically designed sample. The dichroic signal is obtained when making the difference between the two reflectivities measured with the magnetic field applied in two opposite directions parallel to the sample surface. The simulations show that the existence of magnetic dead layers at the interfaces between the Co and Mg layers leads to an important increase of the dichroic signal measured in the vicinity of the third Bragg peak that otherwise should be negligible. The measurements are in agreement with the model introducing 0.25 nm thick dead layers. This is attributed to the Co atoms in contact with the Mg layers and thus we conclude that the Co-Mg interfaces are abrupt from the magnetic point of view.

1. Introduction

The Co/Mg periodic multilayers are promising for optics applications in the EUV range between 20 and 30 nm [1,2]. Indeed, we have measured an experimental reflectivity of 0.43 has been reported and up to 51% for Co/Mg-based multilayers [3,4]. Due to the high quality of the Co/Mg interfaces [1], this value is about 75% of the simulated value for a perfect stack, *i.e.* with perfect interfaces without roughness or interdiffusion. The quality of the interfaces was checked from the physico-chemical point of view by x-ray emission spectroscopy (XES) and nuclear magnetic resonance (NMR) spectroscopy. Both were in agreement and did not indicate any noticeable interdiffusion between the Co and Mg layers. However, given the magnetic character of Co, it is interesting to probe the internal structure of the system through a magneto-optic effect, particularly to check if the interfaces are also abrupt from the magnetic point of view.

Verna *et al.* [5] have demonstrated the possibility to determine the magnetic profiles at manganite surfaces with a monolayer resolution by using resonant magnetic reflectivity. Using the same kind of methodology and taking advantage of the magnetic character of the Co layers, we expect to be able to obtain a precise determination of the magnetization profile within the Co/Mg multilayer. This is made by measuring the reflectivity as a function of the incident photon energy around the Co L_3 and L_2 edges, *i.e.* around 778 and 793 eV, and as a function of the Bragg angle, *i.e.* the grazing angle, for a given circularly polarized radiation and for two opposite magnetizations of the sample.

2. Experiment simulation

Simulations of the reflectivity are performed using the Pythonic Programming for Multilayer (PPM) [6]. The simulations are carried out for a previously studied multilayer: [Co (2.55 nm) / Mg (5.45 nm)]$_{\times 30}$ and considering a Si substrate [1]. The reflectivity simulations in θ-2θ mode are performed for two opposite helicities of the incoming radiation and at two photon energies: 777.6 eV, corresponding to the maximum of x-ray absorption coefficient; 776.2 eV, off the absorption maximum, but where the dichroism is still appreciable. Results for the two energies and for θ up to 26° are presented in figure 1. In this case we report the average of the reflectivities obtained for right and left circularly polarized light. Bragg peaks are much more pronounced at 776.2 eV. This is because out of the absorption maximum, low absorption case, the radiation penetrates deep within the stack and all the 30 bilayers can diffract the incident radiation. Also fringes due to the finite thickness of the overall multilayer are visible. At 777.6 eV, the light is absorbed in the first layers of the stack and then the Bragg peaks are much more smoothed.

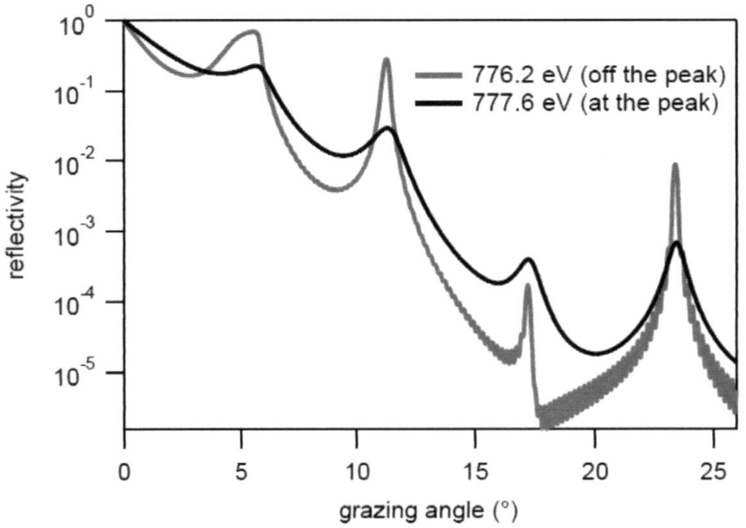

Figure 1. Reflectivity simulations of a Co/Mg multilayer at two photon energies corresponding to the maximum of the absorption peak off this peak.

In order to test the sensitivity of the technique on the magnetization profile of the Co layers, we consider three different magnetization profiles of the Co layer as indicated in figure 2. Profile A corresponds to a fully magnetized Co layer. Profile B considers two 0.25 nm thick dead layers at both Co/Mg interfaces, while for profile C the thickness of the dead layers are 0.5 nm thick. The same profile is taken for all the 30 Co layers of the multilayer. The reflectivity simulations are performed as a function of the photon energy around the Co L_3 absorption edge and at the fixed grazing angles corresponding to the first three Bragg peaks (n=1, θ=5.72°; n=2 θ=11.50°; n=3 θ=17.39°). The dichroic signal, which is the difference between the reflectivity obtained with the two opposite helicities of the incident radiation, is reported in figure 3 for the three different magnetization profiles of figure 2.

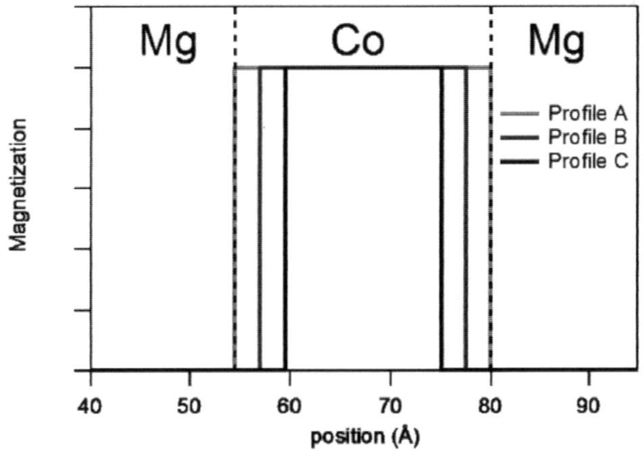

Figure 2. Magnetization profiles (in arbitrary units) of the Co layers considered to simulate the magnetic resonant reflectivity of Co/Mg multilayer.

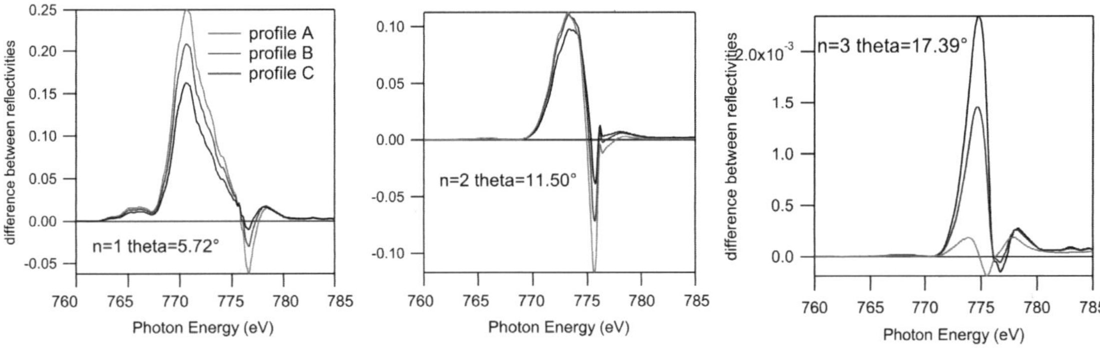

Figure 3. Dichroic signal simulated for Co/Mg multilayer around the Co L_3 edge, for three different grazing angles corresponding to the first three Bragg angles and for the three magnetization profiles of the Co layers described in the figure 2.

For the Bragg peaks diffracted at the first and second diffraction orders, the dichroic signal is not very sensitive to the magnetization profile. In particular n=2 peak remains almost unchanged, while n=1 peak changes the amplitude but not its sign. A large difference is instead evident for the third Bragg peak. With no dead layer the dichroic signal is weak because in this case, the thickness of the Co layer (2.55 nm) is close to a third of the period thickness (8.0 nm). The existence of a dead layer induces a departure to this ratio and leads to a 10 times increase of dichroic signal.

3. Sample design and characterization

Considering the simulations, we design a Co/Mg multilayer which should not present 3rd diffraction order, in fact 3n diffraction orders. Thus if any magnetic dead layer exists at the Co layer interfaces it should give a relatively intense dichroic signal at the 3rd order diffraction peak. So the aimed values of the Co and Mg thicknesses are 2.55 and 5.10 nm respectively, giving a period of 7.65 nm. A multilayer following this design having 30 Co/Mg bilayers was prepared by magnetron sputtering. The deposition was made on a silicon substrate. A 3 nm thick Cr adhesion layer was first deposited onto the substrate. Finally a 3.5 nm thick B_4C capping layer was deposited on top of the stack to prevent its oxidation. After its, preparation the stack was characterized by x-ray reflectivity at 0.154 nm. The reflectivity curve is shown on figure 4. It is observed that the 3rd Bragg peak is almost extinguished and that the 6th Bragg peak is suppressed. This demonstrates that the actual layer thicknesses are closed to the designed ones. This is confirmed by the fit of the reflectivity curve. The thicknesses of the Co and Mg layers deduced from the fit, 2.60 and 5.16 nm, very slightly larger than the designed ones, and the interfacial roughnesses are around 0.5 nm. These values represent the *rms* thickness variation of the Co and Mg layers.

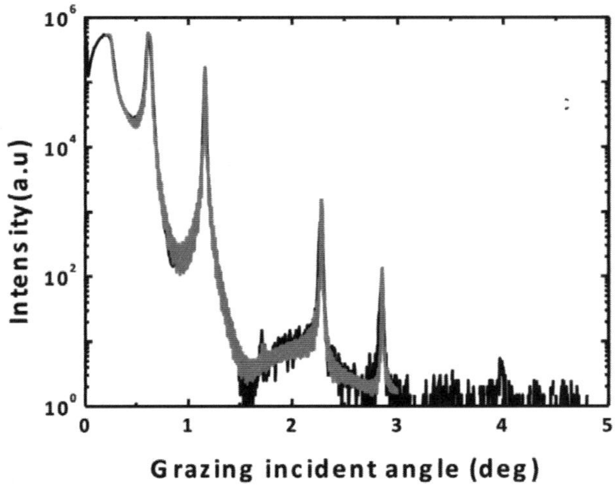

Figure 4. Measured (black line) and fitted (red line) hard x-ray reflectivity curves of the prepared Co/Mg multilayer.

Then MOKE (magneto-optic Kerr effect) experiment is made at the wavelength of the He-Ne laser, *i.e.* 632.8 nm. The sample is a 10 mm x 5 mm rectangle. Figure 5 shows the result when the surface of the sample is perpendicular to the applied magnetic field. The signal on the ordinates, is proportional to the rotation of the light polarization. Its relative variation during a cycle is proportional to the magnitude of the magnetization variation after a linear background subtraction. On the range of the hysteresis loop, this relative intensity variation is 1%. The width of the hysteresis loop allows the determination of a coercitive field of about 600 Oe.

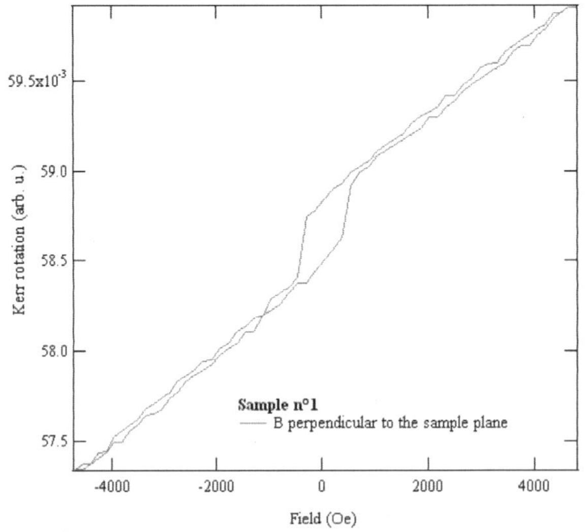

Figure 5. MOKE hysteresis curve of the Co/Mg multilayer. The magnetic field is oriented in the direction perpendicular to the sample plane.

Figure 6 (left) shows the MOKE hysteresis when the surface of the sample is parallel to the applied magnetic field and in our setup the sample is fixed horizontally. The coercitive field is about 90 Oe. The hysteresis curve is not centered around 0 G because of the remanent field of the electro-magnet (about 40 G). In this case the variation of the magnetization is much larger with respect to the perpendicular configuration: on the range of the hysteresis loop, the relative intensity variation is 6%. So the easy magnetization axis is parallel to the sample surface. Figure 6 (right) shows the result when the surface of the sample is parallel to the applied magnetic field and in our setup the sample is fixed vertically, *i.e.* rotated by 90° with respect to the previous case. The curve is almost the same as in the previous case, that is to say there exists no easy magnetization axis for one of the two sides of the rectangle.

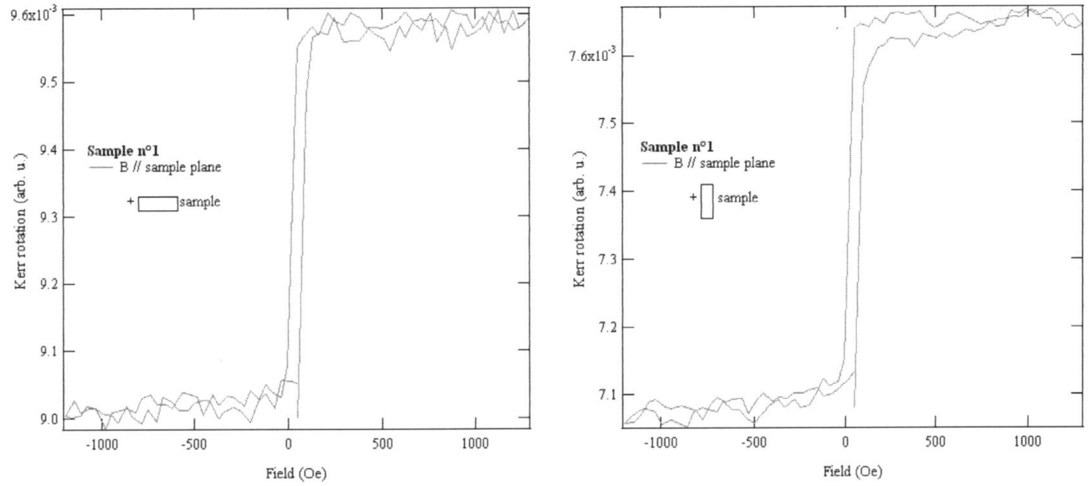

Figure 6. MOKE hysteresis curve of the Co/Mg multilayer. The magnetic field is oriented in the direction parallel to the sample plane. Left: horizontal sample; right: vertical sample.

4. Results and discussion

The magnetic resonant reflectivity measurements are made by scanning the incident photon energy around the Co L_3 and L_2 edges and the grazing angle around the five first Bragg peaks. The incident is almost circularly polarized. The measured degree of circular polarization is 0.8. The magnetic field is applied parallel to the surface of the sample, *i.e.* along the easy magnetization axis as determined by MOKE. The intensity of the applied magnetic field is about 400 Oe, sufficient for a complete saturation of the magnetization. To obtain the dichroic signal, the difference between reflectivities measured with opposite magnetizations is calculated. This kind of experiment is equivalent to the simulated one: use two opposite helicites of the radiation and one magnetization of the sample. Experiments were performed at the BEAR beamline of the Elettra synchrotron centre.

First, we show in figure 7, the reflectivity curves and the dichroic signal around the Co $L_{2,3}$ edge at an angle corresponding to the first Bragg peak. The reflectivity curves are presented without applied magnetic field and in the two cases where the magnetic field is applied in opposite directions. The differences between these three cases are small but can be observed because of the anomaly close to the absorption edge and of the high brightness of the synchrotron radiation. The dichroic signal presents a peak followed by a dip at the L_3 edge and the same at the L_2 edge but with different amplitudes of the peak and dip. The same kind of signal is obtained when making dichroism of the x-ray absorption.

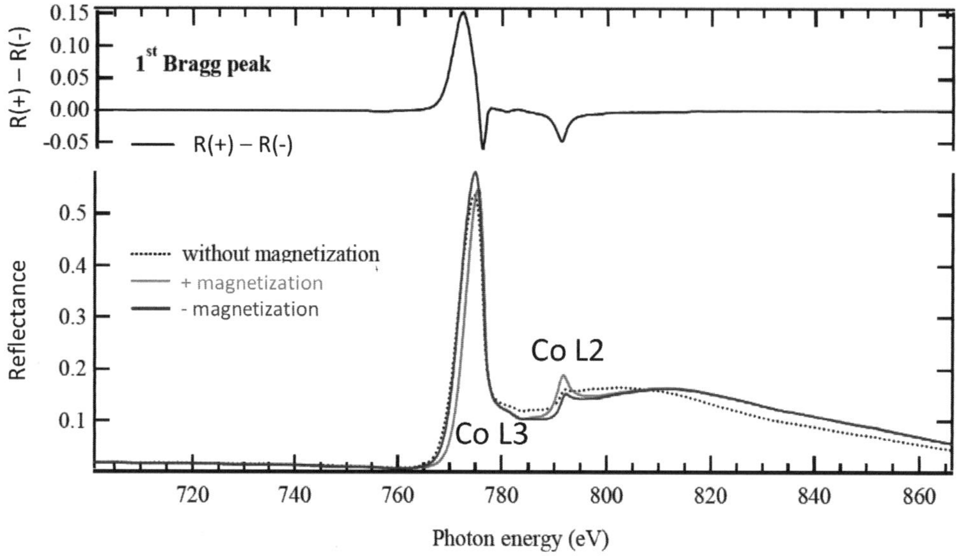

Figure 7. Reflectivity curves, with and without applied magnetic field, and dichroic signal around the Co L edges and at an angle corresponding to the first Bragg peak.

Then, we present on figure 8 the 3D maps obtained by varying both the photon energy (only around Co L_3 absorption edge) and the grazing angle. For sake of clarity we only present the result obtained around the 1st and 3rd Bragg peaks. Around the 1st Bragg peak the dichroic signal is rather intense and the structures such as the ones observed in figure 7 are recognized. For the 3rd Bragg peak, the curves obtained at an angle corresponding to the 3rd Bragg peak being presented in figure 9, no negative value is present in the dichroic signal whose intensity is quite weak and no dip is present in the spectra as a function of energy for every fixed grazing angle. This indicates the presence of a dead layer at the Co interfaces with Mg. The thickness of this dead layer cannot be comparable with the thickness of a Co layer, because this should imply a breaking of the 2:1 thickness ratio and a strong increase of the 3rd Bragg peak with respect to the non-magnetized case, that is not observed. The amplitude of the signal around the 3rd Bragg peak, peak, $1.6\ 10^{-3}$, is compatible with the simulation with a 0.25 nm thick dead layers, and for which only a very small dip is calculated. Also the maximum of the dichroism in the 1st Bragg peak is close to the one simulated with the 0.25 nm thick dead layer.

Thus considering the closeness of the simulated and deposited stacks, we assume the existence of a 0.25 nm thick magnetic dead layer. This thickness represents about the thickness of one Co atomic plane (0.21 nm). Thus the dead layer only results from the Co interfacial atoms in contact with the Mg atoms of the neighbouring layers. These Co atoms have already been observed by NMR spectroscopy [1], giving a NMR signal in the ratio of number of interfacial Co atoms with respect to number of bulk Co atoms. Then, we conclude that the interfaces within the Mg/Co stacks are abrupt from both magnetic and chemical point of views.

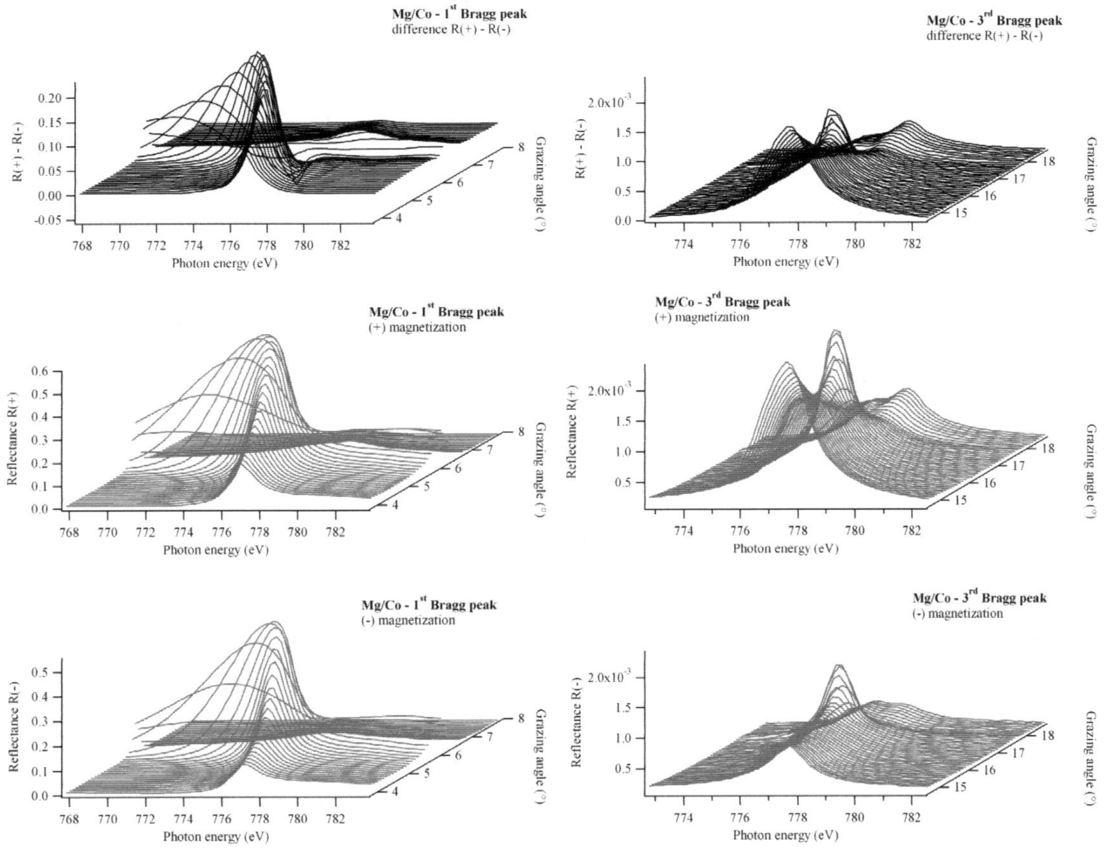

Figure 8. Resonant magnetic reflectivity maps of the Co/Mg multilayer, obtained for incident photon energies around the Co L_3 edge and grazing angles around the 1st (left) and 3rd (right) Bragg peaks. The reflectivity curves are in red and for the two opposite directions of the applied magnetic field, R(+) and R(-). The dichroic signal is plotted in black.

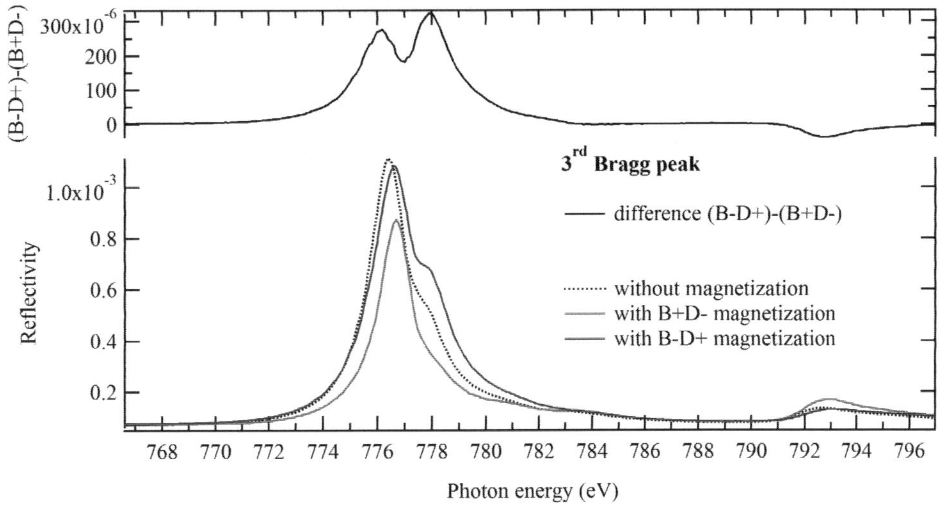

Figure 9. Reflectivity curves, with and without applied magnetic field, and dichroic signal around the Co L edges and at an angle corresponding to the thrid Bragg peak.

Acknowledgments
Funding from ANR through project COBMUL N°ANR-10-INTB-902-01 is acknowledged. A.V. acknowledges support from the FVG Regional Project SPINOX funded by Legge Regionale 26/2005 and the decreto 2007/LAVFOR/1461. Authors are thankful to George Kourousias and Alessio Curri, Scientific Computing team of the IT group at Elettra synchrotron, for their help in the installation and use of PPM program.

References
[1] Le Guen K, Hu M-H, André J-M, Jonnard P, Zhou S K, Li H C, Zhu J T, Wang Z S et Meny C 2010 Development and Interfacial Characterization of Co/Mg Periodic Multilayers for the EUV Range *The Journal of Physical Chemistry C* **114** 6484–90

[2] Zhu J, Zhou S, Li H, Huang Q, Wang Z, Le Guen K, Hu M-H, André J-M et Jonnard P 2010 Comparison of Mg-based multilayers for solar He II radiation at 30.4 nm wavelength *Appl. Opt.* **49** 3922–5

[3] Le Guen K, Hu M-H, André J-M, Jonnard P, Zhou S K, Li H C, Zhu J T, Wang Z S, Mahne N, Giglia A et Nannarone S 2010 Introduction of Zr in nanometric periodic Mg/Co multilayers *Appl. Phys. A* **102** 69–77

[4] Zhu J, Zhou S, Li H, Wang Z, Jonnard P, Le Guen K, Hu M-H, André J-M, Zhou H et Huo T 2011 Thermal stability of Mg/Co multilayer with B4C, Mo or Zr diffusion barrier layers *Opt. Express* **19** 21849–54

[5] Verna A, Davidson B A, Szeto Y, Petrov A Y, Mirone A, Giglia A, Mahne N et Nannarone S mai Measuring magnetic profiles at manganite surfaces with monolayer resolution *Journal of Magnetism and Magnetic Materials* **322** 1212–6

[6] Spiga D, Mirone A, Ferrero C, Cotroneo V, Pareschi G, del Rio M et Vernani D 2004 Fitting X-ray multilayer reflectivities by means of the PPM code *Advances in Computational Methods for X-Ray and Neutron Optics* vol 5536, éd M DelRio (Bellingham: Spie-Int Soc Optical Engineering) p 71–80; https://forge.epn-campus.eu/projects/ppm/wiki

A nanoporous composite by dealloying rapidly quenched $Co_{75}Pd_{20}Si_5$

Yingmin Wang[1], Wei Zhang[1], Akihisa Inoue[2]

[1] Institute for Materials Research, Tohoku University, Katahira 2-1-1, Sendai, Japan
[2] WPI, Tohoku University, Katahira 2-1-1, Sendai, Japan

Corresponding author E-mail: wym@imr.tohoku.ac.jp

Abstract. In this work we report the fabrication of a nanoporous composite by electrochemical dealloying a rapidly quenched $Co_{75}Pd_{20}Si_5$ alloy in a 0.5M H_2SO_4 aqueous solution. The melt-spun alloy exhibited a two-phase microstructure with the precipitation of an unknown secondary phase on the FCC solid solution matrix. Upon dealloying, a nanoporous composite was formed, which consisted of nanoligaments of two different length scales, namely, ~ 5-10 nm and ~ 100-150 nm. The sample surface was free of cracks and the as-dealloyed ribbons exhibited good mechanical integrity. The secondary phase was found to be more corrosion-resistant than the FCC solid solution phase under the same electrochemical conditions, and after dealloying the framework structure of the phase preserved to reinforce the dealloyed microstructure.

1. Introduction

Nanoporous (NP) metals have unique three dimensional structures comprised of interconnected nano-ligaments and nano-pores. Electrochemical dealloying has proven very effective to turn alloys into NP metals [1, 2]. In a dealloying process, the less noble constituents are selectively dissolved out from the alloy, leaving behind a spongy residue consisting mainly of noble metals. Extensive solid solution alloys are the widely used precursor materials for dealloying, from which NP structures with pore size ranging from a few to hundreds of nanometers are readily made. The unique nanostructures deliver intriguing material properties, for instances the anomalous surface plasmon resonance [3] and the surface-enhanced Raman scattering (SERS) effect [4, 5]. The unusual optical properties would give rise to their potential applications as sensors in molecular diagnostics. Some NP metals are ideal electrode or substrate materials for electrocatalysis [6-8]. In a nutshell, the exotic nanostructured materials are currently believed to have a wide range of applications in catalysis, absorption, sensing, and nanomechanics [9].

For device applications, the mechanical integrity and the robustness of NP metals are of critical significance. Usually, a dealloyable extensive alloy is enriched in the less noble metal, and volume contraction of the grains may occur when the majority constituent is dissolving out from the alloy. As a consequence, massive cracking is commonly observed in the as-dealloyed samples [4, 5]. Recently, there appeared several accounts on the making of crack-free bulk NP samples. Senior et al. [10] have successfully obtained NP-Au bulk samples of good mechanical integrity by optimizing the precursor binary alloy composition and the dealloying condition control. Sun et al. used multi-step dealloying strategy to make robust bulk NP-Au samples [11]. In the two cases, the precursor materials are single phase Au-Ag extensive solid solution alloys. Jin et al [12], on the other hand, have realized the fabrication of macroscopically strong NP-Pt by dealloying several two-phase Pt-Ag alloys, in which one solid solution phase appears as a secondary dendrite phase embedded on the matrix solid solution phase. NP structures with good mechanical integrity have attracted much attention in the recent development activities of functional NP metal devices.

Hakamada and Mabuchi [13, 14] have made a comprehensive study on the different dealloying behaviors of Pd-transition metals (Fe, Co, or Ni) solid solution alloys. $Co_{80}Pd_{20}$ was found to be a favourable precursor to obtain NP-Pd with ultrafine nanopores of a few nanometers. In our previous study on the dealloyed microstructure of rapidly quenched $Co_{80}Pd_{20}$ alloy, massive surface cracking were found to occur inevitably in the samples. In the present work, we first attempt to make a two-phase microstructure by alloying $Co_{80}Pd_{20}$ with a glass-forming element Si. It is expected that a small amount of Si addition into the alloy would result in the formation of a more corrosion-resistant secondary phase than the matric FCC solid solution phase, which plays the role of reinforcement in the dealloyed sample for improved mechanical integrity.

2. Experimental

Alloy ingots of $Co_{80}Pd_{20}$ and $Co_{75}Pd_{20}Si_5$ were prepared by arc melting the mixtures of appropriate amount of Si (purity: 99.999 wt.%), Co (99.99 wt.%), and Pd (99.5 wt.%) under a vacuum level of about 5×10^{-4} Pa. Using the master alloys, ribbon samples with a cross section of ~ 0.02 × 1.0 mm were made by single-roller melt-spinning. A HZ-3000 potentiostatic was employed to drive the dealloying of the two kinds of alloys with a platinum counter electrode and a saturated calomel electrode (SCE) reference electrode at room temperature. The dealloying experiments were conducted in a 0.5M H_2SO_4 aqueous solution. As-dealloyed samples were first soaked in deionized water to remove the electrolyte from the internal pores, and then allowed to air dry for subsequent observation. Phase identification for as-quenched and as-dealloyed samples was conducted via a standard θ-2θ scan on a Rigaku RINT-ultima IIIsp diffractometer (Cu-K_α, $\lambda = 0.15406$ nm). The scan rate was set to be 0.01°/s in the 2θ range of 20-100°. The morphology and composition of alloy phases before and after dealloying were analyzed with a field emission scanning electron microscope (Hitachi S-4800 SEM) equipped with an Oxford energy dispersive X-ray spectroscope (EDX).

3. Results and discussion

3.1 Component phases and microstructure

The X-ray diffraction patterns of rapidly quenched $Co_{80}Pd_{20}$ and $Co_{75}Pd_{20}Si_5$ alloys are shown in Fig.1. The $Co_{80}Pd_{20}$ alloy exhibits a single phase FCC structure. In the $Co_{75}Pd_{20}Si_5$ alloy, a secondary phase is found to coexist with FCC solid solution phase, while the diffraction peaks cannot be thoroughly indexed by any of the known Pd-Si and Co-Si phases. The typical microstructure of $Co_{75}Pd_{20}Si_5$ alloy is shown by the inserted SEM image, in which the unknown phase appears to be grey white. The extensive EDX analysis gave a composition range of ~ $Co_{39\text{-}42}Pd_{43\text{-}47}Si_{14}$ for the unknown phase. In comparison with the nominal alloy composition, the phase is enriched in Si and Pd, and is depleted in Co. The alloying effect of Si on the phase selection of $Co_{75}Pd_{20}Si_5$ may be understood in terms of the enthalpy of mixing between constituent elements. The enthalpies of mixing between Pd, Co and Si at 1:1 atomic ratio are ΔH_{Pd1Si1}= -38 kJ/mol and ΔH_{Co1Si1} = -21 kJ/mol, respectively [15]. The strong negative enthalpies of mixing characteristics between Si and Pd and Co metals would favor the formation of intermetallic phase.

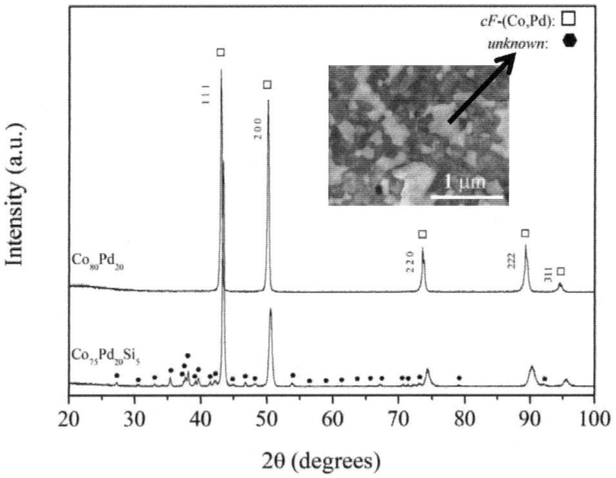

Figure 1. XRD patterns of rapidly quenched $Co_{80}Pd_{20}$ and $Co_{75}Pd_{20}Si_5$ alloys. The inserted SEM image shows the unknown precipitated phase in the FCC solid solution matrix.

3.2 Electrochemical measurement on dealloying behavior

Fig. 2 (a) shows the anodic current-potential behaviors of the two rapidly quenched alloys in 0.5M H_2SO_4. Both alloys exhibit a passivation state at low driving potentials, followed by active dissolution at higher potentials. This electrochemical feature resembles those of dealloyable Au-Ag alloys. It is seen that active dissolution occurs at a relatively lower potential in $Co_{75}Pd_{20}Si_5$ at a sweeping rate of $1x10^{-3}$V/s. To ensure dealloyability, the current-time behaviors of the single and two-phase alloys were studied at constant driving potentials. The typical results are shown in Fig. 2 (b). At potentials of 0.35V ($Co_{80}Pd_{20}$) and 0.15V ($Co_{75}Pd_{20}Si_5$), the current densities are found to increase first rapidly to a peak value and then progressively cease to be zero. All the electrochemical measurements suggest that the two alloys are dealloyable.

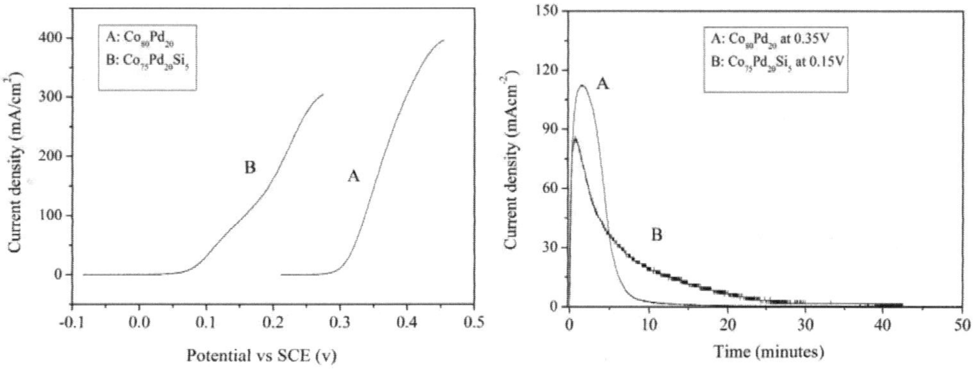

Figure 2. (a) Linear sweep scans at $1x10^{-3}$V/s and (b) Current density-time behaviors of $Co_{80}Pd_{20}$ and $Co_{75}Pd_{20}Si_5$ alloys in 0.5M H_2SO_4, respectively.

3.3 Comparison of dealloyed microstructures

Fig. 3 (a) and (b) compare the surface morphologies of as-dealloyed $Co_{80}Pd_{20}$ and $Co_{75}Pd_{20}Si_5$ ribbon samples. After dealloying, the $Co_{80}Pd_{20}$ sample exhibited massive cracking over the entire ribbon surface, while the as-dealloyed $Co_{75}Pd_{20}Si_5$ was free of cracks. It should be mentioned that after dealloying the $Co_{80}Pd_{20}$ residue exhibited large volume shrinking and the sample became curled, the $Co_{75}Pd_{20}Si_5$ ribbon, however, almost preserved its original shape and dimensions. The latter sample can be broken by bending. The cross-section views of their dealloyed microstructures are shown in Fig. 3 (c) and (d), respectively. Intergranular cracking in $Co_{80}Pd_{20}$ is evidently seen, which can mainly be attributed to the internal tensile stress caused by dissolving-out of Co. In the dealloyed $Co_{75}Pd_{20}Si_5$ bulk, nanoligaments of two different length scales are formed. The fine nanoligaments are of ~ 5-10 nm size and the coarse ones ~ 100-150 nm. The length scale of the fine nanoligaments is comparable to that obtained in $Co_{80}Pd_{20}$. Apart from this, no significant interphase boundary cracking was found in the nanoporous composite.

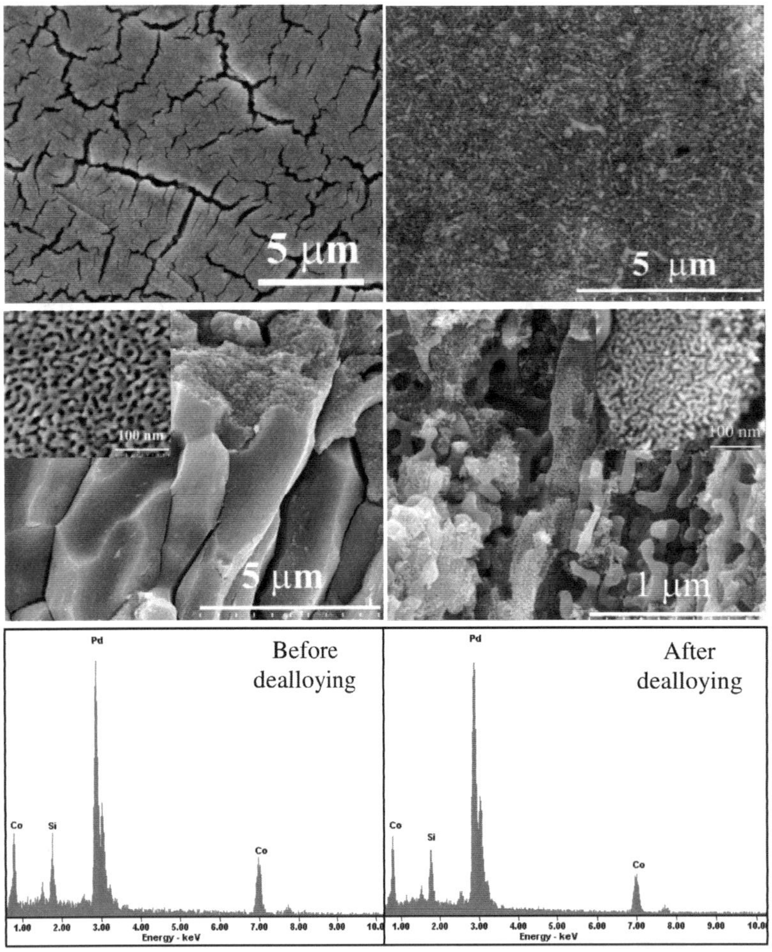

Figure 3. SEM images showing (a) significant surface cracking in dealloyed $Co_{80}Pd_{20}$ ribbon samples (b) Crack-free surface morphology of dealloyed $Co_{75}Pd_{20}Si_5$ (c) Intergranular cracking in dealloyed $Co_{80}Pd_{20}$ (d) the formation of a two-length scale hierarchical nanocomposite in dealloyed $Co_{75}Pd_{20}Si_5$. The inserted SEM images in (c) and (d) compare the finest pore sizes obtained in the two dealloyed microstructures. (e) and (f) EDX spectra taken from the secondary phase in the starting and dealloyed $Co_{75}Pd_{20}Si_5$ samples, respectively. The average compositions of the phase in the two samples are very close.

Fig. 3 (e) and (f) present the EDX spectra taken from the secondary phases existing respectively in the precursor and the dealloyed ribbon samples. The measured composition changes, especially the Si contents in the phase are negligibly small. The compositional evidence indicates that the secondary phase is more corrosion-resistant than the FCC solid solution matrix in the 0.5M H_2SO_4 aqueous solution under a given potential. It is very likely that selective dissolution mainly occurred in the matrix phase. That is, dealloying was locally confined to the regions separated by the secondary phase framework. The interphase boundary structure between the FCC matrix and the secondary phase appears to be a critical issue to understand the good mechanical integrity of the nanoporous composite.

The crystallographic orientational relationship between the two phases needs to be identified for further discussion. Notwithstanding, the fact that there exists no significant interphase boundary cracking in the sample has delivered an interesting implication. Since the interphase boundary is robust against dealloying, it is possible to realize the continuous network formation of the secondary phase along the FCC matrix grain boundaries by optimizing the precursor alloy composition and the melt-spinning conditions such as quenching temperature and cooling rate. Our recent study on the fabrication of novel nanoporous composite in Co-Pd-Si system is along this line, and the results will be reported elsewhere in the future.

4. Conclusion

By substitutional alloying of 5 at.% Si for $Co_{80}Pd_{20}$, a two-phase $Co_{75}Pd_{20}Si_5$ alloy was made by means of melt-spinning. Electrochemical measurement indicated that the $Co_{75}Pd_{20}Si_5$ alloy was dealloyable in 0.5M H_2SO_4. After dealloying, nanoligaments of two different length scales were formed in the alloy. In comparison with the dealloyed microstructure of $Co_{80}Pd_{20}$, the nanoporous composite are free of surface cracking and is of good mechanical integrity. Co-Pd-Si is expected to establish a promising system for the fabrication of various nanoporous structures of good mechanical robustness.

Acknowledgements

The authors acknowledge the financial support of the project of Advanced Materials Development and Integration of Novel Structured Metallic and Inorganic Materials of the Ministry of Education, Science, Sports, and Culture of Japan.

References

[1]Newman R C, Corcoran S G, Erlebacher J, Aziz M J and Sieradzki K 1999 MRS Bull. 24 24
[2]Ding Y and Chen M W 2009 MRS Bull. 34 569
[3]Yu F, Ahl S, Caminade A M, Majoral J P, Knoll W and Erlaebacher J 2006 Anal. Chem. 78 7346
[4]Dixon M C, Daniel T A, Hieda M, Smilgies D M, Chan M H and Allara D L 2007 Langmuir 23 2414
[5]Chen L Y, Yu J S, Fujita T and Chen M W, 2009 Adv. Funct. Mater. 19 1221
[6]Ding Y, Chen M W and Erlebacher J, 2004 J. Am. Chem. Soc. 125 6876
[7]Zhang J, Liu P, Ma H and Ding Y, 2007 J. Phys. Chem. C 111 10382
[8]Zeis R, Mathur A, Fritz G, Lee J and Erlebacher J 2007 J. Power Source 165 65
[9]Erlebacher J and Seshadri R 2009 MRS Bull. 34 561
[10]Senior A and Newman R C 2006 Nanotechnology 17 2311
[11]Sun Y and Balk T J, 2008 Scripta Mater. 58 727
[12]Jin H J, Kramer D, Ivanisenko Y and Weissmüler J 2007 Adv. Eng. Mater. 9 849
[13]Hakamada M and Mabuchi M 2009 J. Alloys Compounds 479 326
[14]Hakamada M, Tajima K, Yoshimura K, Chino Y and Mabuchi M 2010 J. Alloys Compounds 494 309
[15]de Boer F R, Boom R, Mattens W C M, Miedema A R and Niessen A K 1988 Cohesion in Metals—Transition metal alloys (Amsterdam: Elsevier)

Thermal Behavior of Hydrogen Retained in Amorphous-titanium-hydride Thin Films

B. Tsuchiya[1], T. Ishigami[2], S. Nagata[3] and T. Shikama[3]
[1] Department of General Education, Faculty of Science and Technology, Meijo University, 1-501, Shiogamaguchi, Tempaku-ku, Nagoya 468-8502, Japan
[2] The Wakasawan Energy Research Centre, 64-52-1, Nagatani, Tsuruga, Fukui 914-0192, Japan
[3] Institute for Materials Research, Tohoku University, 2-1-1, Katahira, Aoba-ku, Sendai 980-8577, Japan

E-mail: btsuchiya@meijo-u.ac.jp

Abstract. Changes in hydrogen distribution near the surface of non-stoichiometric titanium hydride (TiH_x; x=0.49) thin films possessing an amorphous structure and approximately 170 and 400 nm thick, prepared by a magnetron ion-sputtering device, were investigated by combining an elastic recoil detection (ERD) technique with Rutherford backscattering spectrometry (RBS) one after isochronal annealing at different temperatures (300-500 K) for 10 min. The concentration of hydrogen atoms retained in TiH_x thin films completely vanished at much lower temperatures of 350-400 K, as compared with the decomposition temperature (approximately 473 K) of δ-phase titanium hydride (δ-$TiH_{1.92}$). In addition, the thickness of TiH_x thin films also reduced with an increase in temperature. The rapid decrease of the hydrogen concentration in TiH_x thin films below 400 K is caused by thermal desorption of hydrogen atoms, detrapped from the amorphous structure and associated, in addition, with Ti evaporation.

1. Introduction

The development of a hydrogen-air fuel cell comprising proton-conducting oxides that act as a solid electrolyte and hydrogen storage metals that serve as a hydrogen source has been proposed [1]. This closed-type fuel cell, which does not require hydrogen gas to be supplied from outside, is promising as a compact electric power-generating system that would help to satisfy the growing demand for zero CO_2-emitting energy sources. The rate of hydrogen transport to the fuel cell is one of the most important rate-determinant processes in the closed-type fuel cell. Titanium hydrides (TiH_x) (or titanium deuterides; TiD_x) can potentially be used as hydrogen sources in hydrogen-air fuel cells, since (1) the hydrogen migration rate; that is the thermal diffusion, is considerably high and (2) the decomposition temperature corresponding to the potential energy that develops due to the interface between the topmost surface, made of titanium oxides, and the bulk of the hydrides is approximately 473 K [2]. However, it is necessary to develop new hydrogen sources that can migrate hydrogen to the electrolyte at further lower temperature below 473 K, in order to make it actualized as the fuel cell in the future. In addition, much high energy is consumed for the preparation of TiH_x using a Sieverts

[1] To whom any correspondence should be addressed.

apparatus in hydrogenation conditions wherein a H_2 gas pressure of 1.0×10^5 Pa and a heating temperature of 1073 K [3]. Therefore, we have to discover the new hydrogen storage materials that can prepare at low cost. In this study, TiH_x (x=0.49) thin film specimens having an amorphous structure were prepared using a magnetron ion-sputtering device with flowing H_2 as well as Ar gases at room temperature. The depth profiles of hydrogen atoms near the surface of TiH_x thin films by isochronal annealing have been measured using an elastic recoil detection (ERD) technique which is a form of non-destructive ion beam analysis method with MeV high energy, to investigate the decomposition temperature of hydrogen in TiH_x thin films; that is indicative of the emission of hydrogen atoms from the thin films.

2. Experiments

Two kinds of non-stoichiometric Titanium-hydride (TiH_x) thin films with different thicknesses were deposited onto one face of perfluorosulfonic acid (PFSA) membranes (diameter: 8.0 mm; thickness: 117 μm) which are ones of typical protonic conductive polymers using the magnetron ion-sputtering device. The PFSA membranes have been selected as the substrate in this study to fabricate the closed-type fuel cell utilizing amorphous-TiH_x thin films as hydrogen source. For the deposited conditions, the substrate temperature of the membrane was approximately 296 K, and the gas pressure of argon (Ar) containing hydrogen (H_2), introduced into the magnetron ion-sputtering device, was under a pressure of approximately 0.8 Pa. The gas flow rates of Ar and H_2 were 22 and 40 sccm, respectively. The dimension of the prepared TiH_x thin films was approximately 7.5 mm diameter and 170 nm thick for deposition time of 30 min and 400 nm thick for 90 min.

It can be observed in scanning electron microscope (SEM) images that there are much cracks of approximately 100 μm in the surface of the prepared thin films, as shown in Fig. 1(a) and (b). In addition, much domains of approximately 100 nm are aligned at random directions, as shown in Fig. 1(c) and (d). The X-ray diffraction (XRD) pattern of the TiH_x thin film has a broad peak only; that is indicative of amorphous structure, while that of $TiH_{1.92}$ bulk sample has some keen peaks, related to a face centered-cubic (δ-phase) structure, as shown in Fig. 2.

Figure 1. SEM (SEI: secondary electron image) micrographs of as-prepared amorphous-TiH_x thin film by the magnetron ion-sputtering device.

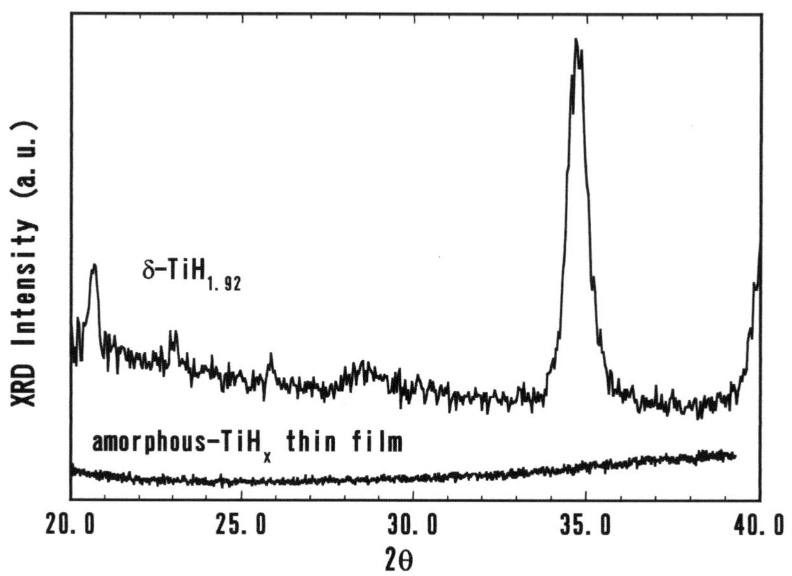

Figure 2. XRD patterns of as-prepared amorphous-TiH_x thin film and δ-$TiH_{1.92}$.

The composition as well as thickness of two kinds of TiH_x thin films were analyzed by means of high-energy ERD, combined with RBS, using 2.8 MeV He^{2+} ion probe beams from a Tandem accelerator, installed at Institute for Materials Research, Tohoku University. The TiH_x thin films were irradiated with He^{2+} ions at an incident angle of 75° to the surface normal of the films. The forward-recoiled hydrogen ions (H^+) by elastic collisions with He^{2+} ions were detected at a scattering angle of 30° to the incident He^{2+} ion direction by a solid state detection (SSD) for ERD, mounted an absorber comprising 12 μm-thick Al film. Simultaneously, the back-scattered He^+ ions by elastic collisions with Ti as well as constituent atoms for the PFSA membranes such as S, O, and C atoms were detected at an angle of 170° to the incident He^{2+} ion direction by a SSD for RBS. From RBS and ERD spectra for as-prepared TiH_x thin films, as shown in Fig. 3(a) and (b), and Fig. 4(a) and (b), the composition, x; that is indicative of ratio of H to Ti atoms, and the thickness for two kinds of TiH_x thin films were determined to be x=0.49 and approximately 400 nm for Fig. 3(a) and (b), and x=0.49 and approximately 170 nm for Fig. 4(a) and (b), respectively.

Subsequently, in order to determine the decomposition temperature of H from amorphous-TiH_x thin films, the depth distribution of hydrogen atoms, retained in the whole TiH_x thin films after isochronal annealing for 10 min at the various temperatures of 296 to 493 K using a tungsten-ceramics heater, was in situ measured in vacuum at room temperature by simultaneous high-energy ERD and RBS measurements. The temperature of the thin films was measured with an alumel-chromel thermocouple in contact with the surface.

3. Results and discussion

Figures 3 and 4 show typical (a) RBS spectra of back-scattered He^{2+} ions and (b) ERD spectra of recoiled H^+ ions from amorphous-TiH_x thin films (first; x=0.49, 400 nm thick, second; x=0.49, 170 nm thick, respectively) after annealing at each temperature of 300-500 K for 10 min, measured using 2.8 MeV He^{2+} ion probe beams. The horizontal axis (Channel Number) corresponds to several energies of the back-scattered He^{2+} ions for RBS spectra and recoiled H^+ ions for ERD spectra, respectively, and represent the distance from the surface. The vertical axis (Counts) corresponds to the hydrogen

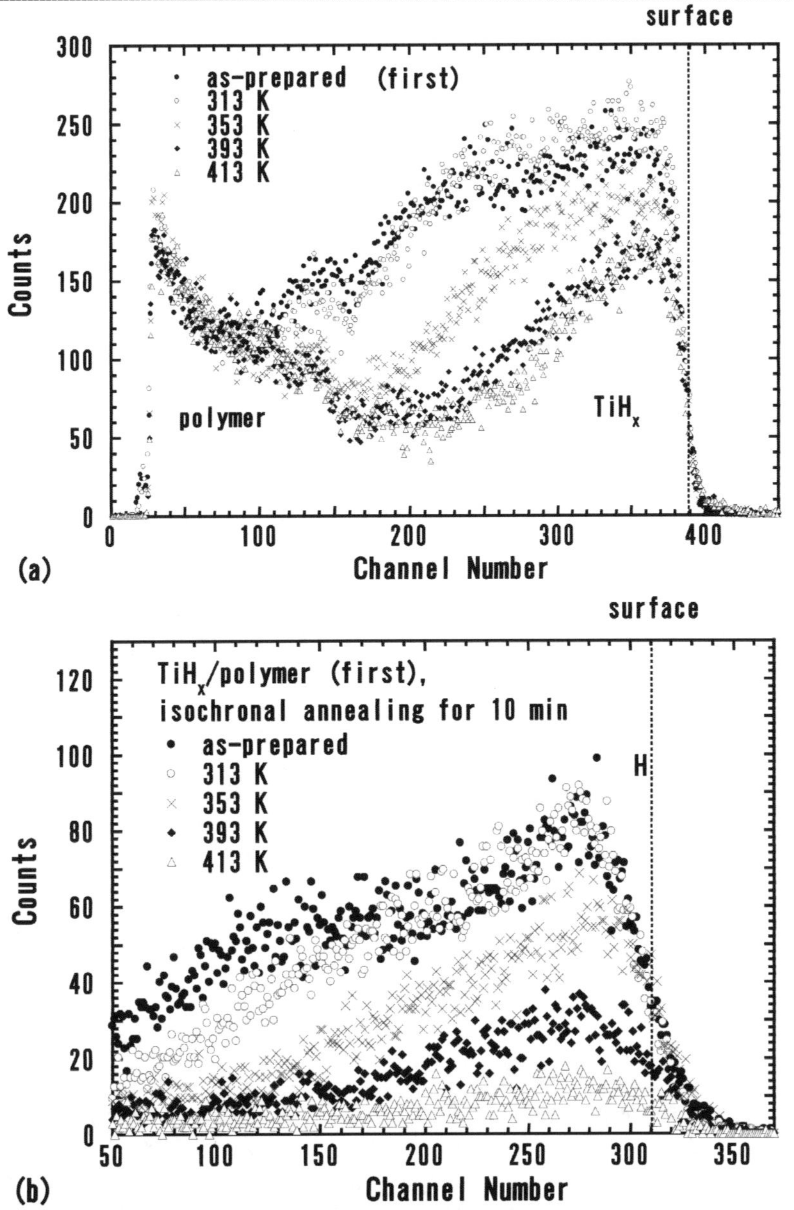

Figure 3. Typical (a) RBS spectra of back-scattered He^+ ions and (b) ERD spectra of forward-scattered H^+ ions from one TiH_x thin film (first; x=0.49, 400 nm thick) before and after isochronal annealing at 313, 353, 393, and 413 K for 10 min.

concentration for ERD spectra and the concentrations of constituent atoms for the PFSA membranes such as S, O, and C atoms under 200 channel numbers in RBS spectra, and Ti atom for TiH_x thin film from the near 400 channel numbers. For ERD spectra, the counts at the near 310 channel numbers corresponds to hydrogen concentration on the top-most surface of TiH_x thin film. The hydrogen distribution shows that the hydrogen concentration is almost constant in each thin film with approximately 170 and 400 nm thick. The counts below approximately 180 channel numbers in Fig. 3(b) and approximately 250 channel numbers in Fig. 4(b), which correspond to water (H_2O), H, and

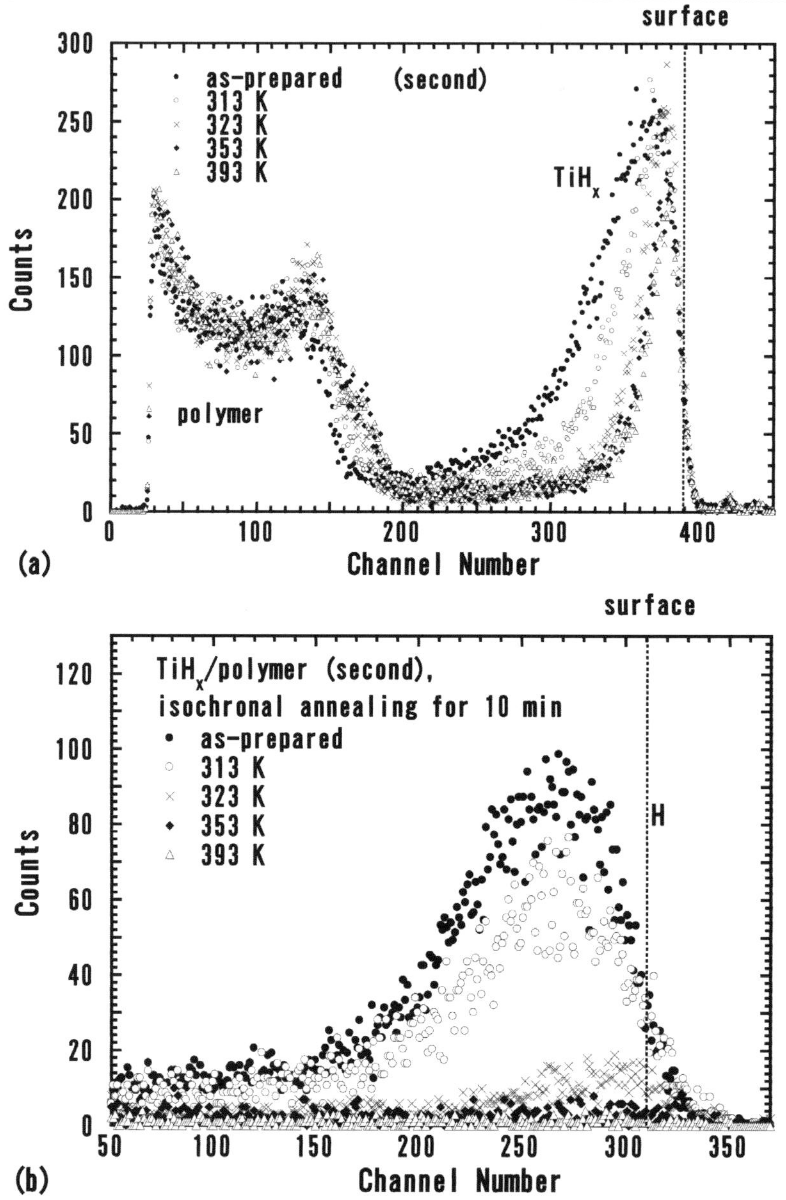

Figure 4. Typical (a) RBS spectra of back-scattered He^+ ions and (b) ERD spectra of forward-scattered H^+ ions from another TiH_x thin film (second; x=0.49, 170 nm thick) before and after isochronal annealing at 313, 323, 353, and 393 K for 10 min.

H_2 at interface between TiH_x thin film and PFSA membrane and H_2O and SO_3H in the PFSA membrane, reduce at 313 K only. The hydrogen concentration at each depth in TiH_x thin film irregularly decreases at 313-413 K. It seems that hydrogen atoms thermally diffuse from interface between the thin film and membrane to the surface of the thin film, and be desorbed with hydrogen molecular formation due to recombination between activated hydrogen atoms on the top-most surface. For RBS spectra, on the other hand, the total integral counts and full width at half maximum of broad peaks between 200-400 channel numbers corresponding to the composition of Ti and the thickness of

TiH_x thin films also reduce with in an increase in temperature. In addition, for thermal desorption spectroscopy (TDS) measurements, the releases of Ti and TiH_x have been also observed at low temperatures of 300-450 K and high temperatures above 450 K. Therefore, the rapid decrease of H in TiH_x thin films may be associated with the evaporation of Ti from top-most surface.

Figure 5 shows the hydrogen releases from δ-phase $TiH_{1.92}$, which has a face-centered cubic structure, and amorphous-TiH_x thin films (first and second) by isochronal annealing at different temperatures of 300-513 K for 10 min. The total integral counts of the ERD spectra corresponding to the whole hydrogen concentration in δ-$TiH_{1.92}$ and TiH_x thin films after annealing at various temperatures were normalized by those in as-prepared δ-$TiH_{1.92}$ and TiH_x thin films. In Fig. 5 it is important to note that the hydrogen concentrations in TiH_x thin films linearly decrease with an increase in temperature, and the decomposition temperatures of 300-400 K for TiH_x thin films are significantly lower than decomposition temperature of 480-500 K for δ-$TiH_{1.92}$. In addition, the hydrogen release rate becomes higher for the thinner TiH_x thin film. These result shows that the amorphous-TiH_x thin films have more potential for use as new hydrogen storage materials that can produce hydrogen gases at low cost. For the hydrogen release process, the hydrogen atoms are released by Ti evaporation. The interstitial hydrogen atoms in the amorphous structure may be probably bonded much weaker with Ti atoms, as compared with hydrogen atoms occupied at tetrahedral site for hydrides. The other surface analyses by means of X-ray photoelectron spectroscopy (XPS) and transmission electron microscope (TEM) and computer simulation are now being further studied for the annealed TiD_x as well as TiH_x, in order to clarify the binding energies between Ti-Ti atoms and between Ti-H atoms in amorphous structure and the mechanism on the thermal behavior of hydrogen atoms in more detail.

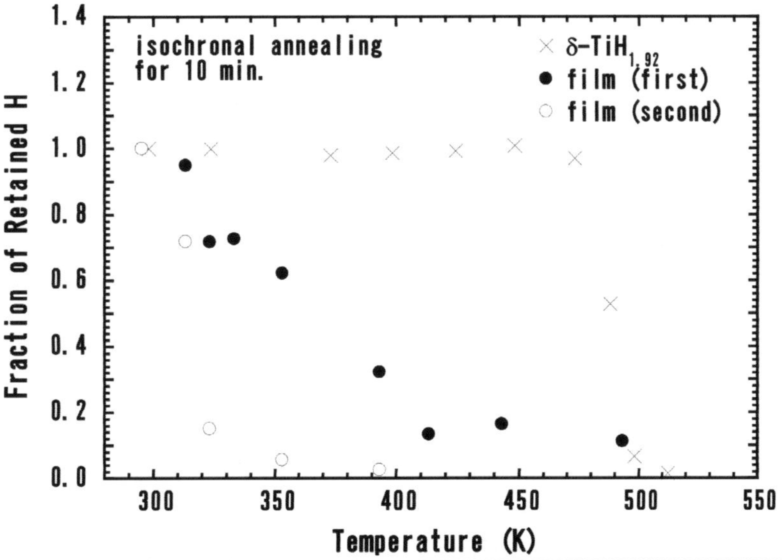

Figure 5. Changes in hydrogen concentrations in δ-phase $TiH_{1.92}$ and amorphous-TiH_x thin films (first and second) after isochronal annealing at different temperatures for 10 min.

4. Summary

The non-stoichiometric TiH_x (x=0.49) thin films with amorphous structure and approximately 170 and 400 nm thick were deposited onto one face of the PFSA membranes at substrate temperature of approximately 296 K and with flowing Ar and H_2 gases by the magnetron ion-sputtering device. The simultaneous high-energy RBS and ERD measurements revealed changes in the hydrogen distribution in TiHx thin films, heated at various temperatures 300 to 500 K for 10 min. It was found from ERD spectra that the decomposition temperatures of 300-400 K for amorphous-TiH_x thin film with approximately 400 nm thick and 300-350 K for amorphous-TiH_x thin film with approximately 170 nm thick are much lower than decomposition temperature of 480-500 K for δ-$TiH_{1.92}$. The experimental result shows the probability of hydrogen production from the hydrogen source materials by expensing with much lower energy. In addition, it was also found from RBS spectra that the thickness of amorphous-TiH_x thin films reduced with an increase in temperature up to approximately 393 K. Therefore, the release of the hydrogen atoms may not only be probably caused due to detrapping from amorphous structure thermally, but also Ti evaporation.

References

[1] C. Folonari C, Lemmi G, Manfredi F and Rolle A 1984 *J. Less-Common Met.* **74** 371
[2] Yamaguchi S, Yugami H and Ikeda S 2002 *J. Alloys Comp.* **330-332** 911
[3] Tsuchiya B, Nagata S, Toh K and Shikama T 2006 *Nucl. Instr. and Meth. in Phys. Res.* B **249** 370
[4] Tsuchiya B, Teshigawara M, Nagata S, Konashi K, Yasuda R, Nishino Y, Nakagawa T and Yamawaki M 2002 *Nucl. Instr. and Meth. in Phys. Res.* B **190** 699

Silicon Oxide Thin Films Prepared by Vacuum Evaporation and Sputtering Using Silicon Monoxide

M Sasaki and T Ehara*
Department of Basic Science, Faculty of Science and Engineering, Ishinomaki Senshu University, Ishinomaki 986-8580 Japan

E-mail: ehara@isenshu-u.ac.jp

Abstract. Dependency of structural properties of silicon oxide thin films prepared using silicon monoxide source material on preparation method has been studied. The structural properties of the silicon oxide films prepared by three methods, vacuum evaporation using e-beam, rf-sputtering, and vacuum evaporation employing resistive heating are studied by IR absorption and Raman spectra. Due to the results, it is found that the structure of the films depends significantly on preparation methods. The films prepared by e-beam evaporation exhibits similar structural properties of SiO_2. The film by rf-sputtering shows Raman spectrum corresponds to significantly random amorphous silicon oxide structure. The film by evaporation employing resistive heating exhibits Raman spectrum that has been explained by silicon-oxygen ring structure.

1. Introduction

Silicon monoxide (SiO) is another stoichiometric silicon oxide material. Because of its vapor pressure characteristics [1], SiO has been used as a source of silicon oxide thin films by vacuum evaporation at relatively high pressure (low vacuum) [2, 3]. Therefore, silicon oxide (SiO_x) thin films formed using a SiO source have been used in multilayer optoelectronic devices [4, 5]. However, there are noteworthy studies indicating that commercially available SiO materials are a composite of two or more compounds [6, 7]. This fact makes structural characteristics study of the SiO_x thin films prepared using SiO problematic. Although some groups have reported the structural properties of the SiO_x films from SiO, their results are sometimes not well consistent each other in details [8-11]. It can be thought that it is because the structural properties of the SiO_x films prepared using SiO material is significantly depend on the preparation conditions.
In the present work, we study the dependency of structural properties of SiO_x thin films prepared from SiO source material on the preparation methods. The preparation methods we used are e-beam vacuum evaporation, rf-sputtering, and vacuum evaporation employing resistive heating.

2. Experimental

The e-beam evaporation of SiO was done by an e-gun voltage of 4 kV for 1 h using SiO source in Ta cell. The films were deposited at pressure of 4 x 10^{-3} Pa and film thickness was 0.25 μm. In the e-beam evaporation, thicker films could not be prepared because of bumping of the source. SiO source for the preparation was grains (M 0.85 – 2.0 mm) with purity of 99.99 % (Kojundo Chemical Laboratory). The SiO_x film growth using rf-sputtering were carried out with a rf-power of 200 W, Ar gas pressure of 2.66 Pa, flow rate of 5sccm, for 1 h. Base pressure of the deposition was 2 x 10^{-3} Pa, and the films with thickness of 2.0 μm were deposited using a sputtering machine (Shimadzu HSR551). The films deposited using SiO target with diameter of 100 mm, purity of 99.99 % (Kojundo Chemical Laboratory). The evaporation by resistive heating was carried out using Ta boat employing an anti-bumping structure for 0.5 h. SiO source was same as that in e-beam evaporation. Base pressure of the preparation was 2 x 10^{-3} Pa and film thickness was 1.5 μm. All the films were deposited on crystalline silicon (100) with thickness of 525 μm. These samples were thermally annealed by the vacuum infrared heating system IVF175 (Thermo Rico) at temperatures of 400, 600, and 800°C, over 30 min at a pressure of 2 x 10^{-3} Pa. The structure of deposited thin films was studied by IR absorption and Raman scattering spectra. Raman spectra of the films were measured using an Ar^+ ion laser of wavelength of 514.5 nm and a power of 50 mW (power density of 2.5 W/cm).

3. Results and Discussion

Fig. 1 shows the IR absorption spectra of SiO_x thin films prepared by e-beam evaporation. The absorption peak with the highest intensity attributed to Si-O-Si stretching vibration [12] is observed at 1050 to 1065 cm^{-1} with a shoulder at higher wavenumber side. The peak position has shifted to higher wavenumber with increase in annealing temperature. Other weak peak is observed around 880 cm^{-1} before and after annealing at 400 and 600°C. At annealing temperature of 600°C, the intensity of the 880 cm^{-1} peak decreased and other peak at 806 cm^{-1} appeared. At annealing temperature of 800°C, the peak around 880 cm^{-1} disappeared and intensity of the peak at 806cm^{-1} increased. The peak around 880 cm^{-1} has been assigned to silicon ring with a non-bridging oxygen hole center (NBOHC) or a silicon ring configurations isolated by oxygen atoms [13]. The peak at 806 cm^{-1} has been assigned to Si-O-Si bending motion [12]. The IR absorption spectrum obtained after the annealing at 800°C is similar to that of thermally oxidized SiO_2 layer on crystalline Si [10]. In addition, all the films prepared by e-beam evaporation were transparent in visible light wavelength. SiO source material used in the present work is not transparent, almost black grains. In addition, it has been reported that the SiO and $SiO_{1.5}$ thin films are not transparent in visible light wavelength [14]. Therefore, it is clear that the films prepared by e-beam evaporation are not SiO films. We think the films are SiO_2 films, formed by the mechanism as follows. At first, the solid phase reaction, $2SiO \rightarrow SiO_2 + Si$ [15-17], occurred by e-beam irradiation to SiO source, then, only the SiO_2 is transferred to the substrate and formed the film.

Fig. 2 shows the IR spectra of SiO_x thin films prepared by rf-sputtering. The peak around 880 cm^{-1} is observed before and after annealing at 400 and 600°C. After annealing at 600°C, intensity of the peak around 880 cm^{-1} decreased and other peak of the Si-O-Si bending appeared at 812 cm^{-1}. At annealing temperature of 800°C, the peak around 880 cm^{-1} disappeared and intensity of peak at 812 cm^{-1} increased. The temperature dependence of these two peaks is similar to that in the e-beam evaporated films. However, the peak of the Si-O-Si stretching is significantly different from the e-beam evaporated films. The peak was observed at 1057 to 1091 cm^{-1} and shifted to higher wavenumber with increase in annealing temperature. The peak shift is more drastically than the e-beam evaporated films. In addition, the peak width is significantly greater and the shape of peak was more asymmetric than the e-beam evaporated films. The results indicate that the structures of the sputtered films are different from the e-beam evaporated films.

Fig. 3 shows IR spectra of SiO_x thin films prepared by resistive heating evaporation. The peak of Si-O-Si stretching vibration is observed at 1000 cm^{-1}. The peak position is lower than films prepared by other methods. The peak position shifted to high wavenumber with annealing temperature to 1080 cm^{-1}. The peak shift is more significant than other preparation method. In addition, peak width is greater than the films prepared by the sputtering or the e-beam evaporation. The peak of Si-O-Si bending motion is also observed at 763 to 800 cm^{-1} with peak shift to higher wavenumber depending on annealing temperature. The peak width of the Si-O-Si bending peaks is wider than that in the films prepared by the other methods. These results indicate that the structure of the SiO_x films by resistive heating evaporation is different from either e-beam evaporated films or sputtered films.

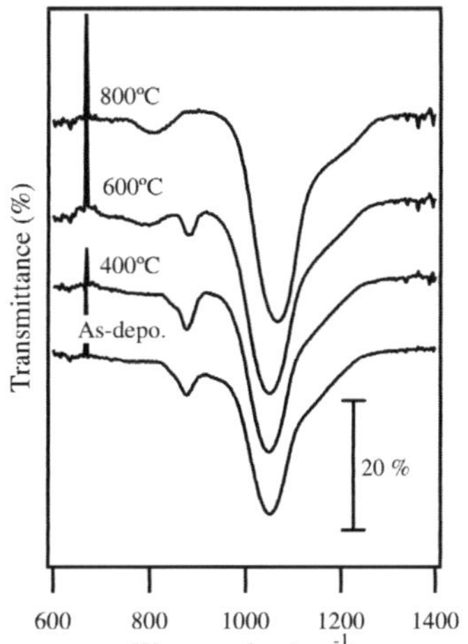

Figure 1. IR absorption spectra of SiO_x thin films using SiO material prepared by e-beam evaporation, before and after annealing at 400 to 800°C.

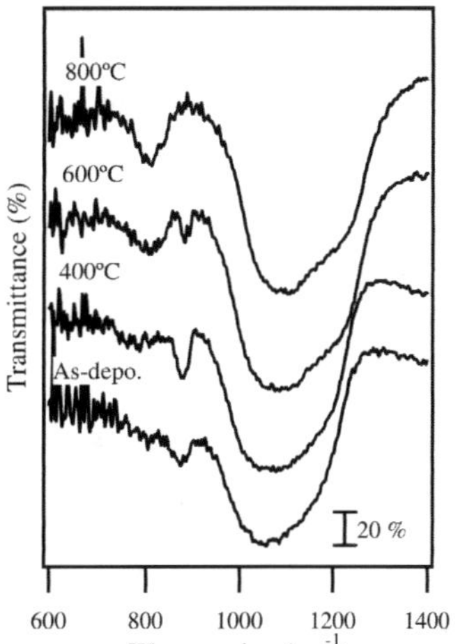

Figure 2. IR absorption spectra of SiO_x thin films using SiO material prepared by rf-sputtering, before and after annealing at 400 to 800°C.

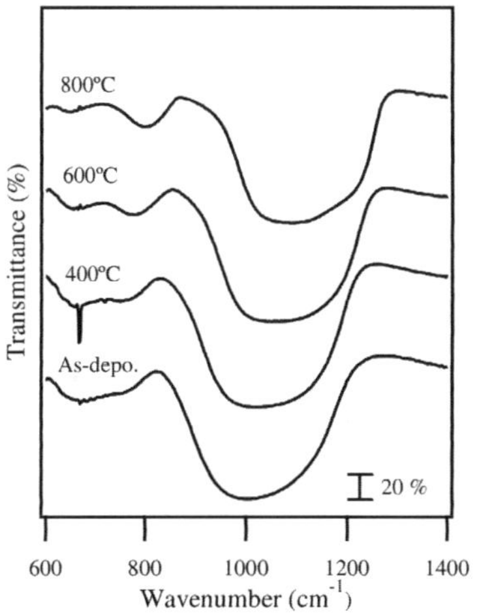

Figure 3. IR absorption spectra of SiO_x thin films using SiO material prepared by evaporation employing resistive heating, before and after annealing at 400 to 800°C.

Fig. 4 shows Raman spectra of the as-deposited SiO_x films prepared by sputtering and resistive heating evaporation using SiO source material. The sharp signal at 520 cm^{-1} is due to the signal from crystalline Si substrate. As shown in the figure, the film prepared by sputtering exhibits almost no peaks at Raman shift of 200 to 700 cm^{-1}. The result indicates that the structure of the film by sputtering is so random that film does not have enough volume of any specific structure to exhibit Raman scattering. In contrast, the film prepared by resistive heating evaporation exhibits a broad signal from 350 to 630 cm^{-1} that has a peak at 566 cm^{-1}. The peak position is different from that of amorphous silicon that has a broad peak at 480 cm^{-1} [18], microcrystalline silicon that has an asymmetric and sharp peak around at 520 cm^{-1} [19], and amorphous SiO_2 films at 450 cm^{-1} [20]. It can be thought that this spectrum is due to the distinctive signal of the SiO. Similar spectrum has been

Journal of Physics: Conference Series **417** (2013) 012028 doi:10.1088/1742-6596/417/1/012028

reported in the previous work, in which samples were prepared by similar evaporation condition [11]. In their work, the structure corresponding to the Raman peak is explained by silicon-oxygen 3-fold (six atoms) or 4-fold (eight atoms) ring structures [11,21].

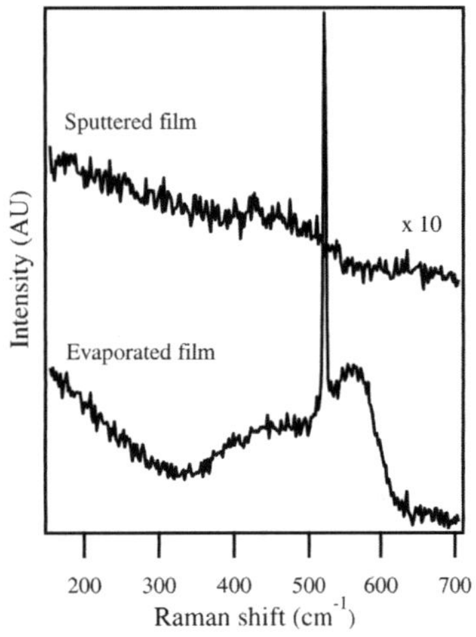

Figure 4. Raman scattering spectra of SiO_x thin films prepared by rf-sputtering and evaporation employing resistive heating.

We think the difference of film structure depending on the preparation methods is caused by the difference in two factors during the evaporation. One is efficiency of the solid phase reaction of SiO to SiO_2 and Si [15-17], caused by energy provided from e-beam irradiation, sputtering, or heating by Ta boat. The other is separation of the ingredients in the SiO source material during the transfer process of source to the substrate. As previously mentioned, the SiO source is composite of more than two ingredients [6,7]. Thus, there are more than two ingredients in the SiO source even before the solid phase reaction occurs. As all the ingredients have different characteristics, some of them transfer to the substrate by evaporation and the others remain in the evaporation source, then, the separation of ingredients occurs. In the case of the e-beam evaporation, it can be thought that the SiO source is changed to Si and SiO_2 with high efficiency, then, only the SiO_2 is transferred to the substrate by the e-beam evaporation. In the case of sputtering, it can be thought that all the ingredients are transferred to substrate by higher energy of sputtering, whether the solid phase reaction occurs or not. Therefore, the films prepared by sputtering are formed from various kinds of ingredients. This means that the sputtered films include wider distribution of the structures than in the films by other preparation methods. Thus, the structure of the films become significantly random, then, the films shows no Raman scattering signals. In the case of the evaporation employing resistive heating, it can be thought that almost only the ingredient that is the precursor of the silicon-oxygen ring structure is evaporated, and is separated from other ingredients. Monticone and co-workers have reported dependency of the Raman spectra on the evaporation pressure in the case of resistive heating evaporation. The intensity of the Raman spectra which is similar to that in the present work decreased with decreasing in evaporation pressure 10^{-2} to 10^{-5} Pa [11]. This result is consistent with the assumption that there are plural ingredients in the SiO source material and separated during the evaporation. At the higher pressure (low vacuum, 10^{-3} Pa), only the precursor of the Si-O ring structure can be transferred to the substrate. In contrast, at lower pressure (10^{-5} Pa), all the ingredients are evaporated and the structure of the films become enough random to show only weak signal in the Raman spectra as well as in the case of sputtered films in the present work. We think the precursor of the Si-O ring structure is the SiO, and it can be transferred to substrate only in the case of resistive heating evaporation because the low efficiency of solid phase reaction of SiO to SiO_2 and Si by Ta boat heating.

Conclusion

SiO_x thin films are prepared by 3 methods, e-beam evaporation, rf-sputtering, and evaporation with resistive heating from SiO source material. The films displayed significant difference in the structure each other. The e-beam evaporated films become transparent SiO_2 films. The sputtered film shows no peaks in Raman spectrum, means significantly random structure. The film by resistive heating evaporation exhibits Raman spectrum that is explained by silicon-oxygen ring structure. The difference is thought to be caused by difference in two factors, efficiency of solid phase reaction, and the separation of the ingredients in SiO source material

Acknowledgement

The authors thank Prof. T. Maeda for his great help with the Raman spectrum measurements.

References

[1] Ferguson F T,Nuth J A 2008 *J. Chem. Eng. Data.* **53** 2824
[2] Brewer L, Edwards R K 1954 *J. Phys. Chem.* **58** 351
[3] Brewer L, Greene F J 1957 *J. Phys. Chem.* Solids **2** 286.
[4] Ohmori Y, Kurosawa Y, Tada N, Fujii A, Yoshino K 1997 *Jpn. J. Appl. Phys.* **36** L1022
[5] Han S, Grozea D, Huang C, Lu Z H, Wood R, Kim W Y 2004 *J. Appl. Phys.* **96** 709
[6] Friede B, Jansen M 1996 *J. Non-Cryst. Solids* **204** 202
[7] Schlmeister K, Mader W 2003 *J. Non-Cryst. Solids* **320** 143
[8] Nesheva D, Raptis C, Perakis A, Hofmeister H *J. Appl. Phys.* **92** 4678
[9] Rinnert H, Vergnat M, Marchal G, Burneau A *J. Appl. Phys.* **72** 3157
[10] Yi L X, Heitmann J, Scholz R and Zacharias M 2003 *J. Phys.: Condens. Matter* **15** S2887
[11] Monticone E, Rossi A M, Rajteri M, Gonnelli R S, Lacquaniti V and Amato G 2000 *Phil. Mag.* B **80** 523
[12] Pai P G, Chao S S, Takagi Y and Lucovsky G 1986 *J. Vac. Sci. Technol.* A **4** 689
[13] Hajnal Z, Deák P, Köhler Th, Kaschner R and Frauenheim Th 1998 *Solid State Commun.* **108** 93
[14] Philipp H R 1971 *J. Phys. Chem. Solids* **32** 1935
[15] Bineva I, Nesheva D, Aneva Z, Levi Z, Raptis C, Hofmeister H and Stavrev S 2003 *J. Mater. Sci.:Mater. Electron.* **14** 799
[16] Feroughi O M, Sternemann C, Sahle Ch J, Schroer M A, Sternemann H, Conrad H, Hohl A, Seidler G T, Bradley J, Fister T T, Balasubramanian M, Sakko A, Pirkkalainen K, Hämäläinen K and Tolan M 2010 *Appl. Phys. Lett.* **96** 081912
[17] Miyazaki H 2008 *Jpn. J. Appl. Phys.* **47** 3766
[18] Ishidate T, Inoue K, Tsuji K Minomura S 1982 *Solid State Commun.* **42** 197
[19] Ehara T, Ikoma T, Akiyama K, Tero-Kubota S 2000 *J. Appl. Phys.* **88** 1698
[20] Galener F L 1979 *Phys. Rev.* **B 19** 4292
[21] Yasaitis J A 1972 *J. Appl. Phys.* **43** 9995

Journal of Physics: Conference Series **417** (2013) 012029 doi:10.1088/1742-6596/417/1/012029

Ni-based amorphous alloy-coating for bipolar plate of PEM fuel cell by electrochemical plating

S Yamaura[1], S C Kim and A Inoue
Institute for Materials Research, Tohoku University,
2-1-1 Katahira, Aoba, Sendai 980-8577, Japan

E-mail: yamaura@imr.tohoku.ac.jp

Abstract. In this study, the Ni-Cr-P amorphous alloy-coated bipolar plates were produced by electro-plating on the Cu base plates with a flow field. The power generation tests of a single fuel cell with those Ni-Cr-P bipolar plates were conducted at 353 K. It was found that the single fuel cell with those Ni-Cr-P bipolar plates showed excellent I-V performance as well as that with the carbon graphite bipolar plates. It was also found that the single cell with those Ni-Cr-P bipolar plates showed better I-V performance than that with the Ni-P amorphous alloy-coated bipolar plates. Furthermore, the long-time operation test was conducted for 440 h with those Ni-Cr-P bipolar plates at the constant current density of 200 $mA\cdot cm^{-2}$. As a result, it was found that the cell voltage gradually decreased at the beginning of the measurement before 300 h and then the voltage was kept constant after 300 h.

Introduction

The bipolar plates are one of the most important components of the proton exchange membrane fuel cells (PEMFC), as they distribute reactant gases (H_2 and O_2/Air) to the electrodes and provide mechanical support and electrical connection between the adjacent cells in the fuel cell stack [1]. Conventionally, bipolar plates have been constructed from carbon graphite because of its high corrosion resistance in the fuel cell environment and high electrical conductivity [2]. However, the brittleness of carbon graphite necessitates the use of thick bipolar plates, resulting in low volumetric power density of the fuel cell stack. Machining of flow fields into carbon graphite plates is also expensive, making carbon graphite impractical for wide-scale commercial uses.

Alternative candidate for bipolar plates is metallic material [3-7]. The metallic materials are suitable to produce thinner bipolar plates since they have many advantages, such as high mechanical strength and excellent machinability to desired shapes. Thinner bipolar plates enable us to make a compact fuel cell stack, leading to enhancing the usefulness of fuel cells. For this purpose, some metallic materials such as the Nb/stainless steel clad material [3], SnO_2:F-coated stainless steel [4] and Cr-N-coated stainless steel [5, 6], were used tentatively to produce thin bipolar plates and the power generation properties of a single fuel cell with those bipolar plates were studied.

Amorphous alloys are known for their advantageous characteristics over crystalline alloys. In particular, they exhibit very high mechanical strength [8] and excellent corrosion resistance [9]. We have tried to develop an amorphous/glassy alloy bipolar plate. In our previous works, the glassy alloy

[1] To whom any correspondence should be addressed.

bipolar plates were produced by hot-pressing the melt-spun glassy alloy sheets [10, 11]. Then the Ni-Cr-P-B glassy alloy-coated bipolar plates were produced by high velocity oxy-fuel (HVOF) spray-coating on Al bipolar plates and the power generation properties of a single fuel cell with those bipolar plates were also examined [12]. As the result, it was found that the single fuel cell with those glassy alloy-coated bipolar plates showed excellent I-V performance. The HVOF spray-coating is suitable for mass-production. However, it needs a special equipment and facility. Then we have adopted electrochemical plating to produce amorphous alloy surface film since it does not need special equipments in a laboratory.

A large number of metal-metalloid amorphous alloys such as Ni-P [13-15], Ni-B [16], Ni-Cr-P [17], Fe-Ni-P [18], Fe-Cr-Ni-P [19] have been developed up to date. For example, electro-less deposited Ni-P amorphous alloy has been widely used in commercial scale. Furthermore, the Ni-Cr-P amorphous alloy was synthesized by electrodeposition and its corrosion resistance was previously studied [17]. In this Ni-Cr-P alloy, P and Cr were added to Ni for amorphization and passivation, respectively. As a preliminary work, we produced the Ni-P amorphous alloy-coated bipolar plates by electro-less plating and examined the I-V performance of a single fuel cell with those bipolar plates [20]. As a result, it was found that the single fuel cell with those Ni-P bipolar plates did not show good I-V performance maybe because the alloy does not contain corrosion resistant elements such as Cr.

In this study, we produced the Ni-Cr-P amorphous alloy-coated bipolar plates by electro-plating and examined the I-V performance and long-time durability of a single fuel cell with those Ni-Cr-P amorphous alloy-coated bipolar plates.

Experimental procedure

The Cu plates (100 mm × 100 mm × 0.3 mm) with a flow field were prepared by hot-pressing. Figure 1 shows a schematic illustration of the flow field designed in this work [12]. The flow field area is 5 cm^2. Surface of the Cu plates with a flow field were cleaned with a nitric acid solution and then with distilled water supersonically before electro-plating.

The electro-plating bath is shown in Fig. 2. The working electrode was the Cu plate and the counter electrode was a Pt net. The electro-plating was carried out at the constant current density of 400 $mA \cdot cm^{-2}$, at the bath temperature of 303-318 K for 24 h. The Cu plate was covered with a polymer film and only the flow field area was exposed to the chemicals. The chemicals for the Ni-Cr-P plating bath were mainly composed of chromium chloride, nickel chloride, citric acid, sodium hypophosphate, formic acid and so on. All the chemicals used in this study were shown in Table 1.

The amorphicity of the electro-plated Ni-Cr-P alloy samples were examined by X-ray diffractometry (XRD) with a monochromatic Cu-Kα radiation at 40 kV and 40 mA. The surface morphology and the

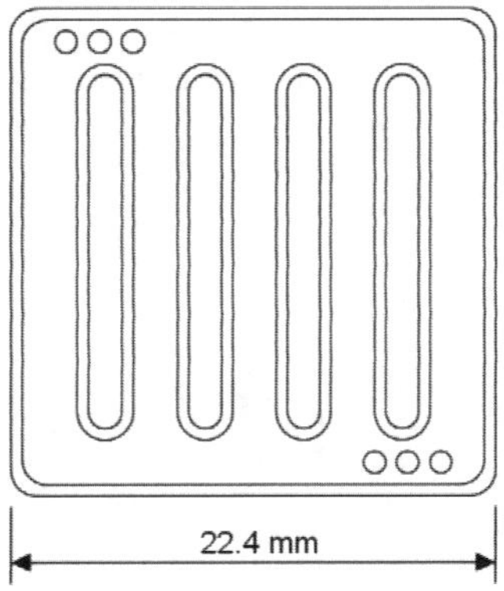

Figure 1. Schematic illustration of the flow field [12].

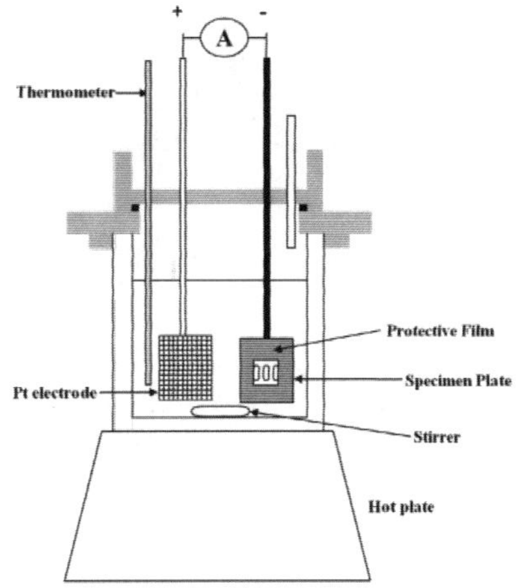

Figure 2. Schematic illustration of plating bath.

Table 1. Chemical reagents used for Ni-Cr-P plating.

Chemicals	Contents
$NiCl_2 \cdot 6H_2O$	25 g/L
$CrCl_3 \cdot 6H_2O$	120 g/L
HCOOH	50 cc/L
KBr	17.3 g/L
NH_4Cl	50 g/L
$Na_3(cit) \cdot 2H_2O$	80 g/L
H_3BO_3	35 g/L
$NaH_2PO_2 \cdot H_2O$	10 g/L
HCl	pH=1.0
NH_4OH	

chemical compositions were obtained by scanning electron microscope with energy dispersive X-ray spectroscope (SEM-EDX).

The I-V characteristics of the single cells with the bipolar plates prepared in this study were measured by an automatic fuel-cell test system (Toyo Corporation, PEMTest8900). A standard fuel cell and membrane-electrode assemblies (MEA) were purchased from Electrochem Inc. Humidification was 100%RH for H_2 and O_2 at 343 K. The cell temperature was 353 K and the gas flow rate was 0.1-0.3 $L \cdot min^{-1}$. After the I-V measurement, the long-time operation test was carried out for 440 h.

Results and discussion

1.1. Bipolar plate preparation

Figure 3 shows an outer view of a Ni-Cr-P alloy-coated bipolar plate produce by electro-plating. The Cu plate was covered with polymer film, but the flow field area was only exposed to the chemicals. So, the Ni-Cr-P alloy was deposited only around a flow field on both sides of the plate.

Figure 4 shows a typical surface morphology of the electro-plated Ni-Cr-P alloy. It was found that

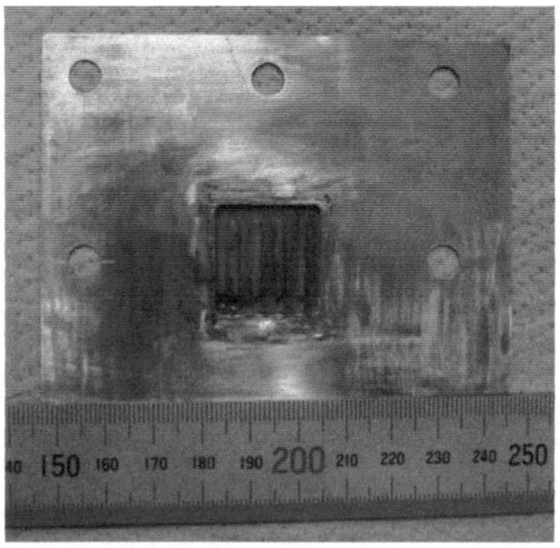

Figure 3. Ni-Cr-P amorphous alloy-coated bipolar plate.

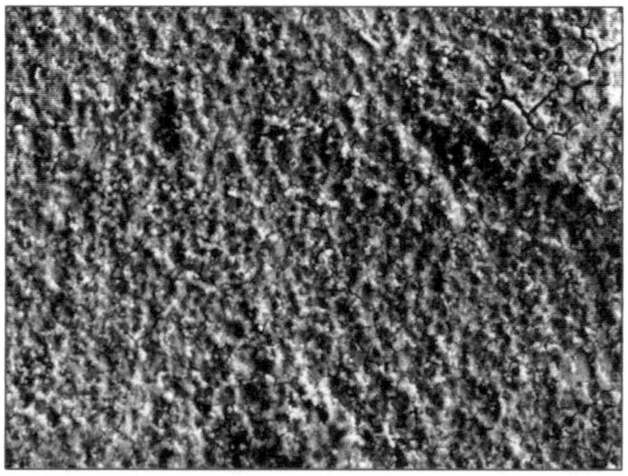

Figure 4. Surface morphology of the Ni-Cr-P alloy deposited on Cu plate by electro-plating.

the surface was not smooth and that some small cracks can be seen. Furthermore, the thickness of the layer was not constant because of a large distribution of electric field depending on the position along the flow field. The thickness of the plated layer was 5-100 μm.

Figure 5 shows the XRD pattern of the electro-plated Ni-Cr-P amorphous alloy. As clearly seen in the figure, the sample showed a broad halo peak and a distinct crystalline peak. The broad halo peak comes from a single amorphous phase and the distinct peak may come from the Cu base plate detected from the outside Cu area around the flow field.

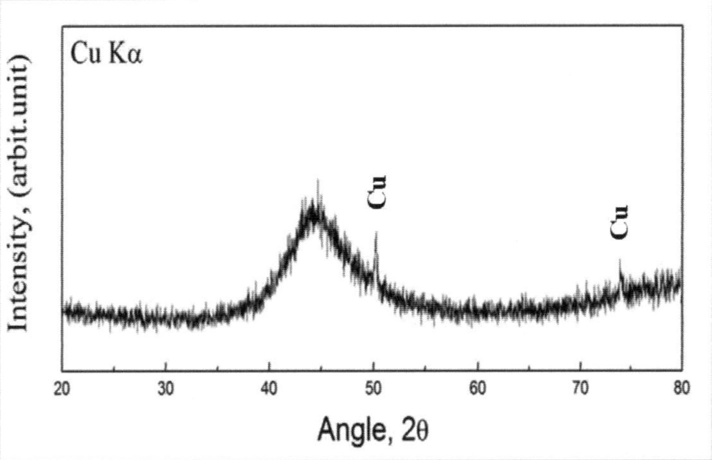

Figure 5. XRD pattern of the Ni-Cr-P alloy electro-plated on Cu plate.

Table 2 shows the detailed alloy compositions of the electro-plated Ni-Cr-P alloy. According to the analysis, the alloy produced in this study was $Ni_{52}Cr_{25}P_{23}$ in at% description. This alloy has P content of more than 20 at%, resulting that the single amorphous phase can be formed. According to the previous papers, it was reported that Ni-P amorphous single phase was formed when P content is as large as 20 at% [13]. So, it was concluded from Fig. 5 and Table 2 that the Ni-Cr-P alloy prepared by electro-plating in this study possessed a single amorphous phase.

Table 2. Alloy compositions of the electro-plated Ni-Cr-P specimen produced in this study.

Element	Wt%	At%
P	14.37	23.45
Cr	25.51	24.80
Ni	60.12	51.75

1.2. Power generation tests

Figure 6 shows the I-V curve of a single fuel cell with two Ni-Cr-P amorphous alloy-coated bipolar plates (hereafter denoted as Ni-Cr-P bipolar plates) measured at 353 K after 50 cycles repetition. The gas flow rate of both H_2 and O_2 was 0.1 $L{\cdot}min^{-1}$. The I-V curves measured with the carbon graphite and electro-less plated Ni-P bipolar plates were reported in our previous paper [20] and were also superimposed on the figure for comparative purpose. We used a new MEA for each single fuel cell assembly. When a new MEA is set in a single fuel cell, it needs activation period for a few tens of I-V cycles to show the maximum performance. So, we showed the I-V curve after 50 cycles repetition. As seen in the figure, the single fuel cell with the Ni-Cr-P bipolar plates shows excellent I-V performance as well as that with the carbon bipolar plates. Furthermore, it was found that the single cell with the

Ni-Cr-P bipolar plates showed better I-V performance than that with the Ni-P bipolar plates. This is because the former one includes high content of corrosion resistant element, Cr. According to the previous paper, the protective chromium oxide passive film can be formed on the surface of the alloy containing both Cr and P [21].

The MEA was fully activated after 50 repetitions of I-V measurement and was ready for long time operation. So, then we started the long time power generation test soon after the I-V measurement shown in Fig. 6.

Figure 7 indicates the result of the long time power generation test conducted at the current density of 200 mA·cm^{-2} at 353 K for 440 h. Final product inside a fuel cell is water. Water should be removed from fuel cell quickly so as not to degrade the cell performance especially in the long time operation. So, the gas flow rate of H_2 and O_2 was increased to 0.3 L·min^{-1} for this test in order to drain off the water produced inside the fuel cell completely from the flow field of the fuel cell. As seen in the figure, the voltage decreased gradually with time at the beginning of the measurement before 300 h and then the voltage was almost kept constant after 300 h. This may be because the protective passive film was formed on the surface of the Ni-Cr-P bipolar plates during the test, leading to fully passivated surface of the bipolar plates. Although some small microcracks were observed in the microstructure as shown in Fig. 4, they did not affect significantly the I-V performance shown in Fig. 6. The result of the long-time operation test in this work was not affected significantly by those microcracks. However, it is not clear whether the microcracks degrade the cell performance after longer operation. We need further investigations. It is also important to optimize the conditions to prepare electro-plated amorphous alloy film having smooth surface without microcracks.

At any rate, the Ni-Cr-P amorphous alloy-coated bipolar plates were produced by electro-plating. The potential of the amorphous alloy-coated bipolar plates produced by electro-plating was successfully shown in this work.

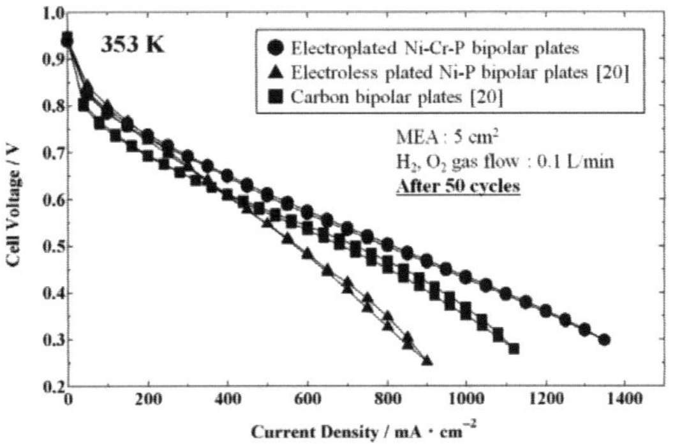

Figure 6. I-V performance of a single fuel cell with the Ni-Cr-P amorphous alloy-plated bipolar plates. Those with the carbon and Ni-P bipolar plates are also shown in the figure for comparative purpose [20].

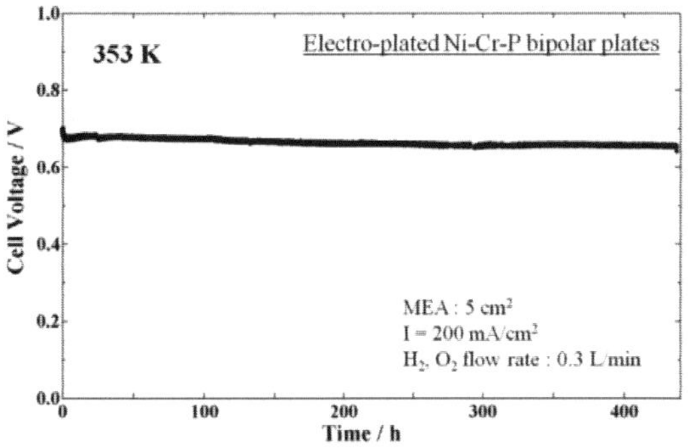

Figure 7. A result of long time power generation at constant current density of 200 mA·cm^2 for 440 h.

Summary

In this study, the Ni-Cr-P amorphous alloy-coated bipolar plates were produced by electro-plating on the Cu base plates with a flow field. The detailed analysis revealed that the alloy composition was $Ni_{52}Cr_{25}P_{23}$.

The I-V characteristics of the single fuel cell with the Ni-Cr-P amorphous alloy-coated bipolar plates produced in this study were examined at 353 K with pure H_2 and O_2 gases. As a result, it was found that the cell with the Ni-Cr-P amorphous alloy-coated bipolar plates showed excellent I-V performance as well as that with the carbon graphite bipolar plates. It was also found that the cell with the Ni-Cr-P amorphous alloy-coated bipolar plates showed better I-V performance than that with the Ni-P amorphous alloy-coated bipolar plates. This may be because the protective passive film was formed on the surface of the Ni-Cr-P amorphous alloy, leading to its high corrosion resistance. Furthermore, the long-time operation test was conducted for 440 h with the Ni-Cr-P amorphous alloy-coated bipolar plates at the constant current density of 200 $mA \cdot cm^{-2}$. The cell voltage gradually decreased at the beginning of the measurement before 300 h. However, the voltage was kept constant after 300 h. This may be because the alloy surface was fully passivated after 300 h.

Acknowledgements

This study was supported by the Ministry of Education, Culture, Sport, Science and Technology (MEXT) of Japan, Grant-in Aid for Scientific Research (C), 21560716, 2011 (S.Y.) and was partly supported by the financial aid from the Global COE program of Tohoku University (S.K.).

References

[1] de las Heras N, Roberts E P L, Langton R and Hodgson D R 2009 *Energy Environ. Sci.* **2** 206
[2] de Bruijn F A, Dam V A T and Janssen G J M 2008 *Fuel Cells* **8** 3
[3] Weil K S, Xia G, Yang Z G, Kim J Y 2007 *Int. J. Hydr. Energy* **32** 3724
[4] Wang H, Turner J A, Li X and Bhattacharya R 2007 *J. Power Sources* **171** 567
[5] Yang B, Brady M P, Wang H, Turner J A, More K L, Young D J, Tortorelli P F, Payzant E A and Walker L R 2007 *J. Power Sources* **174** 228
[6] Pozio A, Zaza F, Masci A and Silva R F 2008 *J. Power Sources* **179** 631
[7] Tawfik H, Hung Y and Mahajan D 2007 *J. Power Sources* **163** 755-767
[8] Inoue A, Shen B L, Koshiba H, Kato H and Yavari A R 2003 *Nature Mater.* **2** 661
[9] Habazaki H, Sato T, Kawashima A, Asami K and Hashimoto K 2001 *Mater. Sci. Eng. A* **304-306** 696
[10] Inoue A, Shimizu T, Yamaura S, Fujita Y, Takagi S and Kimura H M 2005 *Mater. Trans.* **46** 1706
[11] Yamaura S, Yokoyama M, Kimura H M and Inoue A 2009 *J. Phys.: Conf. Ser.* **144** 012001
[12] Kim S C, Yamaura S, Shimizu Y, Nakashima K, Igarashi T, Makino A and Inoue A 2010 *Mater. Trans.* **51** 1609
[13] Cargill III G S 1970 *J. Appl. Phys.* **41** 12
[14] Omi T, Kokunai S and Yamamoto H 1976 *Trans. JIM* **17** 370
[15] Crobu M, Scorciapino A, Elsener B and Rossi A 2008 *Electrochimica Acta* **53** 3364
[16] Watanabe T and Tanabe Y 1976 *Mater. Sci. Eng.* **23** 97
[17] Fu Y Y, Li M and Kowaka M 1993 *Hyoumen-Gijutsu* **44** 450 (in Japanese)
[18] Li M and Kowaka M 1993 *Hyoumen-Gijutsu* **44** 525 (in Japanese)
[19] Li M, Fu Y Y and Kowaka M 1993 *Hyoumen-Gijutsu* **44** 430 (in Japanese)
[20] Kim S C, Yamaura S, Makino A and Inoue A 2011 *Mater. Trans.* **52** 709
[21] Asami K, Hashimoto K, Masumoto T and Shimodaira S 1976 *Corros. Sci.* **16** 909

Journal of Physics: Conference Series **417** (2013) 012030 doi:10.1088/1742-6596/417/1/012030

Synthesis of gold nanoparticles by solution plasma sputtering in various solvents

XL Hu[1, 2, 4*], O Takai[2, 3], and N Saito[2, 3, 4*]

[1]College of Materials Science and Engineering, Nanjing University of Technology, China

[2]EcoTopia Science Research Institute, Nagoya University, Japan

[3]Department of Materials, Physics and Energy Engineering, Graduate School of Engineering, Nagoya University, Japan

[4]Green Mobility Collaborative Research Center, Nagoya University, Japan

E-mail: xiulanhu2010@gmail.com (XL Hu), hiro@eco-t.esi.nagoya-u.ac.jp (N Saito)

Abstract. Gold clusters (average particle size is 1.5 nm) and nanoparticles (average particle size is 3.5 nm) with spherical morphology were fabricated by solution plasma sputtering in solution medium. This process was carried out in an open system under atmospheric pressure. During plasma region, rapid energetic radicals' bombardment, atom vapour diffusion, plasma expansion and medium condensation resulted in the formation of gold nanoparticles from its electrode wires. As-fabricated gold particles have well crystalline even for less than 1 nm cluster. Gold nanoparticles with average size about 3.5 nm gave obviously surface plasmon resonance band at near 511 nm. While surface plasmon resonance band was disappearing for 1.5 nm-cluster formed in liquid nitrogen.

1. Introduction

Metal nanoparticles of a few nanometers in diameter exhibit size-dependent photonic and electric properties that are of interest for applications such as biosensors, catalyst, optics, and electronics [1-8]. The properties of materials change as their size approaches the nanoscale and as the percentage of atoms at the surface of a material becomes significant. However, the changes in physical properties are not always desirable. When gold is prepared as very small particles with diameters less than 10 nm and is highly dispersed on metal oxides, it turns to be a highly active catalyst for many reactions such as CO oxidation and propylene epoxidation in the gas phase. Additionally, gold nanoparticles which were protected by polyallylamine hydrochloride (PAAHC) in the size range of 1.0 nm to 3.0 nm, gave an intrinsic magnetic polarization [9]. Water-soluble gold nanoparticles synthesized by reducing a gold salt in the presence of a designed polymer ligand show good fluorescence [10].Therefore, syntheses of gold nanoparticles have attracted much attention. Various fabrication methods were carried out. Typically, in chemical approaches, reduction of corresponding metal ions or metal complexes is conducted in the presence of additional stabilizing agents (e.g., thiol compounds and polymers). One problem with the chemical methods is the inevitable introduction of byproducts which require subsequent purification steps after the synthesis for its potential application [11].On the other hand, in the physical methods, vacuum metal-vapor-condensation techniques have been developed for the clean preparation of metal nanoparticles dispersed in inorganic/organic solvents without formation of byproducts [12-14].Among recent physical methods, sputtering deposition has better capability to fabricate refractory metals and intermetallic compounds than evaporation and laser ablation. Because the atom vapor is typically generated from targets of pure materials, nanoparticles created by sputtering usually contain fewer impurities than those created by chemical methods. Gold nanoparticles were obtained via aerosol generation by spark discharge at low vapor pressure [15] and using sputtering deposition of gold onto various ionic liquids induced by inert gas [16]. However, the reported physical methods need a special vacuums device and inert gas for obtained gold nanoparticles. Therefore a new technique is urgently required to facilely fabricate size controllable clean gold nanoparticles for their potential applications.

Solution plasma is defined by non-equilibrium plasma in solutions, which provides us a novel reaction field with highly excited energy state. In our research lab, the rapid synthesis of clean nanoparticles without any reducing agents has been succeeded from $HAuCl_4$ aqueous solution [4, 17, 18]. The advantage of this method is the operation in an open system under atmospheric pressure. In the present study, we addressed a novel and one-step route for high-efficient synthesis of gold nanoparticles in a solvent from metal wire electrodes by plasma sputtering. Almost spherical gold nanoparticles with different particle size (including clusters less than 2.0 nm, several nanoparticles) were successfully fabricated in solution medium (such as water, liquid nitrogen, and ethanol) for the first time by solution plasma sputtering without any chemical additions (such as reducing agent and stabilizing polymer).

2. Experimental section

Gold wire (diameter ϕ = 1mm, Aldrich, 99.9%) was used as an electrode. Liquid nitrogen, ethanol and water were selected as a desired solution medium. The experiment of *solution plasma sputtering* technique was carried out in a desired solution and kept a 0.3 mm gap between the electrodes by a screw micrometer during discharging time as shown in scheme 1. Plasma was generated by a high voltage pulsed DC power supply (repetition frequency: 20 kHz, pulse width: 2 μs, Kurita Co. Ltd, Japan), which was shown as the scheme 2. In case of discharge in water / or ethanol, the solution medium temperature was kept at about 0 °C by a cooling system. After discharge, products were collected to characterization. The optical property of colloidal gold nanoparticles was detected by an ultraviolet spectrophotometer UV-3600 (UV-Vis, Shimadzu, Japan) in the range of 200–800 nm. The shape and microstructure of gold nanoparticles were observed in annular dark field scanning TEM (ADF-STEM) and high resolution TEM (HR-TEM) in JEM-2500SE (JEOL, Japan) operated at 200 kV.

3. Results and discussion

Gold nanoparticles with spherical morphology began to form in several microseconds simultaneously with solution plasma sputtering in water in an open system under atmospheric pressure. Figure 1 shows representative UV-Vis spectra of gold nanoparticles in water versus discharge time. A surface plasmon resonance band near 511 nm originated from the gold nanoparticles was detected obviously. The absorbance intensity increased with the prolonging discharge time without peak shift, indicating the amount of gold nanoparticles increased. Note that the production rates of gold nanoparticles are dependent on the power supply parameter (such as pulse width, repetition frequency, and *et al.*) and discharge time. At the same time, the color of water changed from colorless to light red, red dark red with the discharge time. Inset in Figure 1 show a digital photo of colloidal gold aqueous solution.

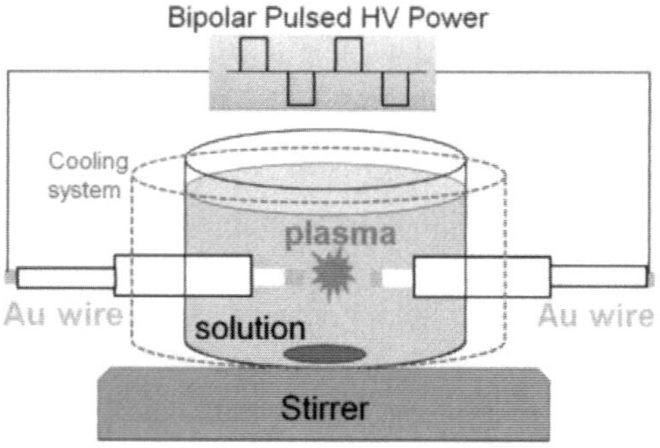

Scheme 1. Schematic diagram of solution plasma sputtering technique.

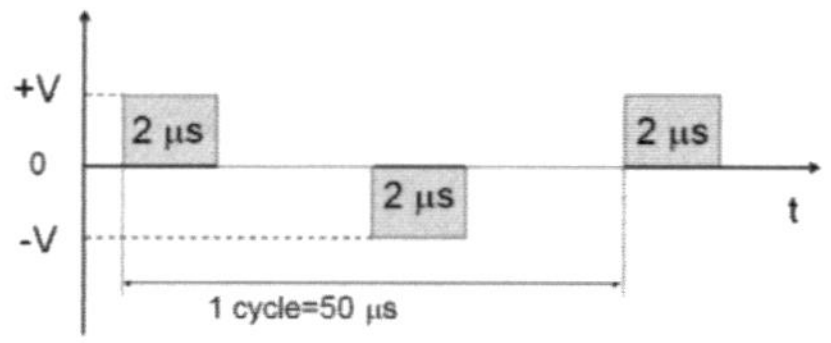

Scheme 2. Pulsed DC power supply (frequency 20 kHz, pulse width: 2 μs)

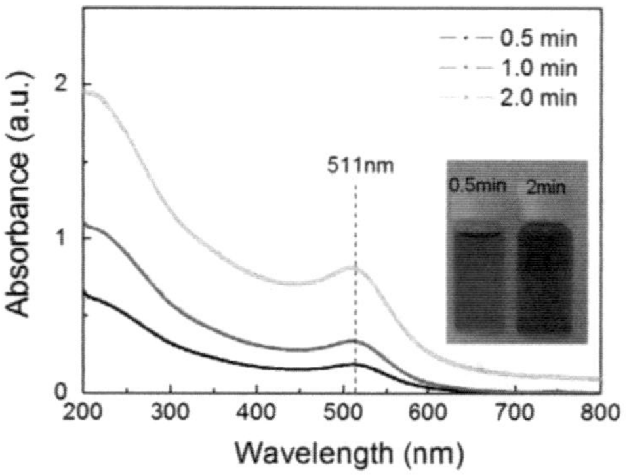

Figure 1. Representative UV-Vis absorbance spectra of gold nanoparticles versus sputtering time. Inset is digital photo for colloidal gold nanoparticles aqueous solution.

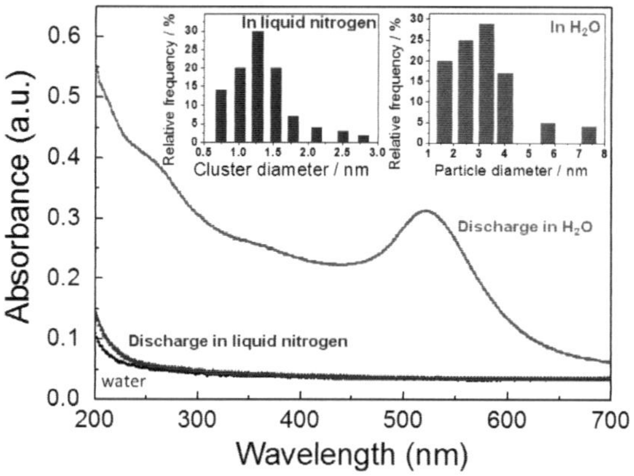

Figure 2. The UV-Vis absorbance spectra of gold clusters and gold nanoparticles fabricated in liquid nitrogen and water, respectively. Insets are their size distribution respectively.

And the effects of solvent on size and morphology were investigated. The plasma sputtering in liquid nitrogen was carried out. Figure 2 show a representative UV-Vis spectra of gold nanoparticles obtained in liquid nitrogen and water. In the case of discharge in liquid nitrogen, almost the same spectra comparing to water were detected without any surface plasmon resonance. The obtained gold cluster aqueous solution is colourless. The calculated size distributions for particles were shown in inset in figure 2. Mean size of 1.25 nm particles was obtained readily in liquid nitrogen by plasma sputtering; while larger mean size of about 3 nm was obtained in water. Therefore, we can conclude that the particle sizes are main dependent on the solvent temperature. That means the particle size can be controllable fabricated by adjusting the temperature of solvent by plasma sputtering. Fabrication of gold clusters by solution plasma sputtering have been discussion at our previous work [19]. Figure 3 show the morphologies of gold clusters obtained in liquid nitrogen and gold nanoparticles obtained in water by solution plasma sputtering, respectively. They are formed as spherical shape. High-resolution TEM image as shown in figure 4 clarifies that these nanoparticles were well crystalline, judging by the well-defined lattice fringes.

When plasma sputtering occurred in ethanol, almost the same particle size gold nanoparticles comparing to in water were formed. It is well known that surface plasmon is frequency-dependent relative permittivity between the external medium and the surface of metal nanoparticles. Thus, the ethanol solution color looks purple as shown in figure 5.

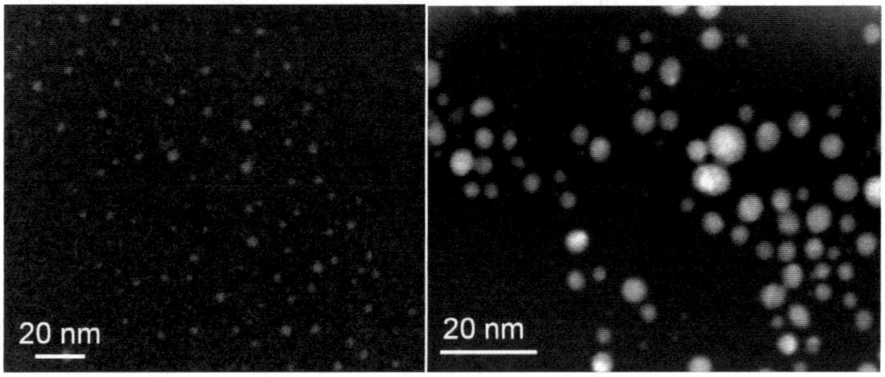

Figure 3. BF-STEM images for (a) gold clusters obtained in liquid nitrogen and (b) gold nanoparticles obtained in water by plasma sputtering.

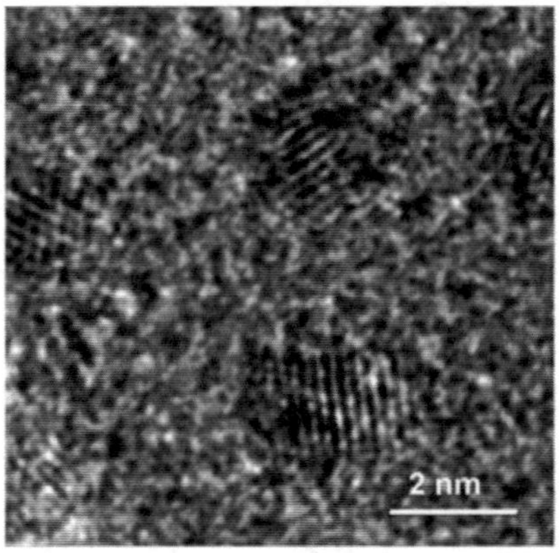

Figure 4. A typical high-resolution TEM image of gold nanoparticles.

Figure 5. Digital photo for colloidal gold solution prepared from plasma sputtering in ethanol.

And we also investigated the effect of polymer surfactant such as polyethylene glycol (PEG) on the particle size. As-reported polymer surfactants were selected to decrease the aggregation of nanoparticles and to obtain smaller particle size. However, PEG is easily decomposed in the discharge region due to the much higher temperature. So black carbon-like compounds was formed readily, and then resulted in fabricated gold particles was easily adsorbed on these carbon-like compounds.

By the way, we investigated the morphology of electrode surface before and after plasma sputtering. Before plasma sputtering, electrode surface was grinded using 800 mesh sandpaper, and looks like smooth and glittering with gold. After discharge in water, the glittering was disappearing. In the case of discharge in ethanol, the surface looks like black. Figure 6 show FE-SEM images of gold electrode surface before and after solution plasma sputtering in different solution medium. Comparing to that of just grinded surface, the surface changed to random rough after plasma sputtering. In the case of sputtering in ethanol, too much spherical-like compounds were observed; it was clarified as carbon-like compounds which resulted from ethanol decomposition. Note that black carbon-like components was formed readily when discharge was happed in an organic solution, and easily adsorbed on the electrode surface.

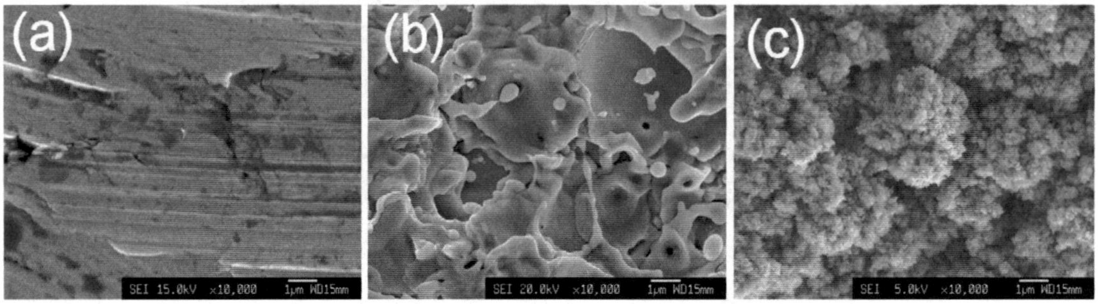

Figure 6. FE-SEM images of gold electrode surface. (a) Before plasma sputtering (just grinded by 800 mesh sandpaper), after solution plasma sputtering in (b) water and (c) ethanol, respectively.

4. Conclusions

Gold clusters (average particle size about 1.5 nm) and nanoparticles (average particle size about 3.5 nm) with spherical morphology were fabricated by plasma sputtering in solution medium in an open system under atmospheric pressure. The particle sizes are dependent on the solvent medium and solvent temperature. During plasma region, rapid energetic radicals' bombardment, atom vapors diffusion, plasma expansion and medium condensation resulted in the formation of gold nanoparticles from gold electrode wires. The as-fabricated gold particles have well crystalline even for less than 1 nm cluster. Gold nanoparticles with average size about 3.5 nm gave obviously surface plasmon resonance band at near 511 nm. While surface plasmon resonance band was disappearing for 1.5 nm-cluster was formed in liquid nitrogen. Thus different size gold nanoparticle will have potential applications in various fields.

Acknowledgement
This research was supported by a grant from the JST-CREST, Japan

References

[1] Bonnemann H *et al* 2001 *Eur. J. Inorg. Chem.***10** 2455
[2] Shishino Y *et al* 2010 *Chem. Commun.* **46** 7211
[3] Zhang Q *el at* 2010 *Adv. Mater.* **22** 1905
[4] Sayan B *et al* 2009 *Adv. Mater.* **21** 4039
[5] Ishida T *et al* 2007 *Angew. Chem. Inter. Edition* **46** 7154
[6] Haruta M *et al* 2001 *Appl. Cataly. A: General* **222** 427
[7] Sau T K *et al* 2004 *J. Am. Chem. Soc.* **126** 8648
[8] Barreto J A *et al* 2011 *Adv. Mater.* **23** H18
[9] Yamamoto Y *et al* 2004 *Phys. Review Lett.* **93** 116801
[10] Schaeffer N *et al* 2008 *Chem. Commun.* **34** 3986
[11] Brust M *et al* 1994 *J. Chem. Soc.-Chem. Commun.* **7** 801
[12] Shishino Y *et al* 2011 *Angew. Chem. Inter. Edition* **50** 703
[13] Tabrizi N S *et al* 2009 *J. Nanopart. Res.* **11** 1209
[14] Valden M *et al* 1998 *Science* **281** 1647
[15] Schwyn S *et al* 1988 *J. Aerosol Sci.* **19** 639
[16] Torimoto T *et al* 2006 *Appl. Phys. Lett.* **89** 243117
[17] Saito N *et al* 2009 *Thin Solid Films* **518** 912
[18] Takai O 2008 *Pure Appl. Chem.* **80** 2003
[19] Hu X L *et al* 2012 *Cryst. Growth Des.* **12** 119

Journal of Physics: Conference Series **417** (2013) 012031 doi:10.1088/1742-6596/417/1/012031

Al(111) nitridation below 473 K induced by supersonic N_2 molecular beam as observed by synchrotron radiation photoemission spectroscopy

Y Teraoka[1,2], M Jinno[1,2], J R Harries[1], A. Yoshigoe[1]

[1] Quantum Beam Science Directorate, Japan Atomic Energy Agency (JAEA), 1-1-1 Kouto, Sayo-cho, Sayo-gun, Hyogo 679-5148, Japan

[2] Graduate School of Material Science, University of Hyogo, 3-2-1 Kouto, Kamigori-cho, Ako-gun, Hyogo 678-1297 Japan

E-mail:yteraoka@spring8.or.jp

Abstract. The Al(111) surface has been directly nitrided using a supersonic N_2 molecular beam at a sample temperature region below 473 K. The translational kinetic energy threshold for nitridation is 1.8 eV. The nitrogen uptake curve obtained at a translational energy of 2.0 eV showed a linear dependence on dose, indicating the formation of a non-protective layer. Nitrogen diffusion into the bulk has been confirmed by a depth profile analysis using angle-resolved photoemission spectroscopy in conjunction with the maximum entropy method. The Al-2p photoemission spectra showed shoulder structures only slightly shifted to higher binding energies, suggesting that sub-nitride components are predominant in the nitride layer. The N-1s photoemission spectra confirm that the three-coordinated nitrogen (N^{3-}) dominates among the four chemical state components (N^{n-}:n=1-4).

1. Introduction

We have developed a supersonic molecular beam (SSMB) apparatus equipped with capabilities for photoemission spectroscopy (PES) at the high brilliance and high energy-resolution soft x-ray synchrotron radiation (SR) beamline BL23SU in the SPring-8 facilities to study surface chemical reaction dynamics [1]. Real-time photoemission spectroscopic measurements during SSMB irradiation are possible using the apparatus. Using a high temperature nozzle and seeded molecular beam techniques, translational energies greater than 1 eV can be achieved for beams of O_2 (2.3 eV) and N_2 (2.0 eV). The use of such high speed molecular beams has contributed to the elucidation of new schemes for many chemical reactions on silicon and metal surfaces [2-7].

Although there have been a few studies previously reported on the surface chemical oxidation reactions of aluminum single crystals, nitridation does not take place upon exposure to N_2 gas near room temperature. We have conducted a study on direct nitridation of the Al(111) surface by irradiation of inert N_2 SSMB from the view point of pure surface science and also from an applications point of view. An ultra-thin film of AlN can be grown by various methods including plasma ion immersion implantation [8], plasma-assisted gas-source molecular beam epitaxy [9] and metal-organic vapor phase epitaxy (MOVPE) [10]. In order to form very thin films on the nanometer scale, the MOVPE method using tri-methyl aluminum and ammonia gases has been reported as a typical method for AlN layer formation on SiC(0001) surfaces [10]. This method, however, requires high temperatures and careful gas handling.

We have conducted direct nitridation of the Al(111) surface by the action of the translational energy of an inert N_2 molecular beam at a surface temperature lower than 473 K. Although aluminum metal does not react with N_2 gas under ambient conditions in this temperature region, and no N-1s photoemission peak could be observed even using high brilliance synchrotron radiation for translational energies lower than 1.6 eV, nitridation of Al(111) was successfully observed using an N_2 SSMB with a translational energy higher than 1.8 eV. In this report, we show evidence for Al(111) nitridation by inert N_2 SSMB at 473 K and discus the mechanisms of the nitridation on the basis of photoemission spectroscopy with high brilliance and high energy-resolution synchrotron radiation.

2. Experimental

All of the experiments described here were conducted at the surface chemistry experimental station (SUREAC2000) of BL23SU in the SPring-8 facilities. The SUREAC2000 apparatus was designed to promote applications of photoemission spectroscopy with synchrotron radiation and SSMB techniques in surface chemical reaction dynamics research. A detailed description of the experimental station can be found in previous reports [1], and only a brief description is given here. The surface reaction analysis chamber is pumped by two turbo-molecular pumps in tandem fashion in conjunction with a titanium sublimation pump so that a base pressure of 7×10^{-9} Pa is obtained. When liquid nitrogen is fed into the manipulator and the chamber trap, a base pressure of 5×10^{-9} Pa can be achieved. The chamber is equipped with a concentric hemispherical electron energy analyzer (Omicron EA125-5MCD).

The N_2 SSMB was continuously generated by the adiabatic expansion of a mixture of N_2, He and Ar. The nozzle, made of pyrolytic boron nitride, can be heated to a temperature of about 1400 K by using a specially-made heater. The translational energy is varied by varying the gas mixing ratio, keeping the nozzle temperature at 1400 K. The upper limit of the translational energy for N_2 molecules is 2.0 eV, as calculated on the basis of the enthalpy conservation law. Some fraction of N_2 molecules must be vibrationally excited since the nozzle temperature is 1400 K. The vibrational distribution is assumed to be constant irrespective of translational energy due to the constant nozzle temperature. Under typical operating conditions, the N_2 beam flux density is estimated to be about 2×10^{14} molecules$\cdot$cm$^{-2}\cdot$s^{-1} on the sample surface. According to rarefied gas dynamics studies, a velocity width ratio $\Delta v/v_p$ is expressed as 1.65/S, where v_p is a peak velocity and S is a velocity ratio. S is expressed as $M(\gamma/2)^{1/2}$, where γ is a specific heat ratio and M is the Mach number. The γ value of 2-atomic molecule is 7/5. Therefore, an energy width ratio $\Delta E/E$ is expressed as $(\Delta v/v_p)^2 + (\Delta v/v_p) \cdot 2v_L/v_p$, where v_L is a lower side velocity in the half maximum of velocity. The $\Delta E/E$ can be approximated as $(\Delta v/v_p)^2 + 2(\Delta v/v_p)$ because the factor of v_L/v_p is nearly one. For example, when M is 30 and γ is 7/5, the $\Delta E/E$ is estimated to be 0.13. This estimation indicates that an energy width ΔE is 13% of a peak energy E. As we used N_2 molecular beam of 2.0 eV, the energy width (FWHM) is estimated to be 0.26 eV. Actually, as mentioned in the section of *3.1.N_2 translational energy threshold for nitridation*, a small amount of N-1s photoemission peak was observed in the N_2 energy setting of 1.8 eV although no signal was detected in the 1.6 eV setting. A clear nitrogen uptake curve was obtained in the 2.0 eV setting. These facts suggest that the FWHM of N_2 molecular beam we used is nearly 0.2 eV as estimated above.

The sample cleaning method is as follows. Firstly, the sample was ultrasonically cleaned in acetone for 5 min, ethanol for a further 5 min, rinsed in de-ionized water, dried with compressed pure nitrogen gas on a filter paper, and placed in the load-lock chamber of the SUREAC2000 apparatus. In order to remove the native oxide layer on the Al(111) surface, the surface was irradiated for 15 min with an Ar^+ ion beam using an acceleration energy of 1 keV, and then thermally annealed up to 773 K in the surface cleaning chamber. The sputtering and anneal cycle was repeated several times. A low energy electron diffraction pattern of 1×1 was observed in the cleaning chamber.

After confirming no detection of C-1s, O-1s, and other contaminant photoemission peaks by the SR-PES, the N_2 SSMB with a fixed translational energy was irradiated on the clean surface. The incident angle of the molecular beam was -10 degrees with respect to the surface normal. The photoelectron detection was +30 degrees. The SR photon energy was selected to be 686.8 eV, and the total spectral resolution was about 0.24 eV. After irradiation with the N_2 beam for several seconds or minutes, photoemission spectra of the Al-2p, N-1s, and O-1s core levels were recorded. This irradiation and measurement cycle was repeated several tens of times. The time evolution, that is, the N_2 dose dependence, of the photoemission peaks were obtained with this successive N_2 beam irradiation and photoemission measurements. The photoelectron detection depth is estimated to be about 3.9 nm (Al-2p), 2.7 nm (N-1s), and 2.0 nm (O-1s) [11], assuming that photoelectrons emitted from depths of up to around three times the inelastic mean free path are detectable. The surface temperature was kept at 300 K, 373 K, or 473 K following flash annealing at 773 K.

3. Results and discussion

3.1. N_2 translational energy threshold for nitridation

An N-1s photoemission peak was not observed for translational energies below 1.6 eV at a surface temperature of 473 K, even following the N_2 dose of 1.12×10^{18} molecules · cm^{-2}. This dose corresponds to 1836 monolayer (ML) of N atoms when the atomic density of the Al(111) plane in the bulk (1.22×10^{15} atoms·cm^{-2}) is defined to be 1 ML. A small N-1s peak was detected following the dose of 1.28×10^{18} molecules·cm^{-2} with a translational energy of 1.8 eV. The N_2 dose of 1.28×10^{18} molecules·cm^{-2} is corresponding to 2098 ML of N atoms. If the detection limit of N atoms on the surface is assumed to be 0.1 ML, the reaction probability is estimated to be 5×10^{-5}. At 2.0 eV, the N-1s peak intensity increased with increasing N_2 dose. Figure 1 shows the uptake of the N-1s SR-PES spectra recorded at various doses between 0 and 1.2×10^{19} molecules·cm^{-2} with the incident energy of 2.0 eV. The energy scales are calibrated assuming that the binding energy of Al-$2p_{3/2}$ component is 72.5 eV [12] for the clean surface. Since very slight nitridation could be observed for a kinetic energy of 1.8 eV, we concluded that the translational energy threshold for the direct nitridation of Al(111) surface by N_2 molecular beams is 1.8 eV.

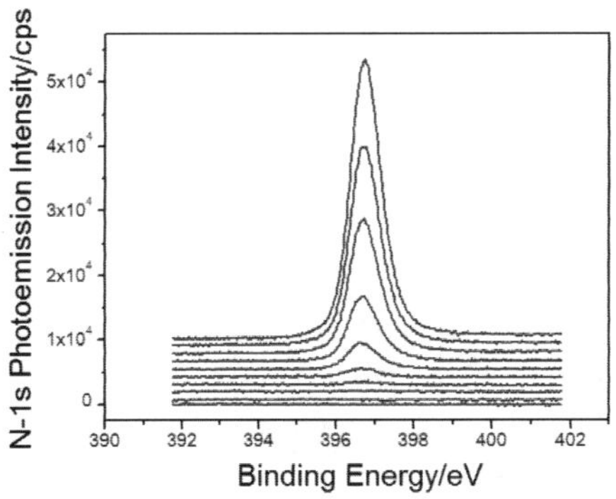

Figure 1. N-1s SR-PES spectra recorded during N_2 beam irradiation from 0 to 1.2×10^{19} molecules·cm^{-2} with the incident energy of 2.0 eV.

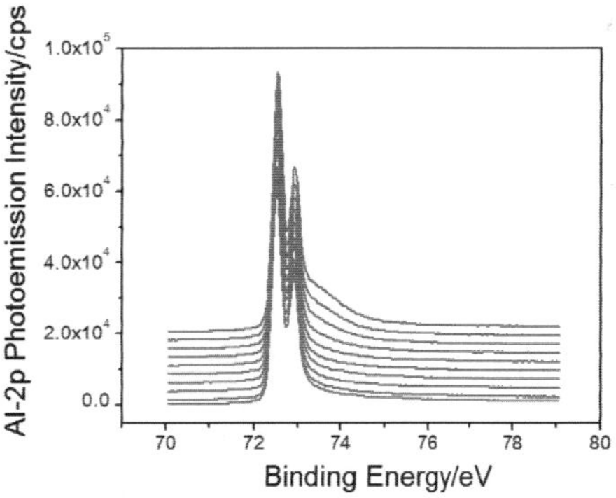

Figure 2. Al-2p SR-PES spectra recorded during N_2 beam irradiation from 0 to 1.2×10^{19} molecules· cm^{-2} with the incident energy of 2.0 eV.

3.2. Chemical bonding states of aluminum

The N_2 dose dependence of the Al-2p photoemission peak profile is shown in figure 2. The total energy resolution is about 0.24 eV, enough to resolve the spin-orbit splitting (0.4 eV) of the Al-$2p_{1/2}$ and Al-$2p_{3/2}$ peaks. The two main peaks are due to contributions from the bulk aluminum under the nitride layer. The Al-2p photoemission peaks show a well-known asymmetric profile tailing to the higher binding energy side due to the induction of electron-hole pairs during photoelectron emission [11]. At increasing N_2 doses a shoulder structure appears on the high binding energy side at around 73 eV to 74.5 eV, overlapping with the high binding energy side tail. This shoulder is attributed to chemical shifts of the Al-2p core levels due to bonding with mainly nitrogen and partly oxygen, as discussed below.

It has been reported [13] that an AlN layer can be formed on an Si(111) substrate by applying the plasma-source molecular beam epitaxy technique. The binding energy values for the Al-N, Al-O (alumina domains and grain boundaries), and Al-O (surface only) bonds were reported to be 73.5 eV, 74.7 eV, and 76.7 eV, respectively. In the spectra recorded here, therefore, we could expect to see satellite peaks or shoulders for these bonds in the region from 73.5 eV to 76.7 eV. In particular, any uptake was not observed in the binding energy region larger than 74.5 eV. Therefore, the shoulder structure shown in the region smaller than 74.5 eV in figure 2 is almost corresponding to Al-N bonds. We thus conclude that Al-N nitride is probably not a complete crystal, but also consists of sub-nitrides in the nitride layer observed here due to the broad shoulder structure.

3.3. Chemical bonding states of nitrogen

The N-1s photoemission peak recorded following a dose of 1.05×10^{19} molecules$\cdot$cm^{-2} is shown in figure 3. For both the zinc-blende and wurtzite crystal structures the maximum coordination number of both nitrogen atoms and aluminum atoms is four. The N-1s peak can thus be deconvoluted into four components, corresponding to n-coordinated sub-nitride (N^{n-} n:1-3) and fully-nitrided AlN local structure (N^{4-}). The minus is due to low electron affinity of nitrogen (-0.07 eV) than that of aluminum (0.45 eV) [12]. Good fits to the data could be obtained using four pure Gaussian line profiles and a linear background. The use of a pure Gaussian line profile is reasonable since the contribution to total energy resolution from the analyzer (0.22 eV) dominates that of the incident synchrotron radiation (about 0.1 eV). As shown in figure 3, the N^{3-} component is the most intense among the four components. This implies that the aluminum surface is not fully nitrided, the surface layers are dominated by sub-nitrides, and that AlN is barely present. This is qualitatively consistent with the observation of broad chemical shift in the Al-2p spectra (figure 2). Experimental evidence supporting the formation of sub-nitrides is also seen in the early stages of nitridation, where the N^{2-} component is dominant, as shown in figure 4(a). It can be concluded that dissociatively-adsorbed nitrogen atoms make chemical bonds between the topmost aluminum atoms to form a sub-nitride state of N^{2-}. As nitridation progresses, the N^{3-} component becomes dominant, as seen in figures 3 and 4(b).

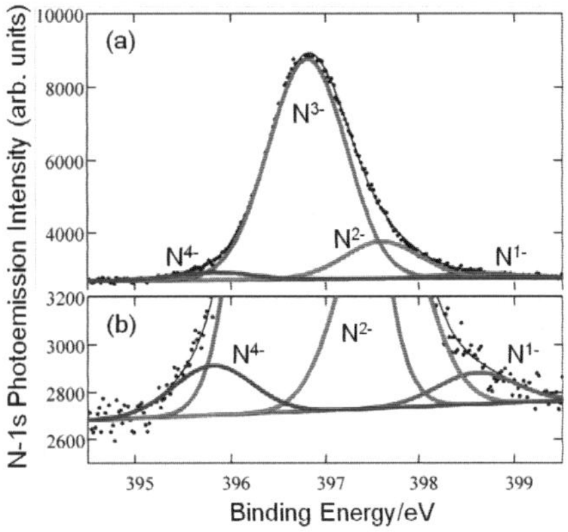

Figure 3. Four-component fit to N-1s SR-PES spectra recorded following a dose of 1.05×10^{19} N_2 molecules$\cdot$ cm^{-2}. The solid line among observed points shows the overall fit, the four lines incide the individual components in (a) and (b). (b) shows in detail the N^{1-} and N^{4-} components.

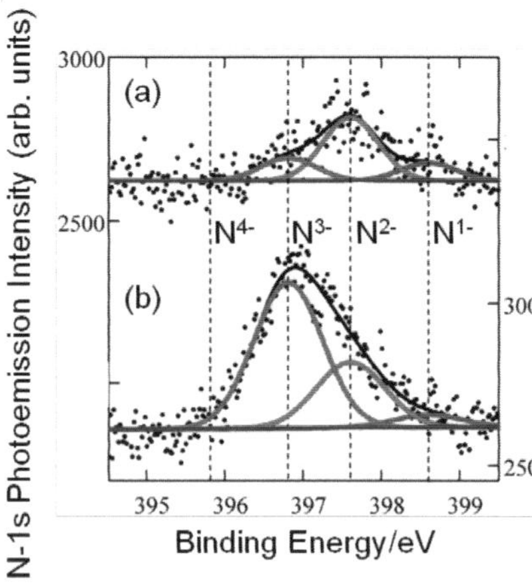

Figure 4. Comparison between fits to N-1s SR-PES spectra recorded in the early nitridation stages after doses of (a) 7.0×10^{16} and (b) 6.5×10^{17} molecules$\cdot$ cm^{-2}.

3.4. Uptake curve of nitrogen adsorption

As can be seen in figure 4, the N-1s peak profile changes with increasing N_2 dose. In order to make clear the evolution of the chemical bonding states of the adsorbed nitrogen, the intensity variations for the deconvoluted components of the N-1s spectra are shown in figure 5 as a function of N_2 dose. For doses less than 10^{17} molecules· cm^{-2} (164 ML of N atoms), the N^{2-} component dominates over the N^{3-} and N^{1-} components, and the N^{4-} component is negligibly small. After a dose of about 10^{17} molecules· cm^{-2}, the N^{3-} component begins to dominate. The N^{2-} component, however, is still larger than the N^{1-} and N^{4-} components. After a dose of about 10^{18} molecules·cm^{-2} (1639 ML of N atoms), the N^{4-} component appears, reaching the same magnitude as the N^{1-} component.

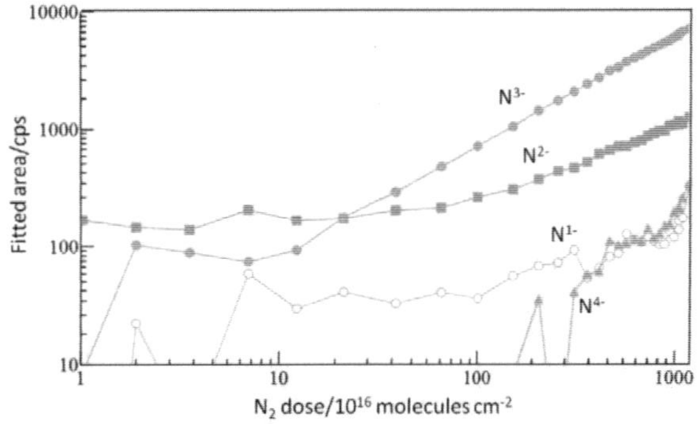

Figure 5. Uptake curves of the four components fitted to N-1s SR-PES spectra, open circles for N^{1-}, solid squares for N^{2-}, solid circles for N^{3-}, and solid triangles for N^{4+}, respectively.

Nitrogen uptake does not saturate even after a dose of 10^{19} molecules·cm^{-2} (16393 ML of N atoms), and linear characteristics of uptake curves are revealed not only for the total N-1s intensity but also for all components. Such linear uptakes generally indicate that the adsorbed layer does not act as a protective layer inhibiting subsequent N_2 adsorption. In other words, the adsorption of an N_2 molecule at a particular site does not protect the atomically local area from further adsorption, and a nitrided local area can again be nitrided by energetic N_2 incidence. As seen in figure 5, surface nitridation proceeds with a constant reaction rate regardless of nitrogen coverage. This finding suggests that a proportion of nitrogen atoms adsorbed on the Al(111) surface migrate into the bulk, with the result that part of the surface remains as a clean Al(111) structure, where further nitridation can take place. The LEED patterns observed after nitridation were not clear suggesting no periodic crystal structure. In order to confirm these facts, depth-profiles of the constituent elements and nitrogen bonding states were generated by applying the maximum entropy method to angle-resolved photoemission measurements.

3.5. Depth profiling

The depth-specific information present in angular-resolved PES spectra can be extracted using the maximum entropy method [14], where the sample structure is modeled as a finite number of layers containing a number of different elements or the chemical states of those elements. By recording O-1s, N-1s, and Al-2p PES spectra at a range of different takeoff angles, we have generated depth profiles of elemental N, O, and Al, and also of the four components fitted to the N-1s spectra for the AlN thin-film sample produced following exposure of the clean Al(111) surface to 1.2×10^{19} N_2 molecules· cm^{-2}. After correcting for the differing photoionisation cross sections [15], the sum of all six components was normalised to 1.0 at each angle. Figure 6(a) shows the atomic concentrations of Al, O, and N as a function of depth. Angular dependencies produced a depth profile of nitrogen concentration which extends to a depth of about 1.5 nm, and the presence of an oxide sub-layer with about 1 nm thickness was suggested. The O-1s photoemission peak was also observed during the N_2 molecular beam irradiation. However the N_2 beam was confirmed free of oxygen atoms and oxygen-containing molecules using the quadrupole mass analyzer, so it must be assumed that oxygen-containing residual gas molecules (for example H_2O) are adsorbed onto the nitride surface. We speculated that oxygen atoms initially-adsorbed on the surface migrated into the bulk to be trapped in the defect sites of the interface region. Incorporated oxygen in the bulk is about 1/3 comparing to nitrogen. The position of oxygen atoms is restricted near the interface so that the depth profile of oxygen is completely different to that of nitrogen. Although oxynitrides may exist near the interface, nitride layers are clearly formed near the surface. The profiles for the four nitrogen components are shown in figure 6(b). The fit suggests that while the N^{3-} component is dominant throughout the AlN film, N^{2-} and N^{4-} are only present close to the surface, and a sub-layer is suggested for N^{1-} in the interface.

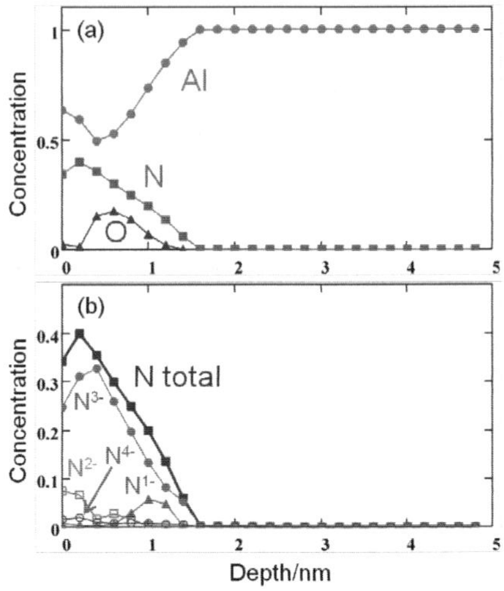

Figure 6. Depth profiles generated using the maximum entropy method. The trace of N in (a) represents the sum of the four components fitted to N-1s SR-PES spectra, the profiles of which are shown separately in (b).

4. Conclusions

Although the Al(111) surface does not react with thermal N_2 gas, we have shown that it can be directly nitrided using an N_2 SSMB at a translational energy larger than 1.8 eV at a sample temperature below 473 K. The nitrogen uptake curve was confirmed to be almost linear with respect to increasing N_2 dose, indicating that the nitridation rate is constant regardless of nitrogen coverage on the surface, and furthermore that a protective layer is not formed on the topmost nitride surface. In order to make clear the reason behind this linear uptake, depth-profiles of the constituent elements and chemical bonding states of nitrogen were calculated from angle-resolved photoemission spectra using the maximum entropy method. The Al-2p photoemission spectra showed only small shoulder structures, suggesting that sub-nitride components are predominant in the nitride layer. The N-1s photoemission spectra confirm that the three-coordinated nitrogen (N^{3-}) dominates among the four chemical state components (N^{n-} n:1-4). We do not mention about penetration of N_2 molecules into the bulk through the Al(111) surface even if the incident energy of N_2 molecule is 2.0 eV. We speculate nitrogen atoms generated on the Al(111) surface by activated dissociative adsorption of high speed N_2 molecules may diffuse into the bulk due to the thermal energy of substrate (473 K). Actually, apparent substrate temperature dependence of nitrogen uptake curve has been observed in our recent experiments. Furthermore, incubation time for nitrogen uptake was also observed when the substrate was at a room temperature. These experimental facts suggest that the nitridation of Al(111) surface by high speed N_2 molecules cannot explained by the penetration of N_2 molecule into the bulk. Consequently, nitrogen diffusion into the bulk and the formation of sub-nitride layers was verified in our high energy resolution photoemission measurements. The study for clarifying the reasons of incubation time and temperature dependence is now on progress.

Acknowledgments
The authors thank Dr. Yuji Saitoh and Dr. Yoshihiro Fukuda for their support of the operation of the monochromator system of BL23SU. The authors gratefully acknowledge the Ministry of Education, Culture, Sports, Science, and Technology of Japan for Grants-in-Aid for Scientific Research (Nos. 20360024, 21760031, 19360336). This work was performed with the approval of the Japan Synchrotron Research Institute (JASRI) (2011A3803, 2008B3802, 2008A3802, 2007A3802, 2006B3801).

References
[1] Teraoka Y and Yoshigoe A 1999 *Jpn. J. Appl. Phys.* 38 Suppl.38-1 642
[2] Yoshigoe A and Teraoka Y 2010 *J. Phys. Chem.* C 114 22539
[3] Okada M, Souwa M, Kasai T and Teraoka Y 2011 *Appl. Surf. Sci.* 257 4257
[4] Okada M and Teraoka Y 2010 Fabrication of Cu Oxides on Single Crystal Cu Surfaces Using Hyperthermal O_2 Molecular Beams *Metal Oxide Nanostructures and Their Applications*

Growth and Properties Part-1, ed Umar A and Hahn Y-B (New York: American Scientific Publishers) chapter 6 pp 1-33

[5] Okada M and Teraoka Y 2010 *Appl. Surf. Sci.* 256 5676

[6] Moritani K, Okada M, Teraoka Y, Yoshigoe A and Kasai T 2009 *J Phys. Chem.* A 113 15217

[7] Takahashi S, Hatta S, Yoshigoe A, Teraoka Y and Aruga T 2009 *Surf. Sci.* 603 221

[8] Valcheva E, Dimitrov S, Manova D, Mändl S and Alexandrova S 2008 *Surface & Coatings Technology* 202 2319

[9] Tanaka S, Kern R S, and Davis R F 1994 *Appl. Phys. Lett.* 66 37

[10] Taniyasu Y, Kasu M and Makimoto T 2007 *J. Crystal Growth* 298 310

[11] Hüfner S 2003 Photoelectron Spectroscopy (Berlin: Springer)

[12] Emsley J 1998 *The Elements* (Oxford: Clarendon Press)

[13] Rosenberger L, Baird R, McCullen E, Auner G and Shreve G 2008 *Surf. Interface Anal.* 40 1254

[14] Chang J P, Green M L, Donnelly V M, Opila R L, Eng J, Sapjeta J, Silverman P J, Weir B, Lu H C, Gustafsson T and Garfunkel E 2000 *J. Appl. Phys.* **87** 4449

[15] Yeh J J and Lindau I 1985 *At. Data Nucl. Data Tables* **32** 1

Journal of Physics: Conference Series **417** (2013) 012032 doi:10.1088/1742-6596/417/1/012032

Development of Advanced Coating Techniques for Highly-durable Casting Dies

S Tanaka [1], M Takagi[1] and T Mano[1]

[1]*Fuji Technical Support Center, Shizuoka Prefectural Industrial Research Institute*, 2590-1, Ohbuchi, Fuji, Shizuoka, JAPAN

E-mail: tanakash@iri.pref.shizuoka.jp

Abstract. In order to improve the durability of aluminum die-casting molds, we applied microstructure-controlled PVD coating techniques. Single-layer and multilayer films consisting of chromium nitride (CrN) or titanium aluminum nitride (TiAlN) were prepared using an ion plating process. Structures of multilayer films were observed using transmission electron microscopy. Pin-shaped mold steel specimens coated with each of the films were soaked in the molten aluminum alloy at 953 K different periods of time, and the amount of weight loss due to erosion was evaluated. The weight losses for the multilayer CrN and TiAlN specimens were found to be less than those for the single-layer specimens. As a practical test, five specimens of core pins used in aluminum die casting of automobile parts were coated with multilayer films, and the number of maintenance operations required to remove aluminum alloy remaining on the specimen surfaces after several thousand castings was counted and compared with six control specimens (core pins treated using a commercial salt bath diffusion process). The number of maintenance operations for CrN- and TiAlN-based multilayer-coated core pins was found to be lower than for the control specimens.

1. Introduction

Die casting, which is a metal casting process that involves the injection of molten metal into a mold cavity under high pressure, can produce large quantities of complex-shaped castings in a short time. However, during the casting process, casting dies that are exposed to molten metal can degrade quickly [1]. In the case of aluminum die-casting, degradation of the casting die causes the aluminum alloy to stick to the mold surface, and dies must be repaired or replaced according to the level of degradation. In order to reduce the cost of repairing or replacing the mold and the associated impact on productivity, a method of extending the life of the casting dies is strongly required.

According to a previous study [2], a metal nitride film formed by PVD on the die component material can suppress damage caused by erosion [3]. However, damage to the base metal is still thought to occur due to flaws in the PVD film. In the present study, multilayer film techniques to inhibit the formation and development of flaws in the film were developed. The practical performance of the developed coating was evaluated using a coated die component intended for use in an aluminum die-casting machine.

2. Experimental Procedures

2.1. Coating Process

Single-layer and multilayer films consisting of chromium nitride (CrN) or titanium aluminum nitride (TiAlN) were prepared by Multi-Arc PVD system (NISSIN ELECTRIC MC-300) using an ion-plating process. Figure 1 shows cross-sectional images of CrN-based multilayer film with 17 layers and TiAlN-based multilayer film with approximately 300 layers.

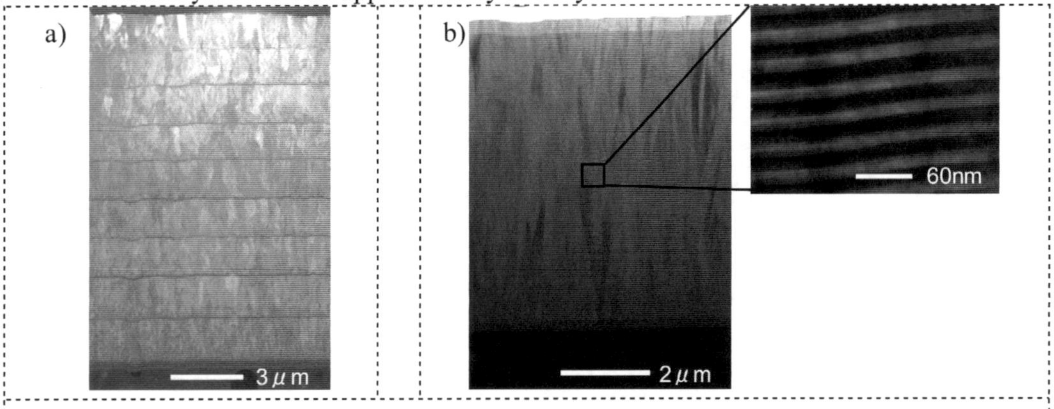

Figure 1. Cross-sectional TEM images of multilayer films: a) CrN-based multilayer film, b) TiAlN-based multilayer film.

2.2. Erosion test

Figure 2 shows the erosion test apparatus used to evaluate the resistance to erosion of the PVD coating materials. The tested specimens were either uncoated, nitrided, or coated with single- or multilayer CrN or TiAlN films. Pin-shaped mold steel (JIS SKD61) specimens (diameter: 10 mm, length: 100 mm) were soaked in molten aluminum alloy (JIS ADC12) at a depth of 50 mm and a temperature of 953 K for different periods of time, and the weight loss due to erosion was measured. Weight measurements were performed after removing the aluminum alloy attached to the surface, and the amount of erosion was then calculated. Erosion pitting of the specimen surface was observed using an optical microscope (KEYENCE: VHX-100).

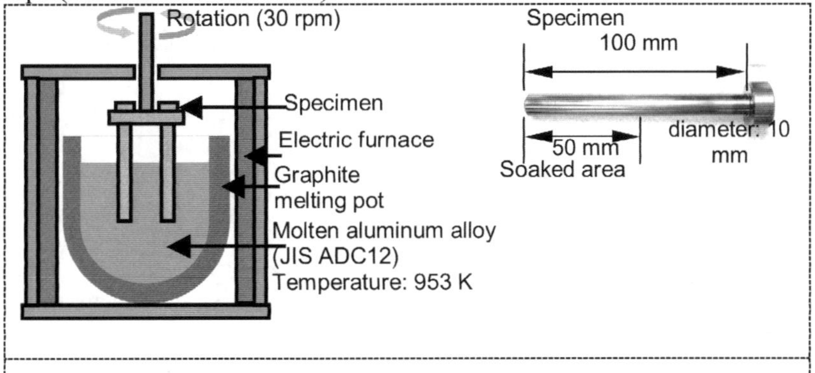

Figure 2. Schematic diagram of erosion test apparatus.

2.3. Practical test for aluminum die-casting

For the practical test, five core pins used in aluminum die casting of automobile parts were coated with multilayer films, and the number of maintenance operations required to remove the aluminum alloy remaining on the die surface after several thousand castings was counted. These results were compared with those for six control specimens (core pins treated using a commercial salt bath diffusion process). Figure 3 shows the cast products and the positions of the core pins used for the casting test. The dimensions of the cast parts were 400 mm × 300 mm × 60 mm (length, width, diameter), and the test material was ADC12 aluminum alloy.

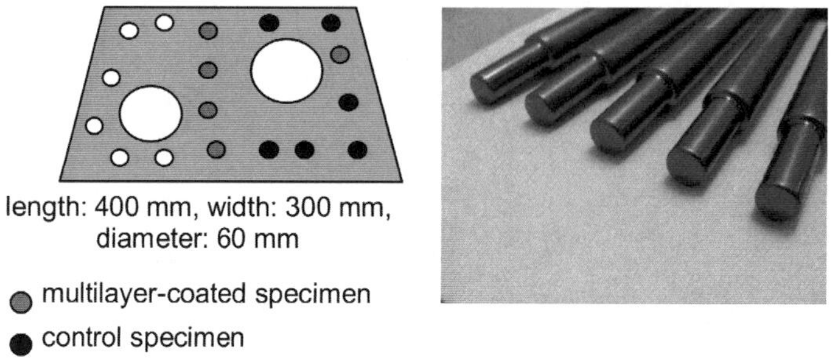

Figure 3. Dimensions of cast product and positions of core pins used for the casting test.

3. Results and Discussion

3.1. Erosion test

When observing the surface state of the untreated specimen after soaking two hours in molten aluminum alloy, wear can be seen in entire of soaked area (Fig. 4). Specimen that treated with nitride was also found to wear over almost all of the soaked area (Fig. 5). The amount of weight loss of each specimens were, 5.60g, 2.92g, respectively. Overall weight of the specimen before the test is 84.4g and weight of soaked area is approximately 30.2g. The weight loss of untreated specimen after soaking for two hours is approximately 19% of the initial weight of the soaked area.

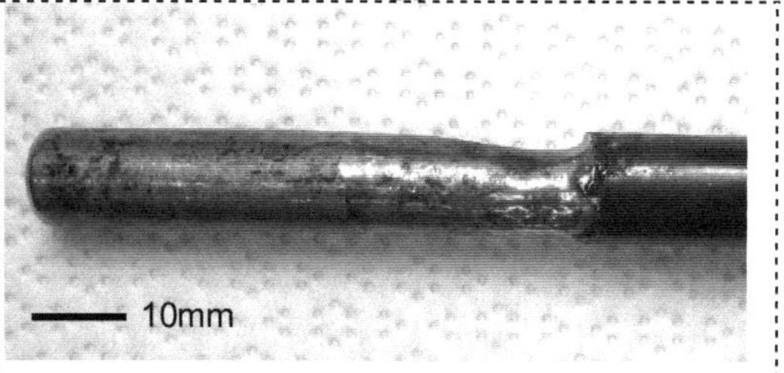

Figure 4. Untreated specimen after soaking for two hours.

Figure 5. Nitride specimen after soaking for two hours.

Figure 6 shows the amount of weight loss of specimens coated with single- or multilayer CrN or TiAlN films after soaking in molten aluminum alloy. Testing time for specimens were, 0.5, 1, 2, 3, 4 hours about single-layer film and 4, 8, 12 hours about multilayer film, respectively.

The weight loss of specimens coated with single-layer CrN or TiAlN films at soaking two hours were, 0.003g, 0.226g, respectively. Compared with specimens untreated or nitrided at test time 2 hours, it was found that the amount of wear significantly reduces about the coated specimens.

For multilayer-coated specimens, erosion was not observed by soaking four hours. Slight erosion was observed at soaking 8 hours, and amount of weight loss of multilayer TiAlN and CrN at soaking 12 hours were 0.130g, 0.061g. Amount of the weight loss of the multilayer-coated specimen after soaking 12 hours were much less than that of single-layer coated specimen after soaking 4 hours, so, it is found that multilayer film can improve the resistant to erosion.

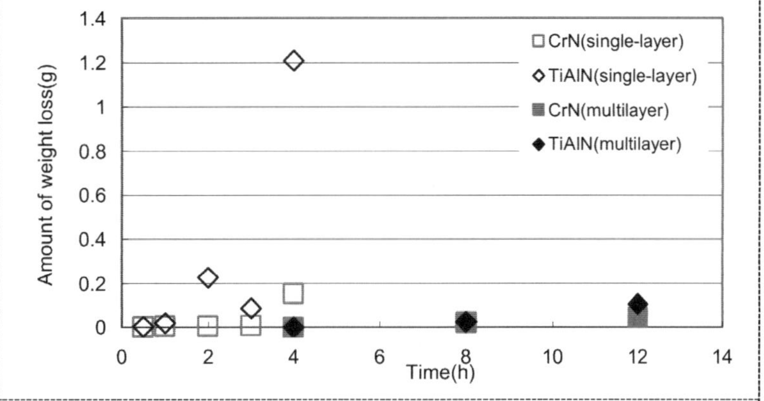

Figure 6. Amount of weight loss of specimens coated with single- or multilayer CrN or TiAlN films for several soaking time.

Figure 7 and 8 shows the surface state of the single-layer coated specimen after soaking four hours. About CrN single-layer specimen, erosion pit can be seen sparsely around soaked area. About TiAlN single-layer specimens, erosion was spread all of soaked area and the coating layer was only partially remained.

In order to pursue a mechanism of damage progression, erosion pit of CrN single-layer film was observed. Figure 9 shows one of typical erosion pit on surface of CrN coated specimen. Shape of erosion pit was nearly perfect circle and CrN film had been left around a pit. Figure 10 shows a cross-sectional image of erosion pit. As cross-sectional shape of erosion pit was spreading inside the substrate and remaining the surface film, the damage of the specimen is considered to progress by erosion of the substrate rather than degradation of the film.

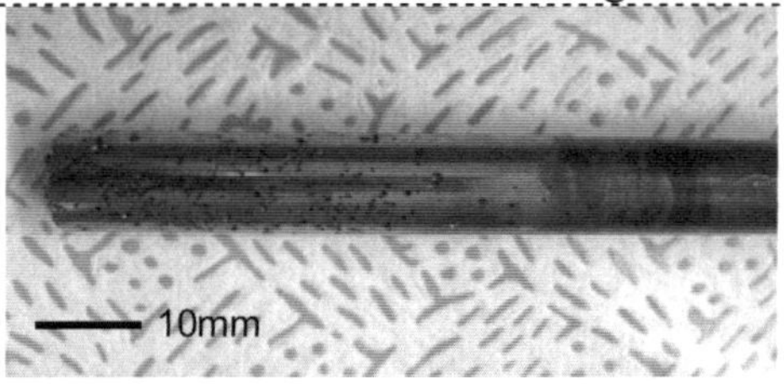

Figure 7. Single-layer CrN specimen after soaking for four hours.

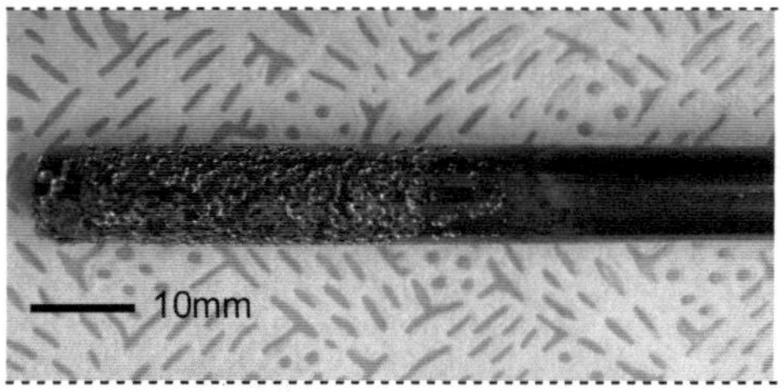

Figure 8. Single-layer TiAlN specimen after soaking for four hours.

Figure 9. Erosion pit of single-layer CrN specimen (surface).

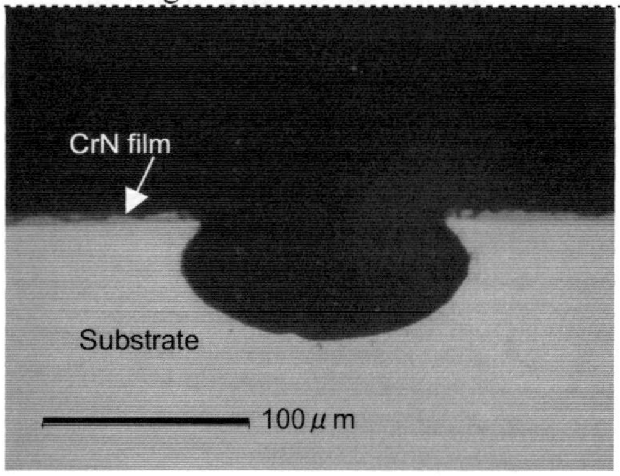

Figure 10. Erosion pit of single-layer CrN specimen (cross section).

Figure 11 and 12 shows the surface state of the multilayer-coated specimen after soaking eight hours. If you watch the surface, in the case of CrN multilayer film is dotted with small defects, in the case of TiAlN multilayer film showed several comparatively large erosion pits (diameter<1mm).

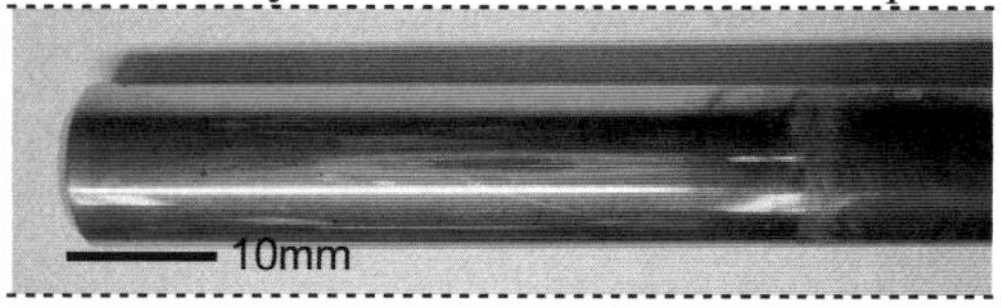

Figure 11. Multilayer CrN specimen after soaking for eight hours.

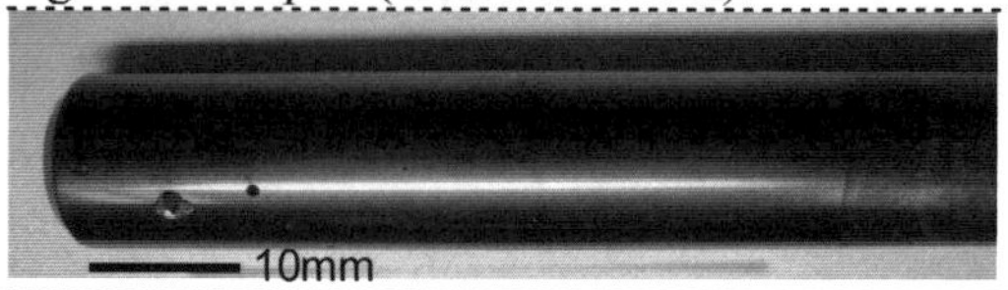

Figure 12. Multilayer TiAlN specimen after soaking for eight hours.

3.2. Practical test for aluminum die-casting

Table 1 shows the results of practical test of multilayer-coated core pins assembled in an aluminum die-casting machine. Core pins coated with multilayer CrN films took four times of maintenance operations during the test casting 2,177 products, whereas control were required to maintain 14 times. Core pins coated with multilayer TiAlN films took none of maintenance operations during the test casting 3,246 products, whereas control were required to maintain 20 times. In both cases that applied CrN or TiAlN multilayer film, the number of maintenance operations was lower than the control specimen of existing product. It is assumed that because the surface damage of core pins was suppressed by multilayer film, adhesion of aluminum was reduced.

Table 1. Results of practical tests

	Number of casting tests	Number of core pins	Number of maintenance operations
CrN (multilayer)	2,177	5	2
Control		6	14
TiAlN (multilayer)	3,426	5	0
Control		6	20

4. Conclusions

CrN and TiAlN film subjected by ion plating can be greatly suppresses damage due to erosion of mold material.

About erosion test the damage of the specimen is considered to progress by erosion of the substrate rather than degradation of the film.

According to the casting test using core pins applied CrN or TiAlN multilayer film, the number of maintenance operations was reduced compared to the control specimen of existing product. It is assumed that because the surface damage of core pins was suppressed by multilayer film, adhesion of aluminum was reduced.

References

[1] C Mitterer, F holler, F Ustel, D heim 2000 *Surface Coatings Technology* **125** 233-239
[2] Hihara M 2003 *Daikasuto yoh kanagatano jumyoutaisaku* (Nikkan Kogyo Shimbun Ltd.)
[3] Yucong W 1997 *Surface Coatings Technology 24th International Conference on Metallurgical Coatings and Thin Films* **94-95** 60-63

Thickness and Composition Measurements of Nanoelectronics Multilayer Thin Films by Energy Dispersive Spectroscopy (EDS)

A. Franquet[1], T. Conard[1], M. Gilbert[1], T. Hantschel[1] and W. Vandervorst[1, 2]

[1] IMEC vzw, Kapeldreef 75, B-3001 Heverlee, Belgium

[2] Instituut voor Kern- en Stralingsfysica, K. U. Leuven, Celestijnenlaan 200D, B-3001 Leuven, Belgium

Alexis.Franquet@imec.be

Abstract. Scaling is one of the main aspects to continuously improve the performance of nanoelectronic devices. This leads to increasingly more complex stacks including new materials; for example SiO_2 is being replaced by high-k materials for gate dielectric applications and metal gates are introduced in place of poly-Si. Moreover, a change in the type of substrate is also considered; Ge and III/V materials are intensively investigated as possible candidates for Si substrate replacement. The introduction of these new materials together with the device downscaling has drastically increased the demand for both composition and thickness metrology methods. Various established techniques providing composition analysis exist such as X-Ray Photoelectron Spectroscopy (XPS) and Rutherford Backscattering Spectrometry (RBS), however, they usually require a large area for the analysis, complex instrumentation and can be time consuming. In this paper, Energy Dispersive Spectroscopy (EDS) coupled to Scanning Electron Microscopy (SEM) is used to determine both composition and thickness of simple and more complex multilayered thin films. The use of EDS for thickness measurements in small scale areas is also investigated.

1. Introduction

The analysis of thin films has become very important nowadays with the downscaling of nanoelectronic devices. In this context, we see in the most recent technology generation for complementary metal-oxide semiconductor (CMOS) applications a tendency to replace the Si substrate by Ge or III/V materials to increase the carrier mobility. This is also accompanied by the replacement in the stack of both SiO_2 and poly-Si by high-k materials as gate dielectrics and by metal gates, respectively. This has lead to a strong increase in the number of elements used in the semiconductor industry, a number that has been multiplied by a factor of around 6 during the last 20 years as shown in Figure 1. Controlling or knowing the real thickness and composition of the films in these multilayered stacks is crucial for electronic research and development. To determine both thickness and composition of thin films, several techniques could be used, such as X-Ray Photoelectron Spectroscopy (XPS), Rutherford Backscattering Spectrometry (RBS) and Medium Energy Ion Scattering (MEIS). These techniques however require large and expensive instrumentations, are usually time consuming and destructive and do not have a small analysis spot. On the other hand, Energy Dispersive Spectroscopy (EDS) coupled to Scanning Electron Microscopy

(SEM) has the potential for fast analysis on small scale areas. In our previous work [1], we used mainly EDS to determine the thickness of both single and multi-layer systems. Very good correlation with other analytical methods was found down to a layer thickness of 0.5nm. In this paper, we show how EDS can be used to determine at the same time with a high accuracy both thickness and composition of complex multi-layer systems with a wide range of elemental composition. Moreover, we will show that EDS when coupled with SEM enables the thickness determination of advanced thin films in small areas.

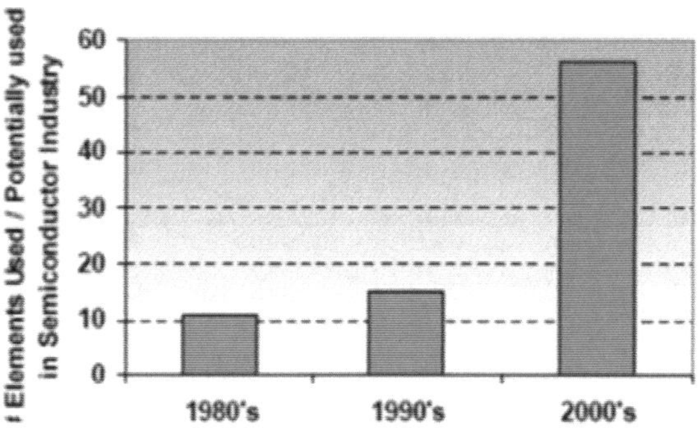

Figure 1. Number of elements used in semiconductor industry (source: sematech / intel).

2. Experimental

The EDS experiments were performed with two Oxford Instruments INCA Energy 350 Microanalysis systems including a 30 mm^2 INCA PentaFET-x3 Si(Li) detector. The resolution of these systems is better than 133eV at Mn Kα for count rates up to 4000. The EDS systems are mounted on two different SEM microscopes (FEI XL30 and FEI Nova200). The obtained EDS data were treated using both INCAEnergy and ThinFilmID software from Oxford Instruments. Calibration standards were measured in the same conditions as the samples under investigation.

3. Results

3.1. EDS for thin film analysis

The basis of the EDS technique is to measure the X-ray line intensities generated from the sample after irradiation by an electron gun. It is well known that these X-rays are generated not only in the top outer part of the sample but are collected from a volume of a few cubic microns of material [2]. The ThinFilmID software is taking advantage of this 'effect' of penetration of the beam in the analyzed sample in order to retrieve more information about the disposition of the elements. For doing this, the collected EDS spectrum by itself is not sufficient, but a description of the sample is also required. The X-ray intensities predicted from the theoretical model of the structure are then compared with the intensities obtained from the X-ray spectra. By using an iterative procedure, the thickness and composition at which the model matches the experimental data are calculated for each layer in the model. The calculation is calibrated using the X-ray intensities from bulk standards.

As an example, ThinFilmID was used to determine the thickness of a Pt metallisation layer deposited on SiO_2 on Si substrate. EDS spectra were recorded using two accelerating voltages of 10 and 20 kV. As shown in Figure 2, both SiKα and PtMα lines are observed under both conditions, the ratio of these lines being different due to the difference in beam penetration. Using a simple model where only the Pt layer thickness is allow to vary, the results reported in Table 1 were obtained. We could see that the thickness calculated from both accelerating voltages is in accordance with the nominal value. To confirm the validity of the ThinFilmID technique, the Pt metallisation layer thickness was also measured with RBS (see Table 1). The obtained result confirms the accuracy of the ThinFilmID procedure.

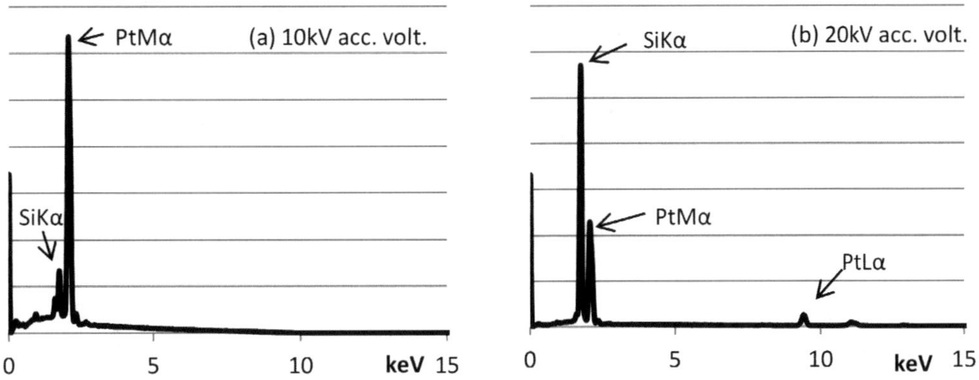

Figure 2. EDS spectra of $Pt/SiO_2/Si$; accelerating voltages (a) 10kV and (b) 20kV.

Table 1. Thicknesses measured for $Pt/SiO_2/Si$ stack.

Nominal thickness	100 nm
EDS thickness	105 nm (10kV) ; 100 nm (20kV)
RBS thickness	110 nm

3.2. EDS analysis of multilayered systems

The use of ThinFilmID has been applied to more complex systems, i.e. multilayered systems where both thickness and composition have to be determined at the same time. Figure 3 describes the stack that has been measured. It consists of an intermediated SiO_2 layer on a Si substrate where the thickness has to be determined. The top layer consists of TiN where both composition and thickness should be measured. Additionally to that, Al and Cl impurities were intentionally introduced in the TiN layer. Various impurity levels were used; this leads to the definition of three series of samples, i.e. A, B and C with no, low and medium Al contents, each series having also three different Cl contents, as indicated in Figure 3. In this context, ThinFilmID has been applied to determine at the same time the thickness of both SiO_2 and TiN layers together with the composition (including impurity levels) of the TiN layer.

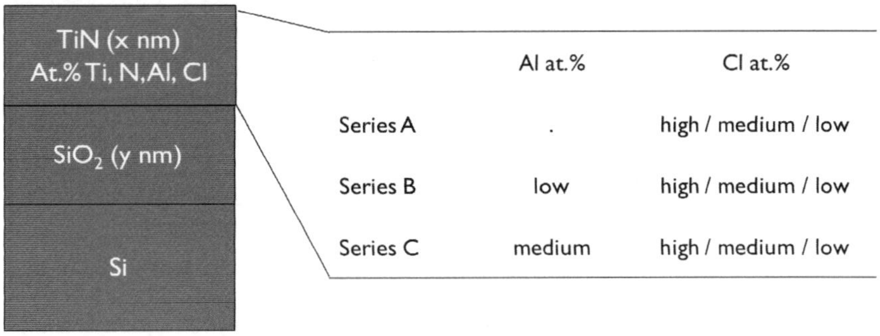

Figure 3. Description of the investigated multilayered system.

The results obtained from ThinFilmID are presented in Figure 4. It shows that the SiO_2 thickness is quite similar for every sample of each series and very close to the nominal values (20nm). On the other hand, we could see more variations in the TiN layer thickness. The three different Cl levels incorporated in the samples of series A (no Al) are clearly identified, i.e. a low, medium and high concentration is measured. A minimum value of 1at.% Cl was found. In series B, a low amount of Al is measured, i.e. between 0.5 and 1at.%, and the three Cl levels are again determined. Finally, similar results are observed for series C, i.e. for the samples with a medium Al content. To evaluate the accuracy of the EDS results, independent RBS analyses were performed on the same series. A relatively good correlation between the values obtained from both methods is found, as displayed in Figure 5.

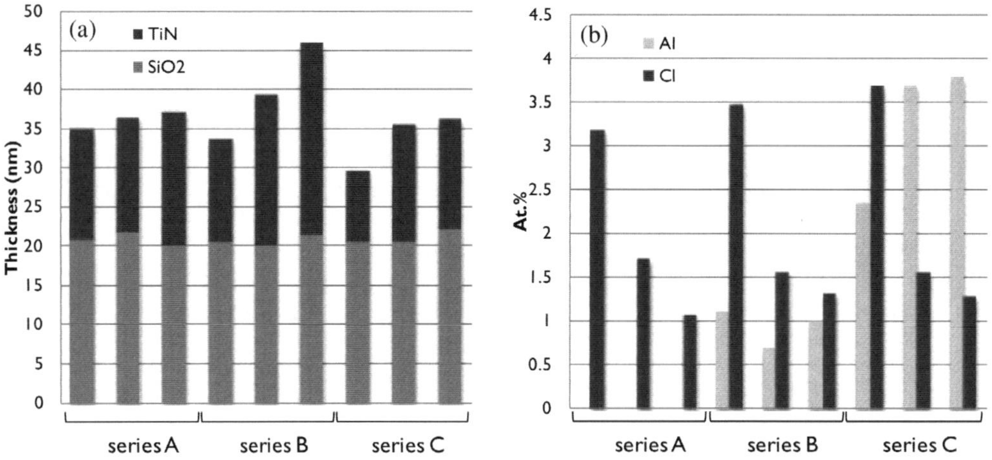

Figure 4. (a) TiN and SiO_2 thicknesses and (b) Aluminium and Chlorine concentrations determined from ThinFilmID.

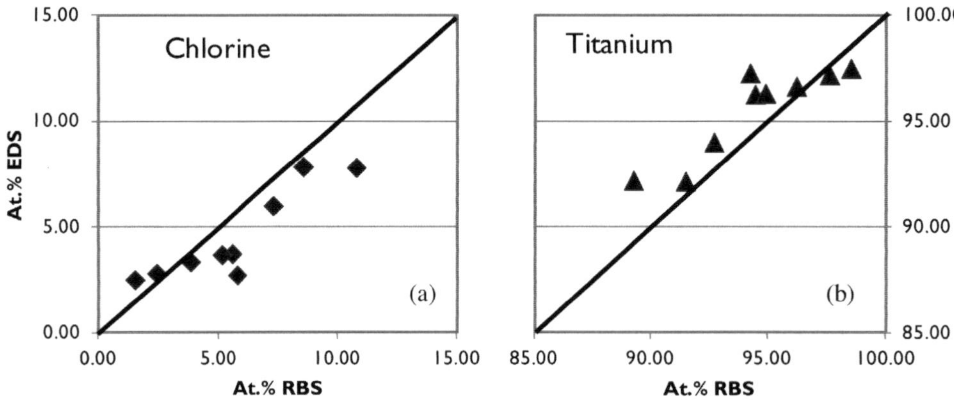

Figure 5. EDS and RBS correlation for (a) Chlorine and (b) Titanium concentrations.

3.3. Thickness measurements of patterned thin film structures

The ability of EDS to measure the thickness and composition of thin films has been demonstrated before. However, as indicated in the introduction, the size of the devices is not only decreasing in thickness but also in lateral dimension. It is therefore important for small patterned structures to evaluate the lateral size limitation for an accurate thickness measurement. To test the practical application of ThinFilmID to this localised measurement, a test wafer has been used (Figure 6). The wafer consists of nominally 128nm thick SiGe (nominally 24at.% Ge) structures on a Si substrate in a range of sizes, including square structures of length 10µm, 5µm, 1µm, 500nm, 300nm, 200nm and 180nm. The SiGe structures are surrounded by Si_3N_4/SiO_2 areas.

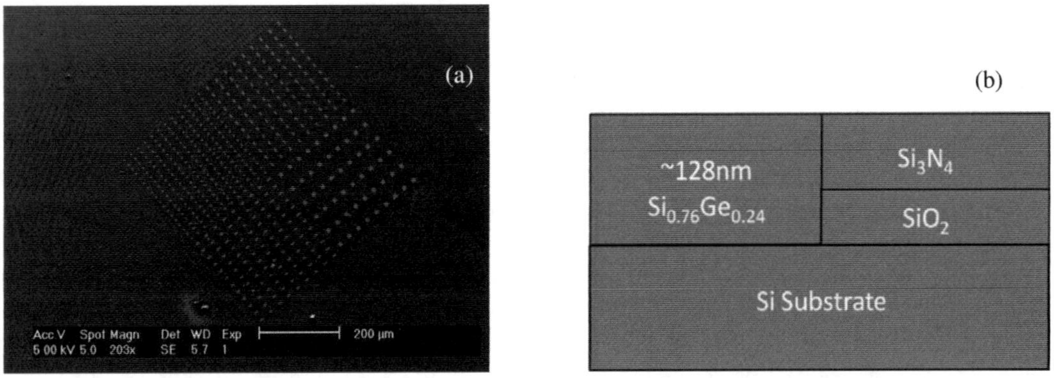

Figure 6. (a) SEM top image showing the array containing the different SiGe structures (bright intensity) and (b) schematic cross-section of the test wafer structures.

EDS spectra were collected from the centre of each square SiGe structure. The thicknesses of these structures as measured by ThinFilmID are plotted against the lengths (lateral dimensions) of the squares in Figure 7(a). The full line presents the results obtained with the nominal value of 24at.% Ge in the model. It shows that for the structures of 10μm, 5μm and 500nm an accurate thickness measurement of ~128nm is obtained. These values are in good agreement with the TEM results, as shown by the example of the 5μm SiGe structure displayed in Figure 7(b). The error on the calculated thicknesses for smaller structures is however larger, i.e. up to 10% for the 300nm and even more for the 200nm and 180nm square structures. This error is due to the generation of X-rays from the Si_3N_4/SiO_2 stack adjacent to the structure. The proportion of this surrounded area in the EDS spectrum is increasing with reducing structure sizes and thus explains the increase of the error observed on the thickness value.

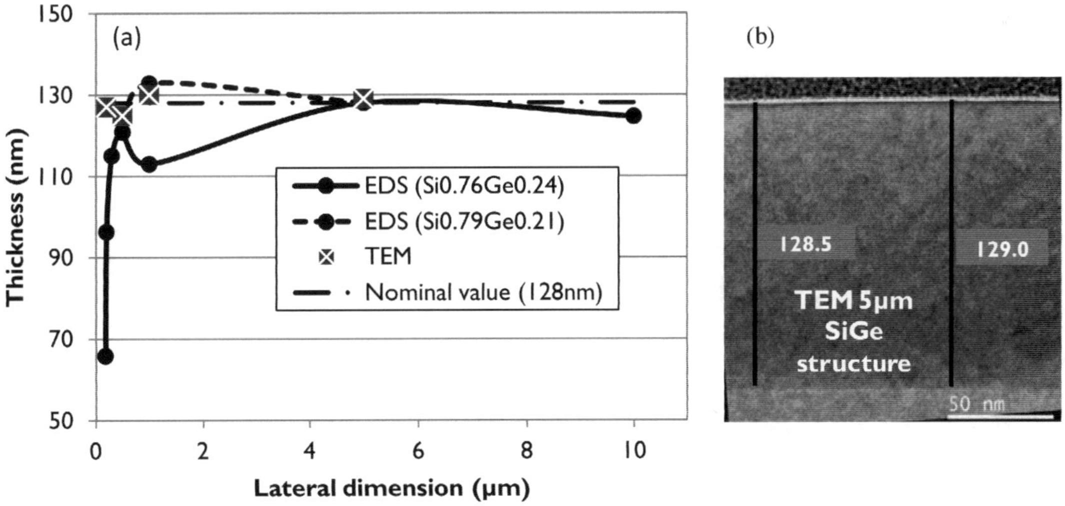

Figure 7. (a) Measured thickness of SiGe vs square structure length and (b) TEM image of 5μm SiGe square structure.

The measured thickness for the 1μm structure is however far from the expected value and could not be explained by the structure size limitation. To overcome that problem, Atom Probe (AP) analysis of this specific size has been performed. In AP, a strong electric field induced by applying a bias of several kilovolts is needed to allow the ejection of atoms from the sample surface; however, to achieve this high electric field the sample has to be manufactured in a tip-shape structure. Figure 8(a) shows the tip prepared from the 1μm square SiGe structure using Focused Ion Beam (FIB) milling, the Pt cap layer is used for sample preparation. The 3D AP reconstruction of the tip is displayed in Figure 8(b). The various components of the tip are clearly observed, i.e. Pt cap layer, SiGe structure and Si substrate. Because of Pt incorporation in the first nm of the SiGe film during FIB preparation (indicated by the intermediate layer), the thickness of the SiGe layer could not be determined precisely. However, the SiGe composition could be easily determined. A value of 21at.% Ge could be measured, which is lower than the nominal 24at.% value. Previous work has demonstrated that AP is quantitative in that range of SiGe concentrations [3]. This difference of concentration could of course be the source of error in the EDS thickness determination. A new ThinFilmID calculation was therefore made using 21at.% Ge in the model for the 1μm square structure; the results are represented by the dotted line in Figure 7(a). A SiGe thickness of around 132nm is found which is in agreement with the TEM results.

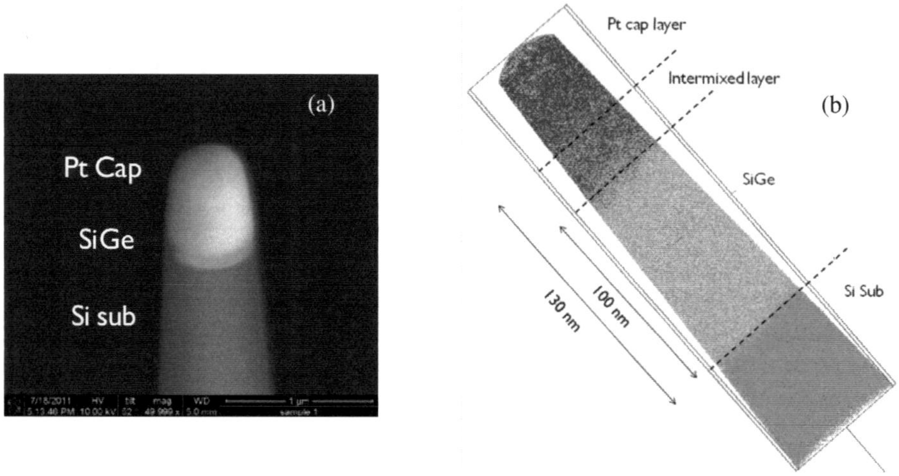

Figure 8. (a) SEM image of the tip fabricated for Atom Probe analysis and (b) 3D Atom Probe reconstruction of the 1μm SiGe square structure.

4. Conclusions

In this work, we show that using proper modelling of the investigated systems, EDS can measure with high accuracy both thickness and composition of multilayered stacks. This is demonstrated by the good correlation existing between the EDS results and independent measurements by RBS, TEM and Atom Probe. It was also shown that the thickness of patterned structures can be measured accurately using ThinFilmID down to lateral sizes of at least 500nm. This is significantly smaller than the 0.5-1mm window sizes often required for other techniques.

5. Acknowledgments

S. Burgess of Oxford Instruments (UK) is acknowledged for his support with the ThinFilmID application. O. Richard and H. Bender (imec, Belgium) are thanked for their help in TEM analyses.

References

[1] T. Conard, K. Arstila, T. Hantschel, A. Franquet, W. Vandervorst, E. Vecchio, S. Burgess and F. Bauer, Mater. Res. Soc. Symp. Proc. Vol. 1184, HH08-08 (2009).

[2] Y. Leng in "Materials Characterization: Introduction to Microscopic and Spectroscopic Methods", JohnWiley & Sons (Asia) Pte Ltd (2008).

[3] S. Koelling, M. Gilbert, J. Goossens, A. Hikavyy, O. Richard and W. Vandervorst, Appl. Phys. Letter 95, 144106 (2009).

Time-evolution of oxidation states at the Ni(111) surface: O_2 incident translational energy dependence

Keisuke Inoue[1,2], Yuden Teraoka[1,2]

1 Quantum Beam Science Directorate, Japan Atomic Energy Agency, 1-1-1 Kouto, Sayo-cho, Sayo-gun, Hyogo 679-5148, Japan

2 Graduate School of Material Science, University of Hyogo, 3-2-1 Kouto, Kamigori-cho, Ako-gun, Hyogo 678-1297, Japan

E-mail:yteraoka@spring8.or.jp

Abstract. Metallic Ni is expected to be a substitution candidate for Pt and Pd in catalytic materials. Generally speaking, catalytic reactions take place on a metal oxide layer. In this study, therefore, the oxidation states of the Ni(111) surface, which were made by irradiation with a supersonic O_2 molecular beam, were analyzed by soft x-ray photoemission spectroscopy with synchrotron radiation. The oxygen uptake curve and initial sticking rate were found to show remarkably strong dependence on the O_2 incident energy for energies of up to 2.3 eV. The intermediate plateau seen in the oxygen uptake curve for low incident energies was found to disappear with increasing incident energy due to a change of the dissociative adsorption mechanism from a two-dimensional island growth model to a direct activated adsorption model. Due to this activated adsorption, the formation rate of NiO and peroxide nickel increased as compared to backfilling oxidation by O_2 gas.

1. Introduction

The Ni(111) surface reacts with O_2 molecules to form a thin oxide layer on the surface at 147 K [1] showing a Langmuir-type oxygen uptake curve. When the clean Ni(111) surface is exposed to O_2 gas at higher temperatures, a plateau appears in an intermediate region of the oxygen uptake curve [1]. These characteristics are also observed in the oxidation of the Ni(001) surface [2]. This behaviour of the oxygen uptake curves can be explained by a two-dimensional (2D) oxide island growth model [2, 3]. Generally speaking, the surface chemical reaction rate of incident atoms and molecules depends also on their translational kinetic energy. Indeed, the oxygen uptake curve for Ni(111) oxidation by an O_2 molecular beam with a hyperthermal energy of 0.6 eV at a surface temperature of 300 K was observed to change from the 2D island growth model to the Langmuir type [4]. This implies that the kinetics of oxygen adsorption at the Ni(111) surface depends actually on the translational kinetic energy of the O_2 molecules, with higher energies allowing a direct activated adsorption mechanism [4]. Here the oxygen uptake showed an intermediate plateau of 0.2 monolayer (ML), and finally saturated at 3.0 ML [4] at 300 K. Translational kinetic energy effects have also been shown for the initial sticking coefficients of O_2 molecules on the Ni(111) surface [5]. Although a trapping-mediated adsorption mechanism was apparent at 123 K for incident energies lower than 0.05 eV, activated adsorption was found to take place in the higher incident energy region, with energies of up to 0.4 eV [5].

The incident energy of O_2 molecules in a beam can be increased to 2.3 eV by using our supersonic molecular beam (SSMB) technique. Using this technique, the effects of incident energy on O_2 adsorption at the Ni(111) surface have been studied using soft x-ray photoemission spectroscopic surface analysis with high brilliance and high energy-resolution synchrotron radiation (SR-PES). The surface chemistry experimental station (SUREAC2000) of BL23SU, one of Japan Atomic Energy Agency's (JAEA) contract beamlines at the SPring-8 facility was used in these experiments [6,7]. Both an O_2 SSMB and the soft x-ray monochromatic synchrotron radiation beam can be irradiated on a sample surface simultaneously at SUREAC2000. Real-time in-situ photoemission spectroscopic observations of chemical bonding states at the oxidized Ni(111) surface can be performed during O_2 SSMB irradiation to elucidate changes in oxidation kinetics by the incidence of hyperthermal O_2 molecules possessing translational energies larger than 0.6 eV. Furthermore, the incident energy dependence of the initial sticking rate can also be estimated from the initial slope of the oxygen uptake curve. In this work, the influence of the incident energy of O_2 molecules on the oxidation of Ni(111) surface is discussed in the energy region up to 2.3 eV.

2. Experimental

All the experiments were conducted using the SUREAC2000 apparatus. The surface reaction analysis chamber of the SUREAC2000 is pumped by two turbo-molecular pumps in tandem fashion in conjunction with a titanium sublimation pump so that a base pressure less than 5×10^{-9} Pa is achieved. A hemispherical electron energy analyzer (Omicron EA125-5MCD) and a conventional twin anode (Al/Mg-K_α) x-ray source are also available for photoemission experiments.

The O_2 SSMB is continuously generated by the adiabatic expansion of a mixture of O_2, He and Ar gas. The free expansion nozzle can be heated up to about 1400 K, corresponding to a calculated upper limit for the incident energy of O_2 molecules of 2.3 eV. The incident energy is controllable by varying the gas mixing ratio and the nozzle temperature. Usually, the incident energy is changed by varying the gas mixing ratio keeping the nozzle temperature constant at 1400 K. The O_2 SSMB was irradiated onto the sample surface with an incident angle of -10 deg with respect to the surface normal. The photoelectron detection was +30 deg. Incident energies less than 0.2 eV were achieved with a nozzle temperature of 300 K. A backfilling experiment (exposure to O_2 gas) was also conducted to obtain data for an incident energy of 0.03 eV.

A clean Ni(111) surface was made by cyclic treatments of Ar^+ ion sputtering (applied voltage;1kV, ion current;1μA) and thermal annealing (1273 K flashing followed by 923 K for 20 min). The cleanliness of the surface was confirmed by measuring an SR-PES wide spectrum and the observation of a 1×1 low energy electron diffraction (LEED) pattern.

After timed irradiation of the O_2 SSMB onto the clean Ni(111) surface, Ni-2p, Ni-3p and O-1s core level photoemission spectra were recorded. Cycles of O_2 SSMB irradiation and SR-PES measurements were repeated at each incident energy until saturation of the O-1s peak intensity. Oxygen uptake curves were obtained from the spectra. The surface temperature was kept at 300 K during both O_2 SSMB irradiation and SR-PES measurements. SR energies of 680 eV and 1150 eV were used.

3. Results and discussion

3. 1. Oxygen uptake curves

The time evolution of the O-1s photoemission spectra towards oxidation saturation is shown in figure 1 for incident O_2 energies of 0.7 eV, 1.4 eV, and 2.3 eV, respectively. All O-1s photoemission spectra show a broad and asymmetric profile which changes with incident energy. The spectral profiles imply that the O-1s peak consists of several components. Figure 1 also demonstrates that the oxygen content at saturation also depends on incident energy.

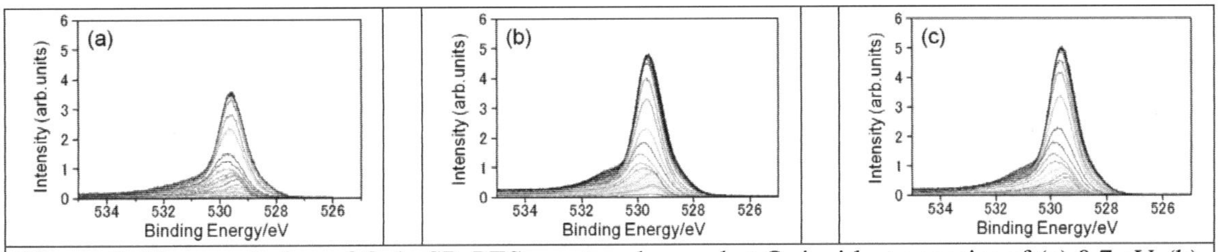

Figure 1. Time-evolution of O-1s SR-PES spectra observed at O_2 incident energies of (a) 0.7 eV, (b) 1.4 eV, and (c) 2.3 eV, respectively.

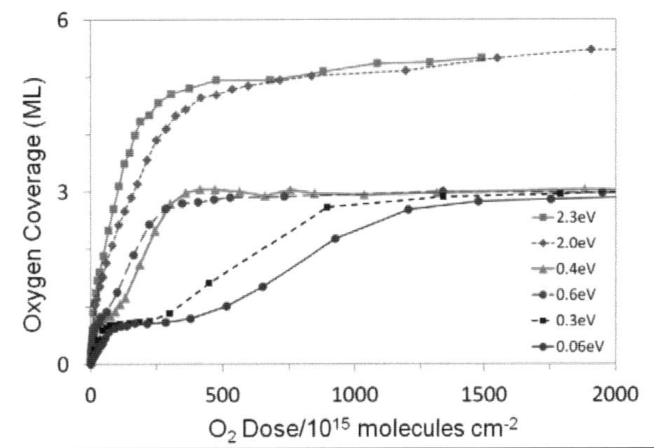

Figure 2. O_2 incident energy dependence of oxygen uptake curve in the oxidation of Ni(111) surface by an O_2 SSMB.

If the area of an O-1s peak is assumed to be proportional to the oxygen content at the Ni(111) surface, an oxygen uptake curve can be obtained by plotting the areas of O-1s peaks as a function of O_2 dose. Figure 2 shows the incident O_2 energy dependence of the oxygen uptake curves. For low incident energy (e.g. 0.06 eV), the uptake curve shows an initial uptake (fast chemisorption stage), an intermediate plateau (oxide onset) followed by a re-increase (fast oxide growth stage) of oxygen coverage. These characteristics can be seen in previous reports [1-4]. The oxygen content at saturation has been reported to be 3 ML [4]. For a higher incident energy region than 0.6 eV, the intermediate plateau region shortened, and the oxygen coverage is higher in the intermediate O_2 dose region. For incident energies higher than 1.0 eV, the plateau almost disappeared as indicated in the 2.0 eV and 2.3 eV uptake curves in figure 2 and the oxygen content at saturation is about 5 ML, larger than that for an incident energy of 0.06 eV. This implies a change in the O_2 dissociative adsorption mechanism from a 2D island growth model to the Langmuir-type direct activated adsorption model. As mentioned below, the increase of oxygen saturation content can be attributed to the increase of the main components shown in figure 4. These are the Ni_xO_y ($x<y$) peroxide component (c) and the nickel oxide component NiO (d).

3.2. Initial sticking rates of O_2 molecules

In order to investigate the influence of incident energy on the dissociative adsorption of O_2 molecules, the relative initial sticking rates for a clean Ni(111) surface were evaluated from the initial slope of each oxygen uptake curve. Figure 3 shows the relative initial sticking rates as a function of O_2 incident energy. The initial sticking rate increased gradually as incident energy increased, and reached a maximum at 1.0 eV. After that the sticking rate decreased slightly and a remarkable re-increase was observed in

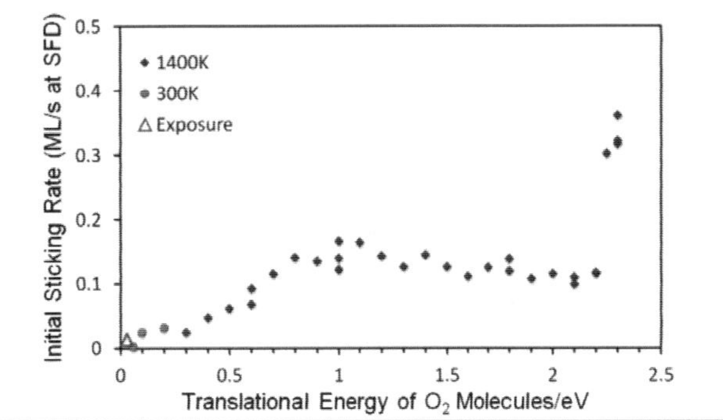

Figure 3. O_2 incident energy dependence of initial sticking rate, ◆ for nozzle temperature of 1400 K, ● for nozzle temperature of 300 K, and △ for exposure to O_2 gas. The initial sticking rate is expressed in units of ML· s^{-1} at a standard O_2 SSMB flux (10^{15} molecules· cm^{-2}).

the region around 2.3 eV. Such behavior is typical for quantum mechanical barrier penetration and over-barrier scattering. The first increase seen for incident energies below 1.0 eV implies that the activated dissociative adsorption of O_2 molecules takes place by penetrating a potential energy barrier of 1.0 eV height. The abrupt increase around 2.3 eV implies that high speed O_2 molecules adsorb by a mechanism which is distinctly different to that which occurs at energies of around 1.0 eV. That is, another potential energy barrier possessing a barrier height larger than 2.3 eV exists in a position closer to the surface than the first potential energy barrier.

3.3. Peak deconvolution of O-1s photoemission spectra

As seen in figure 1, the time-evolution of the O-1s peak profile is not monotonic at each incident energy. The variation of the peak profile during dosage of O_2 molecules implies that the oxidation states of nickel change depending on oxygen coverage and incident energy. We assume here that the O-1s peak consists of six components. A typical peak deconvolution result is shown in figure 4. The main component is (d) NiO. The sub-oxide (f) Ni_xO_y (x>y) and the peroxide (c) Ni_xO_y (x<y) components are incorporated in the deconvolution in addition to (e) an oxygen chemisorption component, which corresponds to initially-chemisorbed oxygen atoms at a three-fold site on the top surface of Ni(111) [8]. The peak labelled (b) near 531 eV may be due to $Ni(OH)_2$ generated from dissociative adsorption of residual H_2O gas molecules during the measurements. Residual H_2O gas molecules also adsorb gradually at the oxidized Ni(111) surface during experiments to form the H_2O component (a). Previous studies indicated that peaks related to H_2O- and OH- appear on the higher energy side of the main O-1s peak [9,10].

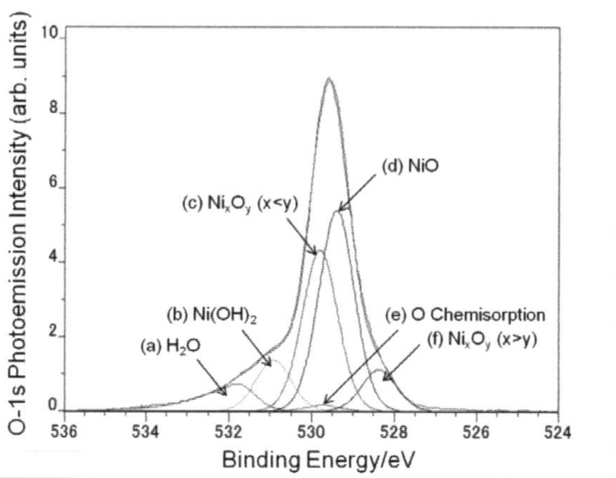

Figure 4. Peak deconvolution of O-1s SR-PES spectrum in the oxygen-saturated Ni(111) surface formed by irradiation of 2.3-eV-O_2 SSMB; (a) adsorbed residual H_2O, (b) nickel hydroxide $Ni(OH)_2$, (c) nickel peroxide Ni_xO_y (x<y), (d) NiO, (e) chemisorbed O atom on the 3-fold site on the Ni(111), and (f) nickel sub-oxide Ni_xO_y (x>y), respectively.

The peak deconvolution enables us to derive the time-evolution of each component during the growth of the oxide layers. The time-evolutions of the four major O-1s components are shown in figure 5. In the early oxidation stages, O_2 chemisorption on the topmost Ni(111) surface initially takes place as the uptake of (e) O-chemisorption component occurs primarily, as shown in figure 5(a). Subsequent collisions of incident O_2 molecules with the initially-formed O-chemisorbed local structures make the O-chemisorption sites change to nickel peroxides Ni_xO_y (x<y). If oxygen diffusion and migration occurs from the peroxide sites, NiO structures may be formed near the peroxide sites. The sub-oxide nickel should be formed in the interface region between the Ni(111) substrate and the oxide over-layer.

With increasing O_2 incident energy, the rate of activated dissociative adsorption through the first potential energy barrier (1.0 eV height) also increases, leading to an abrupt appearance of the uptake of the peroxide component. The uptake of NiO also follows, as shown in figure 5(b). For incident energies higher than 1.0 eV, the NiO content becomes larger than the peroxide content, as shown in figure 5(c). This tendency is prominent at an incident energy of 2.3 eV, as shown in figure 5(d), where the second dissociative oxidation channel is opened through the second potential energy barrier. Collisions of high speed O_2 molecules with the peroxide local structures or another nickel oxide local structure enhance the formation rate of NiO. According to prediction based on the Cabrera-Mott model on the growth of NiO layer by O_2 gas, potential barrier heights of 2.36 eV [11] and 2.56 eV [12] have been reported. Our experimental results are consistent with these theoretical predictions.

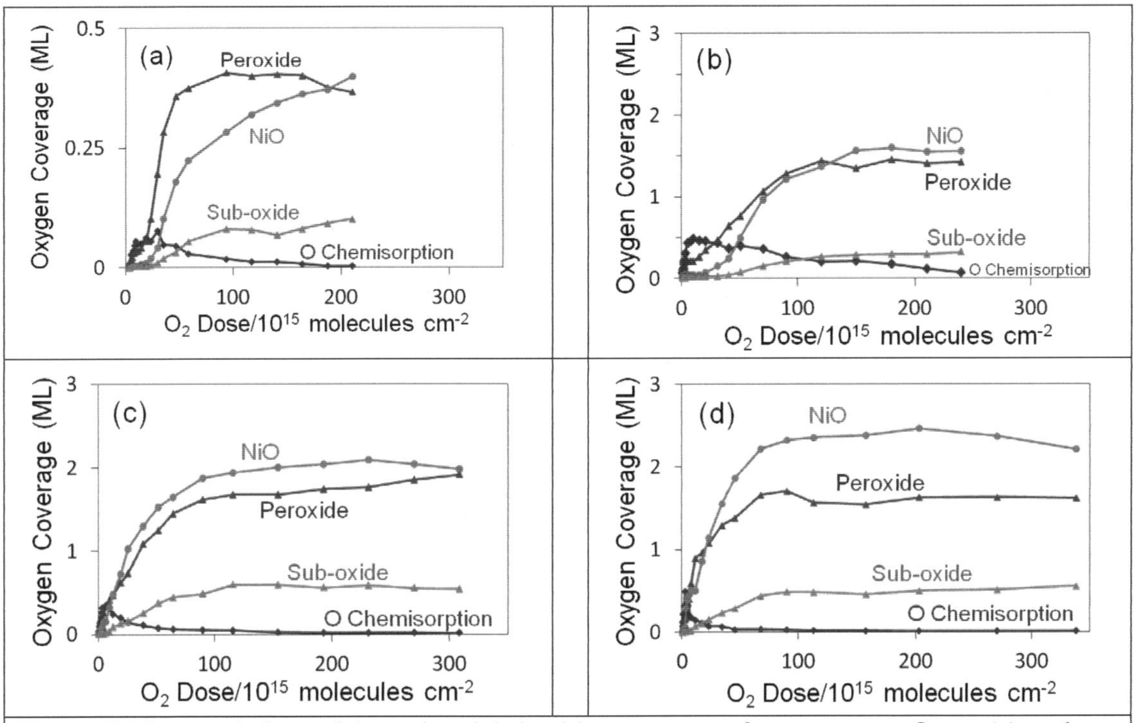

Figure 5. Time-evolutions of the major nickel oxide components for exposure to O_2 gas (a), and to a molecular beam with incident O_2 energy of 0.7 eV (b), 1.2 eV (c), and 2.3 eV (d), respectively.

4. Conclusions

Oxidation processes of a clean Ni(111) surface with hyperthermal O_2 molecular beams have been studied by using soft x-ray photoemission spectroscopy with synchrotron radiation. Oxygen uptake curves have been observed at various O_2 incident energies, extending to energies higher than those of previous reports. The first plateau, which appears at an intermediate region of the oxygen uptake curve, was found to disappear with increasing O_2 incident energy. Furthermore, an oxygen content at saturation of 5 ML was observed for an incident energy of 2.3 eV. This is 1.7 times higher than that observed for energies of 0.06 eV to 0.6 eV. These observations can be attributed to the influence of activated adsorption mechanisms. The initial sticking rate continued to increase as incident energy increased up to 1.0 eV, and a remarkable re-increase was observed in the region around 2.3 eV. The first increase is due to the activated dissociative adsorption of O_2 molecules through a potential energy barrier of 1.0 eV, and the second increase around 2.3 eV implies that another potential energy barrier exists with a height larger than 2.3 eV. Higher O_2 incident energy leads to higher formation rates and higher contents at saturation for the NiO component.

Acknowledgments

This work was supported in part by a Grant-in-Aid for Scientific Research from the Japan Society for the Promotion of Science (JSPS) (20360024). The synchrotron radiation experiments were performed at BL23SU, SPring-8 with the approval of the Japan Synchrotron Radiation Research Institute (JASRI) (Proposal Nos.2011A3802, 2010B3802, 2010A3802, 2008B3803, 2008A3803, 2007A3801, and 2006B3802).

Journal of Physics: Conference Series **417** (2013) 012034 doi:10.1088/1742-6596/417/1/012034

References

[1] Holloway P H and Hudson J B 1974 *Surf. Sci.* **43** 141
[2] Holloway P H and Hudson J B 1974 *Surf. Sci.* **43** 123
[3] Holloway P H 1981 *J. Vac. Sci. Technol.* **18** 653
[4] Zion B D, Hanbicki A T and Sibener S J 1999 *Surf. Sci.* **417** L1154
[5] Beutl M, Rendulic K D and Castro G R 1997 *Surf. Sci.* **385** 97
[6] Teraoka Y, Yoshigoe A and Moritani K 2008 *Electrical Engineering in Japan* **164** 60
[7] Teraoka Y and Yoshigoe A 1999 *Jpn. J. Appl. Phys.***38** Suppl.38-1 642
[8] Schwennicke C and Pfnür H 1996 *Surf. Sci.* **369** 2483-fold O-chemisorption
[9] Gallagher M E, Haq S, Omer A and Hodgson A 2007 *Surface Science* **601** 268
[10] Norton P R, Tapping R L and Goodale J W 1977 *Surface Science* **65** 13
[11] Okazawa T, Nishizawa T and Kido Y 2007 *Phys. Rev.* B **75** 033413
[12] Atkinson A 1985 *Rev. Mod. Phys.* **57** 437

Effects of magnetic flux density and substrate temperature on Ni films prepared by means of unbalanced magnetron sputtering assisted by inductively coupled plasma

K Otani, N Sakata, T Ozaki and K Kawabata

Graduate School of Science and Technology, Hiroshima Institute of Technology, Hiroshima 731-5193, Japan

E-mail: m111003@cc.it-hiroshima.ac.jp

Abstract. We prepared Ni films on glass substrates by means of Unbalanced Magnetron Sputtering Assisted by Inductively Coupled Plasma (UM-ICP). The effects of the magnetic flux density B_C and the substrate temperature T_S on their structures were investigated. We found that the surface grain size observed with an atomic force microscope became large as B_C and T_S increase. We also found that the (111) plane measured by X-ray diffraction was preferentially orientated on the films under B_C=5 mT at T_S=60°C. These indicate that the energetic particle bombardment under B_C and the increase of T_S promote the surface diffusion on the Ni films. The ferromagnetic domain patterns observed with a magnetic force microscope showed the stripe domain structure having the domain width increasing with increasing T_S and the fractal dimension of 1.88.

1. Introduction

Magnetron sputtering is an essential technology to deposit films in electronic devices. The sputtering enables us to deposit films of high melting point materials and compounds with complex compositions. However, highly ferromagnetic materials can be difficult to sputter, because a considerable portion of magnetic flux is shunted by ferromagnetic target [1]. Kawabata *et al*. developed the sputtering system called Multipolar Magnetic Plasma Confinement (MMPC) [2], which made up for the shunted magnetic flux by externally arranged multipolar magnets. MMPC enables the high-speed deposition of Ni films with the rate twenty times larger than the other sputtering methods [3-5]. MMPC combined with a high-frequency coil, which is called MMPC assisted by Inductively Coupled Plasma (MMPC-ICP) [6], promotes the ionization of sputtering gas and the preferred orientation of the (111) plane in the Ni films. The replacement of the multipolar magnets by a solenoid in MMPC-ICP enables to vary the magnetic flux toward a substrate B_C. This system is called Unbalanced Magnetron Sputtering Assisted by Inductively Coupled Plasma (UM-ICP) and able to increase the number of the charged particles incident on the substrate and enhance the energetic particle bombardment, by increasing B_C [7].

The energetic particle bombardment promotes the surface diffusion on the substrate and affects the adhesion, the density and the grain size of films [8, 9]. The bombardment of the particles with suitable energies increases the grain size [10]. It is known that the surface diffusion is also promoted by thermal energy and makes the film highly dense in the temperature region of $0.3<T_S/T_M<0.5$ [11], where T_S and T_M are the substrate and the melting temperatures, respectively. This indicates that Ni films with T_M=1455°C become highly dense in the region of 245°C<T_S<591°C, which includes the ferromagnetic phase transition temperature T_C=358°C. The purpose of the present study is to clarify the effects of the magnetic flux density B_C and the substrate temperature T_S on the structure of the Ni films prepared by UM-ICP. The surface morphology and the magnetic domain structure were observed as their macroscopic structures and the lattice constant was measured as their microscopic structure at room temperature.

2. Experimental

The sputtering system UM-ICP used to deposit the Ni films is schematized in Figure 1. The pumping system is composed of a liquid nitrogen baffle, a turbo molecular pump and a rotary pump. The disposition of the magnets on the reverse of the target is symmetrical as shown in Figure 1. The solenoid was set up in the outside of the chamber. In order to enhance the ionization of sputtering gas and sputtered atoms, the high-frequency coil made of SUS304 was set up 60 mm above the target. Radio frequency power was applied to the high-frequency coil through a matching circuit. DC power was applied to the target through a low-pass filter. The heater was installed in the substrate holder. The target to substrate distance was fixed at 150 mm.

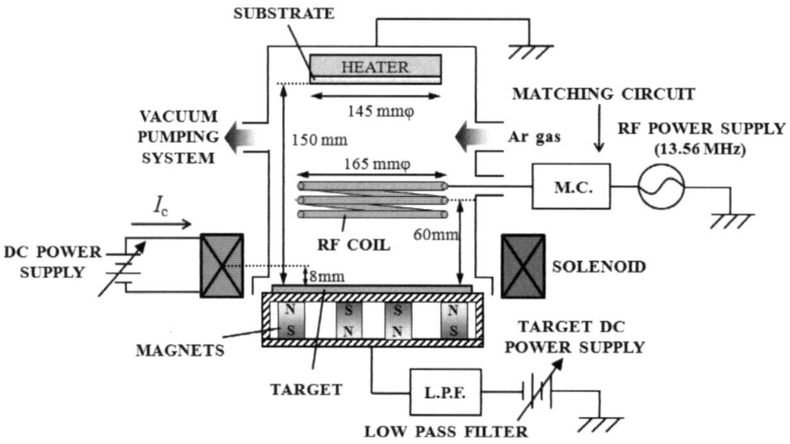

Figure 1. Schematic diagram of the present sputtering system UM-ICP.

The Ni target with the purity of 99.95%, the thickness of 5 mm and the diameter of 200 mm and the Ar gas with the purity of 99.9999% were used. The pressure and the flow rate of the Ar gas were maintained at 1.33×10^{-1} Pa and 1 sccm, respectively. The high-frequency power and the target DC power were maintained at 30 W and 600 W, respectively. The magnetic flux density at the center of the external solenoid B_C was maintained at 0 mT, 3 mT and 5 mT. Figure 2 shows the magnitude of the magnetic flux density measured at the position distant from the center of the substrate $B_\perp$. When B_C is maintained at 0 mT, 3 mT and 5 mT, respectively, $B_\perp$ has the averaged values of 1.1 mT, 2.3 mT and 3.3 mT on the wide area of the substrate.

The Ni films were deposited on the glass substrates with the area of 15 mm × 15 mm (Corning 1737), which were supersonic-cleaned by acetone and ethanol. The film thickness d was about 500 nm. The deposition rate varied between 26.5-32.5 nm/min. T_S was maintained at 90°C, 200°C, 300°C, 380°C under 0 mT, 75°C, 200°C, 300°C, 380°C under 3 mT and 60°C, 200°C, 300°C, 380°C under 5 mT. The lattice constants a of the Ni films were measured by X-ray diffraction (XRD) with the CuKα_1 radiation. The averaged value of the grain size D_G, that is, the diameter of the grain, and the RMS roughness R_q were obtained from the surface morphology observed with an atomic force microscope (AFM). The ferromagnetic domain patterns were observed with a magnetic force microscope (MFM). The average domain width l was obtained from the Fourier transformed figures of the domain patterns.

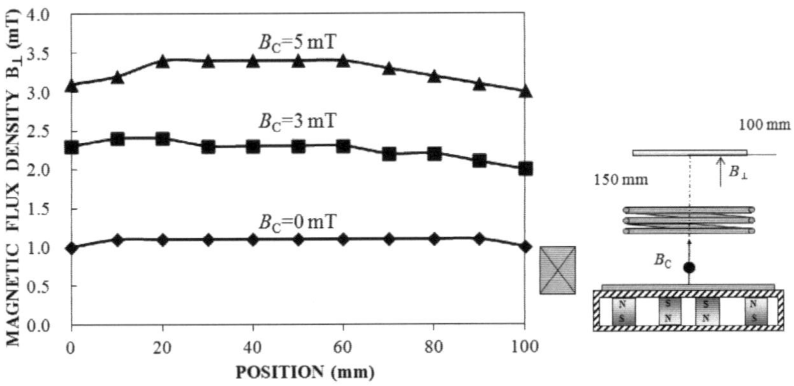

Figure 2. Magnetic flux density at the position distant from the center of the substrate $B_{\perp}$.

3. Results and Discussion

3.1. Surface morphology of Ni films

The AFM images of the Ni films are shown in Figure 3. The figure shows that the surface morphology strongly depends on the magnetic flux density B_C and the substrate temperature T_S. The average surface grain size D_G increases as B_C and T_S increases.

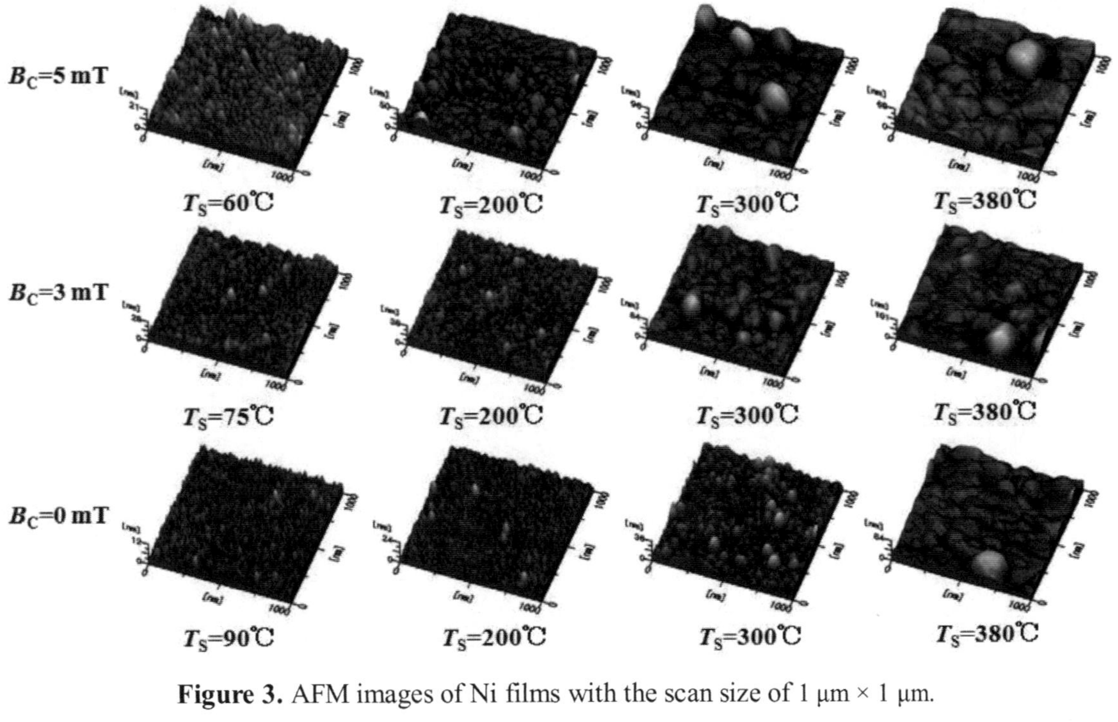

Figure 3. AFM images of Ni films with the scan size of 1 μm × 1 μm.

Figure 4(a) shows that D_G under B_C=5 mT at T_S=380°C is 4.5 times larger than D_G under B_C=0 mT at T_S=90°C. The RMS roughness R_q becomes large according to the increase in D_G, as shown in Figure 4(b). These show that the grains well grow under strong B_C and at high T_S. Therefore, the surface diffusion on the substrate was promoted by the energetic particle bombardment under strong B_C and by the heat distributed at high T_S.

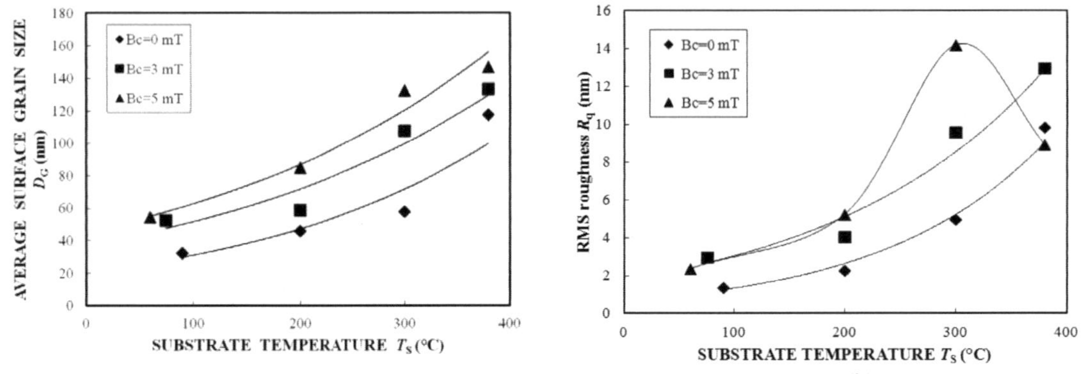

(a) (b)

Figure 4. Effects of the magnetic flux density B_C and the substrate temperature T_S on the average surface grain size D_G (a) and the RMS roughness R_q (b) in Ni films.

3.2. X-ray diffraction patterns of Ni films

The XRD patterns of the Ni films are shown in Figure 5. The clear peaks of the (111) and the (200) reflections show that the Ni films have the fcc structure and are composed of polycrystals. The increase in the peak intensity with the increase of B_C from 0 to 5 mT below T_S=200°C in Figure 5 coincides with the increase in the grain size below D_G=80 nm with increase of B_C in Figure 4(a).

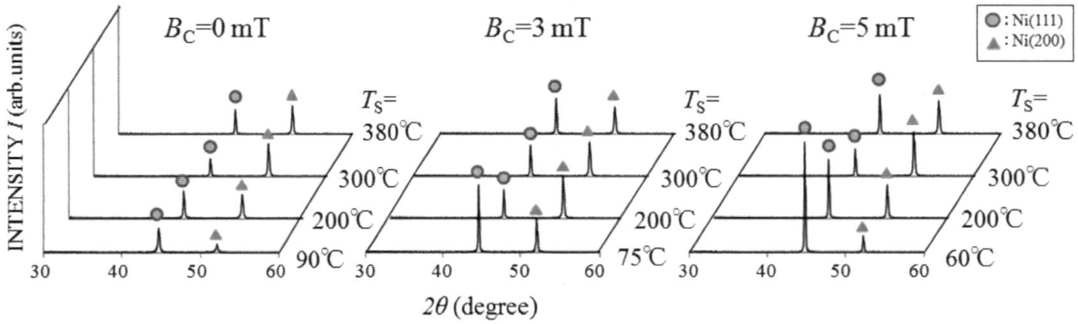

Figure 5. X-ray diffraction patterns of Ni films.

The lattice constants a of the Ni films deposited on the glass substrate were obtained from the XRD patterns shown in Figure 5. All values of a were smaller than the bulk lattice constant a_{bulk}=3.5241 Å [12], as shown in Figure 6. This shows the compressive stresses along the direction perpendicular to the Ni films. The substrate temperature dependence of the ratio between the integrated intensities $I(200)/I(111)$ is shown in Figure 7. The comparison of $I(200)/I(111)$ between the Ni films and the Ni powder brings us the information on the preferred orientation of the lattice plane in the Ni films deposited on the glass substrate. The preferred orientation of the (111) plane is shown under B_C=5 mT at T_S=60°C, indicating that the surface diffusion is promoted under B_C. This is because the preferred orientation of the (111) plane, of which the lattice spacing is smaller than that of the (100) plane, is enhanced by the energetic particle bombardment [6]. It is expected, therefore, that the surface diffusion would be highly promoted under B_C larger than B_C=5 mT. The preferred orientation of the (100) plane in the Ni films deposited at high temperatures would be due to a thermal activation process additional to the effect of B_C. The lowering of the preferred orientation of the (100) plane at T_S=380°C could be attributed to the lowering of the compressive stress due to the absence of magnetostriction above T_C=358°C.

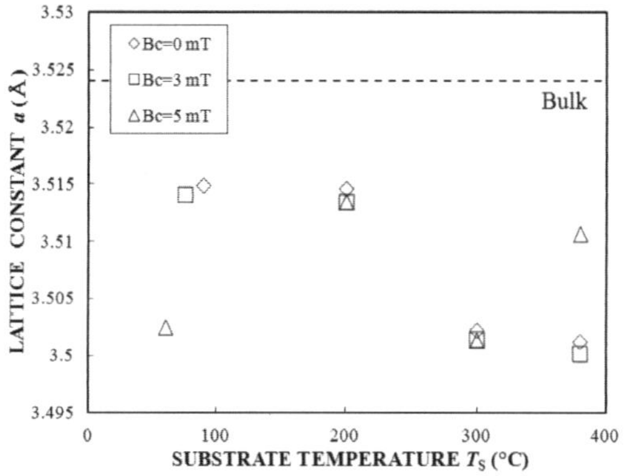

Figure 6. Effects of the magnetic flux density B_C and the substrate temperature T_S on the lattice constant a in Ni films.

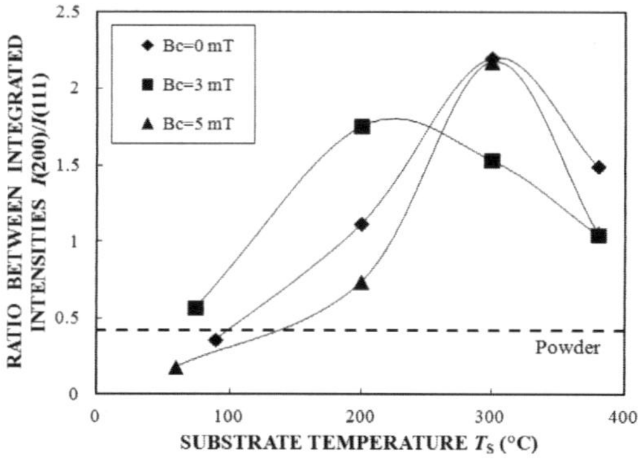

Figure 7. Effects of the magnetic flux density B_C and the substrate temperature T_S on the ratio between the integrated intensities $I(200)/I(111)$ in Ni films.

3.3. *Ferromagnetic domain patterns in Ni films*

The MFM images of the Ni films are shown in Figure 8. The bright and the dark regions correspond to the north and the south poles of the ferromagnetic domains, respectively. The domain patterns show that the lamellar domain structure, which is called "stripe domain structure" [13], formed in the Ni films.

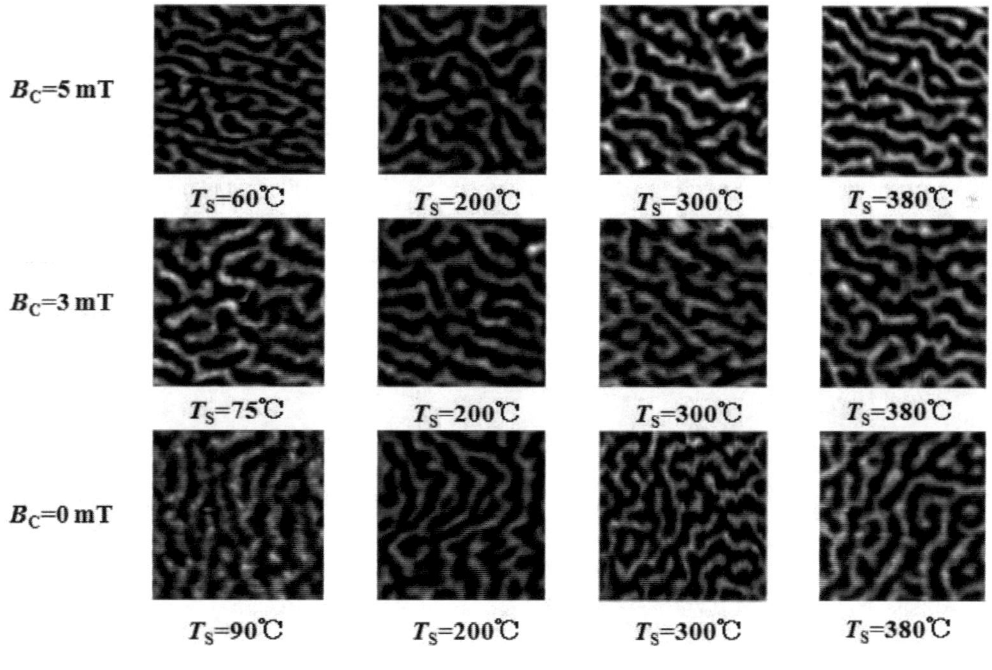

Figure 8. Ferromagnetic domain patterns of Ni films of 5 μm × 5 μm.

The average domain width $l=\lambda/2$ was obtained from the wave number $2\pi/\lambda$ which characterized the Fourier transformed figure of each domain pattern. The substrate temperature dependence of l is shown in Figure 9. The width l increases as T_S increases below T_S=200°C. This indicates that the spontaneous magnetization M_S decreased as T_S increased. This is because M_S of the stripe domain structure is proportional to $\sqrt{d}/l^{3/2}$ [13]. The width l of the Ni film deposited at B_C=5 mT and T_S=60°C is the smallest in Figure 9, showing that the value of M_S is the largest in the domain patterns in Figure

8. This is due to that the magnetization easy axis normal to the (111) plane is nearly perpendicular to the Ni film having the preferred orientation of the (111) plane in Figure 7. The decrease of l under low B_C and high T_S corresponds to the lowering of the preferred orientation of the (100) plane in Figure 7. All domain patterns showed the fractal dimension of 1.88, which is nearly equal to that of the percolation cluster 1.89 [14]. Thus the ferromagnetic domain structures observed at room temperature conserves the characteristics of the structures formed at T_S, though l should become large while the films are aged at room temperature.

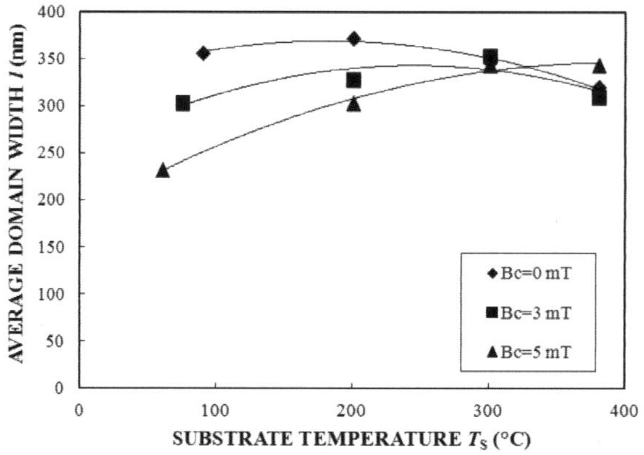

Figure 9. Effects of the magnetic flux density B_C and the substrate temperature T_S on the average domain width l in Ni films.

4. Conclusions

We prepared the Ni films on the glass substrates by means of the sputtering system UM-ICP and investigated the effects of the magnetic flux density B_C and the substrate temperature T_S on the structures of the films. We found that the surface grain size became large as B_C and T_S increased and the preferred orientation of the (111) plane occurred under B_C=5 mT at T_S=60°C. This indicates that the increases of B_C and T_S promote the surface diffusion on the Ni films. We also found that the characteristics of the ferromagnetic domain structures formed at T_S were conserved after cooling to room temperature. Thus we conclude that the structure of the Ni films prepared by the sputtering system UM-ICP is controllable by B_C and T_S.

References

[1] Zhang H, Poole J, Eller R and Keefe M 1999 *J. Vac. Sci. Technol.* A **17** 1904
[2] Kawabata K, Tanaka T, Kitabatake A, Yamada K, Mikami Y, Kajioka H and Toiyama K 2001 *J. Vac. Sci. Technol.* A **19** 1438
[3] Ghebouli B, Layadi A and Kerkache L 1998 *Eur. Phys. J. AP* **3** 35
[4] Howell R S, Sarcona G, Saha S K and Hatalis M K 2000 *J. Vac. Sci. Technol.* A **18** 87
[5] Yu M, Qiu H, Chen X, Wu P and Tian Y 2008 *Thin Solid Films* **516** 7903
[6] Takesako K, Okayama Y and Kawabata K 2006 *J. Vac. Soc. Jpn.* **49** 156
[7] Eboshi R, Muraoka Y, Kurokawa S, Tachibana A, Hasui Y and Kawabata K 2008 *J. Vac. Soc. Jpn.* **51** 182
[8] Mattox D M 1989 *J. Vac. Sci. Technol.* A **7** 1105
[9] Rossnagel S M and Cuomo J J 1989 *Thin Solid Films* **171** 143
[10] Asahara H, Inokuchi A, Watanuki K, Hirayama M, Teramoto A and Ohmi T 2009 *Jpn. J. Appl. Phys.* **48** 04C135
[11] Thornton J A 1986 *J. Vac. Sci. Technol.* A **4** 3059
[12] Ha K, Ciria M, O'Hndley R C, Stephens P W and Pagola S 1999 *Phys. Rev.* B **60** 13780
[13] Saito N, Fujiwara H and SugitaY 1964 *J. Phys. Soc. Jpn.* **19** 421, 1116
[14] Stauffer D 1979 *Phys. Rep.* **54** 1

Preferential adsorption of C_{60} molecules to step edges of the Si(110)-16 × 2 single domain surface

Y Yokoyama[1,2], Y Yamada[1], H Asaoka[2] and M Sasaki[1]

[1]Institute of Applied Physics, University of Tsukuba, Tsukuba, Ibaraki 305-8573, Japan
[2]Quantum Beam Science Directorate, Japan Atomic Energy Agency, Tokai, Ibaraki 319-1195, Japan

E-mail: yokoyama@mac62.bk.tsukuba.ac.jp

Abstract. The Si(110) surface has a quite unique one-dimensional (1-D) stable structure with 16 × 2 reconstruction. Since the perfectly straight rows extend over micrometers, the 16 × 2 structure can be a superior template for 1-D nanostructures. However, controlled nanostructures using the Si(110)-16 × 2 template has not been fabricated yet. In this study, we have investigated the adsorption of C_{60} molecules on the Si(110)-16 × 2 surface at various surface temperatures, and found that the C_{60} preferable adsorption sites are substantially changed, depending on the surface temperature. The obtained results indicate that the organic 1-D nanostructure can be fabricated on the Si(110)-16 × 2 surface by controlling the surface temperature.

1. Introduction

The Si(110) surface has been recently attracting the interests as a candidate for the next-generation metal-oxide-semiconductor field effect transistor (MOSFET) devices because of its high hole mobility compared with other low-index Si surfaces [1-3]. At the same time, it is also known that Si(110) has a quite unique one-dimensional (1-D) 16 × 2 reconstructed structure [4,5]. The 16 × 2 reconstructed structure consists of straight rows of Si adatom units (pair of pentagons; PPs), which run along two equivalent directions, [-112] and [1-12] [6-8]. Therefore, the clean Si(110)-16 × 2 surface usually exhibits a double-domain structure. Furthermore, various reconstructed structures other than 16 × 2 are generated due to small amounts of impurities [9].

Well-defined single domain Si(110)-16 × 2 surface is fabricated by means of controlled dc resistivity heating [10,11]. The dc heating can bring about the electromigration of surface Si atoms to line up the reconstruction rows when the electric current direction matches the orientation of the rows. The annealing temperature, time and current direction are dominant parameters for the 16 × 2 single domain preparation. In this study, we have prepared the surface under the optimized condition; the dc heating along [-112] orientation at 1473 K for 30 sec and prolonged annealing at 873 K for 20 min followed by quenching to room temperature (RT). Figure 1(a) displays a set of the low energy electron diffraction (LEED) and the scanning tunneling microscope (STM) images of the Si(110) surface after dc heat treatment. Figure 1(a) clearly shows that the 16 × 2 single domain is formed throughout the specimen. Figures 1 (b) and (c) show the magnified STM images of the 16 × 2 surface at different sample biases (V_s). At the positive V_s, Si atoms that constitute PPs are resolved (lower right of fig. 2 (b)). On the other hand, at the negative V_s, the electronic states originating from step edges are visualized.

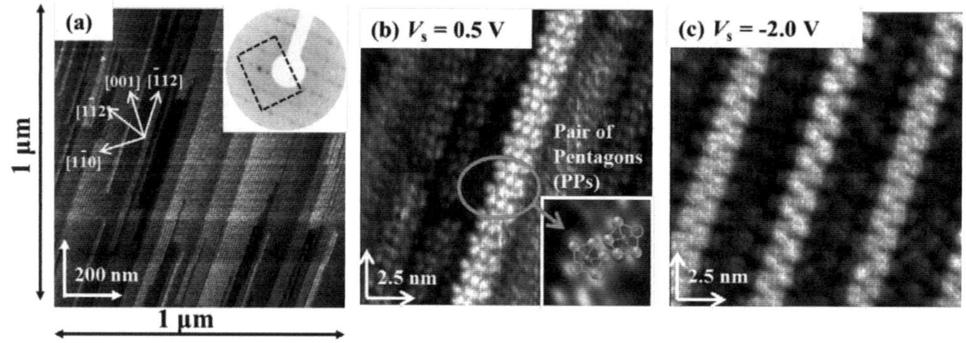

Figure 1: (a) STM and LEED images of the clean Si(110)-16 × 2 single domain surface (1 μm × 1 μm, V_s = 2.0 V). Magnified images (15 nm × 15 nm) obtained at different sample biases of (b) V_s = 0.5 V and (c) V_s = -2.0 V.

Since the perfectly straight rows can extend over micrometers, the 16 × 2 structure can be used as a template of low-dimensional nanostructures. However, the controlled synthesis of nanostructures using the Si(110)-16 × 2 template has not been achieved yet. Especially, the initial process of organic nanostructure formation on the Si(110)-16 × 2 surface has not been studied. Investigating the initial process of adsorption is of great importance, which is needed for controlled nanostructure fabrication. In this study, we have deposited a small amount of C_{60} molecules on the Si(110)-16 × 2 single domain surface at various temperatures to investigate the initial process of nanostructure formation. C_{60} is a typical n type organic semiconductor material, and is a very important molecule for organic semiconductor devices. Moreover, since the size of C_{60} is suited to the width of stripe structure, we expect to align well it on the 16 × 2 substrate.

2. Experimental

The experiments have been carried out by using an ultrahigh vacuum (UHV) STM (JEOL Ltd. JSPM 4500A) system equipped with molecular deposition sources and LEED. The base pressure in the measurement chamber is ~ 2 × 10^{-8} Pa. A medium-doped p-type Si(110) of 0.01 ~ 0.02 Ω cm (size of 1 × 7 × 0.38 mm^3) has been used for the specimen. A single domain 16 × 2 reconstruction surface has been prepared by the dc heating procedure. The deposition of C_{60} molecules are carried out by using a Ta crucible. The deposition rate is calibrated by using a quartz microbalance located at the sample position. All the STM measurements have been carried out at RT.

3. Results and Discussion

First of all, we deposited a small amount of C_{60} molecules on the clean Si(110)-16 × 2 single domain surface at RT. The obtained STM image is shown in fig. 2 (a). The C_{60} molecules about 2 nm in diameter are found on the surface. Note that the apparent size of isolated molecule is measured bigger than the real size of the molecules in STM image. We classify the C_{60} adsorption sites on the clean Si(110)-16 × 2 surface into three sites as shown in fig. 2(e). The labeled A site is the upper terrace, B site is the lower terrace and C site is the step edge. In fig. 2 (a), it is found that the C_{60} are preferentially adsorbed onto the upper terrace (A site) at RT.

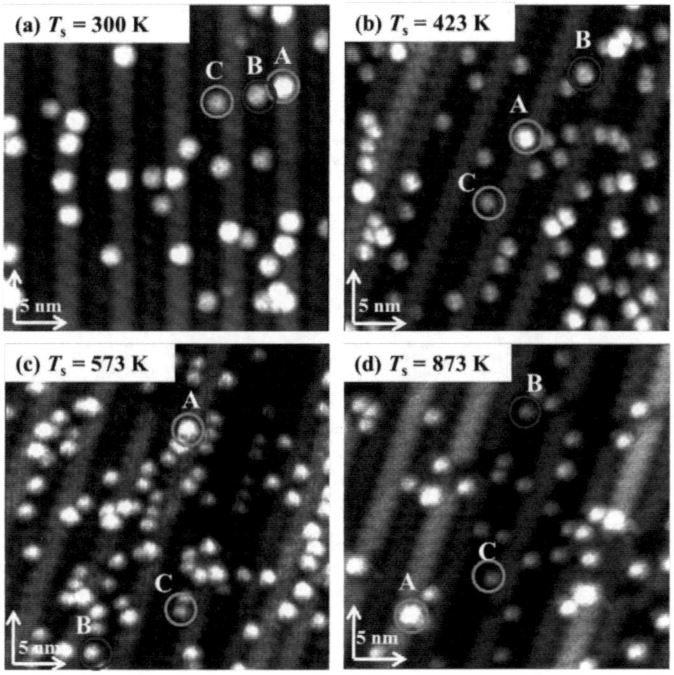

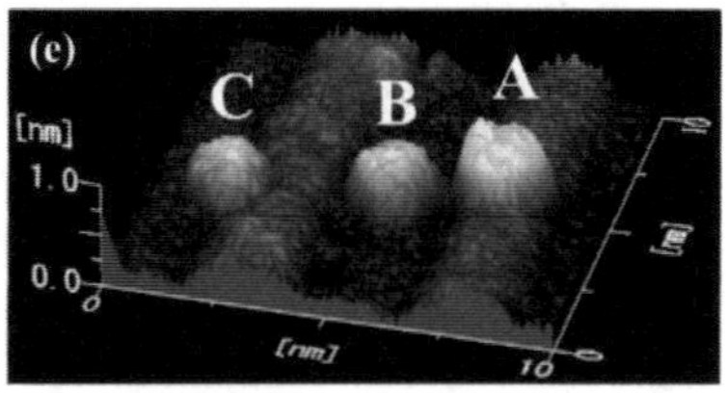

Figure 2: (a) ~ (d) STM images of a small amount of C_{60} deposited on the Si(110)-16 × 2 surface at various temperatures (30 nm × 30 nm). (e) 3D image of C_{60} adsorption sites (A; upper terrace, B; lower terrace, C; step edge).

Next, to investigate the C_{60} adsorption features at different surface temperatures, we deposited the C_{60} molecules at 423 K, 573 K and 873 K. Figures 2 (b) ~ (d) show the STM images after depositing at different temperatures. In the case of C_{60} deposited on the Si(110) surface at 423 K, it seems that the number of C_{60} molecules adsorbed at the step edge largely increases. Preferential adsorptions of C_{60} molecules at step edge are found more obviously in the cases of 573 K and 873 K. Moreover, it is observed that several C_{60} molecules are aligned on the step edges at 573 K. It is known that the C_{60} is polymerized by ultraviolet or electron irradiation [12]. However, the STM image suggests that the polymerization has not occurred in this system since the shapes of C_{60} are maintained the same in fig. 2 (c). According to the recent studies of Feng *et al.*, the electronic state of C_{60} nanowires formed on the Cu surface are connected each other and new electronic states called SAMOs (superatom molecular orbitals) are formed [13]. Therefore, detailed electronic states should be investigated on this surface.

The ratios of classified adsorption sites as a function of adsorption temperature are shown in fig. 3 (a). This result confirms that the ratio of molecules adsorbed onto the upper terrace is substantially larger than that of the other sites at RT. That is, the upper terrace is a preferential adsorption site at RT. However, this is strange in terms of the site reactivity because the upper and lower terraces have the same structure. Now, we have considered the Ehrlich-Schwoebel barrier (ES barrier) [14] to explain the results. It is known that, in order for a surface adsorbate to diffuse from the upper terrace to the lower terrace, the adsorbate must overcome a large barrier located at step edge (ES barrier). The ES barrier on the Si(110)-16 × 2 surface are illustrated in figs. 3 (b) and (c). Since the C_{60} molecules that adsorbed on the upper (lower) terrace cannot diffuse to the lower (upper) terrace at lower temperatures, C_{60} molecules remain within the 1-D terrace where C_{60} molecule impinges (fig. 3 (b)). The C_{60} molecules impinging on the upper terrace shift to the center of the upper terrace, which is probably preferable adsorption sites within the terrace, since the C_{60} cannot reach the step edge of the upper terrace due to the ES barrier. On the other hand, C_{60} molecules impinging on the lower terrace are divided into the center and the edge of the lower terrace. Although the step edge of the lower terrace may be potentially reactive, an activation barrier partly blocks the adsorption at the edge. Because of this reason, the A site adsorption ratio is almost equal to the sum of the B + C site adsorption ratios at RT.

At higher surface temperatures, the features of C_{60} adsorption are drastically changed. Contrary to about 16 % of C_{60} molecules adsorbed at the step edge at RT, about 69 % of C_{60} molecules adsorbed at the step edge at 873 K. At higher substrate temperatures, it is considered that the C_{60} molecules can overcome the ES barrier and surface diffusion from the upper terrace to the step edge is allowed (fig. 3 (c)). Moreover, C_{60} can also overcome the activation barrier at the step edge adsorption from the lower terrace. As a result, while the A site adsorption ratio decreased, the C site adsorption ratio increased remarkably. The similar adsorption ratios of A and B sites at higher temperatures correspond to the similar reactivities of the centers of the upper and lower terraces.

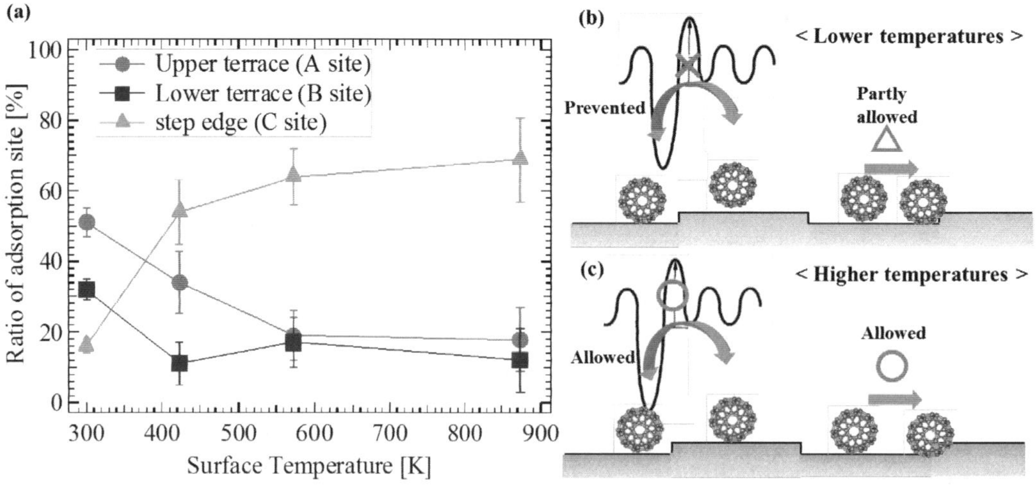

Figure 3: (a) Ratio of adsorption sites deposited at various temperatures which are normalized for the same length of the stripe. Schematic images of ES barrier on the Si(110)-16 × 2 surface at (b) lower temperatures and (c) higher temperatures.

Moreover, it is noticed that the apparent sizes of C_{60} molecules become different for adsorption sites at high surface temperatures. Figures 4 (a) and (b) show the magnified STM image at 873 K and line profiles along the C_{60} molecules adsorbed on the upper terrace (red line) and the step edge (green line), respectively. It is shown that the diameter and the height of C_{60} molecules on the step edge are smaller than that of the upper or lower terrace. It is speculated that the smaller apparent size of adsorbed C_{60} is due to the strong interaction between Si substrate and C_{60} molecule. In the previous study, Yao *et al.* has reported that the apparent size of the chemisorbed C_{60} on Si(100) surface becomes smaller [15]. In the Si(110)-16 × 2 surface, it is known that the step edge has a high reactivity because the electronic states originating from the step edge exist near the Fermi level [16]. Figure 4 (c) shows the magnified STM image of step edge taken at the negative V_s. It is found that the C_{60} molecules adsorbed in-between of the protrusions (white circles in fig. 4 (c)) reflecting the electronic states of the step edge.

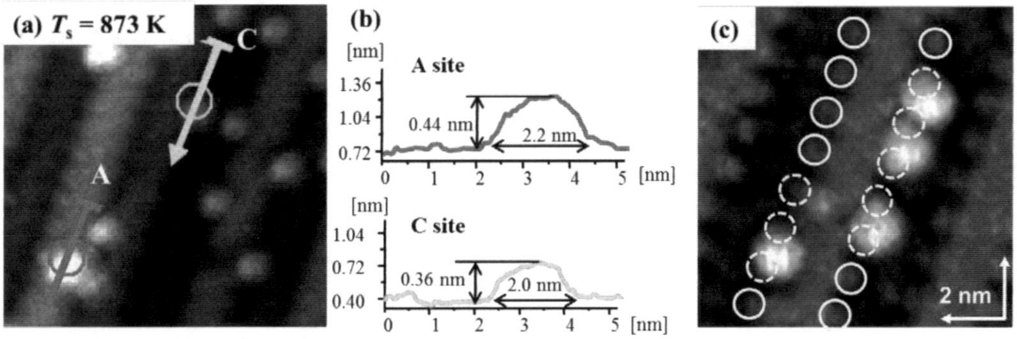

Figure 4: (a) Magnified STM image of C_{60} deposited at 873 K. (b) Line profiles along the C_{60} molecules adsorbed on the upper terrace (red line) and at the step edge (green line) as shown in (a). (c) Magnified STM image taken at the negative bias (V_s = -2.3 V).

4. Conclusion

The C_{60} molecules are deposited on the Si(110)-16 × 2 single domain surface at various temperatures. We found that the C_{60} preferable adsorption sites are substantially changed depending on the surface temperature. At RT, it has been found that the C_{60} molecules preferentially adsorbed onto the upper terrace. It is consider that the adsorbed C_{60} are not able to be diffused from the upper terrace sites to lower terrace sites and vice versa due to the high potential barrier originating from ES barrier at the step edge. In contrast, at higher surface temperatures, it has been found that the C_{60} molecules are preferentially adsorbed on the step edge, which is the most reactive on the surface because the C_{60} molecules can overcome the potential barrier. Moreover, the apparent size of C_{60} which adsorbed at the step edges becomes small at higher surface temperatures. It is speculated that the charge transfer from Si atom to C_{60} molecule occurs at the step edge. The obtained results indicate that the organic 1-D nanostructure can be fabricated on the Si(110)-16 × 2 surface by controlling the surface temperature.

Acknowledgements

Experimental support from T. Yamazaki and the fruitful discussions with S. Shamoto are gratefully acknowledged. This work is financially supported by the Grant-in-Aid for Science Resrach (B), contract No. 20310064, 2008-2011, from the Ministry of Education, Culture, Sports, Science and Technology.

References

[1] Masahara M *et al* 2006 *Appl. Phys. Lett.* **88** 072103
[2] Cheng W *et al* 2006 *Jpn. J. Appl. Phys.* **45** 3110
[3] Sato T *et al* 1971 *Phys. Rev.* B **4** 1950
[4] Yamamoto Y *et al* 1980 *Proceeding of the Annual Meeting of the Physical Society of Japan* (Fukui: Japan) p 300 (un published)
[5] Yamamoto Y *et al* 1986 *Jpn. J. Appl. Phys.* **25** L331
[6] An T *et al* 2000 *Phys. Rev.* B **61** 3006
[7] Stekolnikov A A *et al* 2004 *Phys. Rev. Lett.* **93** 136104-1
[8] Sakamoto K *et al* 2009 *Phys. Rev.* B **79** 045304-1
[9] Ichinokawa T *et al* 1985 *Phys. Rev.* B **31** 5183
[10] Yamada Y *et al* 2007 *Phys. Rev.* B **76** 153309
[11] Yamada Y *et al* 2008 *Phys. Rev.* B **77** 153305
[12] Ueda S *et al* 2006 *J. Phys. Chem.* B **110** 22374
[13] Feng M *et al* 2008 *Science* **320** 1155866
[14] Kyuno K and Ehrlich G 1997 *Surf. Sci.* **394** L179
[15] Yao X *et al* 1996 *Surf. Sci.* **366** L743
[16] Setvin M *et al* 2011 *Phys. Rev.* B **84** 115317

Morphology change of multi-walled carbon nanotubes with SiC coating by electron irradiation

T Taguchi[1], A K M Fazle Kibria[2] and S Shamoto[1]

[1] Quantum Beam Science Directorate, Japan Atomic Energy Agency
2-4 Shirakata-Shirane, Tokai, Ibaraki 319-1195 Japan

[2] Chemistry Division, Atomic Energy Centre
P. O. Box 164, Ramna, Dhaka-1000, Bangladesh

E-mail: taguchi.tomitsugu@jaea.go.jp

Abstract. The different behaviour between morphology changes of multi-walled carbon nanotube (MWNT) and MWNT with SiC coating was investigated by electron irradiation. High resolution TEM images revealed that deformation of graphitic shells occurred in both MWNT and MWNT with SiC coating by electron irradiation. Electron irradiation began changing the outer and inner diameters of MWNT within 3 minutes exposure, whereas it began changing the diameters of MWNT with SiC coating after 10 minutes exposure. The outer diameter of MWNT increased up to irradiation time of 9 min and then decreased rapidly. On the other hand, the outer diameter of MWNT with SiC coating increased up to irradiation time of 50 min and then was almost saturated. These results lead that the morphology stability of MWNT against electron irradiation can be improved by coating with the SiC layer on MWNT.

1. Introduction

Since carbon nanotubes (CNTs) were discovered in 1991 [1], they are promising materials for nanometer scale electronic-devices, energy storage and nanocomposites because of their unique anisotropic structures and physical properties [2-4]. In particular, CNTs are also expected to be used in radiation environments such as nuclear plants and aerospace. Many researchers have reported electron or ion irradiation studies of CNTs [5-7]. Defects can be generated in CNTs and the morphology of CNTs can be changed by electron or ion irradiation [5-7]. Because these generated defects and changed morphology disrupt the ideal hexagonal network of graphitic shells, the mechanical, electric and optical properties of CNTs alter [8-10]. However, there are few reports about improving the morphological stability against electron and/or ion irradiation.

On the other hand, silicon carbide (SiC) is well known to have excellent morphological stability against electron and/or ion irradiation [11, 12]. So far, we have reported about the synthesis of multi-walled CNTs (MWNTs) coated with SiC layer by the heat treatment of MWNTs with Si powder in a vacuum [13, 14]. In this study, the different behaviour between morphology changes of MWNTs with and without SiC coating by electron irradiation was observed by transmission electron microscopy (TEM). The improvement method of morphological stability of CNT against electron irradiation was investigated.

2. Experimental procedure

MWNTs obtained from Furuuchi Chemistry Corporation (Tokyo, Japan) were used in this study. MWNTs with SiC coating were synthesized by heating MWNTs with Si powder (The Nilaco Corporation, Tokyo, Japan) at 1200 °C for 5h in a vacuum (~10^{-3} Pa). The details of fabrication process and characterization of MWNTs with SiC coating were described elsewhere [11, 12]. Electron beam of energy 200 keV was applied directly from TEM (Model JEM-2100F, JEOL Ltd., Tokyo, Japan) on as-received MWNTs and MWNTs with SiC coating samples at room temperature. The flux of electron beam was 4.2×10^{18} e/cm^2/s which corresponding to a beam intensity of 0.67 A/cm^2. The incident electron beam covered an area of 9.6×10^{-10} cm^2. Images of the irradiated MWNTs and MWNTs with SiC coating were recorded by a CCD-camera connected with TEM.

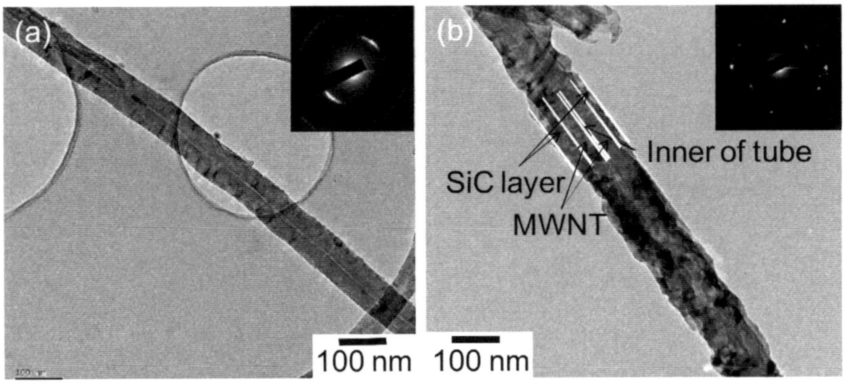

Figure 1. TEM images and the corresponding selected area electron diffraction patterns of as-received MWNT and MWNT with SiC coating before electron irradiation.

3. Results and discussion

Figure 1 shows the TEM images and the corresponding selected area electron diffraction patterns of as-received MWNT and MWNT with SiC coating before electron beam irradiation. TEM observation revealed that SiC layer was produced on the all regions of outer surface of CNT part in the MWNT with SiC coating. The SiC layer was polycrystalline. The thickness range of SiC layer was

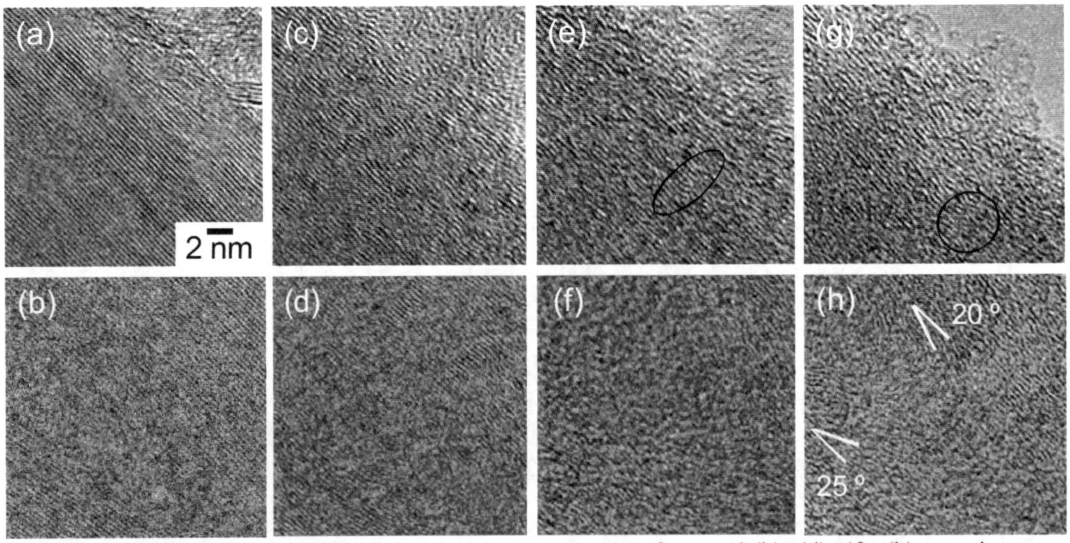

Figure 2. HR-TEM images of (a), (c), (e), (g) near outer surface and (b), (d), (f), (h) near inner tube of as-received MWNT before and after electron irradiation. Irradiation time: (a), (b) 0 min, (c), (d) 8 min, (e), (f) 20 min, (g), (h) 60 min.

approximately 20 to 40 nm. The as-received MWNT and MWNT with SiC coating irradiated in this study had almost the same outer diameter of CNT part, which were approximately 78 and 79 nm, respectively.

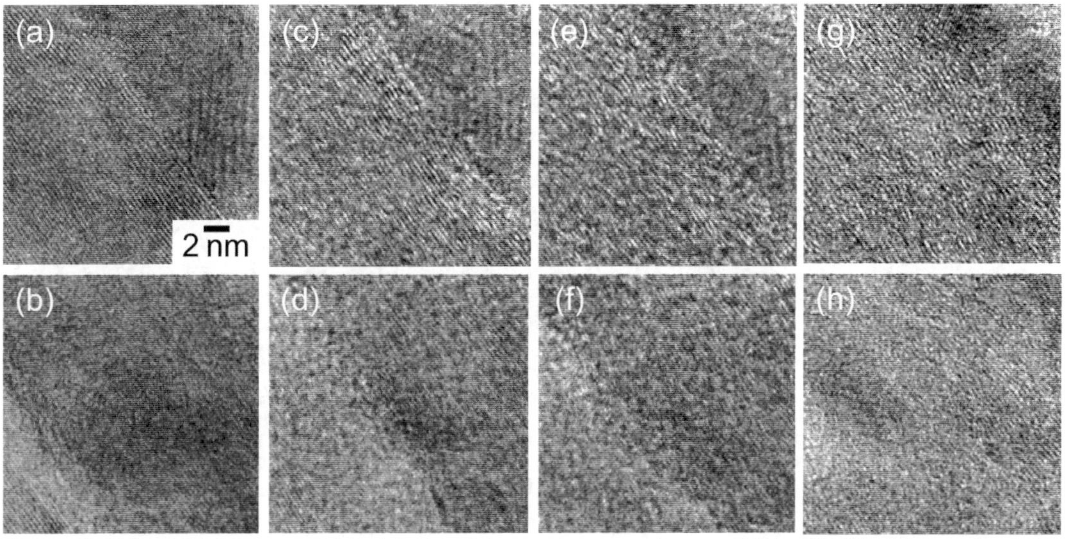

Figure 3. HR-TEM images of (a), (c), (e), (g) near outer surface and (b), (d), (f), (h) near inner tube of MWNT with SiC coating before and after electron irradiation. Irradiation time: (a), (b) 0 min, (c), (d) 8 min, (e), (f) 20 min, (g), (h) 140 min.

High resolution TEM (HR-TEM) images of the near outer surface and the near inner tube of as-received MWNT before and after electron irradiation are shown in figure 2. According to these TEM images, deformation of graphitic shells occurred by electron irradiation. Interstitial loops were induced between different basal planes. The number of interstitial loops increased with increasing electron irradiation time and seemed to be then saturated. It is well known that temperature effects are of particular importance about morphology change of CNTs [5-7, 15]. At room temperature, an accumulation of defects is observed up to a saturation density which depends on the electron energy and beam intensity [15]. Basal planes were also broken to small segments. The size of segments decreased with increasing irradiation time. It was often observed that different basal planes that were broken to small segments were cross-linked each other as shown in figure 2 (e) and (g). Bending of graphitic walls towards the interior hole near inner tube occurred. The curvature near inner tube by electron irradiation was larger than that near outer surface. The curvature angle near inner tube after 60 min irradiation was over 20° as shown in figure 2 (h).

Figure 3 shows HR-TEM images of the near outer surface and the near inner tube of MWNT with SiC coating before and after electron irradiation. The crystal structure and morphology of SiC layer did not change by electron irradiation because of its excellent morphological stability against electron irradiation. Deformation of graphitic shells in CNT part occurred and interstitial loops were induced as well as as-received MWNT. Basal planes were also broken to small segments. Both of deformation and bending curvature in CNT part of MWNT with SiC coating were smaller than those in as-received MWNT. In particular, the deformation of outer layers in CNT part was constricted because the outer surface in CNT part was bound with the SiC layer. Surprisingly, the bending of inner layers, which were not contacted with SiC layer directly, was also suppressed. The average size of segments of broken basal planes in the MWNT with SiC coating seemed to be larger compared to that in as-received MWNT.

Figure 4 shows the changed ratio of outer and inner diameters of CNT parts in as-received MWNT and MWNT with SiC coating by electron irradiation. Electron irradiation began changing the outer and inner diameters of as-received MWNT within 3 minutes irradiation, whereas it began changing the

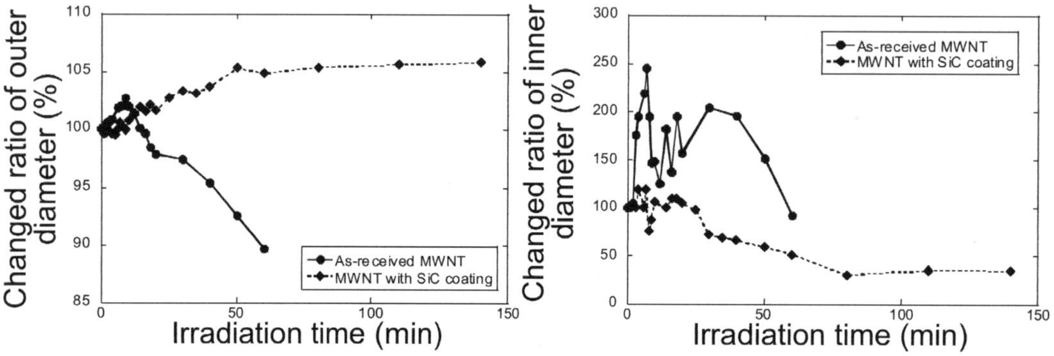

Figure 4. Changed ratio of outer and inner diameters of CNT parts in as-received MWNT and MWNT with SiC coating by electron irradiation.

diameters of MWNT with SiC coating after 10 minutes irradiation. In the as-received MWNT, the outer diameter increased up to the irradiation time of 9 min and then decreased rapidly. The outer diameter swelled at the beginning of electron irradiation because the lattice distance of CNT was widened due to the deformation of graphitic shells and the induction of interstitial loops between different basal planes. It has been reported that all C atoms of single-walled CNT could be ejected by the electron with higher accelerated energy than 139 keV [16]. The energy of electron beam applied in this study was 200 keV. Therefore, the electron irradiation could knock C atoms out from the surface of MWNT. The MWNT contracted rapidly after 9 min irradiation because the effect of knocking out was dominant compared to the widening effect of lattice distance. On the other hand, the outer diameter of MWNT with SiC coating increased up to the irradiation time of 50 min and then was almost saturated. At electron energy of 200 keV, every C atom is susceptible to ballistic ejection [16]. However, in the MWNT with SiC coating, SiC layer that had excellent morphological stability against electron irradiation was coated on the all regions of outer surface of MWNT. It might thereby become hard to knock C atoms out from the outer surface of CNT part. The lattice distance of CNT was widened due to the deformation and the induction of interstitial loops as well as as-received MWNT. These results lead that the outer diameter of MWNT with SiC coating increased with increasing the irradiation time. The changed ratio of outer diameter in MWNT with SiC coating was smaller than that in as-received MWNT.

The inner diameter of as-received MWNT increased rapidly, whereas that of MWNT with SiC coating decreased up to the irradiation time of 80 min and was then saturated. The inner diameters of as-received MWNT and MWNT with SiC coating were approximately 3 and 8 nm, respectively. Banhart et al reported that the stability of nanotubes with diameters below 1-2 nm decreases with decreasing diameter [17]. However, the inner diameters of MWNTs applied in this study were larger than 3 nm. The threshold energy needed to displace C atoms from nanotubes with inner diameters over 2 nm is comparable to that from flat graphite and almost constant against the inner diameter over 2 nm [17]. In spite of these results, the morphological change behaviour of inner diameter in as-received MWNT differed from that in MWNT with SiC coating. The reason is considered that in as-received MWNT it might be easier that C atoms were knocked out from both outer and inner surfaces compared to MWNT with SiC coating. These results lead that the morphological stability of MWNTs against electron irradiation may be improved by coating with SiC layer on the surface of MWNTs.

The changed ratio of wall thickness of CNT parts in as-received MWNT and MWNT with SiC coating is shown in figure 5. The changed ratio of as-received MWNT decreased monotonically. On the other hand, the changed ratio of MWNT with SiC coating increased and was almost saturated after 50 min irradiation. In the as-received MWNT, it might be easy that C atoms were ejected from both the outer and inner surfaces by electron irradiation. Therefore, the wall thickness of as-received MWNT decreased monotonically. On the other hand, it might be harder that C atoms were knocked out from both surfaces of CNT part in the MWNT with SiC coating due to the existence of SiC coating that had excellent morphological stability against electron irradiation. TEM observation revealed that deformation of graphitic shells occurred and many interstitial loops were induced. Therefore, the lattice distance between basal planes increased with increasing the irradiation time. These results lead that the wall thickness of MWNT with SiC coating increased.

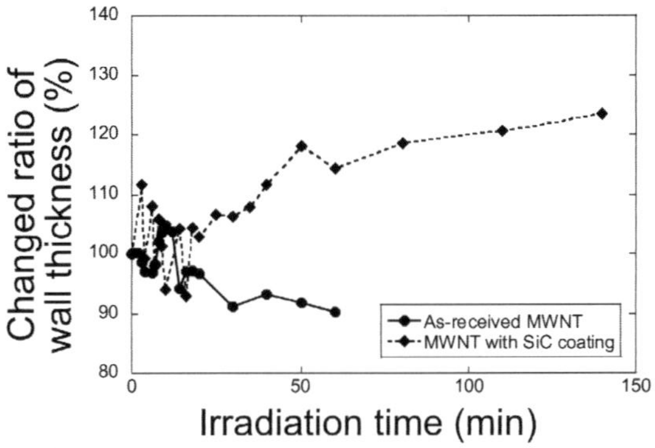

Figure 5. Changed ratio of wall thickness of CNT parts in as-received MWNT and MWNT with SiC coating.

As mentioned above, coating with SiC layer on outer surface of MWNT can bring the facts that the deformation and bending of graphitic shells, the ejection of C atoms from both inner and outer surfaces of CNT part are restrained. Thus, the morphological stability of MWNTs against electron irradiation can be improved by coating with SiC layer.

4. Conclusion

The different behaviour between morphology changes of as-received MWNT and MWNT with SiC coating was investigated by electron irradiation. HR-TEM images revealed that the deformation and bending of graphitic shells occurred in both as-received MWNT and MWNT with SiC coating by electron irradiation. Both of deformation and bending curvature in CNT part of MWNT with SiC coating seemed to be smaller than those in as-received MWNT. Electron irradiation began changing the outer and inner diameters of as-received MWNT within 3 minutes exposure, whereas it began changing the diameters of MWNT with SiC coating after 10 minutes exposure. The outer diameter of as-received MWNT increased up to irradiation time of 9 min and then decreased rapidly. On the other hand, the outer diameter of MWNT with SiC coating increased up to irradiation time of 50 min and then was almost saturated. It might become hard to knock C atoms out from both inner and outer surfaces of CNT part by coating with SiC layer on MWNT. These results lead that the morphology stability of MWNT against electron irradiation can be improved by coating with the SiC layer on MWNT.

Acknowledgement

This work was partly supported by a Grant-in-Aid for Young Scientist (B) (No. 23760646) from the Ministry of Education, Science, Sports and Culture of Japan.

References

[1] Iijima S, 1991, *Nature*, **354**, 56-58.
[2] Baughman R H, Zaknidov A A, de Heer W A, 2002, *Science*, **297**, 787-792.
[3] Yoo Y, Tuck M, Kondakindi R, Seo C Y, Dehouche Z, Belkacemi K, 2007, *J. Alloys Comp.*, **446-447**, 84-89.

Journal of Physics: Conference Series **417** (2013) 012037 doi:10.1088/1742-6596/417/1/012037

[4] Kibria A K M F, Mo Y H, Park K S, Nahm K S, Yun M H, 2001, *Int. J. Hydorgen Energy*, **26**, 823-829.
[5] Ajayan P M, Ravikumar V, Charlier J C, 1998, *Phys. Rev. Lett.*, **81**, 1437-1440.
[6] Hashimoto A, Suenaga K, Gloter A, Urita K, Iijima S, 2004, *Nature*, **430**, 870-873.
[7] Sun L, Banhart F, Krasheninnikov A V, Rodriguez-Manzo J A, Terrones M, Ajayan P M, 2006, Science, 312, 1199-1202.
[8] Gupta S, Patel R J, Smith N, Giedd R E, Hui D, 2007, *Diamond and Relat. Mater.*, **16**, 236-242.
[9] Gomez-Navepro C, de Pablo P J, Gomez-Herrero J, Biel B, Garcia-Vidal F J, Rubio A, Flores F, 2005, *Nature Mater.*, **4**, 534-539.
[10] Salvetat J P, Bonard J M, Thomson N H, Kulik A J, Forro L, Benoit W, Zuppiroli L, 1999, *Appl. Phys. A*, **69**, 255-260.
[11] Taguchi T, Igawa N, Wakai E, Miwa S, Jitsukawa S, Snead L L, Hasegawa, A, 2004, *J. Nucl. Mater.*, **335**, 508-514.
[12] Matsunaga A, Kinoshita C, Nakai K, Tomokiyo Y, 1991, *J. Nucl. Mater.*, **179-181**, 457-460.
[13] Taguchi T, Igawa N, Yamamoto H, Jitsukawa S, 2005, *J. Am. Ceram. Soc.*, **88[2]**, 459-461.
[14] Taguchi T, Igawa N, Yamamoto H, Shamoto S, Jitsukawa S, 2005, *Physica E*, **28[4]**, 431-438.
[15] Banhart F, 1999, *Rep. Prog. Phys.*, **62**, 1181-1221.
[16] Smith B W, Luzzi D E, 2001, *J. Appl. Phys.*, **90[7]**, 3509-3515.
[17] Banhart F, Li J X, Krasheninnikov A V, 2005, *Phys. Rev. B*, **71**, 241408.

Preparation of Copper Nanoparticles in Liquid by Matrix Sputtering Process

Ko Nakagawa, Takashi Narushima, Satoshi Udagawa, and Tetsu Yonezawa*

Division of Materials Science and Engineering, Faculty of Engineering, Hokkaido University, Kita 13, Nishi 8, Kita-ku, Sapporo, 060-8628 Japan

Email: tetsu@eng.hokudai.ac.jp

Abstract. As a new method for nanoparticles preparation, magnetron sputtering of metal atoms and clusters into organic liquids has been intensively used recently. In this study, metallic copper nanoparticles dispersed in pentaerythritol ethoxylate were prepared by this process. Their size control was achieved by controlling the sputtering current. Specific absorption at *ca.* 580 nm was detected by UV-Vis measurement which is attributed to the specific plasmon absorption of metallic copper. TEM observation also revealed the formation of metallic nanoparticles.

1. Introduction

Metal nanoparticles (MNPs) have been attracted significant attention from the scientific and engineering community as their specific physicochemical properties which are different from bulk materials and atomic materials, and it is considered that they play an important role in developing new scientific approaches. MNPs are now considered as a key material of modern nanotechnology. Their unique properties, including nonlinear optics, plasmon phenomena, photo luminescence, catalysis, electronics, biomedicine, magnetic phenomena, as well as energy conversion, are varied with their sizes, especially in a single digit nm region, and with their shape and surface condition. [1-4] For optical properties, extremely small MNPs with their sizes close to Fermi wave length of electron have photoluminescence characterized by quantum size effect. [5-7]

In most cases, recently, especially in laboratories, MNP dispersions were prepared by chemical reduction or decomposition of the corresponding metal compounds by using external reduction agents or solvents. [1, 8-10] Initial cost of wet chemical processes is usually much smaller than vacuum processes and the particle sizes can be controlled by the selection of metal sources, protective reagents, reducing reagents as well as their concentrations.

On the other hand, NP preparation by vacuum process is considered as an industrial process which generates naked MNPs or oxide coated MNPs. Initial cost of vacuum processes is usually much more expensive than chemical reduction processes, but vacuum processes are usually continuous manufacturing ones, whose running cost can be smaller than that of chemical processes, which are batch processes, generally. From this point of view, a new route to prepare MNPs by using a relatively non-expensive apparatus has been demanded. Recently, magnetron sputtering, which is usually focused on the preparation of thin films on solid substrates, has been applied for NP preparation. Ye et al. reported the preparation of rough thin films of Ag on silicon oil. [11] After this work, this magnetron sputtering procedure to prepare MNPs in liquid was extended to other liquids, such as ionic liquids, [12-15] oils, [16] liquid polymers [17] and so on. Size and structure control of Au NPs by changing sputtering parameters has been also discussed. [17-19] The Authors' group proposed to use a thiol organic salt, being solid at room temperature, as the matrix to sputtering preparation of Au NPs by heating the matrix molecules. [20] Other thiol liquid molecules were also used as the matrix for Au NP preparation by sputtering, and this NP dispersion could be polymerized to form a transparent fluorescent resin.[21]

However, these papers published so far reported only Au or Ag NPs as MNPs, which are stable in ambient condition. Quite recently, Suzuki et al. reported metal oxide NPs by sputter deposition into ILs and heat treatment under air. [22,23] In this study, we firstly prepared metallic Cu NPs in liquid organic molecules by magnetron sputtering process.

2. Experimental Section

2.1. Materials

As an organic liquid, pentaerythritol ethoxylate (PEEL, Aldrich, $C(CH_2(OCH_2CH_2)_nOH)_4$) was selected and used as purchased. It is one of the monomers of urethane. Target copper (99.96 %, diameter = 50 mm) was purchased from Nilaco, Japan.

2.2. Preparation of copper nanoparticle dispersion by sputtering process

In order to remove water molecules and other ingredient molecules in PEEL, it was kept at 90 °C under vacuum for more than 2 h. Some bubbles formed from PEEL. The vacuum chamber was evacuated first to ~ 10^{-3} Pa and Ar gas was admitted in until the system was stabilized at a pressure of 2 Pa. Before sputtering, this gas exchange procedure was repeated more than twice in order to evacuate oxygen gas completely. The vacuum pressure was kept at 2 Pa during the sputtering. All samples were obtained by dc-magnetron sputtering of a Cu target over a cylindrical glass petri plate (~ 6 cm diameter × ~ 1 cm height) containing 7.0 g of PEEL. The temperature of PEEL was kept at 20 °C by using a cooling apparatus under the petri plate. Sputtering current was varied in the range of 10 – 100 mA, and sputtering time was 8 or 60 min.

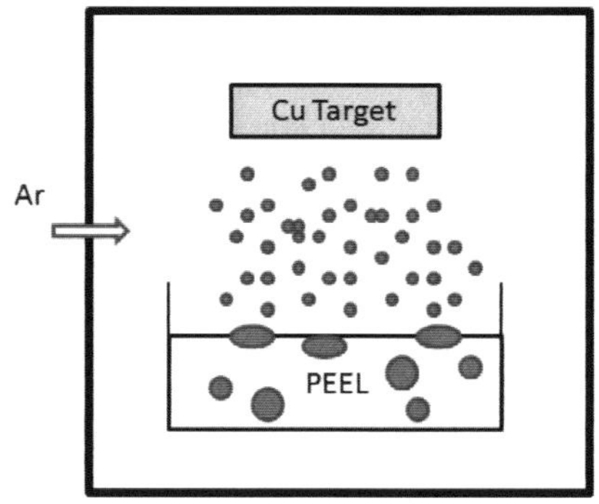

Fig. 1 Schematic illustration of Cu NPs preparation by magnetron sputtering on PEEL.

2.3. Characterization

UV-Vis absorption spectra were obtained with a Jasco V-630 using a 1-mm or 10-mm quartz cuvette. The particle structure and size were observed by TEM with a Hitachi H-9500 at the acceleration voltage of 300 kV. The sample grid (Cu, 3 mmϕ) was prepared by putting a drop of Cu nanoparticle dispersion in PEEL and followed by dipping into ethanol for 10 min to wash excess PEEL.

3. Results and Discussion

3.1. Preparation of Cu nanoparticles by Sputtering

As previously reported, Au NPs can be readily obtained by magnetron sputtering into ionic liquids, oils, *etc*. Discussions of the relationships between sputtering parameters and sizes and shapes of the obtained Au NPs were also discussed. Here, we tried to obtain Cu NPs by sputtering. First, we evacuate the sputtering chamber to 10^{-3} Pa and then filled it with Ar gas up to 2 Pa. This evacuation/Ar introduction process was carefully repeated at least twice in order to remove oxygen gas. Othewise, CuO or Cu_2O NPs can be obtained.

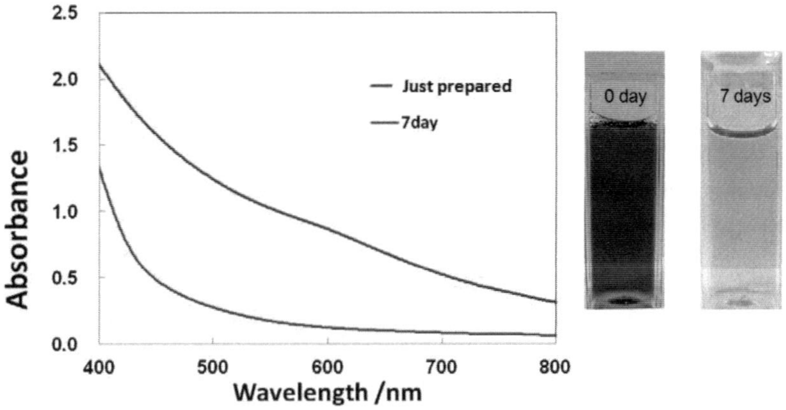

Fig. 2. UV-Vis spectra of Cu NPs in PEEL prepared by sputtering (50 mA, 8 min). Red: Just prepared, Brown: 7 days after preparation. Specific plasmon absorption peak can be observed in the spectrum just after preparation. Images (right) show the Cu NP dispersion in PEEL (just prepared and 7 days after preparation) in a cuvette.

When we carefully changed the chamber gas to Ar, we have successfully obtained CuNPs in PEEL. Figure 2 shows the UV-Vis absorption spectra of Cu NPs obtained by magnetron sputtering at 50 mA for 8 min and the spectra were measured at just after preparation and 7 days after preparation. After preparation, one can find a small absorption shoulder peak at ca. 580 nm, which is corresponding to the specific plasmon absorption of metallic copper NPs. [24] However, after keeping 7 days under ambient condition, the absorbance of the Cu NPs in PEEL dramatically decreased in whole area. In Figure 2, the photographs of the dispersion in a quartz cell are also collected. The color of the dispersion diminished dramatically during 7 days, and the sample after 7 days is slightly yellow. This situation should be attributed to oxidation of Cu NPs in PEEL dispersion. On the other hand, this color change suggests that no aggregation or flocculation of Cu NPs occurred in PEEL.

3.2. Cu NPs in PEEL Prepared by Sputtering with Different Sputtering Parameters

As discussed in the introduction section, precise control of particle size is very important technique to obtain MNPs. Hatakeyama *et al.* changed various sputtering parameters to obtain Au NPs in poly(ethylene glycol) (PEG). [17-19] Temperature, concentration can be parameters which control the size and shape of Au NPs. Here, we have tried to control Cu NPs in PEEL by changing the sputtering current. In order to emphasize the effect of the changing sputtering current, the sputtering time was kept at 60 min. UV-Vis spectra and TEM images of Cu NPs obtained in PEEL by sputtering Cu at various currents in the range between 10 – 100 mA are collected in Figs. 3 and 4, respectively. No plasmon absorption peak is found in the spectrum of Cu NPs in PEEL prepared with sputtering current

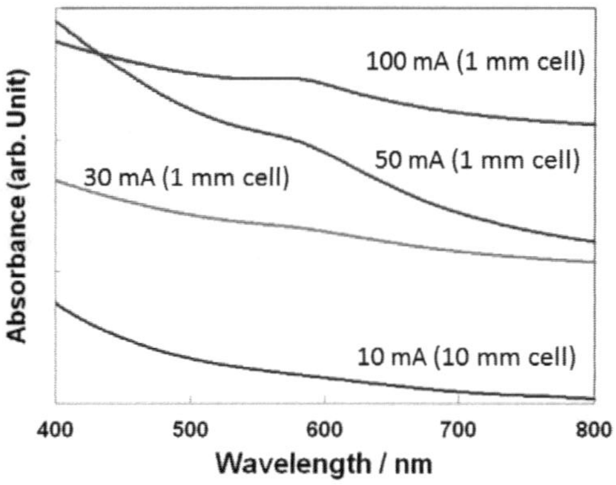

Fig. 3 UV-Vis spectra of Cu NPs prepared by sputtering at various sputtering currents. Sputtering currents used and optical path lengths of the quartz cells are indicated in the Figure (10, 30, 50 and 100 mA; 1 mm or 10 mm cell). The Sputtering time = 30 min.

of 10 mA. More than 30 mA, the plasmon absorption peak at ca. 580 nm can be observed. After being sputtered from the Cu target, Cu atoms and clusters bombarded onto PEEL surface and interact with PEEL molecules. In the surface region, growth of Cu atoms and clusters to Cu NPs proceeds. [16] Therefore, when the sputtering current is higher, the amount of generated Cu atoms and clusters increased. Then, the growth of the particles becomes rapid to generate larger NPs.

In fact, the particle diameter was depending on the sputtering current. Sputtered at 10 or 20 mA, the average particle size is between 2 – 3 nm. However, higher current gave larger particles, especially Cu NPs prepared at 100 mA, showed two peaks in the size distribution, one is 5.5 nm, and the other is larger, about 8 nm.

Precise observation of the UV-Vis spectra in Fig. 3 indicates a slight blue shift in the spectrum of Cu NPs prepared at 100 mA comparing with that prepared at 50 mA, even the particle sizes became larger. It can be explained by the difference of the baselines. However, further experiments with different sputtering conditions are needed.

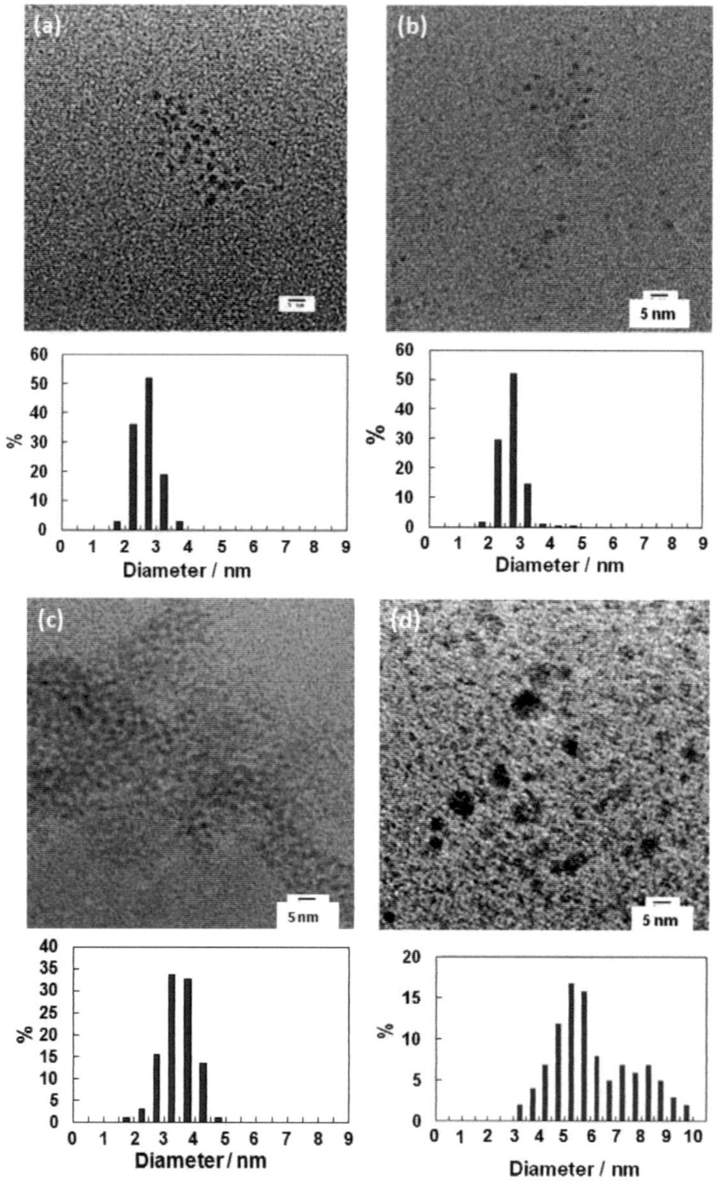

Fig. 4 TEM images and particle size distributions of Cu NPs in PEEL prepared by sputtering process. Sputtering current: (a) 10 mA, (b) 30 mA, (c) 50 mA, (d) 100 mA. Sputtering time = 30 min.

4. Conclusion

Spherical metallic Cu NPs with a diameter of single digit nm were obtained by Ar sputtering onto organic oil, PEEL and Cu NP dispersions were obtained. No aggregation or flocculation of NPs has been observed but oxidation of Cu NPs proceeded when the dispersion was kept under ambient condition. Their particle size has been controlled by sputtering current, and plasmon absorption corresponding to metallic Cu NPs has been clearly detected.

Acknowledgement
This work is partially supported by Grant-in-Aid for Scientific Research in Priority Area "Molecular Science for Supra Functional Systems (477)" to TY (22018001). Authors thank Dr. Y. Shishino (Tokai Opt. Co., Ltd.) for his kind suggestions to our experimental work.

References

[1] Toshima N and Yonezawa T 1998 New J. Chem 1179
[2] G. Schmid Ed., *Clusters and Colloids*, VCH, Weinheim 1994.
[3] T. Pradeep, "NANO: The Essentials", McGraw-Hill Education 2007.
[4] Daniel M.-C. And Astruc D. 2004 Chem Rev **104** 293.
[5] Siwach OP and Sen P 2008 J Nanoparticle Res **10**, 107
[6] Kawasaki H, Kosaka Y, Myoujin T, Narushima T, Yonezawa T, and Arakawa R 2011 Chem Commun **47** 7740
[7] Wei W, Lu Y, Chen W, and Chen S 2011 J Am Chem Soc **133** 2060
[8] Tomonari M, Ida K, Yamashita H, and Yonezawa T 2008 J Nanosci Nanotechnol **8** 2468.
[9] Yonezawa T, Nishida N, and Hyono A 2010 Chem Lett **39** 548
[10] Slatemann C, Lisiecki I, Urban J, and Pileni MP 2004 Langmuir **20** 11772
[11] Ye G, Zhang Q, Feng C, Ge H, and Jiao Z 1996 Phys Rev B **54** 14754
[12] Torimoto T, Okazaki K, Kiyama T, Hihara K, Tanaka N, and Kuwabata S, 2006 Appl Phys Lett **89** 243117
[13] Kameyama T, Ohno Y, Kurimoto T, Okazaki K, Uematsu T, Kuwabata S, and Torimoto T 2010 Phys Chem Chem Phys **12** 1804
[14] Kuwabata S, Tsuda T, and Torimoto T 2010 J Phys Chem Lett **1** 3177
[15] Wender H, de Oliveira LF, Migowski P, Feil AF, Lissner E, Prechtl MHG, Teixeira SR, and Dupont J 2010 J Phys Chem C **114** 11764
[16] Wender H, Gonçalves RV, Feil AF, Migowski P, Poletto FS, Pohlmann AR, Dupont and J, Teixeira SR 2011 J Phys Chem C **115** 16362
[17] Hatakeyama Y, Morita T, Takahashi S, Onishi K, and Nishikawa K 2011 J Phys Chem C **115** 3279
[18] Hatakeyama Y, Okamoto M, Torimoto T, Kuwabata S, and Nishikawa K 2009 J Phys Chem C **113** 3917
[19] Hatakeyama Y, Takahashi S, and Nishikawa K 2010 J Phys Chem C **114** 11098
[20] Shishino Y, Yonezawa T, Kawai K, and Nishihara H 2010 Chem Commun **46** 7211
[21] Shishino Y, Yonezawa T, Udagawa S, Hase, and K, Nishihara H 2011 Angew Chem Int Ed **50**, 703
[22] Suzuki T, Okazaki K, Suzuki S, Shibayama T, Kuwabata S, and Torimoto T 2010 Chem Mater **22** 5209
[23] Suzuki T, Suzuki S, Tomita Y, Okazaki K, Shibayama T, Kuwabata S, and Torimoto T 2010 Chem Lett **39** 1072
[24] Papavassiliou GC and Kokkinakis TH 1974 J Phys F, Metal Phys **4** 67

Journal of Physics: Conference Series **417** (2013) 012039 doi:10.1088/1742-6596/417/1/012039

On the Structural and Optical Properties of SnS Films Grown by Thermal Evaporation Method

P A Nwofe[1], K T Ramakrishna Reddy[1,2], J K Tan[1], I Forbes[1], and R W Miles[1*]

[1]Northumbria Photovoltaics Applications Centre, Northumbria University, Ellison Building, Newcastle Upon Tyne, NE1 8ST, UK.

[2]Thin Films Laboratory, Department of Physics, Sri Venkateswara University, Tirupati-517502, India.

Email: robert.miles@northumbria.ac.uk

Abstract. Tin sulphide (SnS) has a direct energy band gap of 1.35 eV and it consists of abundant, non-toxic elements. It is therefore of interest for use as an absorber layer material in thin film photovoltaic solar cells. In this work, SnS layers with thicknesses in the range 2-3.6μm, were thermally evaporated onto glass substrates using substrate temperatures in the range 280°C to 360°C, and the way the structural and optical properties of the layers varied with the deposition conditions investigated. X-ray diffraction spectra showed a strong (040) reflection as the most prominent peak for films formed between 320°C to 360°C. The peak intensity ratio, crystallite size and grain size were observed to increase with increasing substrate temperature whilst the strain decreased. All the layers were highly light absorbing with the optical absorption coefficient, $\alpha > 10^4$ cm^{-1}. The optical energy band gap was found to be in the range, 1.30-1.34eV, it not changing substantially with substrate temperature. Other optical parameters such as the refractive index and optical conductivity were also evaluated for the layers grown at different substrate temperatures.

1. Introduction

Recently, the need to harness the potentials of SnS for use in various applications especially in solar photovoltaic cells has witnessed a dramatic increase due to the favourable inherent properties of this material. Though thin film solar cells based on the use of polycrystalline inorganic thin film compounds dominates the market share at present, with efficiencies >20% for cells from copper indium gallium diselenide (CIGS), efficiencies > 16.7% using cadmium telluride (CdTe), and modules with efficiencies >10% using both technologies [1], the lack of abundance of the core elements (In and Ga) in the CIGS is expected to limit the large scale production of modules using this technology and to keep material costs high. Also there remain concerns with respect to the use of cadmium, a highly toxic element in some countries. In the countries where the technology is acceptable then it is expected there will be the need for the controlled disposal of CdTe modules after use. These concerns have led to the investigation of other inorganic materials that do not have these problems, one such material being tin sulphide (SnS).

Tin sulphide is a semiconductor with a direct energy bandgap of 1.35 eV and it consists of abundant non toxic elements. The near optimum energy bandgap makes it of interest for application as an absorber layer in thin film solar cells. SnS is also amphoteric and this means that the grain boundaries could be in principle, be passivated by counterdoping p-SnS using an n-type dopant.

Although there have been some studies of the properties of SnS, work on this material to date is limited [2-3].

In this work, structural and optical properties of SnS films grown by evaporation method with a range of substrate temperatures have been investigated with respect to the suitability of the grown films as suitable absorber layers for use in photovoltaic solar cell applications.

2. Experimental

SnS layers with thicknesses in the range (2-3.6μm) were deposited using the thermal evaporation method, performed in an oil pumped vacuum system, operating in the10^{-5} to 10^{-6} Torr range of vacuum pressures. The substrate and source material were heated via resistive elements and the deposition time controlled using a shutter that was incorporated in the thermal evaporation plant. The source material used for absorber was 99.99% purity SnS obtained from Testbourne Ltd. A quartz bottle-shaped crucible served as the evaporation source and was heated by a cylindrical tantalum heating element supplied by the R.D Mathis Company. Spattering (i.e. the ejection of molten lumps of material) was eliminated by placing quartz wool in the crucible and by using a source-substrate distance of 40mm. The substrate temperature was varied between 280°C to 360°C while the deposition time was between 1-5 minutes. The temperatures were measured by thermocouples in direct contact with the crucible and the substrate respectively.

The as-deposited films were characterized using appropriate techniques. A Siemens D5000 X-ray diffractometer (XRD) with a CuK_α radiation source (λ = 0.15406 nm) was used to determine the structural properties of the films. The surface topography of the layers was observed using an FEI Quanta 200 ESEM (Environmental Scanning Electron Microscope) while the surface morphology and elemental composition of the films were determined using an Oxford Instrument EDAX system. The optical transmittance (T) versus wavelength measurements was recorded using a Shimadzu SolidSpec 3700 spectrophotometer.

3. Results and Discussion

The visual observation indicated that the layers are pinhole free and appear bluish black in colour. Fig.1 shows the Sn/S ratio for SnS films grown at different substrate temperatures, determined using EDAX. It can be seen from the figure that Sn/S ratio for the films initially raised sharply up to the deposition temperatures of 300 °C, became relatively stable with stoichiometric composition in the temperature range, 300 -340 °C and then increased gradually afterwards. Similar behaviour has been observed by other authors [4]. The observed non- stoichiometry at growth temperatures < 300° C and > 340° C was due to the effect of differential vapour pressures of tin and sulphur; the higher vapour pressure of sulphur leads to its re-evaporation from the film surface at higher substrate temperatures.

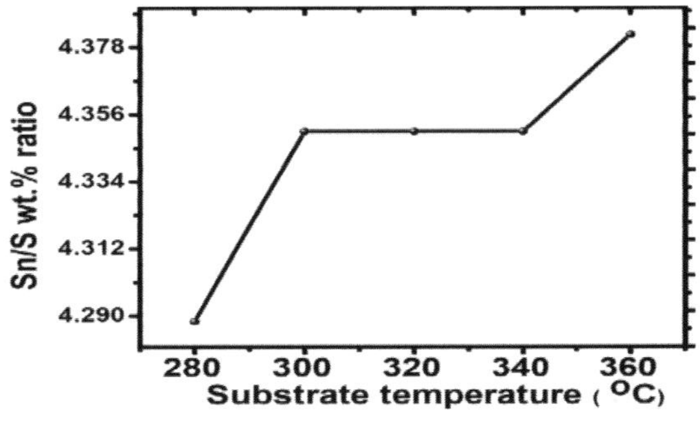

Fig.1. Variation of Sn/S wt.% ratio with substrate temperature in SnS films.

Fig.2 shows the XRD profiles of the SnS films grown at different temperatures. The X-ray diffraction studies showed that all the films were polycrystalline, containing only SnS phase. All the peaks observed from the XRD spectra corresponds to the orthorhombic structure of the SnS and the evaluated lattice parameters were a = 0.433nm, b = 1.119 nm, c = 0.398 nm. The layers formed at lower temperatures (280 - 300°C) had the (111)/ (040) orientation appearing almost simultaneously along with the (110), (120), (021), (101) and (131) peaks. However the intensity of the latter peaks reduced drastically with the increase of substrate temperature. Between the temperatures 320 °C and 360 °C, the intensity of these smaller reflections along the (111) peak was largely suppressed so that the (040) plane became the most prominent peak.

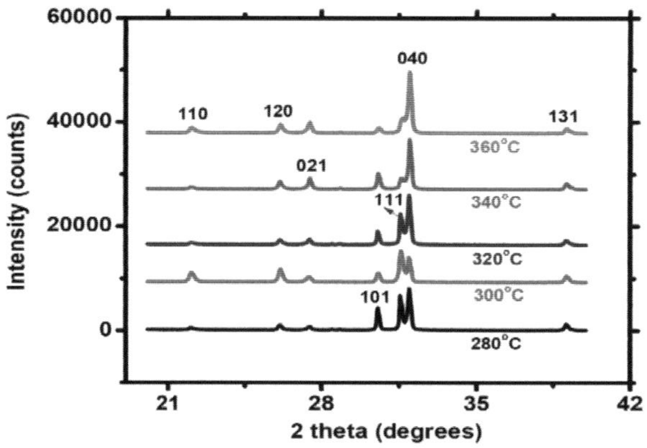

Fig. 2. XRD profiles of SnS films formed at different substrate temperatures.

Information gleaned from the XRD profiles was used to evaluate the crystallite size (D) using the Scherrer's formula [5-9]

$$D = \frac{k\lambda}{\beta \cos\theta} \tag{1}$$

where k is a constant (0.94), β is the (full width half maximum) FWHM value, λ is the wavelength of CuK_α radiation source (λ = 0.15406 nm) and θ is the Bragg angle. The strain (ε) developed in the film was estimated using the following relation [9] in order to evaluate the quality of the grown films and is shown in Fig. 3.

$$\varepsilon = \frac{\Delta\theta}{2(\tan\theta)} \tag{2}$$

where θ is the diffraction angle and Δθ is the fringe width at half maximum (FWHM). The results showed a decrease of strain in the films with substrate temperature. This is due to the increase of crystallite size with temperature that led to a reduction in the inter-crystalline barriers or grain boundary area in the layers. This is also shown in Fig.3.

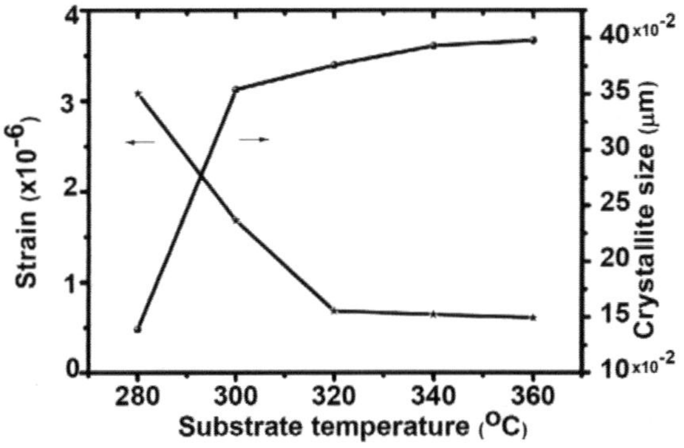

Fig.3. Dependence of stress and crystallite size on substrate temperature.

The XRD data was also used to calculate the peak intensity ratio of the SnS films along the SnS (111) and SnS (040) plane with respect to the SnS (131) plane on the same scale to estimate the crystallinity. Fig. 4 shows the variation of the peak intensity ratio and grain sizes (evaluated from SEM) with substrate temperature. As expected, the peak intensity ratio increased with increase of substrate temperature, which is an indication of an improvement in the crystallinity of the films. This is because at higher growth temperatures, the surface mobility increases, allowing the films to decrease its total energy by growing larger grains thereby decreasing its grain boundary area [5].The grain appeared to increase slowly up to a growth temperature of 340 °C and then steeply afterwards. A similar behaviour has been observed in literature in SnS films deposited by spray pyrolysis [7].

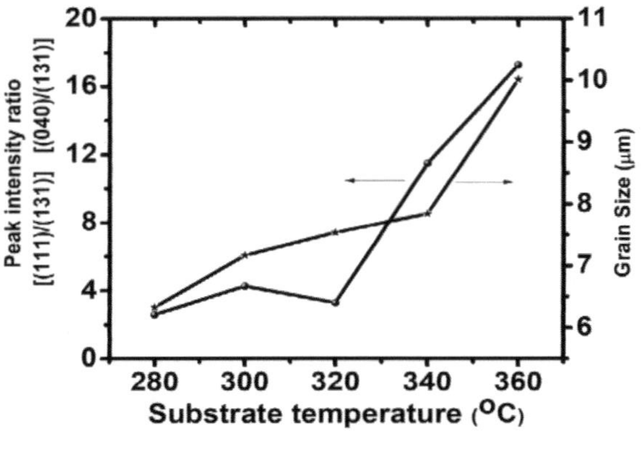

Fig. 4. Peak intensity ratio and grain size vs substrate temperature.

The SEM studies indicated that the layers contain very large grains that were densely packed. Fig. 5 shows the surface topography of SnS layers deposited at two different substrate temperatures. An evaluated grain size of 10μm was obtained for the highest substrate temperature investigated in the study. Some authors [8] attributed the increase of grain size with substrate temperature to the formation of clusters in the crystallites in SnS layers. However the observed variation of grain size in this study has been attributed to the change in the peak intensity ratio from (111)/(040) to (040) plane as observed in the XRD spectra.

(a) 

(b) 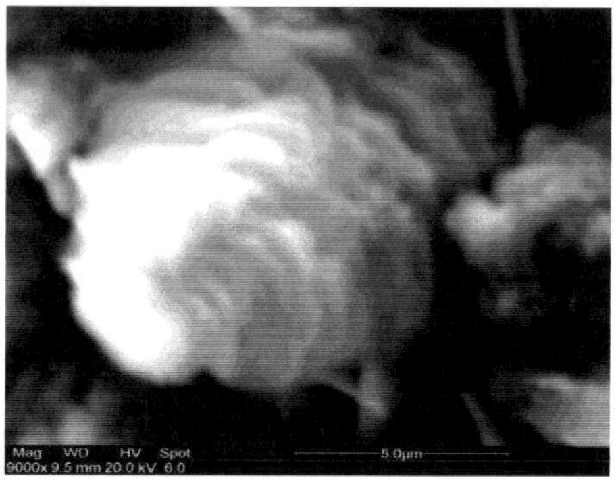

Fig.5. SEM micrographs of SnS films formed at (a) 300°C and (b) 360°C.

The optical transmittance with wavelength of SnS films was measured in wavelength range of 400-1800 nm. Fig. 6 shows such a plot for SnS films. Although a low optical transmittance was observed due to the film thicknesses, however, a clear interference fringes can be observed from the spectra, indicating uniformity of film thickness and optical homogeneity in the grown films. This implies that either scattering or absorption did not occur at the long wavelength region. The spectra also indicated a steep fall in the transmittance spectra near the fundamental absorption edge indicating the absence of other phases in these films as evidenced in the literature [10].

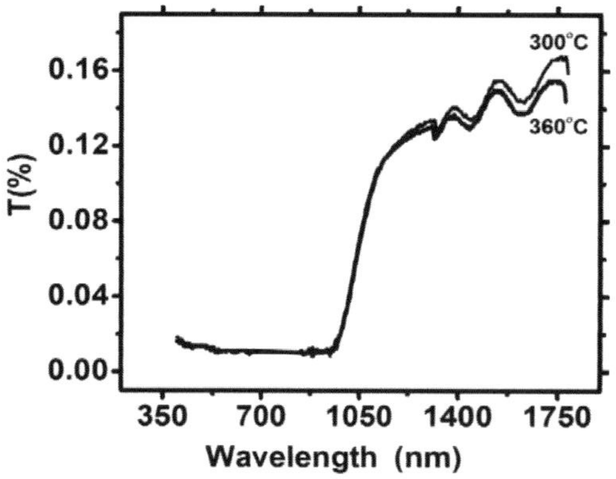

Fig.6. Optical transmittance versus wavelength spectra of SnS films.

From the transmittance versus wavelength data, the optical absorption coefficient (α) was evaluated for all the films and found to be $> 10^4 cm^{-1}$ above the fundamental absorption edge. The optical bandgap was determined using the $(\alpha h\nu)^2$ versus ($h\nu$) plots by extrapolating the plots on the $h\nu$ axis and found to vary in the range, 1.30-1.34 eV for layers grown at different substrate temperatures. The refractive index was calculated in the region where the effect of free carrier absorption is most pronounced in the (α) versus ($h\nu$) plots, using Manifacier's theory for weak absorption region [11]. Fig. 7 shows the change of refractive index with substrate temperature. It can be seen from the figure that the refractive index increases with temperature up to 340°C and then decreased. The optical conductivity was calculated using the relation [12]

$$\sigma = \frac{\alpha n c}{4\pi} \quad (3)$$

where α and n retain their meanings and c is the speed of light. The optical conductivity depends directly on the absorption coefficient and the refractive index of the material. However, it was observed to decrease marginally with substrate temperature up to 340°C and then increased. This might be due to the difference in orientation in the layers grown at lower substrate temperatures.

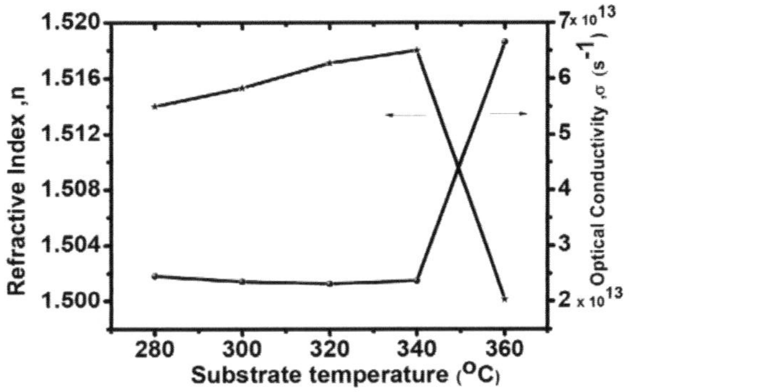

Fig.7. Variation of n and σ with substrate temperature

4. Conclusion

The structural and optical properties of SnS thin films have been investigated with respect to the suitability of the grown films as good absorber layers for solar cell applications. The elemental composition of the films exhibited a non-stoichiometric behaviour. However, high quality films of good crystallinity and very large grains of up to 10μm were recorded. Large values of absorption coefficient $>10^4 cm^{-1}$ were obtained in all the films. The appearance of interference fringes in the optical transmittance spectra indicates thickness uniformity, optical homogeneity and good surface states of the grown films, implying that the films could potentially be useful in various optoelectronic and photovoltaic solar cell applications.

REFERENCES

[1] Green M A, Emery K, Hishikawa Y, Warta W and Dunlop E D 2011 *Prog. Photovolt: Res. Appl.* **19** 565-572
[2] Ramakrishina Reddy K T, Purandar Reddy P, Miles R W and Datta P K 2001 *Optical Materials* **17** 295-298.
[3] Ramakrishna Reddy K T, Prathap P and Miles R W Thin films of tin sulphide for application in photovoltaic solar cells' in "Photovoltaics: Developments, Applications and Impact", Eds. H. Tanaka and K. Yamashita, *Nova Science Publishers, Inc., New York* 2010, 1-27.
[4] Sajeesh T H, Jinesh K B, Sudha Kartha C and Vijayakumar K P 2012 *Appl. Surf. Sci.* **258** 6870-6875.
[5] Maissel L I and Glang R 1970 *Handbook of Thin Film Technology* (New York: McGraw Hill)
[6] Warren B E 1990 *X-ray diffraction (*New York: Dover)
[7] Ramakrishna Reddy K T, Koteswara Reddy N and Miles R W 2006 *Solar Energy Materials & Solar Cells* **90** 3041-3046.
[8] Devika M, Reddy K T R, Reddy N K, Ramesh K, Gopal E S R and Gunasekhar KR 2006 *J.Appl.Phys.* **100** 023518
[9] Qiao Z, Latz R and Mergl D 2004 *Thin Solid Films* **466** 250-258
[10] Orton J 2004 *The story of Semiconductors(*New York: Oxford University press)
[11] Manifacier J C, Gasiot J and Fillard J P 1976 *J.Phys.E:Sci.Instrum.* **9** 1002-1003
[12] Pankove J I 1975 *Optical Processes in Semiconductors* (New York: Dover publishers)

Journal of Physics: Conference Series **417** (2013) 012040 doi:10.1088/1742-6596/417/1/012040

Thickness dependence of the dielectric properties of thermally evaporated Sb_2Te_3 thin films

K Ulutas D Deger, S Yakut
Istanbul University, Science Faculty, Physics Department 34459 Vezneciler-Istanbul, TURKEY

E-mail: deger@istanbul.edu.tr

Abstract. Sb_2Te_3 thin films of different thickness (23 - 350 nm) were prepared by thermal evaporation technique. The thickness dependence of the ac conductivity and dielectric properties of the Sb_2Te_3 films have been investigated in the frequency range 10 Hz- 100 kHz and within the temperature range 293-373K. Both the dielectric constant ε_1 and dielectric loss factor ε_2 were found to depend on frequency, temperature and film thickness. The frequency and temperature dependence of ac conductivity ($\sigma_{ac}(\omega)$) has also been determined. The ac conductivity of our samples satisfies the well known ac power law; i.e., $\sigma_{ac}(\omega) \propto \omega^s$ where s<1 and independent of the film thickness. The temperature dependence of ac conductivity and parameter *s* is reasonably well interpreted by the correlated barrier hopping (CBH) model. The activation energies were evaluated for various thicknesses. The temperature coefficient of the capacitance (TCC) and permitivity (TCP) were determined as a function of the film thickness. The microstructure of the samples were analyzed using X-ray diffraction (XRD). This results are discussed on the base of the differences in their morphologies and thicknesses. The tendency for amorphization of the crystalline phases becomes evident as the film thickness increases.

1.Introduction

Antimony Telluride is a semiconducting chalcogenide of the VB and VIB groups of elements. Due to its attractive photo conducting properties and its high Seebeck coefficient, low thermal conductivity, low electrical resistivity, low band gap, and long-term stability Sb_2Te_3 has wide industrial applications such as target material in television camera, microwave devices, switching devices and optoelectronic devices [1-4]. Sb_2Te_3 thin films have received considerable attention because of their potential application in the fabrication of integrated thermoelectric devices. To obtain Sb_2Te_3 thin films, a number of techniques have been used in the literature, such as thermal evaporation[5,6], atomic layer epitaxy(ALE) [7], sputtering [2,8], electrochemical method [9], flash evaporation [10] and metalorganic chemical vapor deposition (MOCVD) [11].

The electrical, thermoelectrical and optical properties of bulk samples Sb_2Te_3 were studied [12,13]. However, thickness dependence of the dielectric properties of films prepared from that compound were hardly investigated [5]. Therefore, at this study, the dielectric properties of the thermal evaporated Sb_2Te_3 thin films have been investigated with recpect to frequency, temperature, and especially film thickness in detail, in the frequency range 10 Hz-100 kHz and in the temperature range 293-373 K and 23 - 350 nm film thickness.

2.Experimental Details

The Al $|Sb_2Te_3|$ Al thin film sandwich structures were prepared by the following sequential evaporation procedure for dielectric measurements. The substrates were cleaned with detergent and flushed with copious amounts of de-ionized water. Finally the substrates were sequentially rinsed with acetone, methanol and isopropyl alcohol. The base and counter Al electrodes were deposited by thermal evaporation. The system was pumped to a base pressure of less than $1,333x10^{-3}$ Pa. First Al was evaporated to serve as the bottom electrode on the Corning glass substrates. Sb_2Te_3 ingot of high purity (99.99%) were used as the starting material. The crushed ingot were evaporated from molybdenum boat at a vacuum better than $1,333x10^{-4}$ Pa. Al was thermally evaporated to form the top electrode, finally. The evaporation temperature of Sb_2Te_3 ingot was about 600°C and the evaporation temperature of Al wires was about 1200°C. The resulting Al$|Sb_2Te_3|$Al thin film sandwich structures have a capacitive configuration of effective area 6 mm^2. Films were annelated at 373 K for 1h. A HP 4192A 5Hz-13MHz Lf Impedance Analyzer were used for the capacitance and dissipation factor measurements. The ohmic character of contacts were determined by applying front and reverse biases between bottom and top electrodes. For both of them, the resistance measurements gave same results [2]. These measurements were made in approximately $1{,}333 \times 10^{-3}$ Pa vacuum and were carried out in the temperature range 293-373 K. Temperatures of the samples during the study were measured with a copper-constantan thermocouple. The thickness of films were measured by Tolansky interferometric method. The film thickness ranged from 23-350 nm. X-ray difraction analysis was used to investigated the structure of the films. The total conductivity was calculated from the equation $\sigma(\omega) = L/ZA$, where L is the thickness of the film and A is the cross-sectional area. The dielctric constant was calculated from the equation: $\varepsilon_1 = LC/A\ \varepsilon_0$, where C is the capacitance of the film, and ε_0 is the permitivity of the free space. The dielectric loss was calculated from equation $\varepsilon_2 = \varepsilon_1 \tan\delta$, where ($\delta = 90 - \Phi$), Φ is the phase angle. The XRD measurements were carried out in a Rigaku DMAX 2200, Japan. X-ray diffractometer using CuKα radiation, setting 2θ in the range from 10° to 65°.

3.Results

3.1 Structural investigations

Figs.1-a and b shows the X-ray diffraction patterns (XRD) of the thin films prepared on glass substrates. The X-ray diffraction patterns of the films tell about their cristallinity. It is evident from Figs.1-a and b that increasing film thickness changes the film structure from being crystalline into amorphous. For the thinnest film (60 nm), all the diffraction peaks in Fig.1-a have been identified to belong to Sb_2Te_3 and Sb (JCPDS card no: 15-0874 and 85-1322). The peak values for this X-ray output are given in Table 1 where the 2θ values, calculated from the diffraction pattern of Fig.1-a, together with the standard 2θ values of the Sb_2Te_3 and Sb phase are also shown for comparison. A very good coincidence of these 2θ values with the standard ones is seen. Finally, Fig.1-b shows the completion of amorphisation.

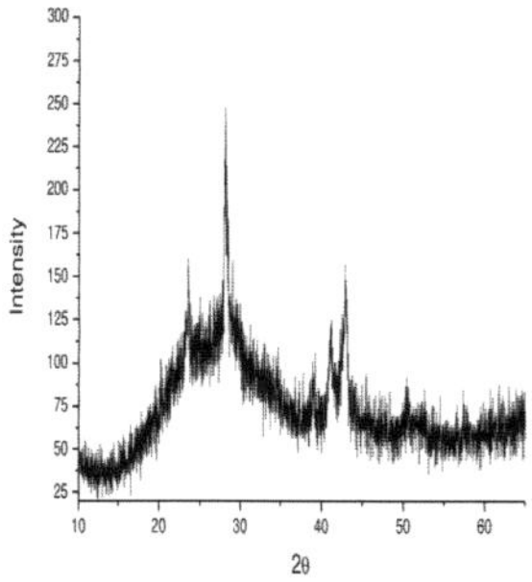

Figure 1a. X-ray diffraction patterns of Sb_2Te_3 as-deposited films. Film thickness: 60 nm

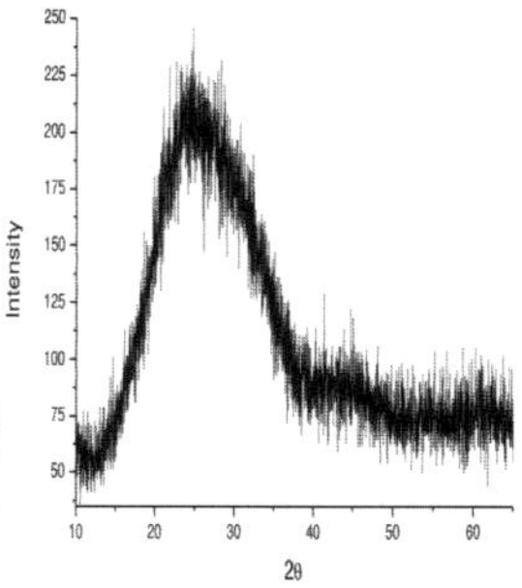

Figure 1b. X-ray diffraction patterns of Sb_2Te_3 as-deposited films. Film thickness: 85nm

Table 1. XRD Diffraction results and JCPDS standarts

Observed 2θ from XRD (Deg.)	Standart 2θ from JCPDS (Deg.)	Standart Intensity	Planes h	k	l	Card Number
23,520	23,656	5,6	0	0	3	85-1322
28,040	28,245	100	0	1	5	15-0874
41,120	41,903	32,7	1	1	0	85-1322
42,880	42,402	25	1	1	0	15-0874

3.2 Dielectric constant

Fig. 2 shows the frequency dependence of dielectric constant, ε_1, at different temperatures for of Sb_2Te_3 film of thickness 42 nm as a representative example. It is clear from the figure that ε_1 with frequency can be attributed to the fact that at low frequencies ε_1, for polar material is due to the contribution of multicomponents of polarizability, deformational (electronic, ionic) and relaxation (orientational and interfacial) as suggested by Farid A.M. et al. [5,15]. When the frequency is increased, the orientational polarization decreases since it takes more time than electronic and ionic polarization. This decreases the value of dielectric constant with frequency reaching a constant value at high frequency due to interfacial polarization.

The increase of ε_1 with temperature can be attibuted to the fact that the orientational polarization is connected with the thermal motion of molecules, so dipoles cannot orient themselves at low temperatures. When the

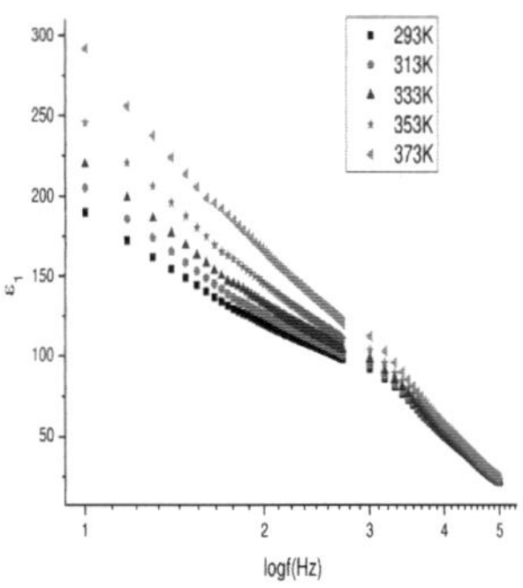

Figure 2. Frequency dependence of the dielectric constant ε_1 for Sb_2Te_3 thin films of thickness 23 nm

Table 2. The thickness dependence of TCC (Temperature Coefficient of Capacitance) and TCP(Temperature Coefficient of Permitivity) for Sb_2Te_3 films at 293K and 1kHz

d(nm)	TCC(ppm/K)	TCP(ppm/K)
23	2666,67	2673,15
42	6357,32	6351,91
120	4898,67	4917,13
240	7871,11	7782,18
270	7362,5	7420,2
350	6981,71	6985,31

temperature is increased the orientation of dipoles is faciliated and this increases the value of orientational polarization and this increases ε_1 with increasing temperatures [16]. The thickness dependence of the dielectric constant, ε_1 at different frequencies and different temperatures in the Sb_2Te_3 thickness range of 23-350 nm are shown in Fig. 3 and Fig. 4. Towards lower thicknesses ε_1 decreases with decreasing thickness. This result is in agreement with the literature [17,18]. The observed decreases in ε_1 is thought to arise from the increase in the density of voids for lower thicknesses. Deepha [19] and Kumar [20] have also reported that the effective thickness of insulators decreases with increasing density of voids causing the dielectric constant to decrease. Hence, the decrease in total polarization can be explained by saying that the effective medium decreases as thickness decreases. The frequency dependence of ε_1 for different film thicknesses is shown in Fig. 5. It is known that the contribution of polarization mechanisms with long relaxation times increases at lower frequencies [17]. Furthermore, the temperature coefficient of the capacitance (TCC) and permitivity (TCP) were determined and are given in Table 2.

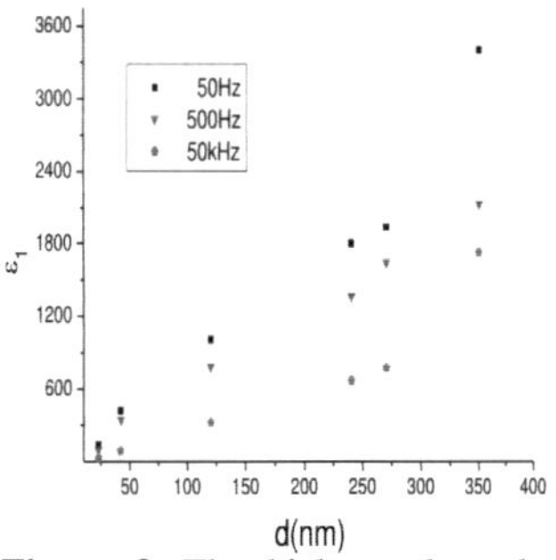

Figure 3. The thickness dependence of the dielectric constant ε_1 at different frequencies in the Sb_2Te_3 thickness range of 23-350 nm.
T = 333 K

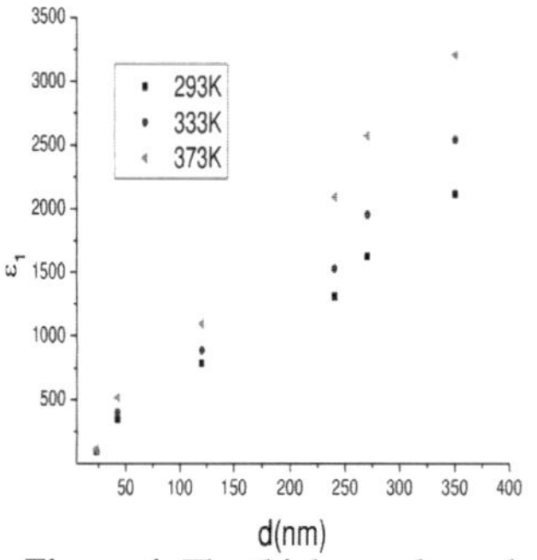

Figure 4. The thickness dependence of the dielectric constant ε_1 at different temperatures in the Sb_2Te_3 thickness range of 23-350 nm.
f = 1kHz

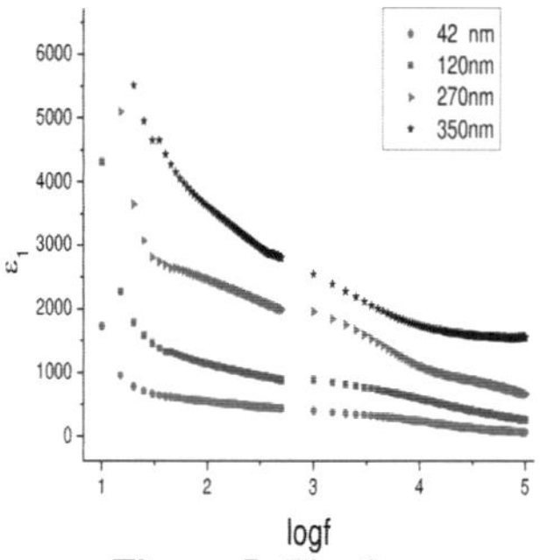

Figure 5. The frequency dependence of ε_1 for different film thicknesses at T = 333 K

3.2 Ac conductivity

A common feature to all amorphous semiconductors is that ac electrical conductivity $\sigma_{ac}(\omega)$ increases with increasing frequency according to the equation [21].

$$\sigma_{ac}(\omega) = \sigma_{tot}(\omega) - \sigma_{dc} = A\ \omega^s \qquad (1)$$

where ω is the angular frequency ($\omega = 2\pi f$), $\sigma_{tot}(\omega)$ is the measured total electrical conductivity,σ_{dc} is the dc electrical conductivity, s is the frequency exponent (s<1) and A is constant dependent on temperature. It is found that the ω^s dependence of ac conductivity $\sigma_{ac}(\omega)$ is a general characteristic for chalcogenide glasses up to a frequency of 10^6 Hz where ω is the angular frequency, s is the frequency exponent and A is a constant independent of frequency [22]. Fig. 6-a shows the frequency dependence of ac conductivity $\sigma_{ac}(\omega)$ for Al_2O_3 film of thickness 270 nm at different temperatures as a representative example. It is clear from the figure that $\sigma_{ac}(\omega)$ increases linearly with frequency according to equation (1). The same behavior of the frequency dependence of $\sigma_{ac}(\omega)$ was obtained for all investigated films. Values of the frequency exponent s were obtained from the slopes of these lines of the figure. The temperature dependence of s for the investigated films is shown in Fig. 6-b. It is seen that s decreases as the temperature increases, independent on film thickness in the investigated range.

According to the quantum mechanical tunneling (QMT) model [23], the exponent *s* is almost equal to 0.8 and slightly increases with increasing temperature or independent on temperature. Therefore, QMT model is considered not applicable to the obtained results. According to the overlapping-large polaron tunneling (OLPT) model [24], the exponent *s* is both temperature and frequency dependent. *s* decreases with increasing temperature from unity at room temperature to a minimum value at a certain temperature then it increases with increasing temperature. Therefore, OLPT model is not applicable to the obtained results.

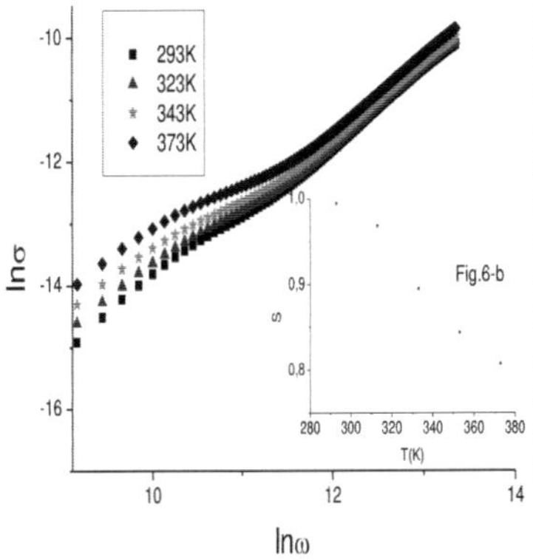

Figure 6a. The frequency dependence of ac conductivity $\sigma_{ac}(\omega)$ for Sb_2Te_3 film of thickness 270 nm at different temperatures.
Figure 6b. The temperature dependence of *s* for the investigated films

Table 3. The thickness dependence of Density of States N(E) for Sb_2Te_3 films at 1kH

	N(E)(1/eVcm^3)(10^{28})						
d(nm)	**293K**	**313K**	**323K**	**333K**	**343K**	**353K**	**373K**
23	0,53	0,52	0,52	0,53	0,54	0,56	0,62
42	1,91	3,14	2,26	2,51	2,81	3,11	3,62
120	2,34	2,58	2,77	3,09	3,45	3,70	4,21
240	3,25	3,70	4,22	4,69	5,04	5,32	5,84
270	5,18	6,01	6,94	7,96	9,04	9,82	11,50
350	3,63	4,38	5,02	6,50	7,24	7,49	7,18

According to correlated barrier hopping (CBH) model, values of the frequency exponent *s* is ranged from 0.7 to 1 at room temperature and is found to decrease with increasing temperature. This is in good agreement with the obtained results, so the frequency dependence of $\sigma_{ac}(\omega)$ can be explained in terms of CBH model. In most chalcogenide glasses the obtained values of s ranged from 0.7 to 1 at room temperature, and have a tendency to decrease with increasing temperature. This means that the obtained experimental results agree with the CBH model. According to the Austin-Mott Formula [25], based on CBH model, ac conductivity $\sigma_{ac}(\omega)$ can be explained in terms of the hopping of electrons between pairs of localized states at the Fermi level. $\sigma_{ac}(\omega)$ is related to the density of states $N(E_f)$ at the Fermi level by:

$$\sigma_{ac}(\omega) = (\pi/3)\ [N(E_f)]\ 2\ K\ T\ e^2\ \alpha^{-5}\ \omega\ [\ln(\upsilon/\omega)]^4 \qquad (2)$$

where α is the exponential decay parameter of the localized states wave function, and υ is the phonon frequency. By assuming $\upsilon = 10^{12}$ Hz and α^{-1}=1nm [26], the density of states is calculated and is given in Table 3. for the Se_2Te_3 at frequency 1kHz. Temperature dependence of the ac conductivity $\sigma_{ac}(\omega)$ at different frequencies was studied for the investigated films. Fig. 7 shows a plot of ln $\sigma_{ac}(\omega)$ against 1000/T of thicknesses 270 nm as a representative example. It is clear from this figure that ln $\sigma_{ac}(\omega)$ increases linearly with the reciprocal of absolute temperature. This suggested that the ac conductivity is a thermally activated process from different localized states in the gap or its tails. The activation energy of conduction, $\Delta E_\sigma(\omega)$ is calculated at different frequencies from the slopes of the straight lines, obtained for all investigated thicknesses using the well known equation
$\sigma = \sigma_o \exp(-\Delta E_\sigma(\omega)/kT)$. The frequency dependence of the activation energy for the investigated films is shown in Fig. 8. It is clear that $\Delta E_\sigma(\omega)$ decreases with increase of frequency. It can also be seen that $\Delta E_\sigma(\omega)$ tends to decrease with increasing frequency as found for other amorphous materials [27].

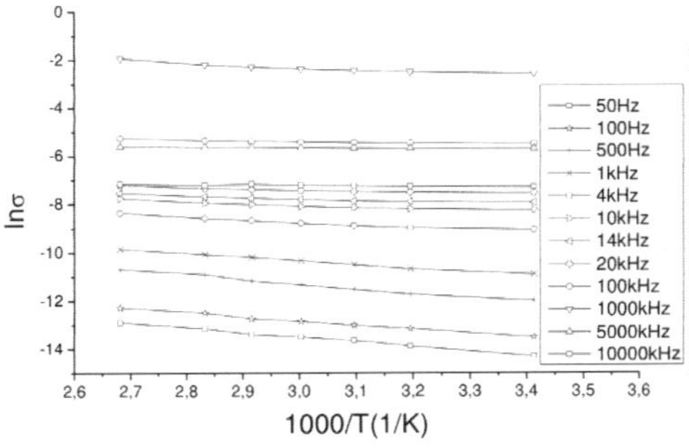

Figure 7. The temperature dependence of ac conductivity $\sigma_{ac}(\omega)$ for Sb_2Te_3 film of thickness 270 nm

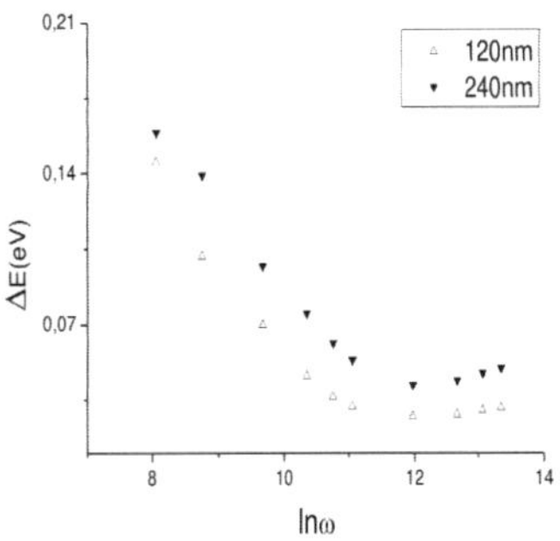

Figure 8. The frequency dependence of the activation energy for the investigated films

The increase of the applied field frequency enhances the electronic jumps between the localized states, consequently the activation energy $\Delta E_{\sigma}(\omega)$ decreases with increasing frequency. The activation energies, which value ~ 0,1 eV can be attributed as the activation energy for electronic polarization. The same behavior was obtained for all investigated films.

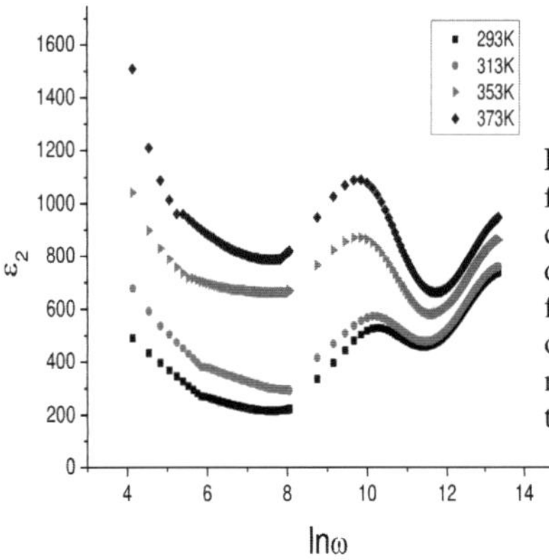

Figure 9. The frequency dependence of dielectric loss for Sb_2Te_3 film of thickness 270 nm at different temperatures.

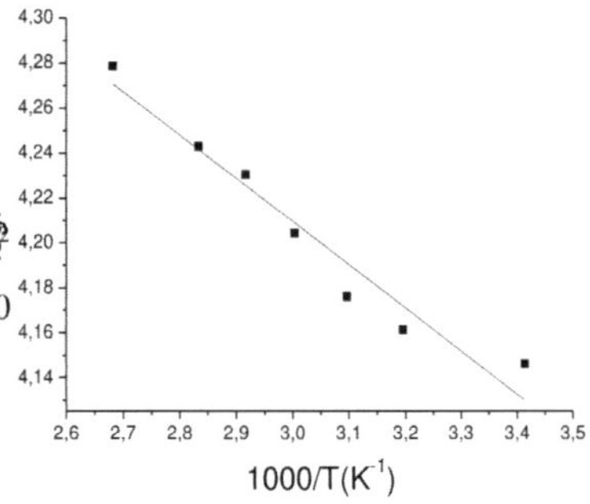

Figure 10. The temperature dependence of frequencies, at which minimum dielectric loss is observed

3.3 Dielectric Loss ε_2

It was observed that ε_2 increases as the frequency increases and reaches a maximum and thereafter decreases as shown in Fig. 9.

As the temperature increases the frequency at which ε_2 reaches a maximum shifted towards higher frequencies. The increase in the dielectric loss may be due to an increase in both the residual current and absorption current. The dielectric loss caused by the dipole mechanism reaches its maximum value at a certain temperature for a definite frequency [28, 29]. The variation of dielectric loss with frequency and temperature represents the effects of various polarization mechanisms at different frequencies. Initially observed mechanism loses effective role with increasing frequency and increasing dielectric loss. At a maximum frequency another type mechanism or one of the mechanism, which the initial mechanism includes, becomes dominant in polarization process.

The activation process was evaluated from;

$$f_{\min} = f_0 \exp\left(-\Delta E/kT\right) \qquad (3)$$

In Fig. 10 variation of the frequencies of minimum dielectric loss with temperature is observed. The activation energy obtained from temperature dependence of dielectric loss 0,06 eV as average.

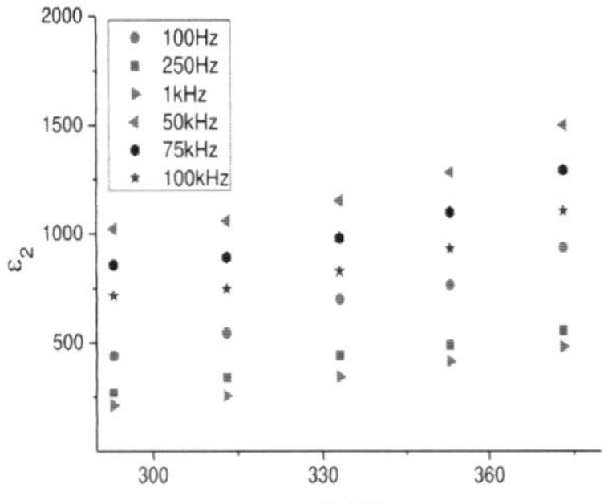

Figure 11. The temperature dependence of dielectric loss ε_2 for Sb_2Te_3 film of thickness 240 nm at different frequncies.

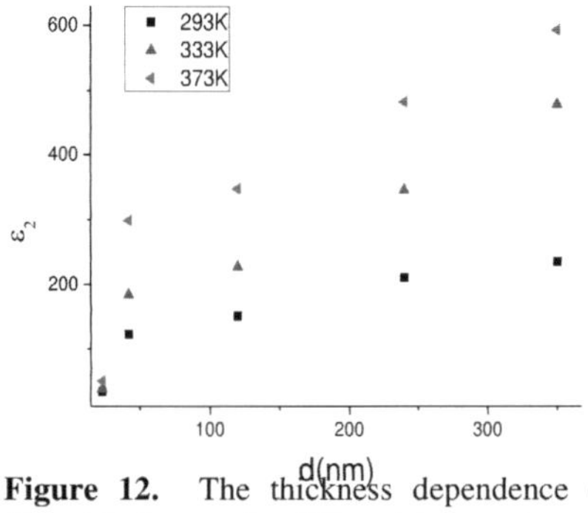

Figure 12. The thickness dependence of the dielectric loss ε_2 for different temperatures in the Sb_2Te_3 thickness range of 23-350 nm at 1kHz.

The activation energy values evaluated from capacitance and dielectric loss factor measurements are in good agreement with each other. Depending on the obtained activation energy, this mechanism may be electronic polarization. Ionic dipole relaxations have been observed with activation energies between 0,5 and 1,0 eV. Electronic processes generally have associated low activation energies (~0,1 eV) and this presents a more favourable mechanism [30]. Temperature dependence of dielectric loss is shown in Fig. 11. Increase in temperature may activate or rise the effectiveness of dipolar(orientational) polarization in net polarization [31]. Fig. 12 is shown the frequency dependence of the ε_2 at different thickness. When the thickness dependence of dielectric loss is investigated, it is found to increase in thickness causes to increase in net polarization in structure. Higher thickness can be thought more charge carriers to be polarized, thus more dielectric loss may occur in structure.

4. Conclusion

Frequency, temperature and film thickness dependence of dielectric constant, dielectric loss factor and ac conductivity of Al| Sb_2Te_3|Al thin film sandwich structures were studied in detail. These factors are found to be frequency, temperature and film thickness dependent. The temperature dependence of ac conductivity and parameter *s* is reasonably well interpreted by the correlated barrier hopping (CBH) model. The activation energies were evaluated for various thicknesses. The temperature coefficient of the capacitance (TCC) and permitivity (TCP) were determined as a function of the film thickness. The microstructure of the samples were analyzed using X-ray diffraction (XRD). This results are discussed on the base of the differences in their morphologies and thicknesses. The tendency for amorphization of the crystalline phases becomes evident as the film thickness increases.

Acknowledgement

This work was supported Research Fund of the University of Istanbul, Project No:2563 and UDP-7203/07052010.

References:

[1] Bätzner D L, Romeo A, Terheggen M, Döbeli M, Zogg H and Tiwari A N 2004 *Thin Solid Films* **451-452** 536-543
[2] Romeo N, Bosio A, Tedeschi R, Romeo A, and Canevari V 1999 *Solar Energy Materials and Solar Cells* **58** 209-218,
[3] Rowe D M 1995 *CRC Handbook of Thermoelectrics* CRC Press
[4] Rajpure K Y, Bhosale C H 2000 *J. Phys. and Chem. Solids* **61** 561
[5] Farid A M, Athia H E and Hegab N A 2005 *Vacuum* **80** 284-294,
[6] Zou H, Rowe D M and Williams S G K 2002 *Thin Solid Films* **408** 270-274,
[7] Yang J Y, Zhu W, Gao X H, Fan X A, Bao S Q and Duan X K 2007 *Electrochimica Acta* **52** 3035-3039
[8] Abken A E and Bartelt O J 2002 *Thin Solid Films* **403-404** 216-222
[9] Erdoğan I Y and Demir U 2009 *Journal of Electroanalytical Chem.* **633** 253-258
[10] Soliman H S, Yaghmour S and Al-Solami H G 2008 *Europan Phys. J. Appl. Phys.* **44** 59-64
[11] Venkatasubramanian R, Colpitts T, Watko E, Lamvik M and El-Masry N 1997 *J. Cryst. Growth* **170** 817–821
[12] Thonhauser T 2004 *Solid State Communications* **129** 249-253
[13] Zou H, Rowe D M, Min G 2001 *J.Vac. Sci. Technol. A* **19** 3
[14] Rajasekar K, Subarayan A, Sathyamoorthy R, Deepa 2004 *J. Ionics* **10** 291
[15] Farid A M, Bekheet A E 2000 *Vacuum* **59** 932
[16] Yildirim S, Ulutas K, Deger D, Zayim E O and Turhan I 2005 *Vacuum* **77** 329-335
[17] Deger D, Ulutas K 2004 *Vacuum* **72** 306-312
[18] Ibrahim A A, El- Sayed N Z, Kaid M A, Ashour A 2004 *Vacuum* **75** 189-194
[19] Dheepa J, Sathyamoorthy R, Sebastian P J 2006 *Solid State Electronics* **50** 1315-1319
[20] Rajesh Kumar B, Subba Raa T 2011 *Chalcogenide Lett.* **8** 83-92
[21] Jonscher A K 1977 *Nature* **267** 673
[22] Durand B, Taillades G, Pradel A, Ribes M, Badot J C, Belhadj-Taher N 1994 *J. Non-Cryst. Solids* **1306** 172-174
[23] Ghosh A 1990 *Phys. Rev. B* **41** 1479
[24] Long A R 1982 *Adv. Phys.* **31** 553
[25] Austin L G, Mott N F 1969 *Adv. Phys.* **18** 41
[26] Agarwal S C, Guha S, Narasimhan K L 1975 *J Non-Cryst. Solids* **18** 429
[27] Angell C 1992 *Ann. Rev. Phys. Chem.* **43** 693
[28] Ulutas K, Yakut S, Deger D 2010 *IOP Conf. Series: Mat. Sci. and Eng.* **15** 012038
[29] Stevels J M 1957 *The electrical properties of glasses Hantbuch der Physik*.350
[30] Lancaster M C 1972 *J. Phys. D: Appl. Phys.* **5** 1133
[31] Stevels J M 1957 *The electrical properties of glasses Hantbuch der Physik* 35

Spectral Changes in Thin Films of Cyclicsilanes and Polysilanes by Heat Treatment

H Tachibana, T Mizuno and Y Akasaka

Electronics and Photonics Research Institute, National Institute of Advanced Industrial Science and Technology (AIST), Central 5, Higashi 1-1-1, Tsukuba 305-8565, Japan

E-mail: h-tachibana@aist.go.jp

Abstract. We report the optical properties of the thin films of cyclic and linear and polysilanes containing phenyl groups by heat treatments. We observe the shift to a longer wavelength of the absorption edge in UV-visible absorption spectra of the thin films, when the cyclic and linear polysilane films were heated at 500 °C after UV irradiation. This behavior is interpreted in terms of the formation of network siloxene films by the dissociation of phenyl groups from silicon structures and the accompanying incorporation of oxygen into the silicon structures. It is shown that UV irradiation before heating and substituent groups attached to cyclic and linear polysilanes affect the spectral changes by heat treatment.

1. Introduction

Polysilanes contains σ-conjugated silicon-silicon bonds in the polymer backbone. Interesting optical properties characteristic of σ-conjugated polysilanes have been known to change by the backbone structures [1, 2]. Polysilanes with various substituent groups have been synthesized as the advance in the synthetic technique. The backbone structure can be controlled by choosing the monomer precursors at the synthetic stage [3, 4]. Recently, we have reported promising results for formation of siloxene-like multilayer by heat treatment of solution-processed thin films using linear and network polysilanes containing anthryl groups as a precursor [5]. We investigate the effect of substituent and backbone structure on heat treatment of polysilane thin films. In this study, we report the optical properties of linear and cyclic polysilanes containing phenyl groups. The effect of UV irradiation before heating and heat treatments of the thin films under vacuum was investigated using UV-visible spectroscopy. The structural changes accompanying heat treatment were analyzed with infrared absorption spectroscopy and atomic force microscopy (AFM).

2. Experimental

2.1. Materials

Poly(methylphenylsilane) (PMPS) and poly[bis-(*p*-*n*-butylphenyl)silane] (PBPS) were polymerized from the corresponding dichlorosilane monomer using dodecane and toluene solution by the Wurtz coupling method, respectively [6, 7]. The decaphenylcyclopentasilane (DCPS) and respective dichlorosilane monomer and were available commercially.

2.2. Film preparation

Solid thin films of cyclic and linear polysilanes were prepared by spin-coating chlorobenzene and toluene solution onto quartz plates for UV-visible absorption measurements, onto Au-evaporated glass plates for IR reflection-absorption (RA) measurements, onto quartz plates for AFM measurements, respectively. Heat treatments were carried out by heating the films for given times under a reduced pressure of 200 Pa. After the heat treatment, the thin films were cooled to room temperature prior to spectroscopic and microscopic measurements. The thin films were irradiated with a 500-W high-pressure mercury lamp in air at room temperature.

2.3. Characterization

UV-visible absorption spectra of the thin films were obtained using a Cary 500 UV-VIS-NIR spectrophotometer. Fourier RA spectra were obtained using a Perkin-Elmer Spectrum 100 FT-IR spectrometer equipped with a liquid-nitrogen-cooled MCT detector. For RA measurements, the incident angle was 80°. AFM measurements were performed using a SEIKO SPA300 microscope operating in a noncontact mode. All the measurements were carried out in air at room temperature.

3. Results and discussions

3.1. Cyclic polysilanes

3.1.1. UV irradiation.

We investigated the effect of UV irradiation on DCPS thin films. Figure 1 shows spectral changes of DCPS thin film on irradiation of UV light. On irradiation with UV light, the absorption band decreases gradually. On UV irradiation more than 30 min, the intensity of the absorption band becomes constant. The structural changes were investigated using FT-IR absorption spectroscopy. Figure 2 shows the variation of the RA spectrum of the DCPS thin film on changing the UV irradiation time. In the RA spectra before irradiation, the absorption bands at approximately 1400 and 700 cm^{-1} are assigned to the phenyl C-Si stretching mode. After UV irradiation for 5 min, an intense absorption band with a broad structure appears at approximately 1100-1200 cm^{-1}, which is assigned to the Si-O-Si stretching mode. On further irradiation with UV light, the intensities of the absorption

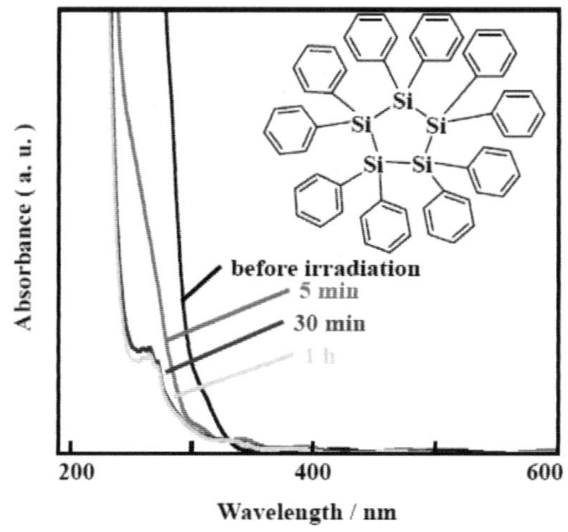

Figure 1. Change of absorption spectra of a DCPS thin film on irradiation with UV light.

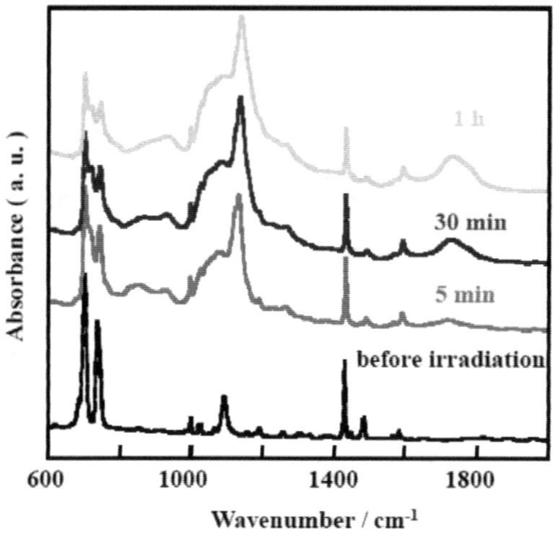

Figure 2. Change of infrared reflection-absorption (RA) spectra of DCPS thin films on irradiation with UV light.

Journal of Physics: Conference Series **417** (2013) 012041 doi:10.1088/1742-6596/417/1/012041

bands increase gradually. In addition, a broad structure appears at approximately 1800 cm^{-1}, which is probably assigned to the phenyl C=C stretching mode. These results suggest that ring opening of DCPS on UV irradiation affords Si-O-Si structures in the thin films with the change in the orientation of phenyl group [8, 9].

3.1.2. Heat treatment.

The thin films after UV irradiation were subjected to heat treatment. The change in absorption spectrum with heating for 10 min under a reduced pressure of 200 Pa is shown Figure 3. The absorption peak at approximately 260 nm increases gradually with increasing heating temperature. With heating at 500 °C, the absorption edge shifts to a longer wavelength. The absorption spectra show a broad absorption up to 600 nm. In the RA spectra of DCPS thin films as shown in Figure 4, the intensities of absorption bands near 1100-1200 cm^{-1} assigned to Si-O-Si stretching band increase appreciably and become broader with increasing heating temperature. In addition, the absorption bands at 700 and 1400 cm^{-1} assigned to phenyl C-Si stretching band decrease at 500 °C. From comparison of two absorptions after irradiation (A_0) and after heating at 500 °C (A_1), the absorption ratio A_1/A_0 of two peaks assigned Si-O-Si and phenyl C-Si stretching band is 1.60 and 0.67, respectively, indicating that the phenyl groups are removed from the thin film. This behavior caused by heat treatment is characteristic of DCPS after UV irradiation, because the absorption decreases gradually and disappears finally with increasing heating temperature in the absorption spectra of thin films of DCPS without irradiation. These results suggest that the shift to longer wavelength of absorption edge is achieved by the ring opening of DCPS, the incorporation of oxygen into silicon chains, elimination of phenyl groups, and the formation of network siloxene structures for the σ-n orbital mixing [5,10, 11].

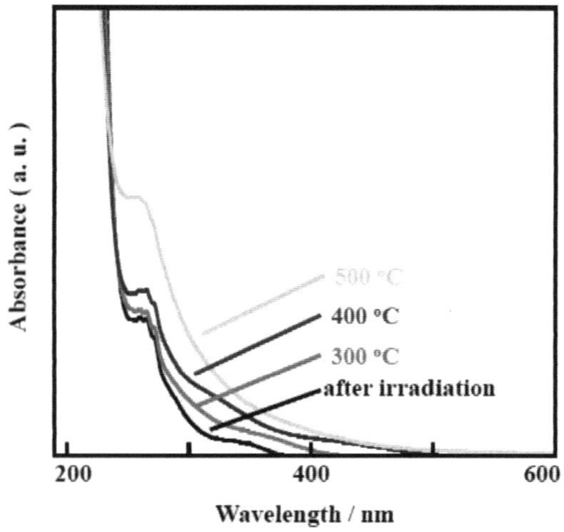

Figure 3. UV-visible absorption spectra of DCPS thin films after UV irradiation for 2 h and after heat treatment at given temperatures for 10 min under a reduced pressure of 200 Pa.

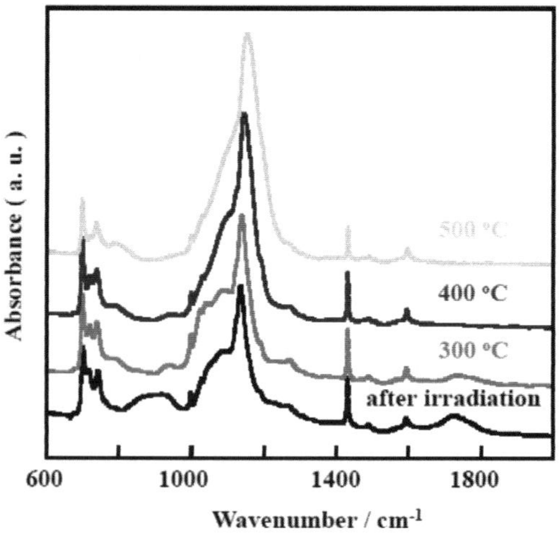

Figure 4. Infrared reflection-absorption (RA) spectra of DCPS thin films after UV irradiation for 2h and heat treatment at given temperatures for 10 min under a reduced pressure of 200 Pa.

3.1.3. Morphology.

The morphological changes before UV irradiation, after irradiation, and after heating were observed by AFM. Figure 5 shows the AFM images of the DCPS thin film before UV irradiation, after irradiation for 2 h in air, and after heating at 500 °C for 10 min under a reduced pressure of 200 Pa. Before UV irradiation, the thin film exhibits rod-like structures. The average width and height of the rod-like structures are 50 nm and 2 nm, respectively. After UV irradiation, the rms surface roughness in the films decreases from 3.3 nm to 1.0 nm and smooth texture is observed. This suggests that the formation of the siloxene structures occurs by ring opening of the DCPS on UV irradiation. On further heating, the films become much smoother with a surface roughness of less than 1 nm, which is probably due to the rearrangement of the network siloxene structures. These behaviors are also supported by the film thickness measurements. The averaged thickness of the thin film decreases by 60 nm after UV irradiation, which may be explained by volume shrinkage accompanying the ring opening of DCPS. On further heating, the thickness decreases to one half after heating at 500 °C, which is probably due to the elimination of phenyl groups.

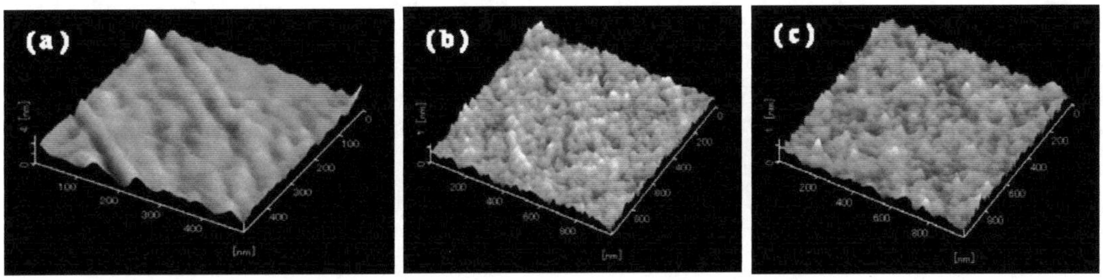

Figure 5. AFM images of a DCPS thin film. The scanned areas of (a), (b), and (c) are 0.5x0.5, 1x1, and 1 x 1 μm^2, respectively: (a) before heating, (b) UV irradiation for 2 h in air and (c) after heating at 500 °C for 10 min under a reduced pressure of 200 Pa.

3.2. Linear polysilanes

To clarify the spectral changes accompanying the siloxene network formation, we investigated the effect of UV irradiation of liner polysilanes on heat treatment. Figure 6 and 7 shows spectral changes of PMPS and PBPS thin film before UV irradiation, after irradiation, and after heating at 500 °C under a reduced pressure of 200 Pa, respectively. After UV irradiation, in both thin films the intensities of the absorption bands decrease appreciably, which is related to the photooxidation of polysilane chains containing Si-O-Si and Si-OH bonds [12, 13]. These structural changes are further supported by the results of the FT-IR measurements. When heating at 500 °C after UV irradiation, the absorption edge shifts to longer wavelength up to 600 nm in the PMPS thin films. On further heating at 500 °C, the intensity of the absorption band increases, but the absorption edge does not shift to longer wavelength. Compared with the absorption spectra with heat treatment after UV irradiation in the PMPS and PBPS thin films, the spectral profile is almost the same, but the absorption edge approaches to 1000 nm in the PBPS thin films, which is probably due to the difference in network siloxene structures formed by the dissociation of phenyl groups [10, 11]. Furthermore, such spectral changes are not observed in thin films of alkyl cyclicsilane and linear poly(di-alkylsilane) without phenyl groups by heat treatment after UV irradiation. These results suggest that The UV irradiation before heat treatment and substituent groups attached to polysilanes play an important role in the changes in electronic structures by heat treatment.

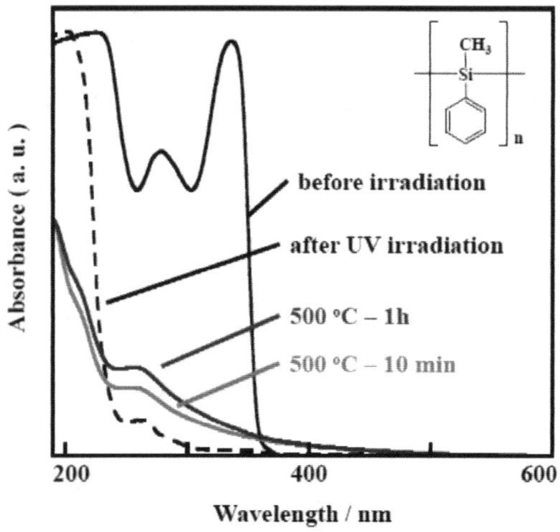

Figure 6. UV-visible absorption spectra of PMPS thin films before UV irradiation, after irradiation for 1 h, and after heat treatment at 500 °C for give times under a reduced pressure of 200 Pa.

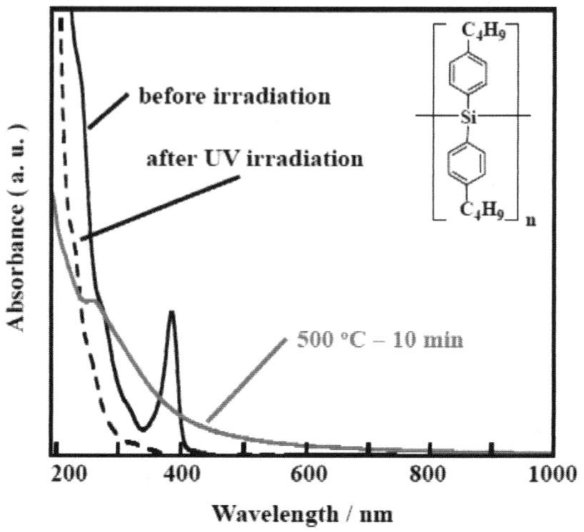

Figure 7. UV-visible absorption spectra of PBPS thin films before UV irradiation, after irradiation for 2 h, and after heat treatment at 500 °C for 10 min under a reduced pressure of 200 Pa.

4. Conclusions

We have demonstrated that the absorption spectra change markedly with heating the thin films at 500 °C under vacuum after UV irradiation, resulting in a decrease in the band gap. The formation of network siloxene structures plays an important role in the decrease in the band gap. The spectral changes by heat treatment depend on UV irradiation before heating and substituent groups attached to cyclic and linear polysilanes.

Acknowledgments

This work was supported by the New Energy and Industrial Technology Development Organization (NEDO).

References

[1] Miller R D and Michl J 1989 *Chem. Rev.* **89** 1359
[2] Tachibana H, Matsumoto M, Tokura Y, Moritomo Y, Yamaguchi A, Koshihara S, Miller R D and Abe S 1993 *Phys. Rev. B* **47** 4363
[3] Koshida N and Matsumoto N 2003 *Mater. Sci. Eng. R* **40** 169
[4] Kyushin S and Matsumoto H 2003 *Adv. Organomet. Chem.* **49** 153
[5] Tachibana H, Toya M, and Ishibe S 2011 *Japan J. Appl. Phys.* **50** 04DK18-1
[6] Trujillo R E 1980 *J. Organomet. Chem.* **198** C27
[7] Acharya A, Seki S, Koizumi Y, Saeki A and Tagawa S 2005, *J. Phys. Chem. B* **109** 20174
[8] Koe J R, Motonaga M, Fujiki and West R 2001 *Macromolecules* **34** 706.
[9] Kusukawa T, Shike A and Ando W 1996 *Tetrahedron* **52** 4995
[10] Takeda R and Shiraishi K 1993 *Solid State Commun.* **85** 301
[11] Brus L 1994 *J. Phys. Chem.* **98** 3573
[12] Nate K, Ishikawa M, Ni H, Watanabe H and Saheki Y 1987 *Organometallic* **6** 1673
[13] Nespurek S, Zakrevskyy Y, Stumpe J, Saplich B and Kadashchuk A 2006 *Macromolecules* **39** 690

The Effect of Substrate on the Low-Temperature Carbon Nanomaterials Growth by Microwave Excited Surface-wave Plasma Chemical Vapor Deposition

R.V. Bekarevich[1,2], S. Miura[1], A. Ogino[1], A.U. Rahachou[2], M. Nagatsu[1]

[1]Graduate School of Science and Engineering, Shizuoka University, 3-5-1 Johoku, Naka-ku, Hamamatsu 432-8561, Japan
[2]Francisk Skarina Homel State University, Sovetskaya 104, Homel 246019, Belarus

rbekarevich@hotmail.com

Abstract. Multi-walled carbon nanotubes (CNTs) have been successfully grown on Si and polyimide substrates at temperature of 230~260°C by the microwave-excited surface wave plasma technique. Graphene-encapsulated nickel nanoparticles have been used as the catalyst for growing CNTs in NH_3/CH_4 plasma. The optimal gas flow mixture ratio and bias voltage for this type of plasma were investigated. High-resolution transmission electron microscopy and Raman spectroscopy investigations show similar results for CNTs grown on the Si and polyimide substrates.

1. Introduction

Carbon nanotubes (CNTs) have become of keen interest to material scientists since their discovery in 1991 [1]. CNTs have different potential applications such as molecular electronics [2], sensors [3, 4], field emission devices [5] and solar cells [6] owing to their combinations of chemical, physical and mechanical properties [7, 8]. The manufacturing of such devices requires growth of CNTs at low temperature where the structure and chemical composition of the functional materials remain unaltered. It is possible to produce CNTs by several methods [9-11] but one of the most promising ways to realize low temperature growth of carbon nanomaterials (CNMs) is the plasma-enhanced chemical vapor deposition (PECVD), because the presence of highly reactive plasma environment decreases the activation energy of the CNT growth process [12]. One of the most popular sources of high density plasma applied in PECVD to grow CNTs is a microwave radiation.

Generally the PECVD methods require a metal catalyst to produce CNTs. Transition metals or their compounds are generally used as catalyst [12-16]. Yamazaki, et al. [14] have grown CNTs at temperature below 400°C by a remote PECVD with Co catalytic nanoparticles deposited by the impactor method on the $TiN/SiO_2/Si$ substrates. At this experiment has been used the gas mixture of methane and hydrogen. In the reference [15] the Co catalyst prepared by sol-gel method has been used to grow CNMs, where the samples were heated up to 280~330°C by Ar/CH_4 or $Ar/He/CH_4$ microwave plasma. Chang, et al. [16] converted Ni catalyst from the film into particles by microwave H_2/N_2 plasma and after that he used CH_4/H_2 plasma to realize the CNTs growth at the temperature of 250°C.

In this study, multi-walled CNTs (MWCNTs) have successfully grown on Si substrates at temperature of 230~260°C by the microwave-excited surface wave plasma technique. Graphite-encapsulated Ni nanoparticles (GNNs) have been used as the catalyst for growing CNTs in NH_3/CH_4

plasma. We chose ammonia as a carrier gas because of its chemical activity. During the growth process, NH_3 may generate atomic hydrogen species providing etching effects for the growing amorphous carbon [17]. Because of these effects NH_3 helps to maintain surface of metal catalyst active [18]. It has been found that the gas mixture ratio NH_3/CH_4 and bias voltage can greatly effect on the CNTs growth. The threshold values of CH_4 concentration and bias voltage of substrate stage for CNTs growth have been investigated. We have also demonstrated the CNTs growth on the polyimide substrate using the present method.

2. Experimental

In this work, we used a microwave surface wave plasma device as shown in Figure 1. The polished (100) n-type silicon wafers and polyimide sheets covered with GNNs produced by DC arc-discharge method [19] have been placed on the stage located at 110 mm from the quartz window of the microwave plasma device to activate GNNs by a ammonia plasma pretreatment, where a 2.45 GHz microwave was introduced through a quartz window via slot antenna [15] at pressure of 20 Pa with the maximum power of 700 W. During the first stage of CNTs growth, the substrate stage was biased negatively by a pulsed squared wave with amplitude of –100 V at a frequency of 500 Hz. After the pretreatment, CNTs growth has been realized by NH_3/CH_4 plasma post-treatment with a low bias voltage at –30 V. Although no heating stage has been applied, due to the heating effect of plasma itself, the substrates were heated up to 230 ~ 260 °C depending on the plasma conditions. The temperature of the sample holder has been measured by a thermocouple during the plasma process. Different NH_3/CH_4 gas flow mixtures (from 100/1 to 100/50 sccm) have been introduced to the microwave plasma system to study the optimal gas flow ratio for CNTs growth.

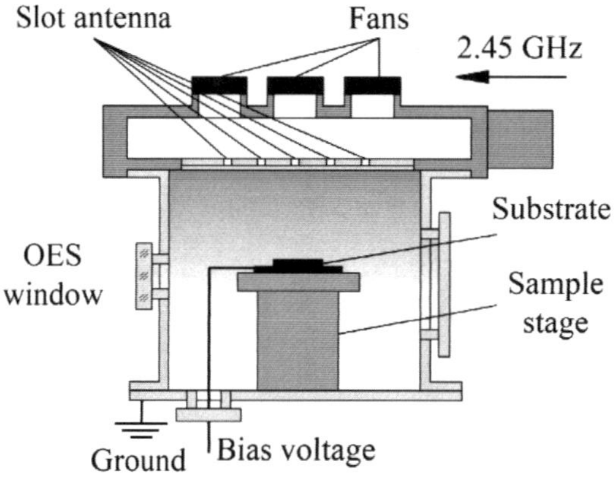

Figure 1. Scheme of the surface-wave plasma device.

Excited atomic and molecular species have been measured during the CNTs growth process by the optical emission spectroscopy (OES) diagnosis, in order to study the correlation between the molecular species and CNTs growth. High-resolution transmission electron microscopy (HRTEM) JEOL JEM-2100F instrument at an operating voltage of 200 kV has been carried out to analyze morphology and detailed structure of the CNTs. The specimens for TEM analysis have been prepared by dissolving the CNTs detached from the substrate in ethanol. Raman spectroscopy has been also used to characterize CNTs grown on silicon and polyimide substrates using JASCO NR-1800 Raman spectrophotometer with Ar^+ ion laser at a wavelength of 514.5 nm.

3. Results and discussion

The typical OES spectrum of NH_3/CH_4 microwave plasma is shown in Figure 2a. It can be seen that line spectra due to methane fragments, such as CH at 430 and 390 nm and the hydrogen Balmer lines H_α at 656 nm, H_β at 486 nm, and H_γ at 434 nm, numerous H_2 molecular bands at 463 and 602 nm [20] are partly obscured by those due to ammonia. Emission band from NH species corresponding to

$A^3\Pi \rightarrow X^3\Sigma^-$transitions at 336 nm, CN species corresponding to $B^2\Sigma^+ \rightarrow X^2\Sigma^+$ transitions have been observed. There is no observation of C_2 radical emission from this spectrum. Hence, it is considered that carbon source for CNTs growth come from the CN violet system which has three strong sequences with heads at 421.6, 388.3, and 359 nm. The strongest CN emission peak at 388.3 nm increases with the CH4 flow rate being increased and reaches a maximum at 30 sccm of methane and after that keep high intensity (Figure 2b). Hydrogen Balmer lines H_α at 656 nm ($3d^2D \rightarrow 2p^2P^0$) and H_β at 486 nm ($4d^2D \rightarrow 2p^2P^0$) are not obscured by ammonia peaks and have been monitored with the variation of plasma conditions. The results of OES studies provided clues to the possible active species present in the plasma during the synthesis of CNTs but there is still a lack of information on the actual surface reaction of the catalysts.

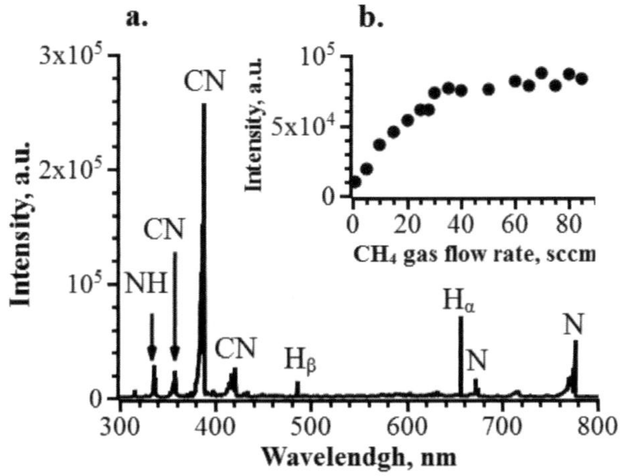

Figure 2. a.– The typical OES spectrum of NH_3/CH_4 microwave plasma (gas flow mixture ratio is 100/30); b. – Effect of CH_4 flow rate on the intensity of CN emission peak at 388.3 nm (NH_3 flow rate fixed at 100 sccm).

It can be seen from HRTEM pictures that MWCNTs with the typical diameters around 5 nm and the typical lengths 50-60 nm start growing when the NH_3/CH_4 gas flow proportion is 100/20 sccm, but optimal gas flow mixtures of NH_3/CH_4 for this type of plasma is 100/30 sccm. In this case straight and curly CNTs are being formed with typical lengths ranged from 200-300 nm to several microns and diameters within 15-25 nm roughly corresponding to the typical sizes of the catalytic Ni grains.

In Figure 3a-b, the product typically exhibits tube-like structures with closed ends for both Si and polyimide substrates. The wall is made of many layers of carbon sheets, between which the lattice fringes of interplanar spacing can be clearly identified in the figure. The average value of spacing is 0.34 nm, which is close to the literature data [21]. In the case of polyimide substrate, MWCNTs with opened ends with diameters within 50 nm were also observed (Figure 3c).

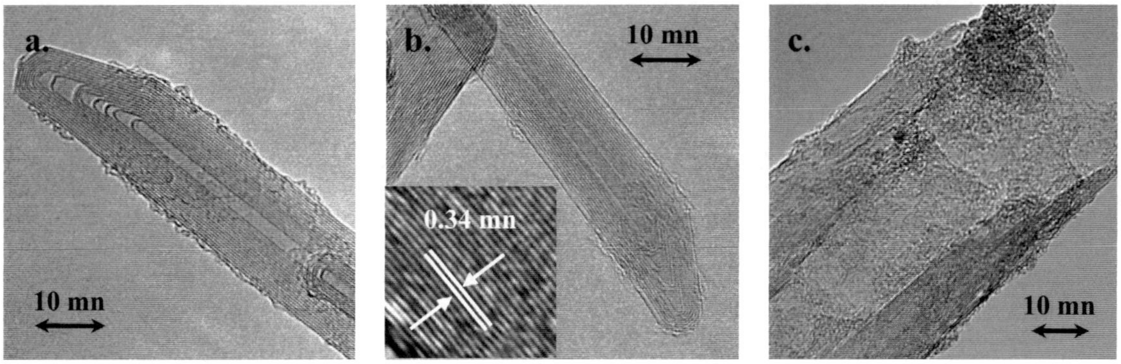

Figure 3. CNTs obtained from NH_3/CH_4 plasma with gas flow ratio 100/30 sccm at -30 V.
a. – CNT with closed ends on the Si substrate; b. – CNT with closed ends on the polyimide substrate; c. – CNT with opened ends on the polyimide substrate.

Raman spectrum of CNTs grown on the Si substrate (Figure 4a) shows strong D and G-bands, which correspond to typical CNTs vibration modes. The bands are ascribed to the disorder-induced (D) mode at 1350 cm^{-1}, the C-C stretching tangential (G) mode at 1582 cm^{-1}, and the disorder-induced (D′) mode at 1620 cm^{-1} [22]. The distinct shoulder of CNTs near 1620cm^{-1} (D′) originates from the A_{1g} tangential Raman mode for MWNTs [23]. The intensity of D-band peak is slightly stronger than that of G-band peak, which might be caused by defects of MWCNTs [24]. The similar results have been observed for CNTs synthesized on the polyimide substrate (Figure 4b).

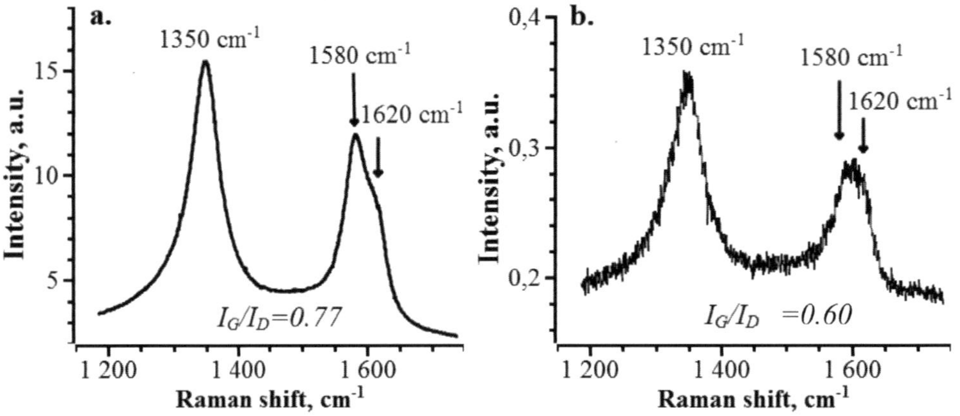

Figure 4. Raman spectrum of the final product:
a. – CNTs grown on the Si substrate; b. – CNTs grown on the polyimide substrate.

4. Conclusions

The low-temperature growth of MWCNTs on the Si and polyimide substrates has been realized by microwave surface wave plasma CVD at temperature of 230 ~ 260 °C, where the graphene-encapsulated nickel nanoparticles were used as the catalyst for growing CNTs in NH_3/CH_4 plasma. The optimal gas flow rate ratio and substrate bias voltage at the stage of post-treatment for CNTs growth have been investigated for ammonia/methane surface wave plasma plasma. HRTEM and Raman spectroscopy investigations show similar results for CNTs grown on the Si and polyimide substrates. Hereby, this work shows the possibility of CNTs growth from NH_3/CH_4 environment at a relatively low temperature directly on polymer substrate.

5. Acknowledgement

This work was supported in part by a Grants-in-Aid for Scientific Research (Grant No. 2110010) from the Japan Society for the Promotion of Science (JSPS).

References

[1] Iijima S 1991 *Nature* **354** 56–8
[2] Feldman A K, Steigerwald M L, Guo X and Nuckolls C 2008 *Acc. Chem. Res.* **41** 1731–41
[3] Arlett J L, Myers E B and Roukes M L 2011 *Nat. Nanotechnol.* **6** 203–15
[4] Choong C-L, Milne W I and Teo K B K 2008 *Int. J. Mater. Form.* **1** 117–25
[5] Jeong H J, Jeong H D, Kim H Y, Kim J S, Jeong S Y, Han J T, Bang D S and Lee G W 2011 *Adv. Funct. Mater.* **21** 1526–32
[6] Umeyama T and Imahori H 2008 *Energy Environ. Sci.* **1** 120
[7] Ayala P, Arenal R, Loiseau A, Rubio A and Pichler T 2010 *Rev. Mod. Phys.* **82** 1843–85
[8] Li T-S, Lee C, Chang S-C and Lin M F 2008 *Solid State Commun.* **145** 451–5
[9] Zhu H W, Xu C L, Wu D H, Wei B Q, Vajtai R and Ajayan P M 2002 *Science* **296** 884–6
[10] Joselevich E, Dai H, Liu J, Hata K and Windle A H 2008 *Top. Appl. Phys.* **111** 101–64

[11] Seah C-M, Chai S-P and Mohamed A R 2011 *Carbon* **49** 4613–35
[12] Hofmann S, Ducati C, Robertson J and Kleinsorge B 2003 *Appl. Phys. Lett.* **83** 135
[13] Assouar M B, Dossot M, Rizk S, Tiusan C and Bougdira J 2010 *Nanotechnology* **21** 065708
[14] Yamazaki Y, Sakuma N, Katagiri M, Suzuki M, Sakai T, Sato S, Nihei M and Awano Y 2008 *Appl. Phys. Exp.* **1** 034004
[15] Lu D, Ogino A, Matsuda T, Ma Q and Nagatsu M 2009 *J. of Plasma and Fusion Res. Series* **8** 544–7
[16] Chang S-C, Lin T-C, Li T-S and Huang S-H 2008 *Microelectron. J.* **39** 1572–5
[17] Wang S, Wang P and Zhou O 2006 *Diam. Relat. Mater.*, **15** 361–4
[18] Juang Z Y, Chien I P, Lai J F, Lai T S and Tsai C H 2004 *Diam. Relat Mater*, **13**, 1203–9
[19] Nagatsu M, Yoshida T, Mesko M, Ogino A, Matsuda T, Tanaka T, Tatsuoka H and Murakami K 2006 *Carbon* **44** 3336–41
[20] Clay K J, Speakman S P, Amaratunga G A J and Silva S R P 1996 *J. Appl. Phys.* **79** 7227
[21] Shen J, Huang Z, Li J, Gao L, Zhou H and Qian Y 2005 *Carbon* **43** 2823–6
[22] Amama P B, Lan C, Cola B a, Xu X, Reifenberger R G and Fisher T S 2008 *J. Phys. Chem. C* **112** 19727–33
[23] Rao A M, Jorio A, Pimenta M A, Dantas M S S, Saito R, Dresselhaus G and Dresselhaus M S 2000 *Phys. Rev. Lett.* **84** 1820–3
[24] Li Q W, Li Y, Zhang X F, Chikkannanavar S B, Zhao Y H, Dangelewicz a M, Zheng L X, Doorn S K, Jia Q X, Peterson D E, Arendt P N and Zhu Y T 2007 *Adv. Mater.* **19** 3358–63

Journal of Physics: Conference Series **417** (2013) 012043 doi:10.1088/1742-6596/417/1/012043

Gas Barrier Properties of Diamond-like Carbon Films Synthesized by Using Remote Type Microwave Plasma CVD under Sub-ambient Atmospheric Pressure

Y Tachimoto, M Noborisaka, A Shirakura, C Kuroyanagi and T Suzuki
Center for Science of Environment, Resources and Energy, Keio University
3-14-1 Hiyoshi, Kohoku-ku, Yokohama-shi, Kanagawa-ken, Japan

E-mail: spand-sofic@hotmail.co.jp

Abstract. Diamond-like carbon (DLC) films were synthesized on polyethylene (PE) substrates with a remote type microwave plasma chemical vapor deposition (MPCVD) apparatus at 10 kPa. We investigated the effect of source gas flow rate and microwave power on oxygen transmission rate of DLC films. Acetylene (C_2H_2) and argon (Ar) gas were used for source gas and carrier gas, respectively. From the result of contact-type profiler and atomic force microscope, oxygen transmission rate decreased as thickness increased and surface roughness decreased with increasing C_2H_2 flow rate. The oxygen transmission rate of DLC film synthesized at the C_2H_2 flow rate of 0.50 l/min and the microwave power of 1.0 kW on 300 μm thick PE substrates was 141 cc/m^2·day, while that of uncoated PE substrates was 247 cc/m^2·day.

1. Introduction

There has been a great attention about the study of diamond-like carbon (DLC). Several studies have reported that DLC films possess many excellent properties: low friction coefficient, low wear rate, high hardness, fine corrosion resistance and low gas permeability [1-3]. Because of these properties, DLC films have yielded practical applications in mechanical, electrical and medical fields [4,5]. This film has been synthesized inside polyethylene terephthalate (PET) bottles to improve gas barrier properties, and plasma enhanced chemical vapor deposition (PECVD) method is applied for commercializing this application [6]. However, since oxygen ingress through plastic caps causes deterioration of the contents as well as through bottles, high gas barrier films should be synthesized on caps to protect the contents.

Conventional PECVD method under low pressure is not suited to synthesis of DLC films on inexpensive products like caps due to unbearable cost resulted from the vacuum process. For this reason, we have studied synthesis methods at sub-ambient atmospheric pressure without high-vacuum technique and recently we have focused on CVD processes using a sub-ambient atmospheric microwave plasma generating device utilizing an annular wave guide.

Microwave plasma chemical vapor deposition (MPCVD) devices use induction coupled plasma and there is no necessity to make electrodes fit along substrates. Therefore MPCVD devices are adapted for application to three dimensional substrates. Over the past few years, many researchers have synthesized DLC films by using MPCVD devices under high vacuums or high temperatures [7,8]. However, there have been no noticeable studies on synthesis of DLC films by MPCVD devices on polymer substrates such as polyethylene (PE), which is common material of plastic caps, carrying out under sub-ambient atmospheric pressures at low temperatures. In recent year, novel MPCVD devices which can generate plasma at sub-ambient atmospheric pressure have been developed by irradiating surrounded microwaves through the annular wave guide to fused quartz cylinders [9,10].

The present study was undertaken in order to synthesize DLC films under sub-ambient atmospheric pressure and low temperature by a remote type MPCVD apparatus and to investigate the effect of source gas flow rate and microwave power on oxygen transmission rate of the DLC films and to clarify the factors influencing the decrease in the oxygen transmission rate.

2. Experiments

We employed a MPCVD device for the synthesis of DLC films under sub-ambient atmospheric pressures. Figure 1 shows a schematic view of the remote type MPCVD equipment with an annular wave guide for generating surface wave plasma (CYRANNUS I-6, iplas, Germany). The equipment excites carrier gases by irradiating the microwave of 2.45 GHz to the fused quartz cylinder whose diameter is 140 mm and height is 143 mm, and generates plasma. The generated plasma was exhausted as a jet stream generated by the pressure difference between an upper chamber with lower pressure and the fused quartz cylinder, and a source gas was introduced into the plasma jet and was excited.

Table 1 shows the experimental conditions. First, nitrogen (N_2) and argon (Ar) gas which have long lifetime of excited states were used as carrier gas and we investigated the effect of the carrier gases on substrate temperatures. The Microwave power was 2.0 kW, and the output wave, which decides the operation modes of the microwave generator, was continuous wave. Second, acetylene gas (C_2H_2) was used as the source gas and its flow rates were 0.10, 0.25, 0.50 l/min, and Microwave power were 1.0, 2.0, 3.0, 4.0 kW to investigate the effect of source gas flow rate and microwave power on oxygen transmission rate of the DLC films. PE sheets with a thickness of 300 μm were used as the substrates for oxygen transmission rate measurement system and single crystal Si (100) wafers were used for all the other analyses.

The substrate temperature was measured by a thermocouple thermometer. The structure of the films was determined by Raman spectroscopy (STR300, Seki Technotron Corp., Japan). The oxygen transmission rate was determined by an equal-pressure method of ISO 15105-2 (OX-TRAN Model 2/21, Mocon, Inc., USA). The film thickness was measured by a contact-type surface profiler (Dektak3030, Veeco, USA). The surface roughness was measured by an atomic force microscope (SPM-9600, SHIMADZU CORPORATION, Japan).

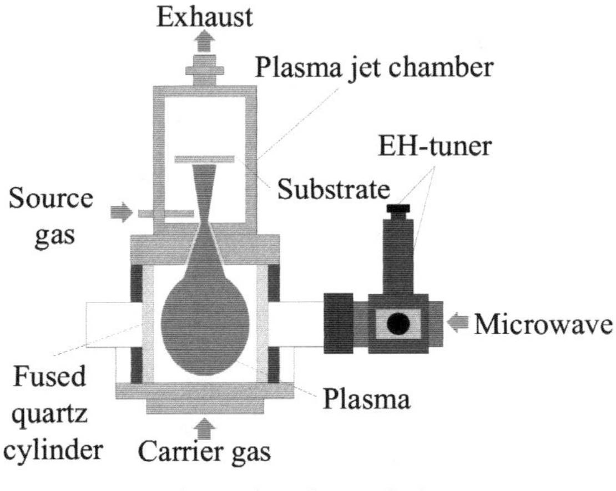

Figure 1. Schematic view of the remote type MPCVD equipment.

Table 1. Experimental conditions.

Microwave power (kW)	1.0, 2.0, 3.0, 4.0
Output wave	Continuous wave
Source gas	C_2H_2
Source gas flow rate (l/min)	0.10, 0.25, 0.50
Carrier gas	N_2, Ar
Carrier gas flow rate (l/min)	10
Substrates	PE, Si
Synthesis time (min)	10
Synthesis pressure (Pa)	10^4

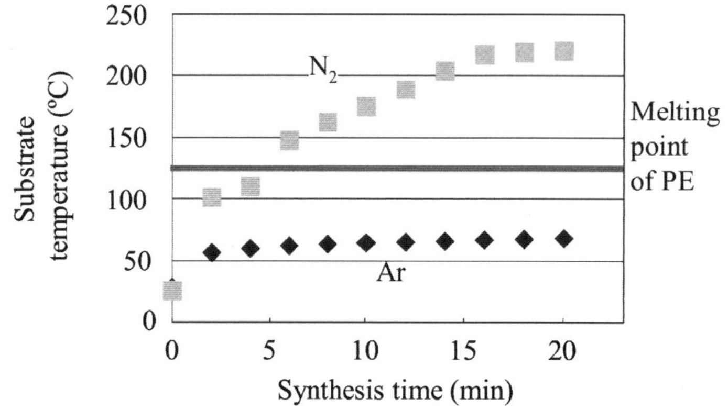

Figure 2. Comparison of N_2 to Ar as carrier gas on substrate temperatures at microwave power of 2.0 kW.

3. Results and Discussions

3.1. Effect of carrier gases on substrate temperatures

Figure 2 shows the substrate temperatures using N_2 and Ar as the carrier gases at the microwave power of 2.0 kW with a function of synthesis time. The substrate temperature was exceeding 200ºC when N_2 was used for a carrier gas though melting point of PE is 110-140ºC. When Ar was used for a carrier gas, the substrate temperature was decreased to about 70ºC. The substrate temperatures at 4.0 kW was kept a little below 110ºC while it was higher than that at 2.0 kW.

3.2. Synthesis of DLC films using Ar carrier gas and measurement of oxygen transmission rate

Figure 3 shows Raman spectra of DLC films using Ar carrier gas when microwave power was 1.0 kW. Raman spectra of DLC-specific which had G band (1540 cm^{-1}) and D band (1380 cm^{-1}) were obtained. From Figure 3, the structure of the films did not depend on C_2H_2 flow rate. The same spectra were observed with different microwave powers, which indicate the structure of the films is not affected by microwave power.

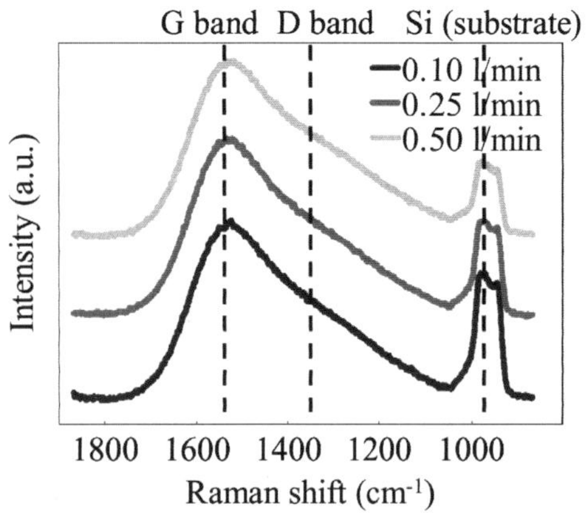

Figure 3. Raman spectra of DLC films on Si substrate (Microwave power: 1.0 kW).

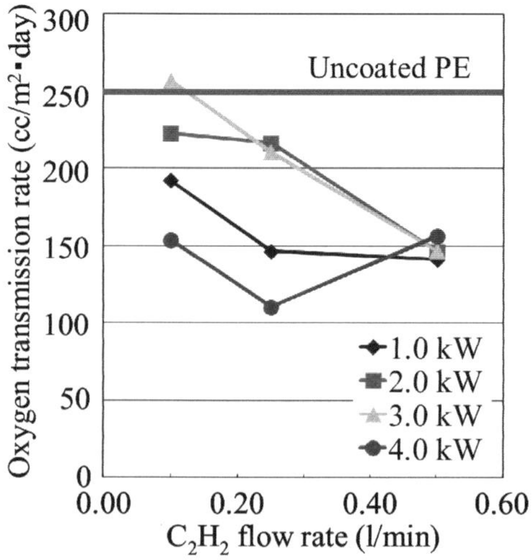

Figure 4. Oxygen transmission rate of DLC films on PE substrates as a function of the C_2H_2 flow rate.

Figure 4 shows Oxygen transmission rate of DLC films on PE substrates as a function of the C_2H_2 flow rate. Oxygen transmission rate decreased with increasing C_2H_2 flow rate except the DLC film obtained at 4.0 kW. When microwave power was 4.0 kW, oxygen transmission rate decreased to about 190 cc/m^2·day in spite of no C_2H_2. This result indicates that higher substrate temperature closed to the melting point of PE accelerated crystallization of PE and decreased oxygen transmission rate.

3.3. Factors influencing the decrease in the oxygen transmission rate

Figure 5 shows thickness of DLC films as a function of the C_2H_2 flow rate at various microwave power. Thickness increased with increasing C_2H_2 flow rate and microwave power in the range of 1.0-3.0 kW. Figure 6 shows Oxygen transmission rate as a function of the thickness. Thickness was correlated with oxygen transmission rate at the microwave of 1.0-3.0 kW. The correlation coefficients (r) were calculated to be -0.90, -0.98 and -0.98 at 1.0, 2.0 and 3.0 kW, respectively. There was no correlation at 4.0 kW (r=-0.23). Increase of thickness led to a decrease of oxygen transmission rate.

Figure 7 shows surface roughness of DLC films as a function of the C_2H_2 flow rate at various microwave power. Surface roughness decreased with increasing C_2H_2 flow rate. Figure 8 shows Oxygen transmission rate as a function of the surface roughness. Surface roughness was correlated with oxygen transmission rate at the microwave of 1.0-3.0 kW (r=0.97, 0.85 and 0.93 at 1.0, 2.0 and 3.0 kW, respectively). There was no correlation at 4.0 kW (r=0.06). Decrease of surface roughness led to decrease of oxygen transmission rate. It was suggested that surface roughness was associated with an internal structure of the DLC films.

The DLC film synthesized at lower microwave power showed nearly the same level of oxygen transmission rate at low thickness. This result suggests low microwave power plasma reduced damage to the PE substrates. The DLC film synthesized at C_2H_2 flow rate of 0.50 l/min and microwave power of 4.0 kW showed abnormality, and requires further investigation.

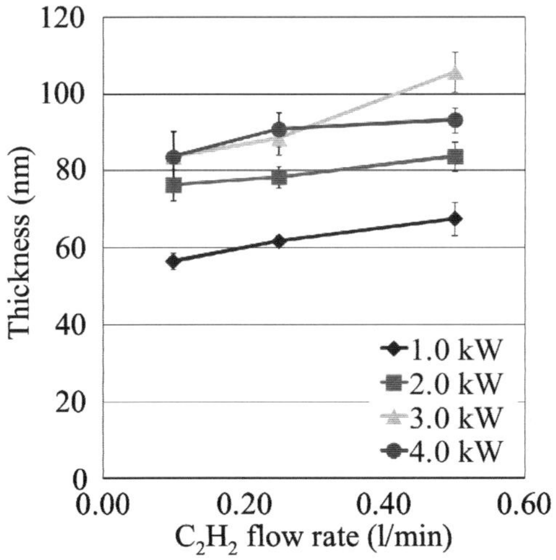

Figure 5. Thickness of DLC films as a function of the C_2H_2 flow rate.

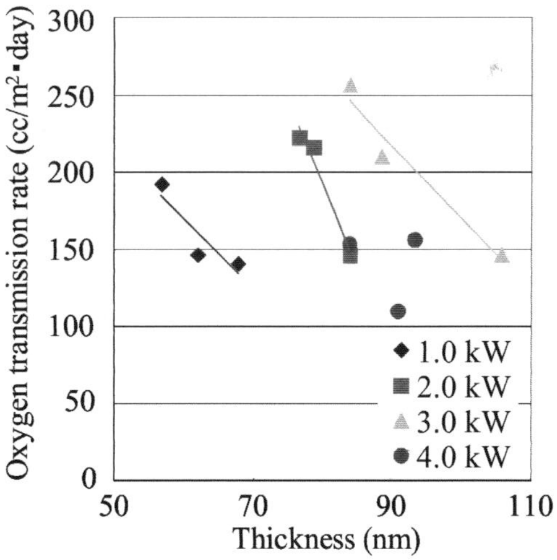

Figure 6. Oxygen transmission rate as a function of the thickness.

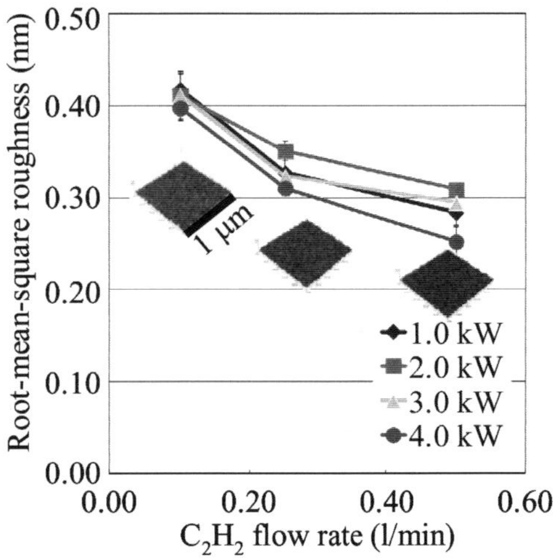

Figure 7. Surface roughness of DLC films as a function of the C_2H_2 flow rate.

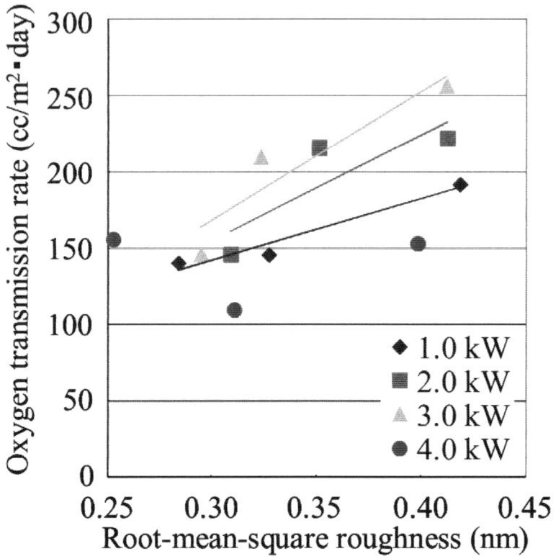

Figure 8. Oxygen transmission rate as a function of the surface roughness.

4. Conclusions

The DLC films were synthesized at the condition of 10 kPa, 70 °C using MPCVD equipment. In terms of carrier gases, Ar was found to be a better candidate for synthesizing DLC films on plastics like PE. By using Ar carrier, lower synthesizing temperatures were obtained. Oxygen transmission rate decreased as thickness increased and surface roughness decreased with increasing C_2H_2 flow rate. The DLC films synthesized at low microwave power have a higher ratio of gas barrier to thickness compared to those at higher microwave power.

References

[1] Grill A 1997 *Surf. Coat. Technol.* **94** 507
[2] Robertson J 2002 *Mater. Sci.Eng.* R **37** 129
[3] Abbas G A, Roy S S, Papakonstantiou P and McLaughlin J A 2005 *Carbon* **43** 303
[4] Piazza F, Grambole D, Schneider D, Casiraghi C, Ferrari A C and Robertson J 2005 *Diamond Relat. Mater.* **14** 994
[5] Hauert R 2004 *Tribol. Int.* **37** 991
[6] Shirakura A, Nakaya M, Koga Y, Kodama H, Hasebe T and Suzuki T 2006 *Thin Solid Films* **494** 84
[7] Umehara Y, Murai S, Koide Y and Murakami M 2002 *Diamond Relat. Mater.* **11** 1429
[8] Aryal H R, Adhikari S, Ghimire D C, Uchida H and Umeno M 2007 *Diamond Relat. Mater.* **16** 1269
[9] Sung-Spitzl H 2001 *Kunstst-Plast. Eur.* **91** 46
[10] Werner F, Korzec D and Engemann J 1994 *Plasma Sources Sci. T.* **3** 473

Effect of dilution gas on hardness of hydrogenated amorphous carbon-based films synthesized by atmospheric pressure plasma enhanced CVD

M Noborisaka, Y Tachimoto, R Horikoshi, A Shirakura and T Suzuki
Center for Science of Environment, Resources and Energy, Keio University
3-14-1 Hiyoshi, Kohoku-ku, Yokohama, Japan

E-mail: okura0306@gmail.com

Abstract. The deposition of hydrogenated amorphous carbon (a-C:H)-based films at atmospheric pressure is a cost-effective process without using a high-vacuum technique. In this study, we synthesized a-C:H-based films by atmospheric pressure-plasma enhanced CVD method from C_2H_2/He and C_2H_2/N_2 mixture gas and investigated the effect of dilution gas on the hardness of the films. The film deposited from C_2H_2/N_2 contained nitrogen elements with forming C–N bonds and the film was softer than a-C:H film deposited from C_2H_2/He. The hardness of the film deposited from C_2H_2/He was 1.1 GPa and twice higher than that from C_2H_2/N_2.

1. Introduction

Hydrogenated amorphous carbon (a-C:H) films have been widely used for many industrial applications because of their excellent characteristics such as high hardness [1], low friction coefficient [2], and gas barrier properties [3]. PECVD is a commonly used technique with simple but powerful control of the deposition and generally operated at low pressure. However, the vacuum technique has proven to be costly because of the expensive pumping equipment and the time-consuming vacuum process. An atmospheric pressure plasma glow (APG) discharge system is cost-effective and employing the system in the PECVD process thus serves as one approach to overcoming the problems. Highly functionalized films are much desired and considerable efforts have been made to synthesize by the atmospheric pressure-plasma enhanced CVD (AP-PECVD) system; however, films synthesized under atmospheric pressure are generally inferior to those synthesized at low pressure in hardness, and shows a similar hardness to polymeric materials around 0.5 GPa [3]. Kondo *et al.* reported on a millimeter-sized deposition of DLC films by atmospheric pressure plasma and obtained films with a high hardness of 20 GPa [4]. However, large-area deposition of carbon-related films under atmospheric pressure has not been achieved. In a previous paper, we reported that hardness of amorphous carbon films synthesized under atmospheric pressure was dependent on the substrate temperature [5]. In this study, we synthesized a-C:H-based films by AP-PECVD method using helium and nitrogen gas as a dilution gas and investigated the effect of dilution gas on the hardness of the deposited films.

2. Experimental Procedures

Figure 1 shows a schematic view of the AP plasma enhanced CVD equipment we built. Acethylene

(C_2H_2) gas diluted with helium (He) or nitrogen (N_2) gas was introduced between the parallel-plate electrodes with a distance of 1 mm. The gas mixture ratio as $[C_2H_2/(C_2H_2 + \text{He or } N_2)] \times 100$ was 1% with a total gas flow rate of 5 l/min. The plasma was generated and sustained between the electrodes. The frequency of the power source and the voltage applied were 5 kHz and 15 kV, respectively. The upper electrode of 50 mm square was covered with dielectric material, alumina, and chilled to 15°C and the lower electrode of 70 mm square was heated to 100°C. Single crystal Si (100) wafers were used as the substrate. The film thickness was measured by a contact-type surface profiler (Dektak3030, Veeco, USA). Hardness of the films was measured using a triboindenter (TI-900, Hysitron, USA) with a Berkovich diamond indenter. The penetration depth was about 50 nm and the load-displacement curve was analyzed using the Oliver and Pharr method [6]. The measurements were performed at 5 different surface locations and the results are expressed as the mean of 5 replicates and the corresponding standard deviation. To investigate the surface chemical bonding structure, X-ray photoelectron spectroscopy (XPS; JPS-9000MX, JEOL, Japan) was used. The RMS (root-mean-square) surface roughness of the films was examined by atomic force microscope (AFM; SPM-9500J3 SPM-9600, Shimadzu, Japan). The cross-sectional surface of the films was observed by scanning electron microscope (SEM; Silion, FEI and S-4700, Hitachi, Japan).

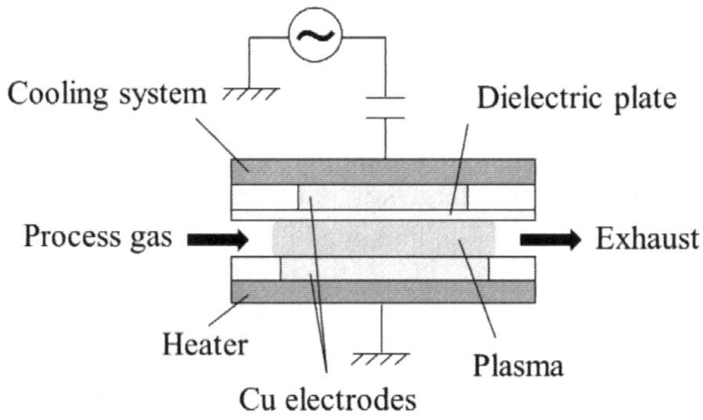

Figure 1. Scheme of AP-PECVD equipment

3. Results and Discussion

3.1. Deposition rate, surface roughness and hardness

Table 1 shows the data of deposition rate, surface roughness and hardness. The deposition rate of the film synthesized from C_2H_2/N_2 was about twice faster than that from C_2H_2/He. The film synthesized from C_2H_2/He showed 1.1 GPa, which was more than about four times harder than polymeric materials and about twice harder than that from C_2H_2/N_2.

Figure 2 and Figure 3 show the AFM surface images and the cross-sectional SEM images of the films deposited from C_2H_2/He and C_2H_2/N_2, respectively. The surface of the film deposited from C_2H_2/N_2 was much rougher than that from C_2H_2/He (Figure 2). The size of the particles that constitute the film deposited from C_2H_2/N_2 was bigger than those from C_2H_2/He (Figure 3). Kodama *et al.* reported that coagulation of particles and falling particles onto substrates lead to a formation of poor density films in AP-PECVD process [3].

Table 1. Deposition rate, RMS roughness and hardness of the films deposited from C_2H_2/He and C_2H_2/N_2.

	Deposition rate (nm/s)	RMS roughness (nm)	Hardness (GPa)
C_2H_2/He	7.6	0.43	1.1
C_2H_2/N_2	13	13	0.56

Figure 2. AFM surface images (scanned area: 1×1 μm^2) of the films deposited from (a) C_2H_2/He and (b) C_2H_2/N_2

3.2. Effect of chemical bonding structure on hardness

Figure 4 shows the XPS spectra in C1s and N1s region of the films deposited from C_2H_2/He and C_2H_2/N_2. Nitrogen incorporation in the films deposited from C_2H_2/N_2 has been confirmed, as shown in Figure 4 (b).

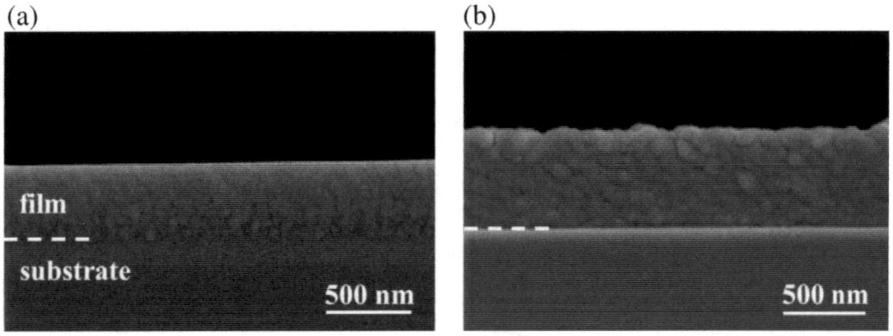

Figure 3. Cross-sectional SEM images of the films deposited from (a) C_2H_2/He (deposition time: 60 sec) and (b) C_2H_2/N_2 (deposition time: 40 sec).

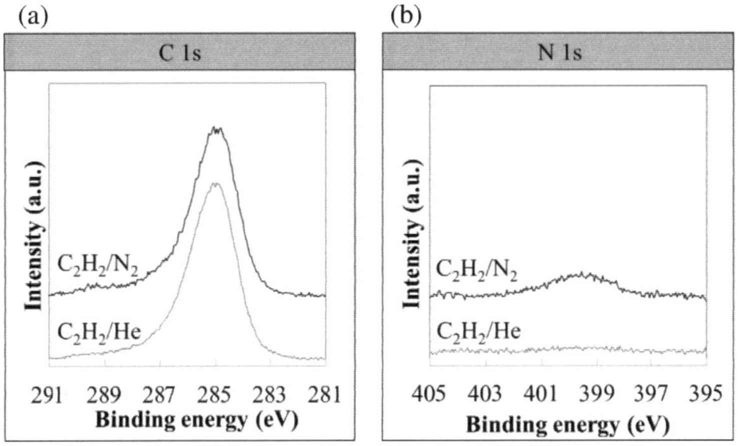

Figure 4. (a) C1s and (b) N1s XPS spectra of the films deposited from C_2H_2/He and C_2H_2/N_2.

Figure 5 shows the curve-fitted XPS spectra in C1s region of the films deposited from C_2H_2/He and C_2H_2/N_2. The data are presented after background subtraction and using Gaussian peak fitting. The C1s peak of the film deposited from C_2H_2/He can be fitted into four components around 284.4 eV,

285.2 eV, 286.5 eV and 289.2 eV, respectively (Figure 5 (a)). The one around 284.4 eV corresponds to sp^2 C–C bonds, that around 285.2 eV corresponds to sp^3 C–C bonds, that around 286.5 eV corresponds to C–O bonds, and that around 289.2 eV corresponds to C=O bonds [7-9]. For the film deposited from C_2H_2/N_2, a new C1s peak near 287.1 eV assigned to C–N bonds was observed (Figure 5 (b)). From these results, we found that the film deposited from C_2H_2/N_2 contained nitrogen elements assigned to dilution gas and that C–N bonds were formed in the films. Many researchers reported that nitrogen incorporated a-C:H (or DLC) films are softer with a rougher surface compared with a-C:H films (or DLC) by AP-PECVD process or electron cyclotron resonance (ESR) plasma CVD process [3,7,10]. Kodama *et al.* reported the C–N structures are likely to prevent carbon atoms from forming three-dimensional carbon structures as for the N_2 dilution gas, resulting in a reduction of the film density. X. Yan *et al.* reported that the N incorporation in carbon increased the fraction of sp^2 carbon bonds, which led to a rough surface. However, the deconvoluted peaks do not show significant changes in the fraction of sp^2 carbon bonds with a change of dilution gas in our study. Increase of surface roughness when using N_2 as a dilution gas is still a question and further investigation is needed.

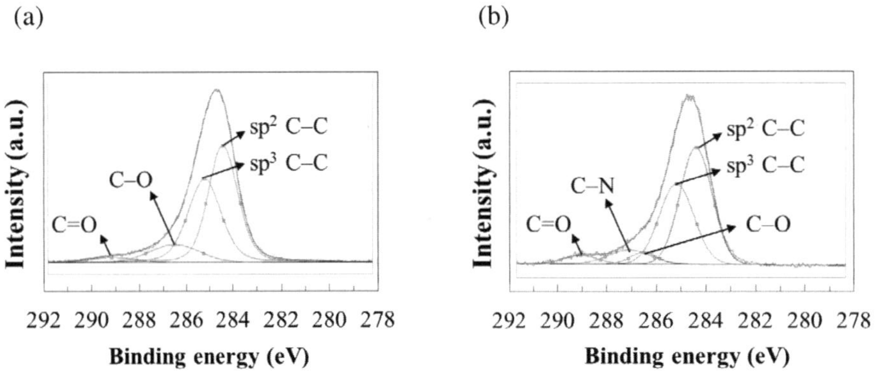

Figure 5. The deconvoluted binding energy spectra of the deposited films: (a) C1s spectrum of the film deposited from C_2H_2/He and (b) C1s spectrum of the films deposited from C_2H_2/N_2.

4. Conclusions

We studied the effect of dilution gas on the hardness of a-C:H-based films deposited by AP-PECVD method from C_2H_2/He and C_2H_2/N_2. The films deposited from C_2H_2/N_2 contained nitrogen elements assigned to dilution gas and C–N bonds were formed in the films. The film deposited from C_2H_2/He was harder than nitrogen incorporated a-C:H film deposited from C_2H_2/N_2. The film synthesized from C_2H_2/He showed 1.1 GPa, which was more than about four times harder than polymeric materials. In this research, it was found that kind of dilution gas influences the hardness of a-C:H-based film and the films deposited with He dilution are harder than those deposited with N_2 dilution. Further detailed studies are ongoing that focus on the effect of the dilution gas ratio on hardness of the films.

References

[1] Grill A 1990 *J. Res. Develop.* **34** 849
[2] Grill A 1997 *Surf. Coat. Technol.* **94** 507
[3] Suzuki T and Kodama H 2009 *Diam. Relat. Mater.* **18** 990
[4] Kondo Y, Saito T, Terazawa T, Saito M and Ohtake N 2005 *Jpn. J. Appl. Phys.* **44** L1573
[5] Sakata T, Kodama H, Hayashi H, Shimo T and Suzuki T 2010 *Surf. Coat. Technol.* **205** S414
[6] Oliver W C and Pharr G M 1992 *J.mater. Res.* **7** 6 1564

[7] Yan X, Xu T, Chen G, Yang S and Liu H 2004 *Appl. Surf. Sci.* **236** 328
[8] Li H, Xu T, Wang C, Chen J, Zhou H and Liu H 2006 *Diam. Relat. Mater.* **15** 1585
[9] Adhikari S, Adhikary S, Omer A M M, Rusop M, Uchida H, Soga T and Umeno M 2005 *Diam. Relat. Mater.* **14** 1824
[10] Saitoh H, Takamatsu H, Tanaka D, Ito N, Shigeo O and Ito H 2000 *Jpn. J. Appl. Phys.* **39** 1258

Formation of C-N-Si Film for Interlayer of Hard Material Coating by Pulsed Discharge Plasma CVD

Mikio Noda
Pulse Plasma Technology GK
Higashisango 222, Owariasahi, Aichi 488-007, Japan

E-mail: m_noda2003@ybb.ne.jp

Abstract. Carbon nitride (C-N) and Si added C-N (C-N-Si) films were deposited on SKD61 steel plate by pulsed discharge (PD) plasma chemical vapour deposition (CVD). When the films were deposited with N_2 and H_2 diluted CH_4 source gases, the deposition rate increased and the hardness decreased with increasing CH_4 concentration. By means of adding mono-methyl-silane (MMS) gas to the source gases, the C-N-Si film having high hardness of 10 GPa and low Yung's modulus of 80 GPa could be deposited. The FT-IR spectrum of the films showed a peak of graphitic ring and peaks of terminating with H and N were observed. Raman spectra showed D and G peaks at around 1350 and 1590 cm^{-1}, respectively. The intensity of D peak compared to G peak decreased when MMS was added. Ball on disk test of the C-N-Si film against SUJ2 ball showed the friction coefficient was about 0.2. These results shows that the C-N-Si film having high hardness and low Young's modulus, which is suitable for interlayer of hard film coating, can deposit by the PD plasma CVD, and suggest that the film is composed of fullerene-like structure which forms resilient and fracture tough materials.

1. Introduction

Formation of high hardness and low Young's modulus film is desired for use as interlayer of hard coatings. Furthermore, it is desirable that the interlayer material is mainly composed of C and N because the typical hard coating materials such as diamond, diamond like carbon (DLC), TiN and c-BN contain C and N. For this point of view, fullerene-like allotrope of carbon nitride (FL-CNx) is suitable for the interlayer material. As reviewed by Broitman et. al. [1], FL-CNx is a predominantly sp^2-hybridized material with nitrogen structurally incorporated either substitution in a graphite sheet or in a pyridine-like manner. The substitution of nitrogen results in bending by formation of pentagon structure and cross-linking. The nitrogen-induced bending and cross-linking between the sheets contributes considerably to the strength of FL-CNx by preventing inter-planar slip, and results in an extremely fracture tough and elastic material, where deforms occur with reversible bond rotation and bond angle deflection.

The N incorporation into the carbon network is a complicated procedure since carbon and nitrogen form multiply bonding configuration [2-4]. Therefore, it is necessary to excite the carbon and nitrogen to form the pentagon structure, where the N atom is three-fold coordinated through three σ bonds and two electrons consist a lone pair orbital. Deposition at low temperature below 550°C is also necessary when the film is deposited on heat treated materials such as SKD 61 and SKD11 steel. In the present experiments, pulsed discharge (PD) plasma chemical vapour deposition (CVD) method was used to resolve these requirements. The film having high hardness and low Young's modulus could be deposited by using high peak power, after glow effect and low mean power of the PD plasma CVD.

2. Experimental procedure

The carbon nitride (C-N) and C-N-Si films were deposited on SKD61 steel by means of the PD plasma CVD method by using pure N_2 gas, hydrogen diluted CH_4 gas (CH_4/H_2) and hydrogen diluted mono-methyl-silane (MMS) gas (MMS/H_2). The concentration of CH_4 of (CH_4/H_2) gas was changed from 10 to 30 %, and that of MMS of (MMS/H_2) was 5 %. These gases were prepared by a company and their purity was assured more than 99.9 %. Figure 1 shows the schematic diagram of the PD plasma CVD system fabricated for the present experiments. The source gases were introduced into the chamber after pre-evacuation by a rotary pump. The flow rate of the N_2, CH_4/H_2 and MMS/H_2 gases were 100, 100 and 10 mL/min, respectively. The total gas pressure during the deposition was kept constant as 1 kPa by adjusting the evacuating speed by the rotary pump. A SKD61 steel plate of about 2 cm in diameter was used as substrate. The distance between the Mo electrode and the substrate used as cathode was about 10 cm. When the discharge was produced between the electrode and substrate by using the pulse power supply, the plasma was produced between the electrode and substrate, and the film was formed on the substrate.

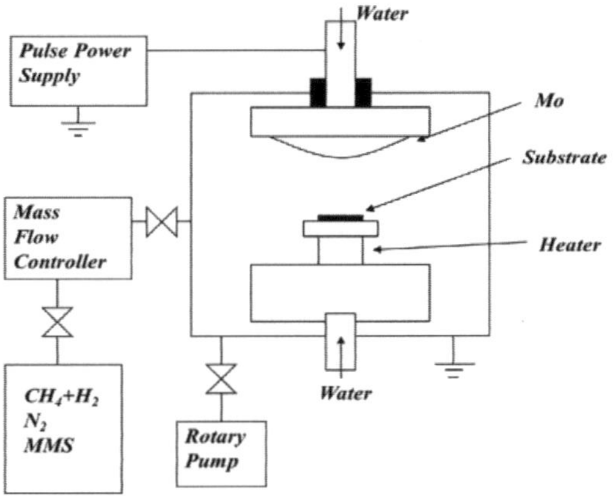

Figure 1. Schematic view of the deposition system.

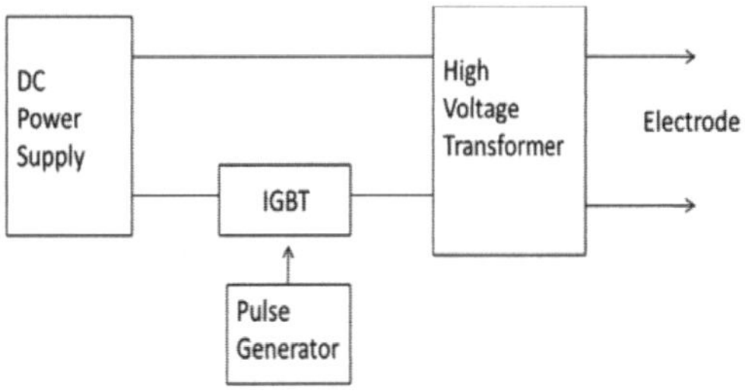

Figure 2. Schematic view of the pulse power supply.

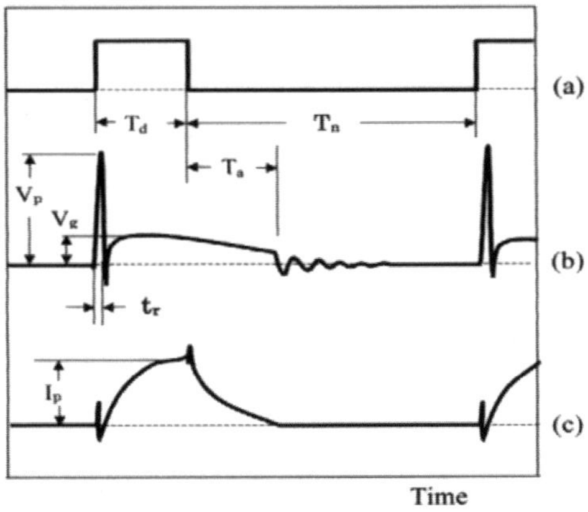

Figure 3. Typical waveforms of the pulse generator (a), discharge voltage (b) and current (c).

The pulse power supply was fabricated by means of chopping the output of the DC power supply with IGBT and elevating the voltage by the transformer as shown in Figure 2. Detailed description of the pulsed discharge was reported before [5-8]. Frequency and duty ratio of the pulsed discharge in the present experiments were 1 kHz and 20%, respectively. Figure 3 shows typical waveforms of the pulse generator (a), discharge voltage (b) and current (c). The peak voltage (V_p) was about 4 kV, and voltage of glow discharge (V_g) was about 700 V. The peak current of the discharge (I_p) was 4 A. The deposition time was changed from 30 to 120 min. The substrate was heated with lamp heater and the substrate temperature was from 400 to 500°C.

The hardness test of the deposited film was performed by using a nano-indenter made in Elionics INC. The hardness was calculated from the road-displacement curves. The load was chosen 1 or 10 mN so that the penetration depth was less than 10 % of the film thickness. Raman analysis of the films was performed using a 532 nm line of a green laser.

3. Results and discussion

Table 1 show changes of deposition rate, hardness and Young's modulus when the source gases are changed. When the CH_4 concentration of (CH_4/H_2) gas is increased from 10 to 30 %, where the flow rate of the (CH_4/H_2) and N_2 gases are 100 mL/ min, the deposition rate increases, and the hardness and Young's modulus decreases as shown for sample 1 to 3 in table 1. These decreases of hardness and Young's modulus are probably due to decrease of elimination of weak bonds by atomic hydrogen as well known for diamond and DLC depositions [9].

Table 1. Changes of deposition rate, hardness and Young's modulus of the C-N and C-N-Si films when CH_4 concentration of the CH_4/H_2 gas is changed and Si is added.

Sample	CH_4 concentration of CH_4/H_2 gas (%)	Deposition time (min)	Thickness (nm)	Deposition rate (nm/min)	Hardness (GPa)	Young's modulus (GPa)
1 (C-N)	10	120	100	0.8	11.1	268
2 (C-N)	15	120	250	2.1	5.3	239
3 (C-N)	30	60	500	8.3	4.1	181
4 (C-N-Si)	30	30	1,140	38.0	10.0	80

When the MMS/H_2 gas was added to the source gases of the sample 3, the hardness of the film is increased from 4.1 to10 GPa and the deposition rate is increased from 8.3 to 38 nm/min as shown for sample 4. Whereas, Young's modulus is decreased from 180 to 80 GPa. These extreme increases of the deposition rate and the hardness, and the extreme decrease of Young's modulus suggest that the structure of the film changes to the FL-CNx by the adding of Si.

In the case of DLC prepared from CH_4, H_2, Ar and $Si(CH_3)_4$ source gases by means of DC plasma CVD, it has been reported that adding of small amount of Si prevent the formation of sp^2 nano-clustering structure and polycarbosilane of Si-CHx-Si bonds are formed [10]. In the present experiments, the flow rate of MMS/H_2 gas was 10 mL/min and considerably smaller than that of N_2 and CH_4/H_2 of 100 mL/min. Therefore, the adding of small amount of Si in the present experiments is probably effective to the formation of basal plane of the FL structure. Furthermore, it can be considered that Si in the basal plane forms cross-linking between the basal planes because Si has sp^3 configuration. Therefore, it can be speculated that the significant increase of hardness and extreme decrease of Young's modulus by adding Si are caused by enhancing the formation of FL structur and cross-linking between the basal plane.

Figure 4 shows optical photograph of the surface of the sample 3 and 4 denoted in table 1. The structure of the sample 3 is grain-like and comes to smooth with adding Si, suggesting that the structure changes to FL by adding Si.

Figure 5 shows FT-IR spectra of the sample 3 and 4. A peak of 1530 cm^{-1} due to C=N graphitic ring denoted by Laskarakis et al [3] is observed. A broad peak at around 2100 cm^{-1} due to terminating with

N appears for the sample 4, reflecting that the size of the aromatic clusters become smaller and more disordered.

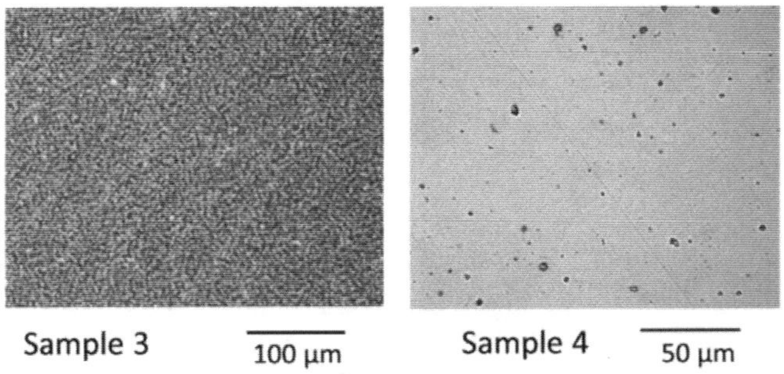

Figure 4. Optical photograph of the surface of the sample 3 denoted in table1 (C-N) and sample 4 (C-N-Si) which is deposited with adding Si to the sample 3.

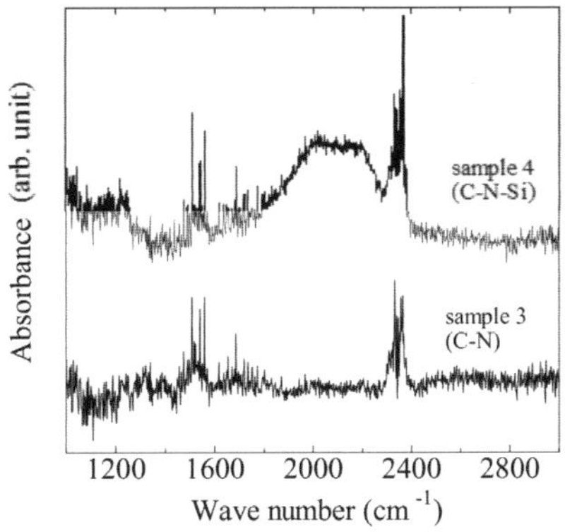

Figure 5. FT-IR spectra of the sample 3 (C-N) denoted in table 1 and sample 4 (C-N-Si) which is deposited with adding Si to the sample 3.

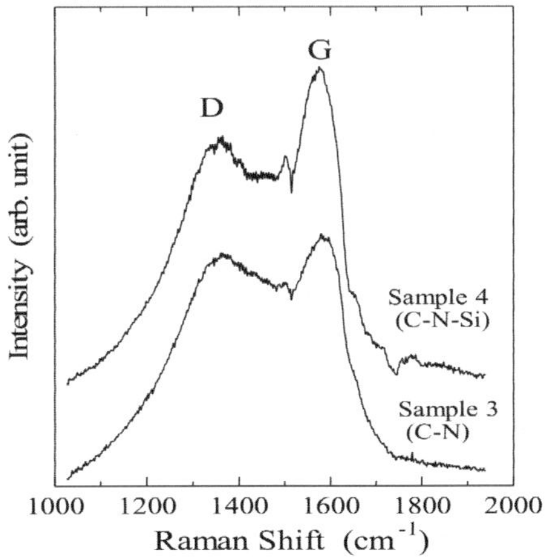

Figure 6. FT-IR spectra of the sample 3 (C-N) denoted in table 1 and sample 4 (C-N-Si) which is deposited with adding Si to the sample 3.

Figure 6 shows Raman spectra of the sample 3 and 4 shown in table 1. Two main peaks positioned at around 1350 and 1590 cm^{-1}, which are denoted by D and G respectively, are observed. The intensity of D peak (I_D) compared to that of G peak (I_G) decreases when Si is added to the sample 3. The decrease of (I_D/ I_G) ratio means the size of the aromatic clusters become smaller and more disordered as reported by Ferrari et al [4]. This result coincides with the results shown in Figure 4, which shows the film surface becomes smooth due to the decrease of the cluster size by adding Si. Furthermore, the G peak is caused by in-plane bond stretching vibration and can be present in aromatic clusters [11]. Therefore, the increase of I_G suggests that FL structure of graphite is enhanced by the adding of Si.

Figure 7 shows ball on disk test of the sample 4 shown in table 1 against SUJ2 ball, where the load was 500 g and the rotational speed was 120 rpm. The friction coefficient is about 0.25 until about 55 m sliding distance. This value of the friction coefficient is almost same as that of FL-CNx films [1].

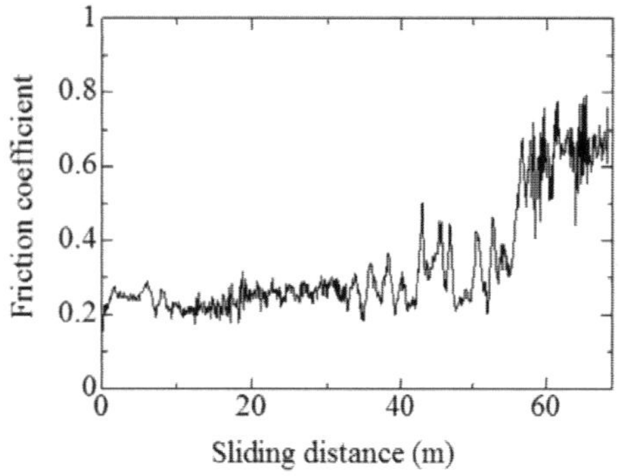

Figure 7. Ball on disk test of the Si added C-N-Si film (sample 4) against SUJ2 ball, where the load is 500 g and the rotational speed is 120 rpm.

4. Conclusion

The C-N and C-N-Si films were deposited by PD plasma CVD by using N_2, CH_4/H_2 and MMS/H_2 source gases. In the case of C-N films deposited by using N_2 and CH_4/H_2 gases, the deposition rate increases with increasing the CH_4 concentration of CH_4/H_2 gas. Whereas, the hardness decreases, and Yaung's modulus also decreases. In the case of C-N-Si films, where MMS/H_2 gas is added to the N_2 and CH_4/H_2 gases, both the deposition rate and the hardness are extremely increased, and significant decrease of Young's modulus is also observed in comparison with those of the C-N films. As a result, the C-N-Si film of high deposition rate of 38 nm/min, high hardness of 10 GPa and low Yaung's modulus of 80 GPa can be deposited. These extreme changes due to adding of Si to the C-N films suggest that the film comes to contain fullerene-like structure composed of pentagon configuration which forms resilient and fracture tough materials by inducing the curvature in the basal plane and inter-planar cross-linking. The results of FT-IR, Raman spectra and friction coefficient measurements support this consideration. The C-N-Si film having high hardness and low Yaung's modulus is suitable for interlayer of hard material coatings.

5. References

[1] Broitman E, Neidhardt J and Hultman L, 2008, *Tribology of Diamond-Like Carbon Films:Fundamentals and Applications,* ed C. Donnet and A. Erdemit (Springer, 2008) P. 620-653.
[2] dos Santos M. C. and Alvarez F, 1998, Phys. Rev. **B 58** 13918.
[3] Laskarakis S, Logothetidis S and Gioti M, 2001, Phys. Rev. **B 64** 125419.
[4] Ferrari A. C., Rodil S. E. and Robertson, 2003, Phys. Rev. **B 67** 155306.
[5] Noda M, 2001, J. Indian Inst. Sci. **81** 627.
[6] Noda M and Umeno M, 2005, Diamond Relat. Mater. **14** 1791.
[7] Noda M, Shinagawa T, Kawai S and Umeno M, 2008, Diamond Relat. Mater. **17** 646.
[8] Kawai S, Shinagawa T, Noda M and Umeno M, 2008, Diamond Relat. Mater. **17** 676.
[9] Spear K. E and Frenklach M, *Synthetic Diamond: Emerging CVD Science and Technology*, ed H. E. Spear and J. P. Dismuhes (John Wiley & Sons, 1994) P. 243-304.
[10] Iseki T, Mori H, Hasegawa H, Tachikawa K and Nakanishi K, 2006, Diamond Relat. Mater. **15** 1004.
[11] Abrasonis G, Gago R, Vinnichenko M, Kreissig A, Kolitsch A and Moller W, 2006, Phys. Rev. **B 73** 125427.

Formation of SiC in DLC/a-Si films as characterized by Raman spectroscopy and XPS

C Srisang[1,2,3,a], P Asanithi[1], K Siangchaew[2], A Pokaipisit[1,3,b] and P Limsuwan[1,3]
[1] Department of Physics, Faculty of Science, King Mongkut's University of Technology Thonburi, Bangkok, 10140, Thailand
[2] Western Digital (Thailand) Company Limited, Ayuthaya, 13160, Thailand
[3] Thailand Center of Excellence in Physics, CHE, Ministry of Education, Bangkok 10400, Thailand

E-mail: [a]Jirawat.Srisung@wdc.com, [b]pokaipisit@gmail.com

Abstract. Bilayer DLC/a-Si films was deposited on germanium substrate. a-Si films was initially deposited as a seed layer on the substrate using DC magnetron sputtering. DLC films was then deposited on the a-Si layer via a pulsed filtered cathodic arc (PFCA) system. *In situ* ellipsometry was used to monitor growth in thickness of a-Si and DLC films, allowing a good control over growth conditions. The thickness of DLC and a-Si layers were 9 nm and 6 nm, respectively. In this work, the interface of a-Si and DLC layers was investigated by Raman spectroscopy and X-ray photoelectron spectroscopy (XPS). It was found that the SiC layer was formed at the interface of a-Si and DLC layers.

1. Introduction

Diamond like carbon (DLC) consists of an amorphous form of the carbon containing both graphitic type bonding (sp^2) and diamond type tetrahedral bonding (sp^3) [1]. The properties of this non-crystalline material cover a wide range and are intermediate between the properties of diamond, graphite, and hydrocarbon polymers [2]. The properties depend on the experimental parameters. Under appropriate deposition condition, chemically inert and optically transparent films can be prepared [3]. As the hardness of DLC films approaches that of diamond, at present, it has been widely used as a protective layer on magnetic recording head and disk media [3, 4]. DLC films are also finding applications as corrosion resistant coating, anti reflection and scratch proof coating on germanium and silicon optics [5]. Another important application of these coatings is for bearing because of their low sliding friction on the metals.

DLC films can be prepared by various techniques, i.e. ion beam assisted deposition [6], sputtering [7], filtered cathodic vacuum arc [8], plasma-assisted chemical vapor deposition [9], and pulsed laser deposition [10]. It was reported that structure and properties of DLC films can be controlled by substrate bias [11, 12]. An important problem appears in DLC film deposition is the ability to control physical properties of the film over a substrate. Different types of substrate or substrates of different roughness may provide different physical properties of the film, e.g. surface roughness [13]. Moreover, the problem of low adherence on the substrate is also significant. The interface layer between the substrate and DLC films may play an important role. Castillo *et al.* improved the adherence of DLC films using amorphous hydrogenated carbon (a-C:H) and titanium nitride (TiN) as interface layer between DLC films and silicon substrate [14].

In current hard disk drive technology, ultra-thin DLC films have been used for coating on both disks and recording heads. However, prior to DLC coating a thin amorphous silicon (a-Si) layer is pre-coated on the substrate to increase the adhesion between the substrate and DLC films.

In this study, a-Si with a thickness of 6 nm was coated on germanium (Ge) substrate as a seed layer. Then, DLC films with a thickness of 9 nm was coated on a-Si film. The formation of SiC at the DLC/a-Si interface was investigated using Raman spectroscopy and X-ray photoelectron spectroscopy.

2. Experimental details

2.1. Pulsed filtered cathodic arc system

The pulsed filtered cathodic arc (PFCA) system used in this work (Veeco, model Nexus DLC-X) is shown by a schematic diagram in Figure 1. The system consists of four vacuum chambers, i.e. load lock chamber, transfer chamber, pre-clean chamber, and PFCA chamber. All chambers were attached together by isolation gate valve. Each has its own turbo molecular pump to evacuate the chamber individually. Ionization gauges were used to measure the pressure in all four chambers. Prior to the deposition processing of the DLC films, all chambers were evacuated to a base pressure of about 5.0×10^{-7} torr. The transfer chamber was equipped with an automated robot arm system for transferring the substrate from load lock chamber to pre-clean and PFCA chambers.

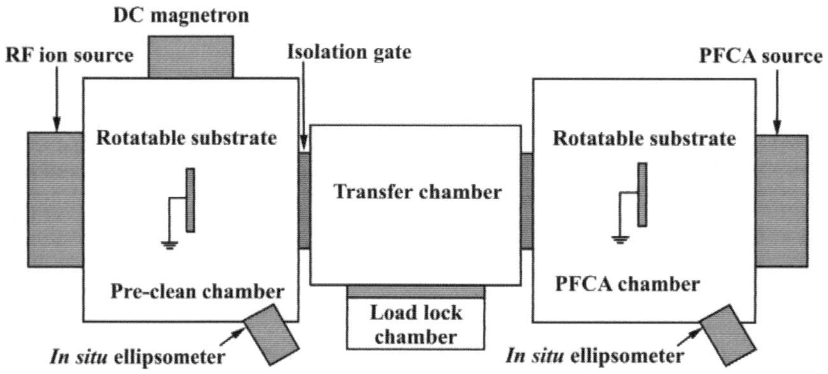

Figure 1. Schematic diagram of PFCA system.

The pre-clean chamber was equipped with radio frequency (RF) ion source, 16 inch in diameter, and DC magnetron sputtering source with a target of 4 inch in diameter. The ion source was used for pre-cleaning of the substrate whereas the DC magnetron source was used for the deposition of a-Si as a seed layer. The substrate with deposited silicon layer was then transferred to substrate fixture in the PFCA chamber by robot arm. The PFCA chamber was equipped with a cathodic arc for the deposition of DLC films.

The substrate fixtures in pre-clean and PFCA chambers are capable of tilt and rotation. Therefore, the a-Si seed layer and DLC films can be deposited onto the substrate at any required incident angle. Pre-clean and PFCA chambers were installed with a multi-wavelength ellipsometry (J.A. Woollam, M-2000) for *in situ* monitoring of the film thickness of a-Si layer and DLC films, respectively. Figure 2 shows a schematic diagram of PFCA chamber and cathodic arc source.

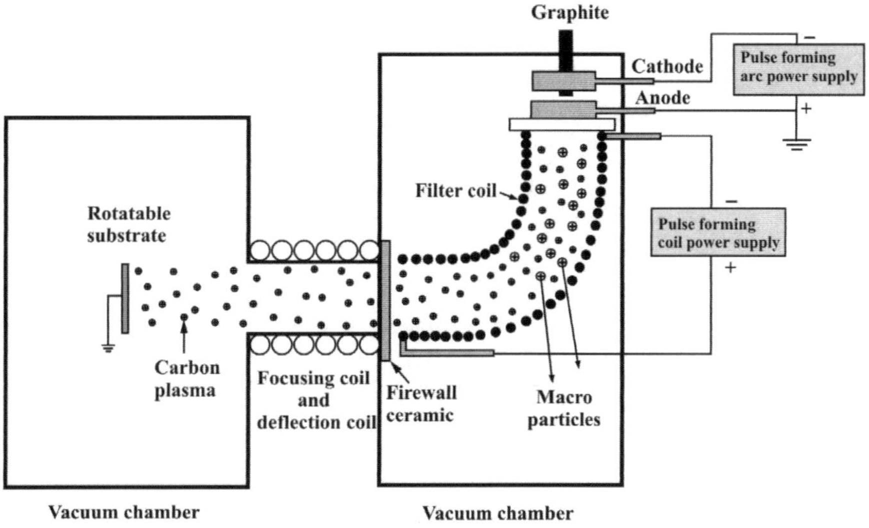

Figure 2. Schematic diagram of PFCA chamber and cathodic arc source.

The cathodic arc source includes a high purity graphite cathode of 0.25 inch in diameter and 8 inch in length (99.999% purity), obtained from Poco Graphite. The arc power supply is operated in pulsed mode. The positive potential of power supply is connected with anode and ground. The voltage between anode and cathode, called arc voltage, can be varied from 0-1000 V with pulse frequency of 1-5 Hz. Since the distance between anode and graphite cathode is about 2 mm, it is high enough to create current discharge between anode and cathode. Then, graphite is vaporized at the cathode. All charge particles are fed into a curved 90 degree filter coil. Electrons are moved in spiral along magnetic field generated by filter coil, and ions will be guided due to they are charge particles. In order to improve the film thickness uniformity, a coil operates as a magnetic lens to focus or defocus the plasma beam. A deflection coil is used to raster the beam before striking the substrate. The macro particles move in strange trajectory so they may either leave the filter through openings gap between the turns of the coil or stick to the turns. Thus, unwanted macro particles and neutral atoms are filtered out and coating species reaching the substrate are pure carbon plasma.

2.2. Films preparation
DLC films were deposited using the PFCA deposition method. Crystalline n-type Ge wafers with (100) orientation, and a dimension of 1 cm × 1 cm with a thickness of 0.5 mm were used as substrates.

The DLC deposition process consists of three steps: (1) the substrate was cleaned with low energy Ar^+ ion plasma etching at an incident angle of 60° with respect to substrate, for 60 seconds, (2) a-Si seed layer was deposited on the cleaned Ge substrate at an incident angle of 44° with respect to substrate, using DC magnetron with a power of 150 W and argon gas flow rate of 40 sccm, and (3) DLC layer was deposited at the normal incident angle with respect to substrate using PFCA with a pulse frequency of 1 Hz, arc voltage of 950 V and coil voltage of 900 V. The first two steps were carried out in pre-clean chamber while the third step was done in PFCA chamber. The film thickness was monitored using *in situ* ellipsometry. Figure 3 shows illustration of deposited films on Ge substrate.

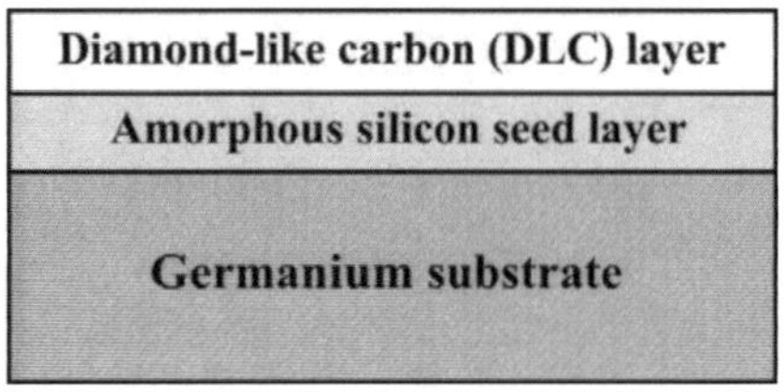

Figure 3. The layers of DLC films and a-Si layer on Ge substrate.

2.3. Characterization
Raman measurements were performed using Renishaw inViaReflex Raman Spectrometer with an excitation wavelength of 514 nm from Ar^+ ion gas laser. The laser output is 20 mW and the objective lens is 50x. The incident power on the sample is approximately 4 mW. The scan range was from 400 to 1900 cm^{-1}.

The XPS depth profiles were recorded by PHI Quantera SXM scanning X-ray using AlK_{α} X-ray source with a spot size of 200 µm. To obtain the depth profile, the DLC/a-Si (9/6 nm) film was etched using ion bombardment from an ion gun of 1 kV for acquiring the XPS spectrum information. Furthermore, the Shirly background type was used for XPS measurements.

3. Results and Discussion
Raman spectroscopy was used to investigate the vibrational modes of carbon and silicon atoms due to chemical bonding, especially in SiC, in the DLC/a-Si film. The longitudinal optical (LO) and the transversal optical (TO) phonon modes were observed at 796 cm^{-1} and 972 cm^{-1}, respectively, as shown in Figure 4.

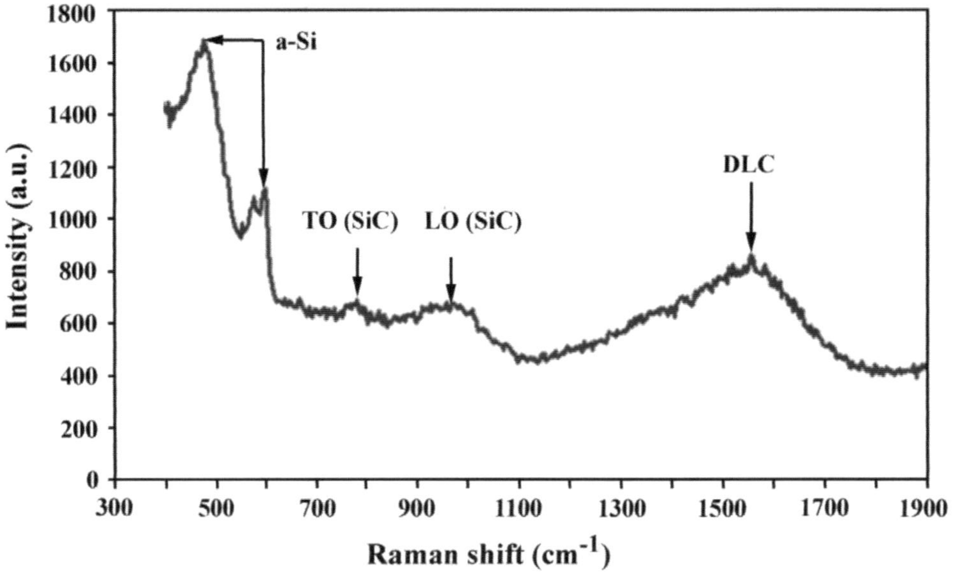

Figure 4. Raman spectra of coated DLC /a-Si films with a thickness of 9/6 nm.

The Raman shifts located at these positions are in agreement with those reported for 3C-SiC (β-SiC) [15]. The prominent peak at 480 cm^{-1} is related to the vibrational mode of a-Si since it was used as a seed layer for depositing the DLC films [16]. However, no peak from crystalline silicon at 520 cm^{-1} is observed. This suggests that during the deposition of a-Si layer on the Ge substrate there is probably no crystalline silicon formation. In the higher wavenumber, the wide band in the 1100-1800

cm^{-1} region is related to the vibrational modes of DLC graphitic characteristics, generally consisting of G and D bands.

Table 1. XPS binding energy of different bonds.

Bond	**Binging energy (eV)**
C-C	284 - 284.5
C-Si (C 1s)	282.3 - 283.7
Si-C (Si 2p)	99.8 - 101.3
Si-Si	99.2 - 99.6

Table 1 shows XPS binding energy after the resolved components in the DLC/a-Si film. The higher-energy range of 284-284.5 eV was assigned to the bonding of carbon atoms in the DLC films. The C 1s and Si 2p configurations, corresponding to the bonding between carbon and silicon atoms in SiC at the DLC and a-Si interface, are in the ranges of 282.3-283.7 eV and 99.8-101.3 eV, respectively [17]. The lowest-energy one in the range of 99.2-99.6 eV corresponds to the bonding of silicon atoms in a-Si layer.

In addition, the XPS sputter depth profile was also used to have further quantitative information on the SiC formation at the DLC/a-Si interface. The DLC/a-Si (9/6 nm) film was etched using ion bombardment for acquiring the XPS spectrum information. The actual depth for each XPS analysis depends upon the etching rate (sputter time) and the material being etched at any given depth.

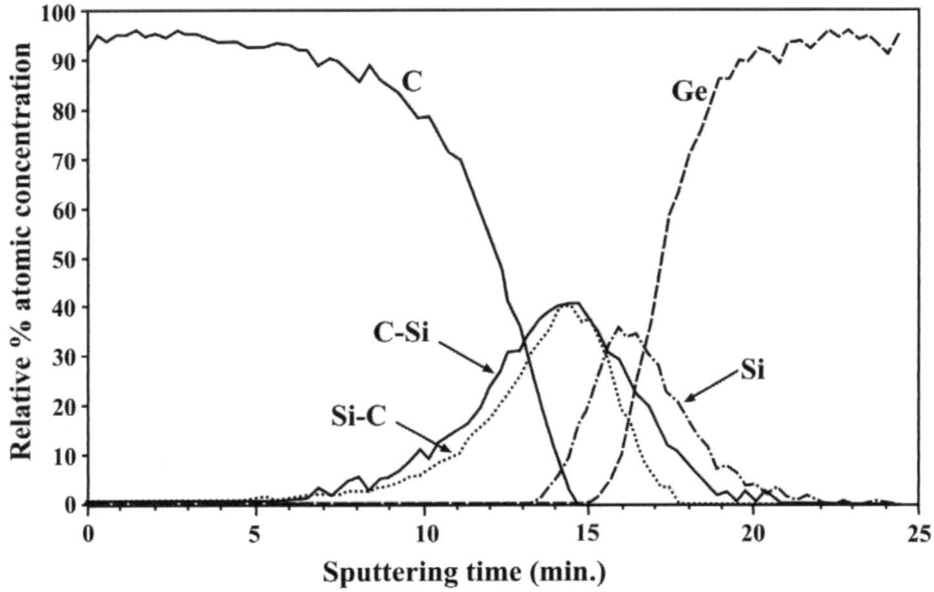

Figure 5. The XPS depth profile of DLC/a-Si films with a thickness of 9/6 nm on Ge substrate.

Figure 5 shows five major spectra of depth profile ranging from DLC films (top layer), SiC (C 1s and Si 2p) at the DLC and a-Si interface, a-Si layer, and Ge substrate. At the sputter time of less than 5 min, the depth profile represents the high percentage of DLC films of more than 90%. The DLC percentage gradually decreases for longer sputter time and becomes 0% at 15 min. At this time, the spectrum corresponding to a-Si layer is initially appeared and reaches the maximum content of 38% at 16 min of sputter time. The peaks obtained from the bonding between carbon and silicon atoms in the C 1s (C-Si) and the Si 2p (Si-C) locating in between those of DLC and a-Si films, are clearly observed as a result of the SiC formation. They reach the maximum contents of 41%. The last spectrum profile indicates the Ge substrate. From Figure 5, the thickness of SiC layer can be estimated and it was found to be approximately 2.8 nm.

4. Conclusions

We presented that there is the formation of SiC layer during the deposition of DLC/a-Si thin film. The LO and the TO phonon modes of SiC were observed at the Raman peaks of 796 cm^{-1} and 972 cm^{-1}, respectively. The result indicates that the SiC occurred between the DLC and a-Si interface is the 3C-SiC. The results were also confirmed using XPS binding energy and XPS depth profile analysis. There are two possible bonding configurations of carbon and silicon atoms at the interface, including the C 1s and the Si 2p which are in the ranges of 282.3-283.7 eV and 99.8-101.3 eV, respectively.

Acknowledgment

This work was supported by Industry/University Cooperative Research Center (I/UCRC) in HDD Advanced Manufacturing, Institute of Field roBOtics, King Mongkut's University of Technology Thonburi and National Electronics and Computer Technology Center, National Science and Technology Development Agency. This work was partially supported by Thailand Center of Excellence in Physics (ThEP) and King Mongkut's University of Technology Thonburi under The National Research University Project.

Reference

[1] Gradowski M V, Ferrari A C, Ohr R, Jacoby B, Hilgers H, Schneider H H and Adrian H 2003 *Surf. Coat. Technol.* **174-175** 246
[2] Robertson J 2002 *Mater. Sci. Eng. R* **37** 129
[3] Kim Y T, Cho S M, Choi W S, Hong B and Yoon D H 2003 *Surf. Coat. Technol.* **169-170** 291
[4] Liu F X and Wang Z L 2009 *Surf. Coat. Technol.* **203** 1829
[5] Patil D S, Ramachandran K, Venkatramani N, Pandey M and Cunha R D 2000 *Pramana J. Phys.* **55** 933
[6] Weissmantel C, Bewilogua K, Dietrich D, Erler H J, Hinneberg H J, Klose S, Nowick W and Reisse G 1980 *Thin Solid Films* **72** 19
[7] Savvides N 1986 *J. Appl. Phys.* **59** 4133
[8] Xu S, Tay B K, Tan H S, Zhong Li, Tu Y Q, Silva S R P and Milne W I 1996 *J. Appl. Phys.* **79** 7234
[9] Shim J Y, Chi E J, Baik H K and Lee S M 1998 *Jpn. J. Appl. Phys.* **37** 440
[10] Scheibe H S, Drescher D, Schultrich B, Falz M, Leonhardt G and Wilberg R 1996 *Surf. Coat. Technol.* **85** 209
[11] Zhang W, Tanaka A, Wazumi K and Koga Y 2002 *Diam. Rel. Mater.* **11** 1837
[12] Sheeja D, Tay B K, Lau S P and Shi Xu 2001 *Wear* **249** 433
[13] Zhong M, Zhang C and Luo J 2008 *Appl. Surf. Sci.* **254** 6742
[14] Castillo H A, Parra E R and Arango P J 2011 *Appl. Surf. Sci.* **257** 2665
[15] Lee C J, Pezzotti G, Okui Y and Nishino S 2004 *Appl Surf Sci* **228** 10
[16] Veres M, Koós M, Tóth S, Füle M, Pócsik I, Tóth A, Mohai M and Bertóti I 2005 *Diam. Rel. Mater.* **14** 1051
[17] Cho W, Oh Y, Kim C, Osada M, Kakihana M, Lim D and Cheong D 1999 *J. Alloys Comp.* **285** 255

Functional Nanowires Array Electrodeposited into Nano-porous Membrane Thin Films

T Ohgai[1] and K Hashiguchi[2]

[1]Division of Chemistry and Materials Science, Nagasaki University, 1-14 Bunkyo-machi, Nagasaki 852-8521, JAPAN
[2]Graduate School of Science & Technology, Nagasaki University, 1-14 Bunkyo-machi, Nagasaki 852-8521, JAPAN

E-mail: ohgai@nagasaki-u.ac.jp

Abstract. Ni, Co, Fe and Co/Cu multilayered nanowires were electrodeposited into a nano-well template to synthesize novel functional ferromagnetic devices. Growth rate of Co/Co multilayered nanowires was around 40 nm•sec^{-1} and the cylindrical shape was precisely transferred from the nanochannels to the nanowires and the aspect ratio reached up to ca. 150. Magnetic hysteresis loops of Ni, Co and Fe nanowires with diameter 40 nm showed typical perpendicular magnetization behavior due to the uni-axial shape anisotropy and the coercive force reached up to around 1 kOe. 2% of anisotropic magnetoresistance effect was observed in Ni nanowires electrodeposited into anodized aluminum oxide template, while 20% of giant magnetoresistance effect was observed in Co/Cu multilayered nanowires.

1. Introduction

One-dimensional nanowires with a large aspect ratio have received much attention due to their shape anisotropy and extremely large surface area [1-3]. Nanowires can be fabricated by manipulating metallic atoms one by one using scanning tunneling microscopy (STM) or stretching a micro-size wire down to nano-size, while they can be also prepared by electrochemically depositing metallic atoms into a nano-well template with numerous nanochannels [4, 5]. Nano-well templates such as anodized aluminum oxide films or polycarbonate membrane films with high density of nanochannels (about 10^8~10^{10} cm^{-2}) can be used in the template synthesis technique [6, 7]. Especially, Ni, Co and Fe nanowires array is the best candidate to be applied for magnetic storage media with high density domain, while Co/Cu multilayered nanowires array can be applied to magnetic field sensor with giant magnetoresistance property [8-10].

In this study, electrodeposition process of Ni, Co, Fe and Co/Cu multilayer into a nano-well template was investigated to synthesize novel functional ferromagnetic devices with anisotropic magnetoresistance and giant magnetoresistance property.

2. Experimental Procedures

Ion track-etched polycarbonate membrane films and anodized aluminum oxide films were used as a template for growing nanowires. On a surface of the membrane filter, a gold layer was sputter-deposited to cover the pores and make a cathode. Aqueous solution containing metal sulfate ($NiSO_4$, $CoSO_4$, $FeSO_4$, $CuSO_4$, H_3BO_3, etc.) was used as electrolyte. To determine the optimum deposition potential for growing nanowires, cathodic polarizarion behavior was investigated in a wide range of cathode potential. Co/Cu multilayered nanowires were electrodeposited by alternatingly changing the cathode potential from -0.6 V to -1.0 V. After growing the nanowires, the polycarbonate membrane films or anodized aluminum oxide films were dissolved in solvent and the remains consisted of nanowires and a gold layer which served as a sample for SEM observation. Magnetic hysteresis loops and magnetoresistive curves of electrodeposited nanowires were obtained using VSM with increasing the magnetic field up to 10 kOe.

3. Results and Discussion

Cathodic polarization curves for electrodeposition of Cu, Co and Ni are shown in **Fig. 1**. The equilibrium potentials of Cu and Co are estimated to be around +0.05 V and –0.48 V (vs.Ag/AgCl) on the basis of the Nernst equation. The cathodic current occurs at the potential region close to the equilibrium potential of Cu as shown in **Fig. 1**. It is well-known that Cu^{2+} ions begin to electrodeposit without an accompanying overpotential from the aqueous solution. Therefore, this cathodic current corresponds to the deposition current of Cu. With increasing cathodic current, at around 10^{-5}A, the potential significantly polarizes to the less-noble region. This phenomenon seems to be caused by the diffusion control of Cu^{2+} ions. In the potential region less-noble than the equilibrium potential of Co, the cathodic current increases again at circa –0.8 V. It is also well-known that the electrodeposition of iron-group metals such as Ni, Co, and Fe is accompanied by the overpotential due to the rate determining multi-step reduction mechanism. Therefore, this increase in cathodic current is mainly caused by the deposition current of Co. At the potential region less than –1.2 V, with increasing cathodic current, the potential polarizes significantly to be less-noble region due to the diffusion limit of Co^{2+} ions. Consequently, the optimum cathode potential for electrodeposition of Cu and Co can be determined to –0.6V and –1.0V.

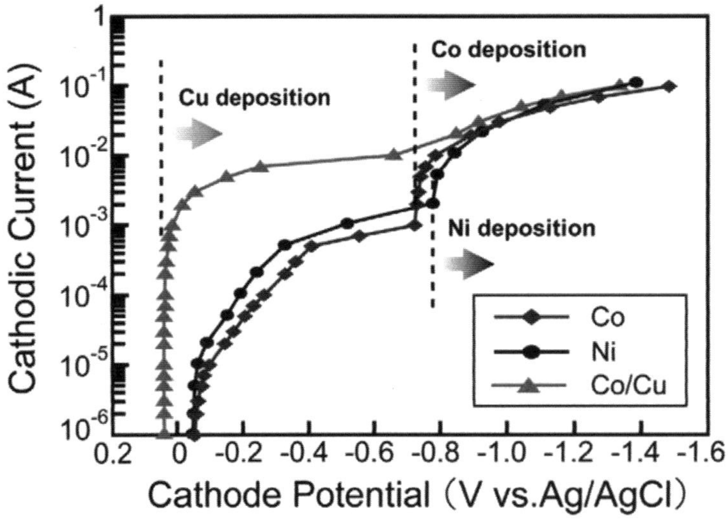

Figure 1. Cathodic polarization curves for electrodeposition of Cu, Co and Ni from aqueous solutions.

Figure 2 shows effect of cathode potential on the time-dependence of cathodic current during electrodeposition of Co-Cu nanowires. To determine the growth rate of nanowires, the filling time of a nanochannel 6000 nm in length was estimated by monitoring the time-dependence of deposition current at each cathode potential as shown in **Fig.2**. When the nanochannels are filled and the nanowires reach the membrane surface, the current will increase drastically due to the formation of hemispherical caps. The growth rates were estimated by dividing channel length by channel-filling time. At –0.6 V, the filling time is around 300 s and the deposition rate is estimated to be about 20 nm

s^{-1}, while the filling time is close to 30 s at –1.0 V and the deposition rate is estimated to be around 200 nm s^{-1}. On the basis of the results shown in **Fig.2**, for example, Co (50nm)/Cu (50nm) multilayered nanowires can be electrodeposited by alternating the cathode potential from -0.6 V (2.5s for Cu layer) to -1.0 V (0.25s for Co-rich layer). However, in this process, deposited Co layer will dissolute during the Cu deposition. Therefore, Co deposition time should be adjusted longer than the estimated time. **Figure 3** shows the time-dependence of cathodic current during electrodeposition of Co (50nm)/Cu (50nm) multilayered nanowires. Co deposition time was adjusted to 0.5s, which is two times of the estimation. The filling time is around 150 s and the deposition rate is estimated to be about 40 nm s^{-1}.

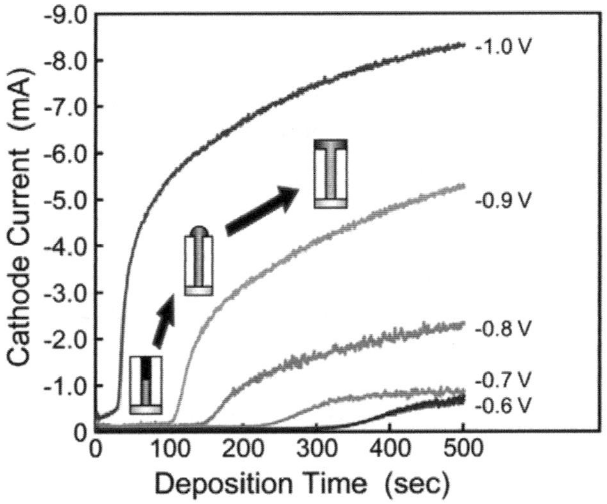

Figure 2. Effect of cathode potential on the time-dependence of cathodic current during electrodeposition of Co-Cu nanowires.

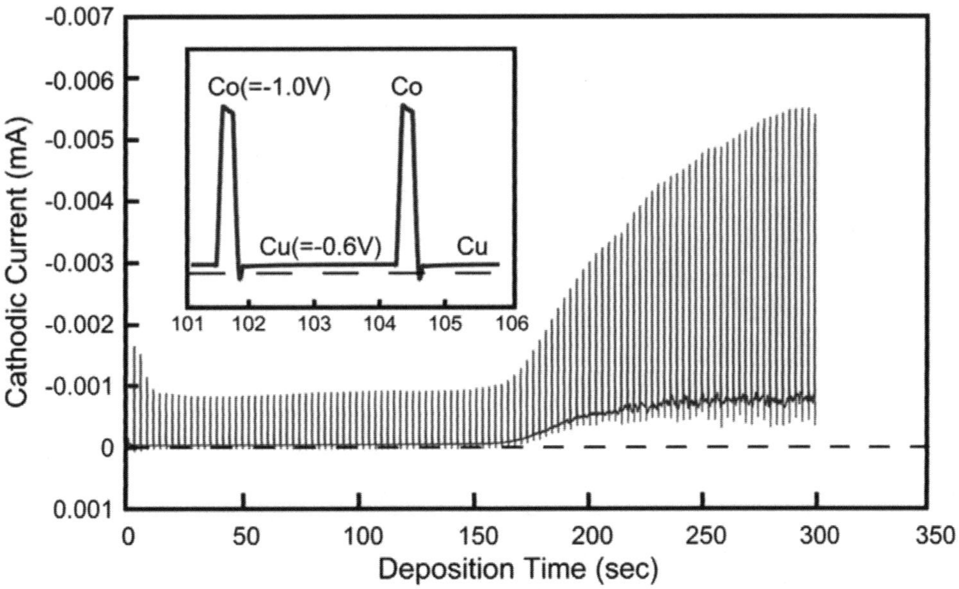

Figure 3. Time-dependence of cathodic current during electrodeposition of Co/Cu multilayered nanowires.

SEM images of electrodeposited Co nanowires separated from polycarbonate membrane filters are shown in **Fig. 4**. The cylindrical shape was precisely transferred from the nanochannels to the nanowires and the aspect ratio reached up to ca. 150.

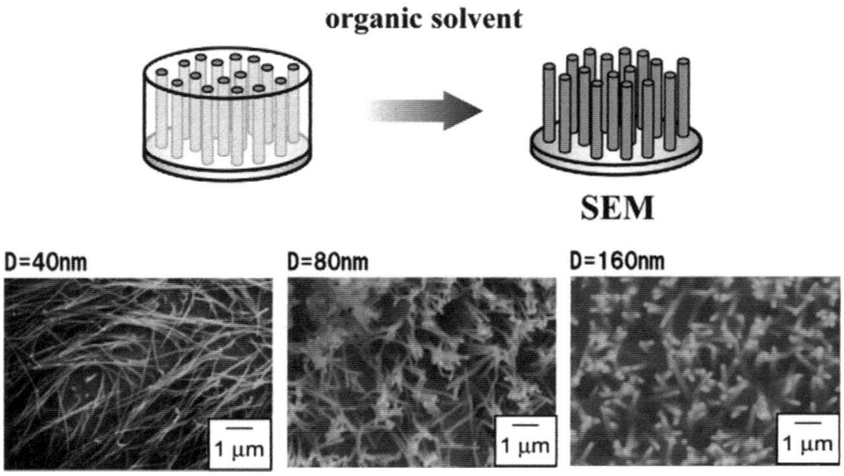

Figure 4. SEM images of electrodeposited Co nanowires separated from polycarbonate membrane filters.

Magnetic hysteresis loops of electrodeposited Ni, Co and Fe nanowires with diameter of 40 nm and length of 6000 nm are shown in **Fig. 5**. Magnetic field was applied to perpendicular direction to the membrane film plane. The perpendicular direction to the membrane film plane corresponds to the parallel direction to the long axis of nanowires. These magnetic hysteresis loops show typical perpendicular magnetization behavior and coercive force of Co nanowires was ca. 1 kOe. This is resulting from the uni-axial magnetic anisotropy and single magnetic domain structure of the nanowires with large aspect ratio.

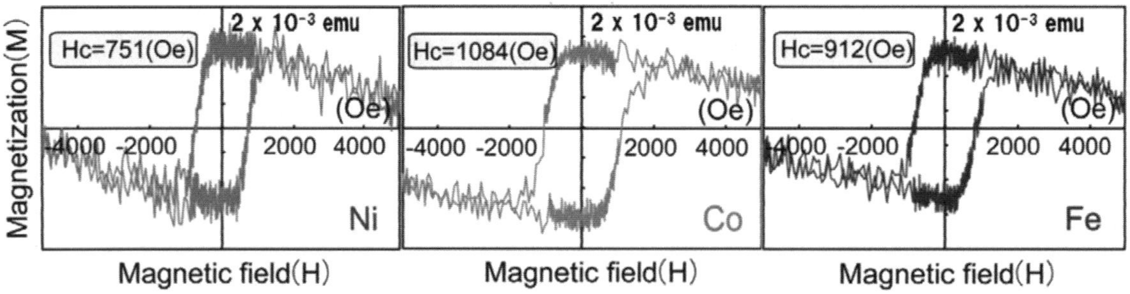

Figure 5. Magnetic hysteresis loops of electrodeposited Ni, Co and Fe nanowires with diameter of 40 nm and length of 6000 nm.

Magnetoresistive hysteresis of Ni nanowires electrodeposited into polycarbonate template (a) and anodized aluminum oxide template (b) are shown in **Fig. 6**. When the magnetic field was applied to perpendicular direction to the membrane film plane, the resistance was almost constant with increase in magnetic field. On the other hand, when the magnetic field was applied to in-plane direction to the membrane film plane, the resistance was decreased with increase in magnetic field and 2% of anisotropic magnetoresistance effect was observed in Ni nanowires electrodeposited into anodized aluminum oxide template.

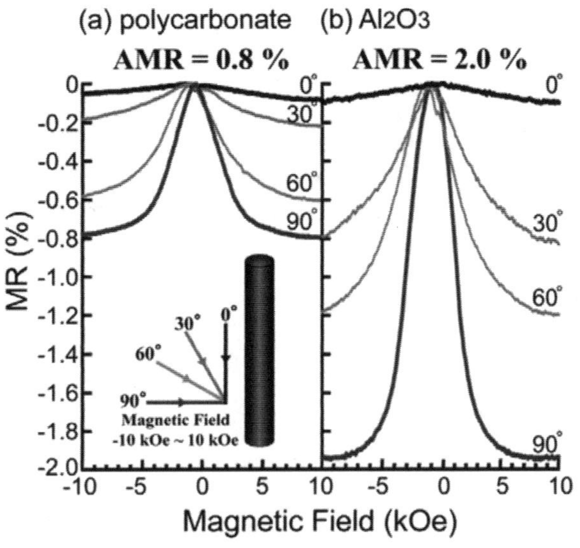

Figure 6. Magnetoresistive hysteresis of Ni nanowires electrodeposited into polycarbonate template (a) and anodized aluminum oxide template (b).

Magnetoresistive hysteresis of Co/Cu multilayered nanowires electrodeposited into anodized aluminum oxide template (a) and polycarbonate template (b) are shown in **Fig. 7**. Independently of the magnetic field direction, the resistance was decreased with increase in magnetic field and 20% of giant magnetoresistance effect was observed in the multilayered nanowires with Co layer 10nm and Cu layer 10 nm electrodeposited anodized aluminum oxide template.

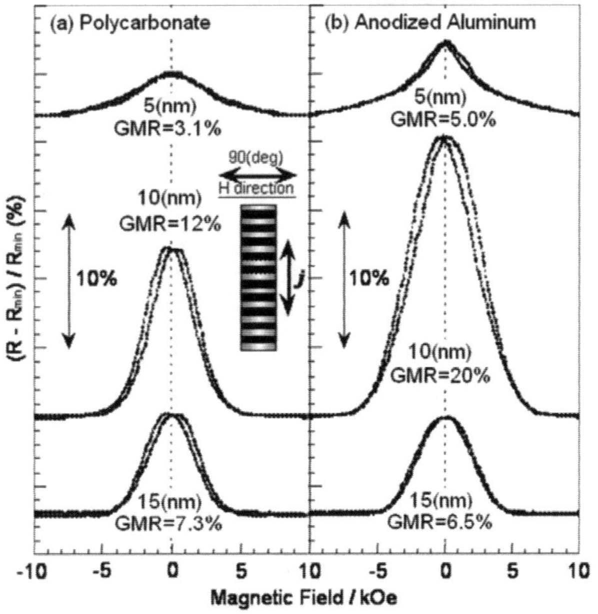

Figure 7. Magnetoresistive hysteresis of Co/Cu multilayered nanowires electrodeposited into polycarbonate template (a) and anodized aluminum oxide template (b).

4. Conclusions

Growth rate of Co/Co multilayered nanowires was around 40 nm•sec^{-1} and the cylindrical shape was precisely transferred from the nanochannels to the nanowires and the aspect ratio reached up to ca. 150. Magnetic hysteresis loops of Ni, Co and Fe nanowires with diameter 40 nm showed typical perpendicular magnetization behavior due to the uni-axial shape anisotropy and the coercive force reached up to around 1 kOe. 2% of anisotropic magnetoresistance effect was observed in Ni nanowires electrodeposited into anodized aluminum oxide template, while 20% of giant magnetoresistance effect was observed in Co/Cu multilayered nanowires.

References

[1] Martin C R 1991 *Adv. Mater.* **3** 457
[2] Whitney T M, Jiang J S, Searson P C and Chien C L 1993 *Science* **261** 1316
[3] Martin C R 1994 *Science* **266** 1961
[4] Nielsch K, Müller F, Li A P and Gösele U 2000 *Adv. Mater.* **12** 582
[5] Sauer G, Brehm G, Schneider S, Nielsch K, Wehrspohn R B, Choi J, Hofmeister H, and Gösele U 2002 *J. Appl. Phys.* **91** 3243
[6] Ohgai T, Gravier L, Hoffer X, Lindeberg M, Hjort K, Spohr R and Ansermet J P 2003 *J. Phys. D: Appl. Phys.* **36** 3109
[7] Ohgai T, Gravier L, Hoffer X and Ansermet J P 2005 *J. Appl. Electrochem.* **35** 479
[8] Ohgai T, Hoffer X, Gravier L, Wegrowe J E and Ansermet J P 2003 *Nanotechnology* **14** 978
[9] Ohgai T, Hoffer X, Fabian A, Gravier L and Ansermet J P 2003 *J. Mater. Chem.* **13** 2530
[10] Ohgai T, Hoffer X, Gravier L and Ansermet J P 2004 *J. Appl. Electrochem.* **34** 1007

MoS_2 sputtering coating for ultrahigh vacuum manipulation

Toshiyuki Maruyama[1], Jun Nakagawa[1], Katsumi Endo[1], Akira Kasahara[2], Masahiro Goto[2], Masahiro Tosa[2]

[1]Toyama Co., Ltd., 4-13-16, Hibarigaoka, Zama-shi, Kanagawa 252-0003, Japan
[2]National Institute for Materials Science, 1-2-1 Sengen, Tsukuba-shi, Ibaraki 305-0047, Japan

E-mail: t-maruyama@toyama-jp.com

Abstract. Lubrication applied to sliding part is one of the key components for the development of high performance driving mechanism and manipulator in ultrahigh vacuum. Some solid lubrication has been widely used for this purpose, especially molybdenum disulfide (MoS_2). Low friction, low wear and low outgassing during the sliding motion on lubrication are required for a high performance manipulation in ultrahigh vacuum. Although the tribology of MoS_2 has been reported to be suitable in a vacuum, the outgassing characteristics while sliding are not clean. In this work, we investigate the characteristics of MoS_2 sputtering coating in order to develop high performance ultrahigh vacuum driving mechanism. Their friction and outgassing characteristics were evaluated and we have found out suitable parameters for manipulation.

1. Introduction

The final target of this research is to produce high-performance manipulation under ultrahigh vacuum environment. Ultrahigh vacuum offers very clean space and its vacuum pressure is 10^{-6}~10^{-8} Pa. Manipulation mechanisms are made of sliding parts, such as bearings and guide rails. In these parts, it is necessary to use lubricant on sliding surfaces for driving smoothly. Low evaporating grease is generally used in high vacuum. But in ultrahigh vacuum, liquid lubricant become an outgassing source and ultrahigh vacuum pressure cannot be maintained. Therefore the solid lubricant, such as coatings and powders, is recommended for use in ultrahigh vacuum. But the solid lubricant has two major disadvantages. The first disadvantage is short lifetime suffered from wear. Lifetime of solid lubricant is not longer than liquid one and often become a weak point for manipulation in ultrahigh vacuum. Additionally, the precision of manipulation is reduced by wear of lubricant.

The second disadvantage is outgassing. Absorbed gases are desorbed from lubricant and ultrahigh vacuum pressure cannot be maintained during sliding. With respect to these two points, lubricant showing great performance has not been developed yet. Therefore, we have aimed to develop an advanced solid lubricant with long lifetime, little wear and low outgassing. In this study, we focused molybdenum disulfide (MoS_2) for solid lubricant. This compound is well known as a good lubricant and provides low friction in a vacuum. The friction characteristics have already been studied by many researchers [1,2,3], but outgassing characteristics during sliding have not been reported so far. We focused on this point and studied optimum coating parameters in order to improve lifetime and to reduce outgassing. As coating method, we selected RF sputtering that has an advantage to control film-thickness well to be suitable for precision manipulators.

2. Experimental

2.1. Coating

In the present study, we used RF magnetron sputtering source and coated MoS_2 on commercial-type 440 martensite stainless steel substrates, which material is typically used for sliding parts. These coatings with different film thickness were evaluated the friction and outgassing characteristics. In addition, we made two more coated samples in order to test whether the outgassing can be reduced by heating the samples. One is heated in a vacuum after sputtering and the other is heated during sputtering process.

2.2. The evaluation of friction characteristics

We measured the sliding friction coefficient of MoS_2 by using a vacuum friction measurement system based on the Bowden-Leben-type system [4]. We evaluated the friction changes with film thickness in a vacuum (10^{-5}~10^{-6} Pa). The sliding indenter is sapphire ball that is 1/8 inch in diameter. The sliding rate is 0.5 mm/s, the sliding length is 5mm and the load is 0.98 N. The measurement was repeated 10 times on the same line in the same direction.

In addition, we performed the long time friction measurement and evaluated the lifetime characteristics of different film-thickness samples. The apparatus was the same measurement system with the exception that the sample stage was replaced translator driving into high speed rotary one. The sliding indenter is type 440 stainless steel ball that is 1/8 inch in diameter. The sliding rate is 140 mm/s and the load is 0.98 N. The measurement was repeated on the same line in the same direction until friction increased significantly.

2.3. The evaluation of outgassing characteristics

We evaluated outgassing value by using an outgassing measurement system based on the throughput method [5]. The outgassing measurement apparatus is shown in Figure 1. This apparatus contains the sliding mechanics on the sample similar to the friction measurement in the vacuum chamber and the outgassing was measured while the indenter slides on the surfaces of samples. Also mass spectrometer is equipped in order to analyze components of outgassing. We evaluated the outgassing variation with respect to film thickness and heating methods. The sliding indenter is type 440 stainless steel cylindrical roller that is 3 mm in diameter and 3 mm in length. The sliding rate is 1 mm/s, the sliding length is 4 ~ 10 mm and the load is 1.96 N, and the measurement was repeated 20 ~ 200 times on the same line in the same direction. The base pressure of vacuum chamber reached 10^{-7} Pa just before outgassing measurement was started.

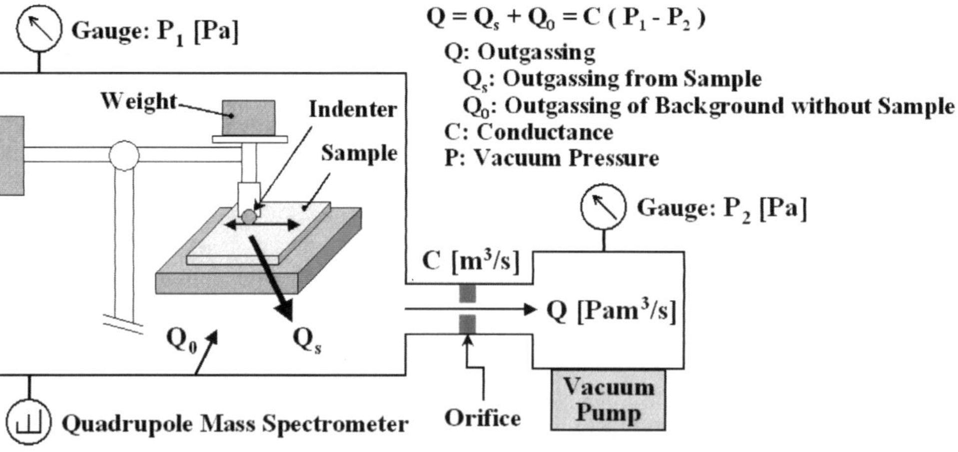

Figure 1. Schematic diagram of outgassing measurement apparatus.

3. Result and discussion

3.1. Friction characteristics

Figure 2 shows the change in friction coefficient of MoS_2 with film-thickness. The friction coefficient in a vacuum is lower than that in atmosphere in common with general characteristics of MoS_2 [1]. In addition, friction coefficients show the tendency to increase with increasing of film-thickness. The similar trend is obtained in both at atmospheric and vacuum conditions.

Figure 3 shows the change in friction coefficient of MoS_2 with number of frictions about three different film-thickness samples. In 0.5 and 1.8 μm thick samples, these friction coefficients increase suddenly when number of frictions exceeds about 250,000 and 350,000 times. It means that these coatings have completely worn and reached the end of lifetime. But in 1.0 μm thick sample, friction coefficient still maintains a constant value around 0.05 even when number of frictions exceeds 1,000,000 times.

It means that there exists optimum thickness of MoS_2 for long lifetime. The thicker coating has higher friction resistance which causes large wearing on the coating. On the other hand, for thinner coating samples, the film is so thin that only a little amount of wearing could destroy the coating completely. The cause of this optimum thickness is considered that crystalline structures or compositions of MoS_2 changes with thickness. The next step would be to analyze these features.

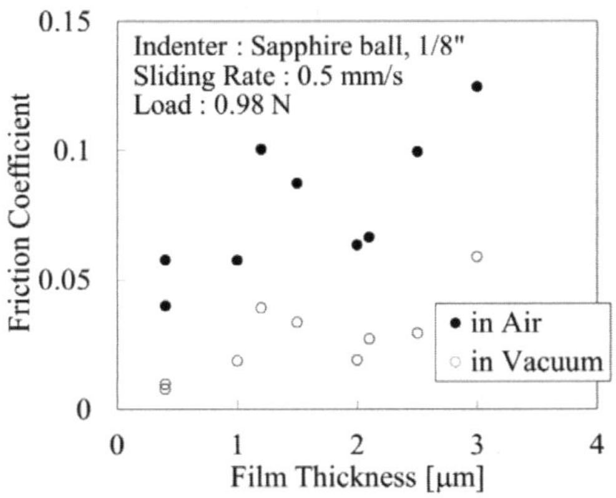

Figure 2. Changes in friction coefficient of MoS_2 coatings with film thickness in the air and vacuum (10^{-5}~10^{-6} Pa).

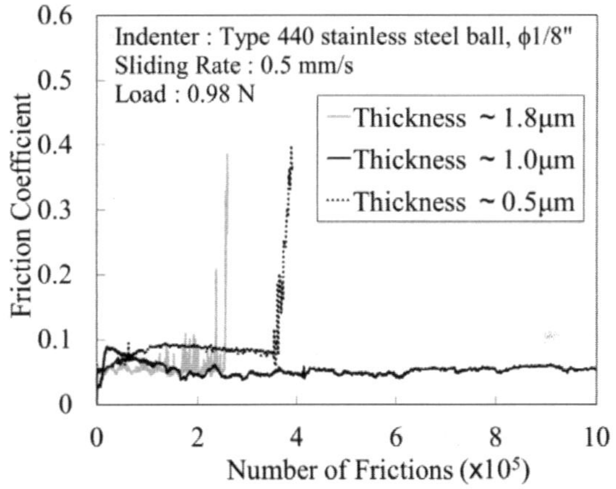

Figure 3. Changes in friction coefficient of MoS_2 coatings with number of frictions in a vacuum (10^{-5}~10^{-6} Pa).

3.2. Outgassing characteristics

Figure 4 shows the outgassing changes with number of frictions about two different film-thickness samples. At beginning of sliding, a large amount of outgassing arises. But the outgassing reduces with increasing frictions. Furthermore, the outgassing values of two different film-thickness samples are almost the same during the sliding procedure. It means that the outgassing of MoS_2 coatings does not depend on thickness. Figure 5 shows the outgassing mass spectra of 1.0 μm thick sample at beginning of sliding process. The major outgas is argon (Ar), the second is carbon dioxide (CO_2) and the third is carbon monoxide (CO), respectively. Since Ar was used as sputtering gas, it can be considered that Ar gas is absorbed inside MoS_2 coatings during sputtering. The Ar outgassing during sliding has been also reported in the case of Ag ion-plated coating where Ar gas was used for the coating process [6].

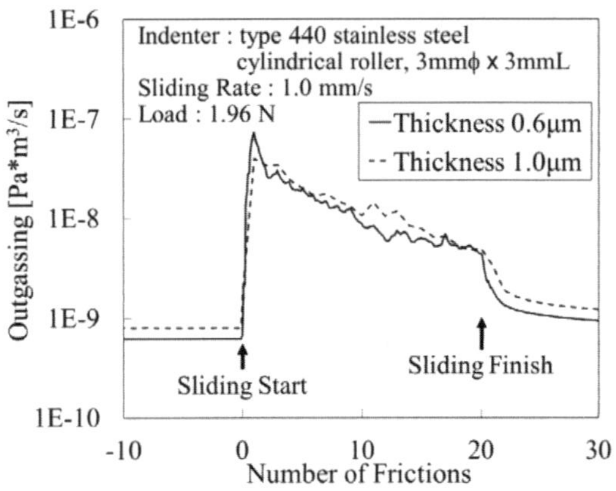

Figure 4. Changes in outgassing of the different film-thickness MoS_2 coatings with number of frictions.

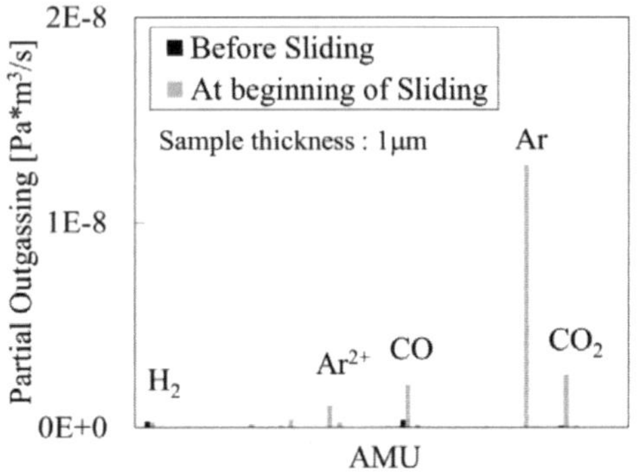

Figure 5. Outgassing mass spectra of MoS_2 coating at the beginning of sliding.

Figure 6 shows the result of similar outgassing measurement about two heated samples and non-heated one with achievement of more frictions. In all samples, outgassing has their peaks at beginning of sliding and reduced with increasing frictions. But the reduction of outgassing is saturated after 100 times frictions. The outgassing of no heating sample is almost the same as that of the sample with heating ones after sputtering. It means that the sample with heating after sputtering is not effective in reducing the outgassing. On the other hand, the heated sample during sputtering shows rather smooth outgassing reduction figure after beginning of sliding. But after saturation, the differences of outgassing of all samples are small and the amounts are nearly background level.

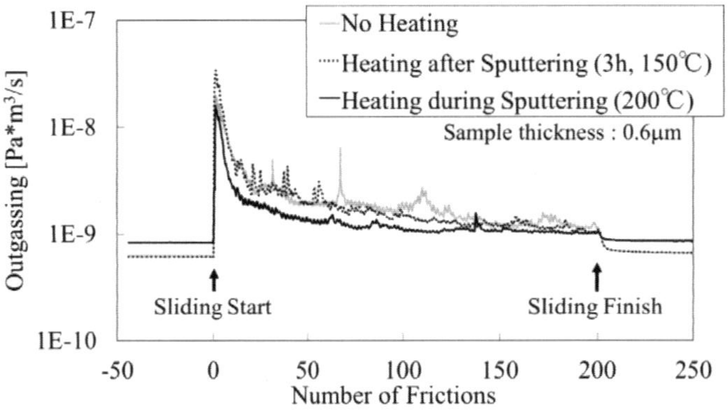

Figure 6. Changes in outgassing of the heated and non-heated MoS_2 coatings with number of frictions.

Figure 7 shows the changes in major outgassing masses with number of frictions about samples without heating (Figure 7 (a)) and with heating during sputtering (Figure 7 (b)). These results indicate that the outgas components of sample with heating is smaller or reduces faster than without heating. Ar gas reduces with increasing frictions in both samples, but the reduction in the sample with heating is faster than that without heating. It shows that Ar reduces further with more increasing frictions. On the other hand, CO and CO_2 gases in the sample with heating are smaller than without heating. Furthermore, CO of the sample with heating shows almost no variation during sliding process. It can be considered that heating process during sputtering can prevent from gaseous adsorption inside MoS_2 coatings. But after saturation, the amounts of CO and CO_2 gases of the sample with heating are nearly background level, so changes are very small during sliding.

In future, we try to measure outgassing precisely by reducing background pressure and want to observe whether the changes of amounts of CO or CO_2 gases occur actually.

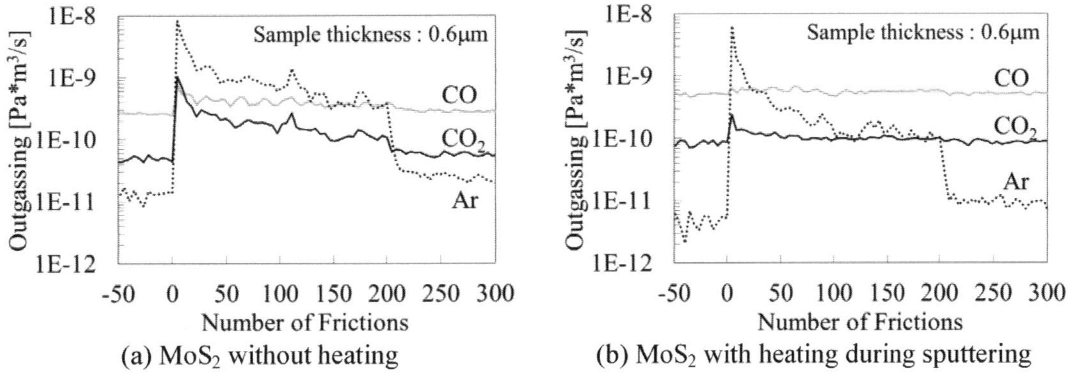

Figure 7. Changes in major outgassing masses of MoS_2 coatings with number frictions.

4. Conclusions

We evaluated friction and outgassing characteristics of MoS_2 coating for optimum solid lubricant under ultrahigh vacuum environment. It was confirmed that optimum thickness coating can offer longer lifetime for ultrahigh vacuum lubricant but film-thickness has little effect to the outgassing ratio. Large outgassing of Ar, which was used as sputtering process gas, was occurred in early stage of sliding. Ar and other absorption gas (CO and CO_2) can be decreased by heating substrate during sputtering.

References

[1] Halter J A 1967 *Wear* **10** 397
[2] Fusaro L R 1976 *ASLE Trans.* **20**-1 15
[3] Spalvins T 1987 *J. Vac. Sci. Technol.* **A 5** 212
[4] Kasahara A, Goto M, Tosa M and Yoshihara K 2003 *J. Electroanal. Chem.* **559** 45
[5] Redhead A P 2002 *J. Vac. Sci. Technol.* **A 20** 1667
[6] Fujii Y and Ono H 2004 *Koyo Engineering Journal English Edition* No.164E (Osaka: Koyo Seiko) p 24

Journal of Physics: Conference Series **417** (2013) 012049 doi:10.1088/1742-6596/417/1/012049

Correlation between local atomic structure and ultraviolet luminescence of AlGdN thin films

Kuniyuki Ichii[1], Shinya Kitayama[1], Shinya Iwahashi[1], Junya Nakamura[1], Vidya Sagar Reddithota[1], Takashi Kita[1], Yoshitaka Chigi[2], Tetsurou Nishimoto[2], Hiroyuki Tanaka[2], Mikihiro Kobayashi[2], Tsuguo Ishihara[3], and Hirokazu Izumi[3]

[1]Department of Electrical and Electronic Engineering, Graduate school of Engineering, Kobe University, Japan.

[2]YUMEX Inc, Japan.

[3]Hyogo Prefectural Institute of Technology, Japan.

E-mail: 118t204t@stu.kobe-u.ac.jp

Abstract. The present study reports on the correlation between the local atomic structure and cathodoluminescent properties of $Al_{1-x}Gd_xN$ thin films grown by reactive rf magnetron sputtering at ultra-high vacuum conditions. Those thin films were characterised using X-ray absorption fine structure (XAFS) and cathodoluminescence (CL). From the CL measurements, we have observed a narrow intense ultraviolet emission at 318 nm which is originated from the intra-orbital *f-f* transition in Gd^{3+} ions. In order to understand the local atomic structure around the $Al_{1-x}Gd_xN$ (x=0.1 to 6.0 mol%) thin film, XAFS measurements have been carried out. Analysis of the local atomic structural results showed that both the large distance among Gd atoms and nitrogen vacancies in $Al_{1-x}Gd_xN$ lattice significantly contribute to the richness in the ultraviolet emission intensity.

1. Introduction

The current interest in using ultraviolet (UV) light source in various areas of science and technology such as lithography for large display panels, UV curing, sterilization and medical care. For the past decades, mercury lamps are widely used as efficient UV light emitting source in contrary mercury toxicity harms human life and environment. Mercury lamps have several issues such as short-life time, low emission efficiency and risk of burst to lamp. Due to these demerits, researchers are searching for new mercury-free UV light sources. Recent times, UV light emitting diode's (LEDs) have great deal of interest due to their potential properties viz. long-life time and wavelength controllability. The efficiency of UV LEDs have significantly low due to the low hole concentrations in *p*-type semiconductor [1, 2]. Thus, UV semiconductor devices without use of *p*-type semiconductor have emerged. Of those devices, the output of 100 mW and power efficiency of 40% at a wavelength of about 240 nm was achieved by using electron beam technique without any hole doping [3]. However, these UV light sources have point source where the emission spectrum was broader (10 nm~), which are not suitable for UV applications of lithography for large display and medical care. It has been reported earlier that the wide band gap semiconductors could be found to be appropriate for UV light emitting source [4-7]. The large band gap of AlN could be the good host material for bright and prominent UV light because the similarity of electronic states of Al and the trivalent rare earth ions and high thermal conductivity, chemically stable material and obviously nontoxic of AlN [8]. Gd ion is extremely narrow and temperature insensitive luminescent element due to the the intra-4*f* electron transition and electro-magnetic shield effects. According to the earlier investigations, luminescence performance of rare earth doped in semiconductors is discussed with help of local atomic structure [9-11]. Our earlier investigations on $Al_{1-x}Gd_xN$ thin films showed the CL emission intensity is drastically changed as function of growth temperature [12,13]. In this paper, we analyze the correlation between local structure around Gd^{3+} ion and the CL properties.

2. Experimental

$Al_{1-x}Gd_xN$ (GdN content 0.1, 0.8, 1.3, 1.8 and 6.0 mol %) thin films were grown on fused silica substrates using reactive radio-frequency magnetron sputtering technique performed at the base pressure of 6×10^{-5} Pa. The 4N Al and 3N Gd were used as the metal target. We have used a 6N pure mixed gas of argon and nitrogen for the reactive growth and the partial pressure ratio was even. A deposition chamber equipped two targets which are separated from a substrate-introduction chamber. During the growth of $Al_{1-x}Gd_xN$ thin films, we set Gd metal tip on the target and the argon and nitrogen flow rates were controlled with a flow rate controller and adjusted to keep the total sputtering pressure of 5.0 Pa. Initially, we deposited a AlN buffer layer of 700 nm to reduce the interface defects then we deposited 700 nm $Al_{1-x}Gd_xN$ thin films. The $Al_{1-x}Gd_xN$ thin films have been characterized using XAFS and CL. The XAFS measurements were performed in the fluorescence mode at the SPring-8 (beam line BL14B2) using synchrotron radiation from the 8.0 GeV storage ring at room temperature [14]. The CL measurements were carried out at the acceleration voltage of 4.0 kV and the injected electron current of 0.2 mA in vacuum of 1.0×10^{-4} Pa using a single monochrometor of the spectrum resolution of 1 nm and the signal is detected by a Si charge-coupled device.

3. Result and Discussion

Figure 1 (a) shows the CL emission spectrum for $Al_{0.999}Gd_{0.001}N$ thin films at room temperature. It is important to note that a narrow (line width ~1 nm) ultraviolet emission at 318 nm is observed which is especially originated from the intra-orbital *f-f* transition in Gd^{3+} ions. From the energy-band diagram of Gd^{3+} ion, the ~318 nm emission is attributed to the transition from the excited state of $^6P_{7/2}$ to the ground state of $^8S_{7/2}$ which is shown in inset of Fig. 1 (a) [4-7,12,13]. Figure 1 (b) shows the CL intensity as function of GdN mole fraction. It is clear that the CL intensity of $Al_{1-x}Gd_xN$ is deteriorates with the increase of GdN-mole fraction. We propose that the difference in the atomic structures of host with GdN content causes the change in CL intensity of $Al_{1-x}Gd_xN$.

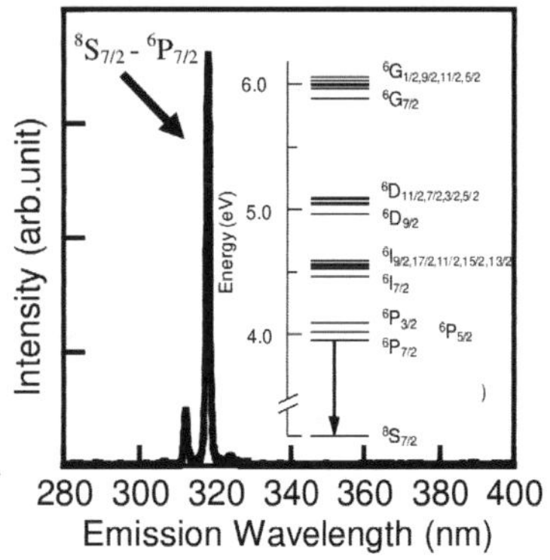

Figure 1. **(a)** CL spectrum of $Al_{0.999}Gd_{0.001}N$ thin film at room temperature. The energy level diagram of Gd is shown in inset of the Figure.

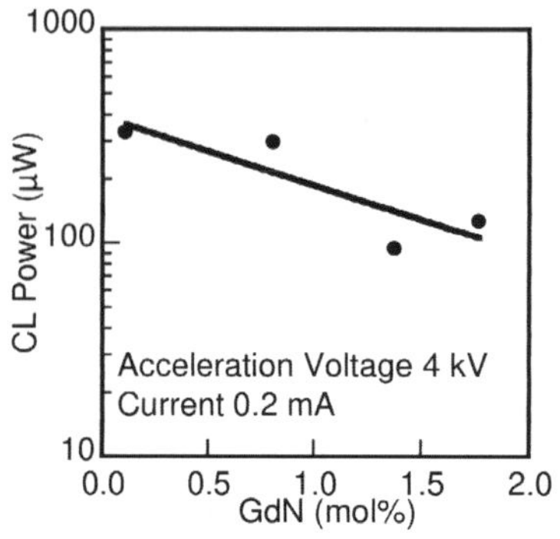

Figure 1. **(b)** The CL intensity of $Al_{1-x}Gd_xN$ thin films as function of GdN mole fraction.

Figure 2 shows the normalised absorption spectrum for $Al_{1-x}Gd_xN$ thin film and the Gd $-L_3$ absorption edge for AlGdN sample is found at 7.238 keV. From the absorption spectrum, it is evident that there is a characteristic series of oscillations (inset of Fig. 2) just starts above the absorption edge. The magnified characteristic series of oscillations can be seen clearly in the inset of Fig. 2. The oscillations could be prevailed from the interference of the emitted and reflected photoelectrons generated within the inner shell of the Gd L_3 edge.

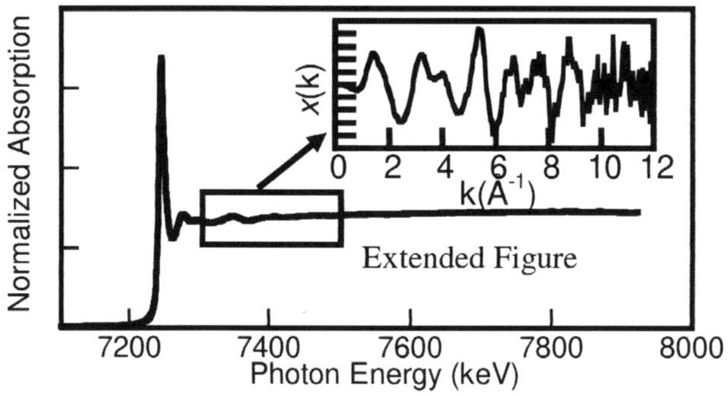

Figure 2 The absorption spectrum for the $Al_{1-x}Gd_xN$ recorded at room temperature. Inset figure shows the magnified fine oscillation above the absorption edge.

Figure 3 shows the Fourier transform of the XAFS oscillations in an absorption spectrum yields the radial structure function for $Al_{1-x}Gd_xN$. Fourier transformed XAFS oscillations are presented in the range of 2.5 ~ 10 Å^{-1} and the oscillations were well fitted in the range Δr = 1.0 ~ 3.3 Å. This range of oscillations explores the 1st neighbour shell and 2nd neighbour shell information. Accordingly, the large random noise and the uncertainties from the background in the fine oscillations have been eliminated. These spectra show two peaks; the 1st peak at 1.9 Å which corresponds to the 1st neighbor shell of N and the 2nd peak at 3.1 Å corresponds to the 2nd neighbor shell of Gd or Al. It is obvious that the 1st neighbor shell peak intensity slightly increased and the 2nd neighbor shell peak intensity decreased as function of GdN mole fraction. The 1st neighbor shell signal gives information in regarding to the coordination number (CN) and the 2nd neighbor shell signal gives the distribution of group-III atoms in $Al_{1-x}Gd_xN$ lattice. The experimental radial structure function is well fit theoretically using FEFF 6 program [15]. From the FEFF 6 program, the back scattering amplitudes and phase shifts are determined. In the fitting of the XAFS data, the interatomic distance (r), the Coordination Number (CN) and the Debye-Waller factor (DWF) were obtained by curve fitting of the 1st and 2nd neighbor shell of $Al_{1-x}Gd_xN$ thin film.

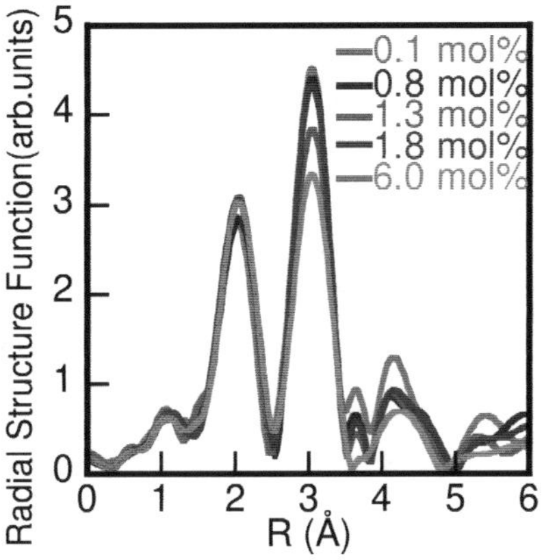

Figure 3 Radial structure functions for $Al_{1-x}Gd_xN$ thin films grown on fused silica substrate.

Figure 4 shows the CN and DWF as function of GdN mole fraction. The 1st neighbor shell CN of $Al_{1-x}Gd_xN$ is enhanced with the increase of the GdN-mole fraction It is found that the disorder of the 2nd neighbor shell is increased with the increase of the GdN-mole fraction; this could be due to the doping of Gd into the large local distortion in AlN lattice.

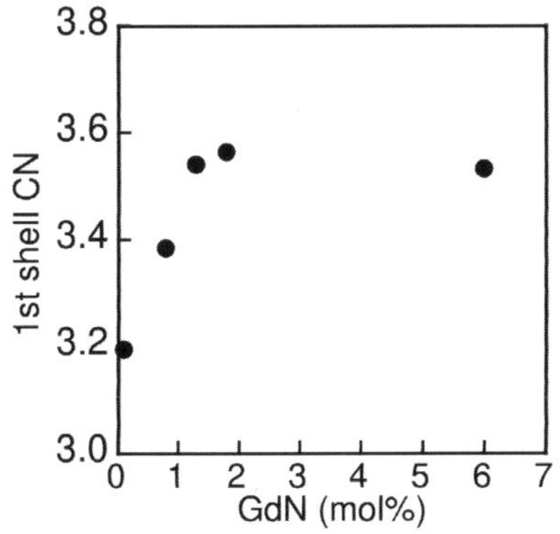

Figure 4. **(a)** The 1st neighbor shell CN with GdN mole fraction in $Al_{1-x}Gd_xN$ thin films.

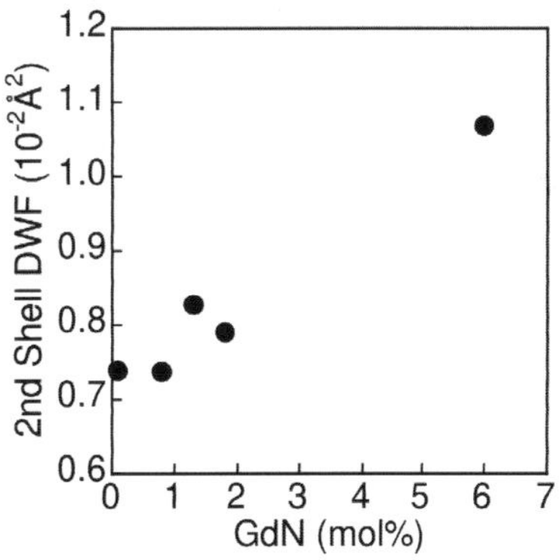

Figure 4. **(b)** The 2nd neighbor shell DWF with respect to GdN mole fraction in $Al_{1-x}Gd_xN$ thin films.

Figure 5 shows the CL intensity with respect to the CN of the 1st neighbor shell and the DWF of the 2nd neighbor shell. The CL intensity is enhanced with the decrease of the CN in the 1st neighbor shell, which indicate the nitrogen vacancies in the 1st neighbor shell and play a vital role in the enhancement of the luminescence intensity. Luminescence from RE ions in solid hosts involves transitions within the 4*f* electronic shell that are strictly forbidden in the free ion but become more or less allowed through the admixture of states of opposite parity induced by the crystal field in solids. Luminescence from the intra-orbital *f-f* transition is allowed in a solid by reduction of the local symmetry and opposite-parity mixing under the crystal effect. [16,17] Nitrogen vacancies lead to lattice distortion of Gd ions and lowering the *f*-orbital symmetry of Gd ions enhance the transition probability. The CL intensity is decreased with the increase of the

DWF of the 2nd neighbor shell and it is clearly shown in Fig. 5 (b). Low DWF indicates that atomic structure is homogeneously distributed in the AlN crystal structure. Thus, the average Gd-Gd distance becomes maximum at the low GdN mole-fraction and inhibits concentration quenching. In other words, the relatively large distance among Gd atoms inhibits the quenching of the luminescence intensity.

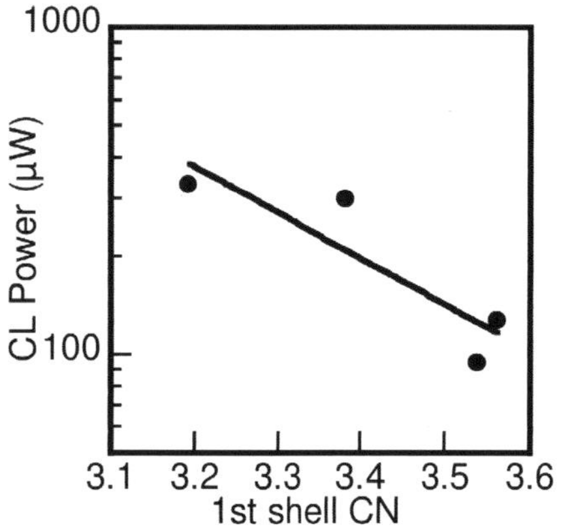

Figure 5. (a) CL intensity with respect to the 1st shell CN for $Al_{1-x}Gd_xN$ on fused silica substrate.

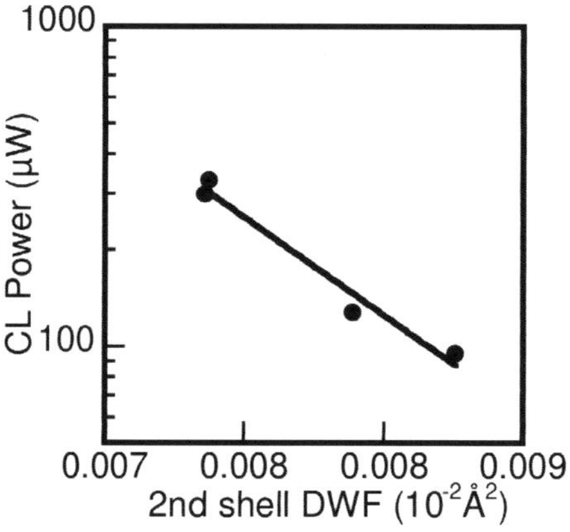

Figure 5. (b) CL intensity with respected to 2nd shell DWF for $Al_{1-x}Gd_xN$ on fused silica substrate.

4. Summary

$Al_{1-x}Gd_xN$ (x=0.1,0.8,1.3,1.8 and 6.0 mol %) thin films were successfully grown on fused silica substrates using reactive rf sputtering at high vacuum condition. The CL measurements confirmed that the films showed UV narrow emission at 318 nm, which is originating from the intra-orbital *f-f* transition ($^6P_{7/2}$ to $^8S_{7/2}$) within Gd^{3+} ions. The absorption spectra showed the absorption edge at 7.238 keV which is assigned to Gd L_3-edge. The CL intensity of $Al_{1-x}Gd_xN$ grown on fused silica substrate is deteriorates with the increase of GdN content. From the XAFS measurements, the 1st neighbor shell CN is decreased with increase of the CL intensity and 2nd neighbor shell DWF is decreased with increase of the CL intensity. The enhancement of CL intensity is the cooperative phenomena of nitrogen vacancies and large distance among Gd atoms.

5. References

[1] Hirayama H, Fujikawa S, Noguchi N, Norimatsu J, Takano T, Tsubaki K, Kamata N 2009 *Phys. Status Solidi* A **206** 1176
[2] Bhattacharyya A, Moustakas T D, Zhou L, Smith D J and Hug W 2009 *Appl. Phys. Lett.* **94** 181907
[3] Oto T, Ryan G, Banal K. Kataoka, Funato M and Kawakami Y 2010 *Nature Photonics*, **4** 767
[4] Zavada J M, Nepal N, Lin J Y, Jiang H X, Brown E, Hömmerich U, Hite J, Thaler G T, Abernathy R C and Pearton J S 2006 *Appl. Phys.* Lett. **89** 152107
[5] Vetter U, Zenneck J and Hofsäss H 2003 *Appl. Phys.* Lett. **83** 2145
[6] Gruber J B, Vetter U, Hofsäss H, Zandi B and Reid F M 2004 *Phys. Rev. B* **69** 195202
[7] Maqbool M, Ahmad I, Richardson H H and Kordesch M E 2007 *Appl. Phys. Lett.* **91** 193511
[8] *Properties of Advanced Semiconductor Materials*, edited by Levinshtein M E, Ramyantsev S L, and Shur M S ~Wiley, New York, 2001, p. 31.
[9] Ofuchi H, Kawamura D, Matsubara N, Tabuchi M, Fujiwara Y and Takeda Y 1998 *Microelectron.Eng.* **43-44** 745

[10] Ofuchi H, Nishikawa T, Takaba K, Ogawa K, Tabuchi M, Takeda Y, Wakahara A, Yoshida A, Oshima T and Ito H 2006 *PhysB*. **376-377,** 496-498
[11] Bang H, Morishima S, Li Z, Akimoto K, Nomura M and Yagi E 2002 *J. Cryst. Growth* **237-239** 1027
[12] Kita T, Kitayama S, Kawamura M, Wada O, Chigi Y, Kasai Y, Nishimoto N, Tanaka H, and Kobayashi M 2008 *Appl. Phys. Lett.* **93** 211901
[13] Kitayama S, Yoshitomi H, Iwahashi S, Nakamura J, Kita T, Chigi Y, Nishimoto T, Tanaka H, Kobayashi M, and Ishihara T 2011 *J. Appl. Phys.* **110** 093108
[14] http://www .spring8.or.jp/en/
[15] Ravel B, Newville M, 2005 *J. Synchrotron Radiat.* **12** 537
[16] Judd B R 1962 *Phys.Rev.***127** 750
[17] Ofelt G S 1962 J. *Chem. Phys*. **37** 511.

Laser Pulse Compensation Applied to Phase-Change Recording Film to Form Nano-Sized Marks

S Furumiya[1,2] **and Y Fujiwara**[2]
[1] Storage Media Group, Digital & Network Technology Development Center, Panasonic Corporation
3-1-1 Yagumo-nakamachi, Moriguchi, Osaka 571-8501, Japan
[2] Division of Materials and Manufacturing Science, Graduate School of Engineering, Osaka University
2-1 Yamada-oka, Suita, Osaka 565-0871, Japan

E-mail: furumiya.shigeru@jp.panasonic.com

Abstract. We have examined the application of the laser pulse compensation to the phase-change recording film of the triple-layer optical disc by using an original writing laser pulse controller, which compensates dynamically the pulse edges in order to form the marks at the correct position. We confirmed 112 nm sized marks smaller than the optical spot formed on the thin recording film. The results of the compensation had some characteristics depending on the data pattern and/or the layer. We discussed the causes by considering about the structure of the recording films and the phase-change process to form the marks.

1. Introduction

An optical disc is well known as the medium of the high reliable data storage system, the recording is realized by forming many small marks on the recording film by the irradiation of the laser beam spot. A kind of the rewritable optical discs is made from the phase-change material of a Ge-Sb-Te for the recording film [1]. The data capacity of the disc has been increasing from 4.7 Gbyte of the digital versatile disc random access memory (DVD-RAM) by the red laser optics [2] to 25 Gbyte of the Blu-ray disc rewritable (BD-RE) by the blue laser optics [3]. The smaller laser spot size depending on the optical diffraction limit is, the larger capacity is obtained, but it seems impractical for the commercial products to use shorter wavelength of the laser or larger numerical aperture (NA) of the objective lens than now.

Recently, two effective approaches to increase the data capacity per one disc were studied. The one way is piling the recording layers. Nishihara *et al.* reported a structure of the rewritable phase-change triple-layer disc [4], which has very thin front recording layer with high transparent and high reflective contrast. The other way is increasing the data density by the recording method. Nakamura *et al.* proposed an adaptive write strategy [5], which decreases the symbol error of the high density recording at the front layer of the triple-layer disc.

In this paper, first of all, we describe an original configuration of the writing laser pulse controller, which was developed to execute the adaptive write strategy. By using this controller, we have examined the application of the laser pulse compensation to the phase-change recording film of the triple-layer disc to form the nano-sized marks. Then, we report that the results of the compensation have some characteristics depending on the data pattern and/or the layer, when the marks are formed on the recording films as the correct mark edges. Moreover, we discuss the causes by the relation between the thermal diffusion of the thin film and the phase-change process of the mark formation.

2. Writing laser pulse controller

The phase-change recording is performed by forming the amorphous marks and the crystalline spaces on the recording film by the irradiation of the laser pulses to the spinning disc. The writing laser pulse controller is for generating the laser pulses irradiated from the optical pick-up unit.

2.1. Writing laser pulses for the phase-change recording

Figure 1 shows the relation between the writing laser pulses and the mark and space train on the recording film. The writing laser pulses to form the mark of 2T length (2Tm), where T means a period of data clock, consist of a top pulse of the write power and a cooling pulse of the bottom power. The phase-change film is melted by the strong irradiation of the top pulse and cooled rapidly by the cooling pulse, and then the region becomes a mark of the amorphous phase. And to form the space of 3T length (3Ts) between the marks, the erase power is irradiated. The region irradiated by constant erase power becomes the space of the crystalline phase by keeping the film at a crystallization temperature.

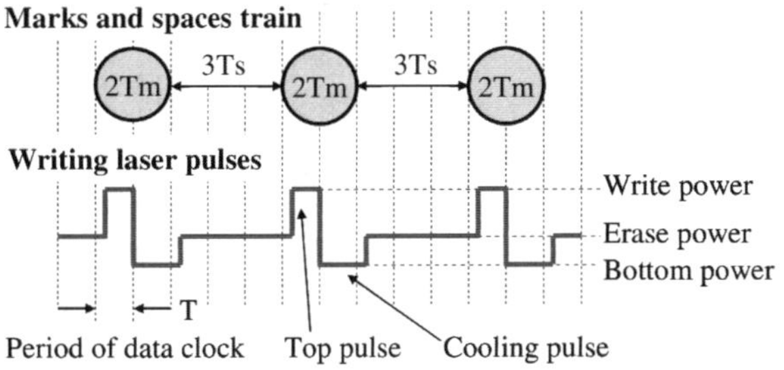

Figure 1. The circles image the mark shapes on the recording film as the length of 2 times periods of data clock T. The pulse waveform indicates the writing laser levels. The horizontal direction of the figure means the edge positions of the marks and also the time scale for the pulses.

2.2. Signal processing for the write laser pulse compensation

We have developed a writing laser pulse controller, which has the special functions of not only generating the pulses at high time accuracy, but also compensating dynamically the pulse edges adapted for the recording data pattern, so as to form each mark at the correct position consequently.

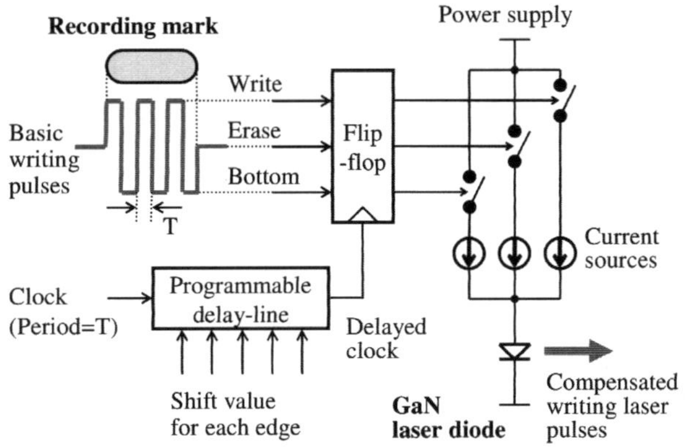

Figure 2. Block diagram for the signal processing of the writing laser pulse controller. The programmable delay-line is a key part to control the timing of pulses.

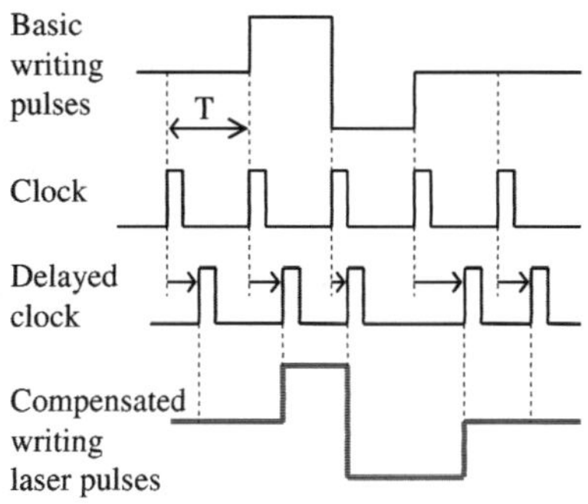

Figure 3. Compensated writing laser pulses with arbitrary duration are made from individually delayed clock pulses.

Figure 2 shows conceptually the block diagram of the signal processing of the writing laser pulse controller. Compensated writing laser pulses are made in the following order, firstly generating the basic writing pulses roughly, wherein the interval of each pulse is T, secondly making the delayed clock through the programmable delay-line loaded the shift value for each clock edge, thirdly latching the basic writing pulses precisely by the trigger of the delayed clock, and lastly driving a GaN laser diode by three switching current sources. Figure 3 indicates the relation between the basic writing pulses and the compensated writing laser pulses.

At the actual circuit, the processing is capable of the clock frequency of up to 800 MHz and the time accuracy of 80 ps by using four programmable delay-lines (Onsemi MC100EP195) for the processing layers divided into four time phases. As the examples, figure 4 indicates the operating pulses shifted by the step of 240 ps, and figure 5 is the optical output of the compensated writing laser pulses.

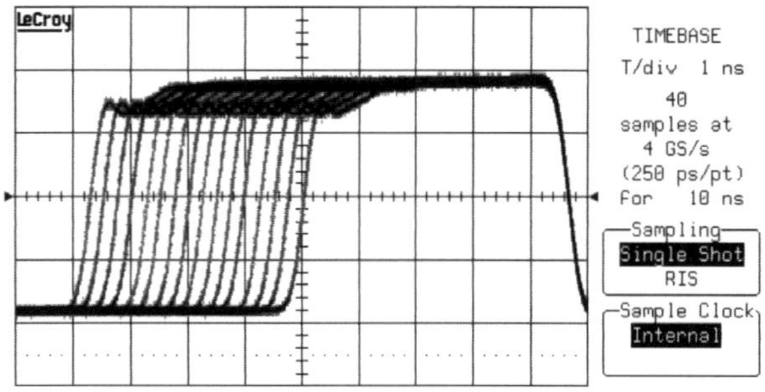

Figure 4. Waveform for overlapped sixteen edges by the step of 240 ps. These internal pulses of the writing laser pulse controller are measured by using the persistence mode of a digital oscilloscope.

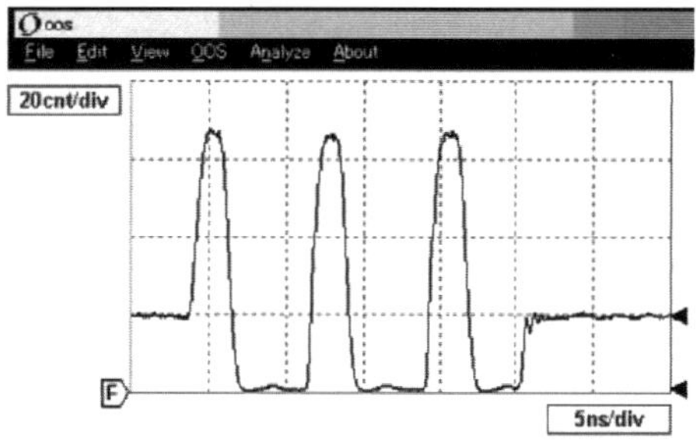

Figure 5. Optical output waveform observed by an optical oscilloscope. The rising time of the laser pulse is obtained less than 1 ns.

3. Experimental results and discussion

In this section, we describe the experimental procedures applying the laser pulse compensation to the phase-change recording films of the triple-layer disc to form the nano-sized marks. Then, we discuss the results and the causes.

3.1. Preparation of the triple-layer disc

Figure 6 shows a schematic image of the layer structure of an experimental triple-layer disc. This structure includes three important designs. First, the GeTe-Sb_2Te_3 materials are adopted as the active recording films for getting high reflectivity contrast, although thickness is very thin. Second, the silver alloy films add for the optical reflection and the thermal radiation. Third, the layer1 and layer2 have high refractive index films on the silver films for obtaining high optical transmittance.

3.2. Experimental procedures and conditions

Figure 7 shows an experimental setup for data recording to the above triple-layer disc. The data is processed at the writing laser pulse controller explained in the previous section. Through the optical pick-up unit with a wavelength of 405 nm and an NA of 0.85, the laser pulses are irradiated to the spinning triple-layer disc. The reading signal measurement was done by the partial-response maximum-likelihood (PRML) detector on the computer [6,7].

The pulse width and the optical output power for writing a 2T mark are in the table 1. The major experimental conditions are listed in the table 2. The data bit length on the disc is 56 nm. By these conditions, the data capacity per one layer is 33.4 Gbyte at the Blu-ray format, so the total capacity of the triple-layer disc becomes 100 Gbyte.

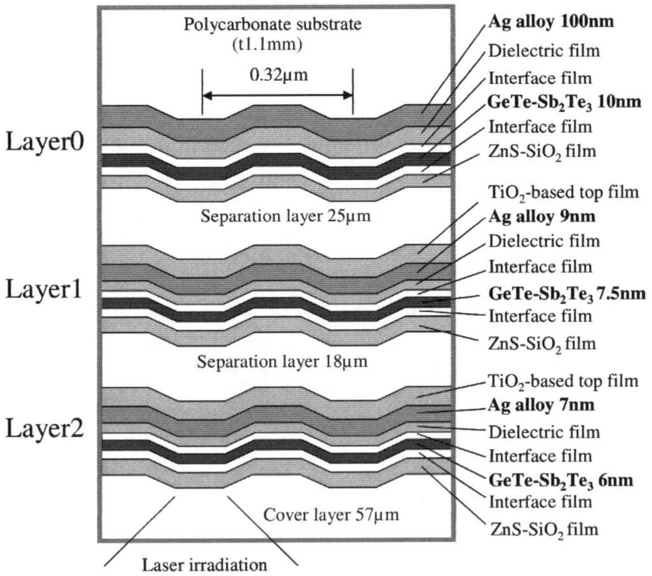

Figure 6. Cross-sectional view of the triple-layer disc. Three recording layers are formed on a polycarbonate substrate. All of the inorganic material films are deposited by the sputtering method. Especially in the layer1 and the layer2, very thin films of Ag and GST material with under 10nm thickness are adopted.

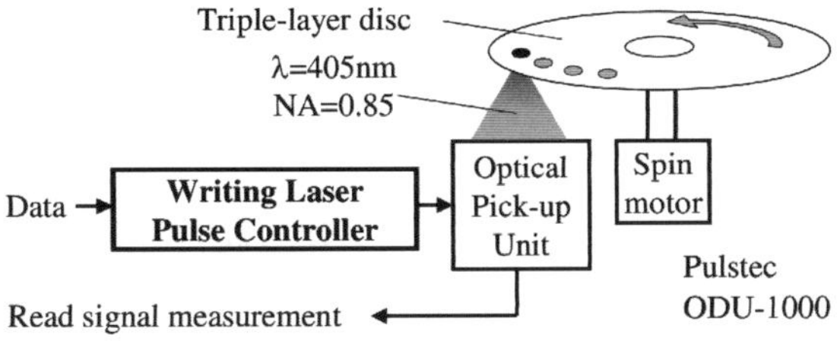

Figure 7. Experimental setup for data recording. An original writing laser pulse controller was added to the Pulstec ODU-1000.

Table 1. Pulse width and the optical output power for writing a 2T mark for each recording layer of the triple-layer disc.

	Layer0	**Layer1**	**Layer2**
Top pulse width	4.26 ns	2.13 ns	2.13 ns
Cooling pulse width	none	4.02 ns	3.79 ns
Write power	25.7 mW	23.5 mW	13.6 mW
Erase power	9.5 mW	6.3 mW	4.2 mW
Bottom power	0.1 mW	0.1 mW	0.1 mW

Table 2. Major experimental conditions for the data recording.

Recording linear velocity	7.38 m/s
Data clock period; T	7.58 ns
Pulse control accuracy; T/32	240 ps
Data bit length on the disc	56 nm
Data capacity per one layer	33.4 Gbyte

3.3. Results

The photograph in figure 8 is a SEM image of recorded marks on the layer2. Amorphous marks could be formed at the minimum size of 112 nm, although the optical recording spot size is large as 250 nm at the half-width. So, we have observed the nano-sized marks smaller than diffraction limit on the phase-change film. There, it was necessary to compensate the pulse edges when the 2Tm was formed on the recording film as the correct mark edges. The characteristic results of the compensation were obtained showing in table 3. There is a dependence on recording data pattern. The compensation value is not zero for the 2Tm between the preceding 2Ts and the succeeding 3Ts or for the 2Tm between the preceding 3Ts and the succeeding 2Ts. And there is a dependence on the three layers, too. The compensation value of the layer2 was larger than the layer0 or layer1.

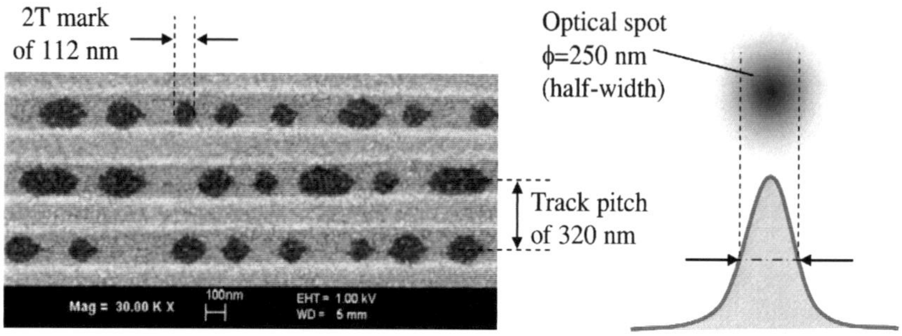

Figure 8. The photograph is a SEM image of recorded marks on the layer2 compared with optical spot size as the schematic. It is observed forming nano-sized marks smaller than diffraction limit.

Table 3. Compensation parameters for forming 2T marks at correct edge positions. The positive value means leading and the negative value means lagging in figure 9.

		Layer0		**Layer1**		**Layer2**	
Compensation values of 2Tm		Succeeding space		Succeeding space		Succeeding space	
		2Ts	3Ts	2Ts	3Ts	2Ts	3Ts
Preceding space	2Ts	0 ps	480 ps	0 ps	480 ps	0 ps	720 ps
	3Ts	-480 ps	0 ps	-480 ps	0 ps	-720 ps	0 ps

3.4. Discussion

Figure 9 represents the relation between the data patterns on the recording film and the writing laser pulses. We estimate a reason of the dependence on the data patterns as follows. In figure 9, (a) is the symmetrical data pattern of the 2Tm-3Ts-2Tm-3Ts-2Tm, (b) is its writing laser pulses, (c) is the asymmetrical data pattern of the 2Tm-3Ts-2Tm-*2Ts*-2Tm just changed succeeding space from the 3Ts to the 2Ts, and (d) is its writing laser pulses. When the center 2Tm and the succeeding 2Tm are formed, the interval of the 2Ts is short enough to cause the optical interference by the overlap of the laser spot and the thermal interference through the recording film. These interferences affect the cooling process of the film, the cooling speed becomes slow. This means easy to crystallize, in other words, insufficient to form amorphous in the phase-change process, therefore, the marks shrink like (c). On the other hand, if the pulse is compensated like (f), the 2Tm can be formed at right edge position like (e). In case of the center 2Tm in the symmetrical pattern as the 2Tm-2Ts-2Tm-2Ts-2Tm, the interference seems to be balanced.

Additionally, we estimate a reason of the dependence on the layers as follows. Each layer has a sliver alloy film shown in figure 6, which is very thin as 7 nm in the layer2 in contrast to 100 nm in the layer0. A thinner film tends to decrease the heat radiation efficiency for the direction of the film depth, on the other hand, to increase the heat spread for the direction of the film plane. Therefore, wide thermal interference occurs in layer2 compared with layer0, the compensation value of the layer2 becomes large as in table 3. Furthermore, there is no cooling pulse for only layer0, although it is necessary for the layer1 and layer2 as in table 1. This reason is explained by high cooling capability of the heavy silver alloy film in the layer0.

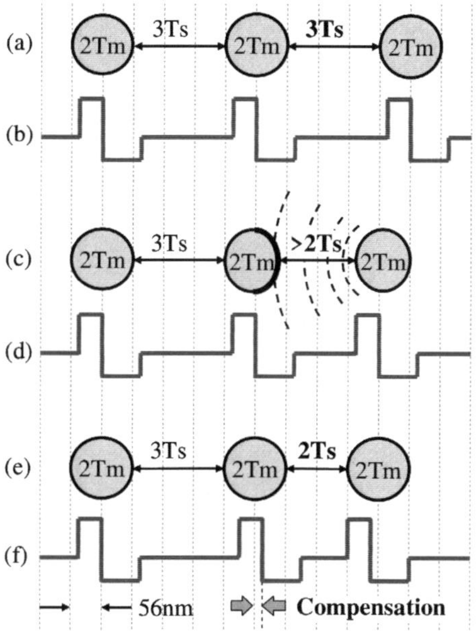

Figure 9. Relations between the data patterns on the recording film and the writing laser pulses. (a) is the data pattern of the 2Tm-3Ts-2Tm-3Ts-2Tm, (b) is its writing laser pulses, (c) is the data pattern of the 2Tm-3Ts-2Tm-*2Ts*-2Tm just changed succeeding space from the 3Ts to the 2Ts, which is formed by the pulses (d) without the compensation. They are shrinking because of the optical and thermal interference. Some arcs by the broken line show the interference schematically. On the other hand, the marks can be formed at right edge position (e) by the pulses (f) with the compensation.

4. Conclusion

New writing laser pulse control, which can compensate so as to form the correct mark edges on the thin recording film, was applied to the phase-change triple-layer disc. It has enabled to form 112 nm-sized marks smaller than the optical spot. The characteristic values of the compensation can be explained as the results of cancelling the optical and thermal interference in the phase-change process of forming the marks by considering about the spot size of the writing laser and the structure of the recording films.

References

[1] Yamada N, Ohno E, Akahira N, Nishiuchi K, Nagata K, and Takao M: Jpn. J. Appl. Phys. Suppl. **26**-4 (1987) 61.

[2] ISO / IEC 16824: Information technology-120 mm DVD-Rewritable disk (DVD-RAM) (1999).

[3] Blu-ray Disc Association: White paper of physical format specifications for BD-RE 3rd edition (2010).

[4] Nishihara T, Tsuchino A, Tomekawa Y, Kusada H, Kojima R, Yamada N: Jpn. J. Appl. Phys. **50** (2011) 062503.

[5] Nakamura A, Kobayashi I, Narumi K, Takaoka T, Furumiya S, and Miyagawa N: Jpn. J. Appl. Phys. **49** (2010) 08KG01.

[6] Miyashita H, Nakajima T, Kimura N, Hino Y, and Ishibashi H: Jpn. J. Appl. Phys. **43** (2004) 4850.

[7] Shiraishi J, Kobayashi S, Miyashita H, and Hino Y: Tech. Dig. ISOM2009, **Tu-F-08** (2009) 74.

Terahertz radiation from the coherent longitudinal optical phonon-plasmon coupled mode in an *i*-GaAs/*n*-GaAs epitaxial structure

Shuichi Tsuruta[1] , Hideo Takeuchi[2] and Masaaki Nakayama[1*]

[1]Department of Applied Physics, Graduate School of Engineering, Osaka City University, Sugimoto, Sumiyoshi-ku, Osaka 558-8585, Japan

[2]Department of Electronic Systems Engineering, School of Engineering, The University of Shiga Prefecture, 2500 Hassaka-cho, Hikone, Shiga, 522-8533, Japan

E-mail: nakayama@a-phys.eng.osaka-cu.ac.jp

Abstract. We have investigated the time-domain terahertz radiation from an undoped GaAs/*n*-type GaAs (*i*-GaAs/*n*-GaAs) epitaxial structure using an optical gating method with a femtosecond pulse laser. We have found from the Fourier power spectra of the terahertz waveforms that two terahertz bands, the frequencies of which obviously depend on the pump power, appear in a high density excitation regime in addition to a broad band due to a surge current and a sharp band due to the coherent longitudinal optical (LO) phonon. Based on a model for the LO phonon-plasmon coupled (LOPC) mode, we reasonably explain the pump-power (photogenerated-carrier-density) dependence of the frequencies of the two bands peculiar to the high density excitation. This fact demonstrates that the two terahertz bands are assigned to the lower and upper branches of the LOPC mode. The frequencies of the LOPC mode are determined only by the pump power. Thus, the LOPC mode originates from the *i*-GaAs layer; namely, the plasmon component of the surge current flowing through the *i*-GaAs layer couples with the coherent LO phonon.

1. Introduction

The generation of terahertz electromagnetic waves has attracted much attention from aspects of physics of ultrafast phenomena and applications in terahertz spectroscopy, sensing, and imaging [1]. There are two well-known generation mechanisms of terahertz waves from a semiconductor surface with use of illumination of a femtosecond pulse laser. One is a surface surge current due to instantaneously generated carriers, which produces a broad terahertz spectrum [2-4]. The other is a coherent optical phonon, which shows a sharp band [5-9]. Here, we emphasize that the surge current is a flow of the collection of photogenerated carriers. Consequently, the collective motion of the carriers, the so-called plasmon, should coexist in the surge current. Thus, it is expected that the plasmon component of the surge current couples with the coherent longitudinal optical (LO) phonon: the formation of the LO phonon-plasmon coupled (LOPC) mode. However, little has been known about the terahertz radiation from the LOPC mode.

[*] To whom any correspondence should be addressed.

In the present work, using time-domain terahertz spectroscopy, we have investigated the LOPC mode in an undoped GaAs/*n*-type GaAs (*i*-GaAs/*n*-GaAs) epitaxial structure. The frequency spectrum was obtained from the Fourier transform (FT) of the time-domain signal. It is found that two terahertz bands appear in a high density excitation regime. The frequencies of the two bands obviously depend on the pump power. The pump-power (photogenerated-carrier-density) dependence of the frequencies of the two bands is reasonably analyzed by a model for the LOPC mode, and the two bands are assigned to the lower and upper branches of the LOPC mode. This fact demonstrates that the LOPC mode is instantaneously formed in the *i*-GaAs layer; namely, the surge current flowing through the *i*-GaAs layer has a plasmon component that couples with the coherent LO phonon. Here, we note that the present result is quite different from the earlier works by Kersting *et al.* [10] and by Hasselbeck *et al* [11]. The earlier works reported on the LOPC mode in *n*-type GaAs and InAs crystals using terahertz spectroscopy. However, it was concluded that the frequency of the LOPC mode is determined only by a background carrier concentration originating from dopants. Thus, the present results of the terahertz radiation from the LOPC mode are peculiar to the *i*-GaAs/*n*-GaAs epitaxial structure.

2. Experimental

The sample was an *i*-GaAs(200 nm)/*n*-GaAs(3 μm and 3×10^{18} cm^{-3}) epitaxial structure grown on a semi-insulating (001) GaAs substrate by metal organic vapor phase epitaxy, where the values in the parentheses denote the individual layer thickness and doping concentration. According to Ref. [12], the surface Fermi level pinning produces a built-in electric field, which was theoretically estimated to be 35 kV/cm, in the *i*-GaAs layer. Thus, the *i*-GaAs layer was completely depleted, and photogenerated carriers were efficiently swept out from the *i*-GaAs layer by the built-in electric field, which is advantageous to the flow of the surge current.

The signals of the time-domain terahertz waves were measured using laser pulses with a duration time of ~60 fs. The repetition rate of the laser pulse was ~90 MHz. The pump beam was focused on the sample with an incidence angle of 45°. The diameter of the spot on the sample surface was about 120 μm. The photon energies of the pump and gate beams were the same: 1.57 eV. The absorption

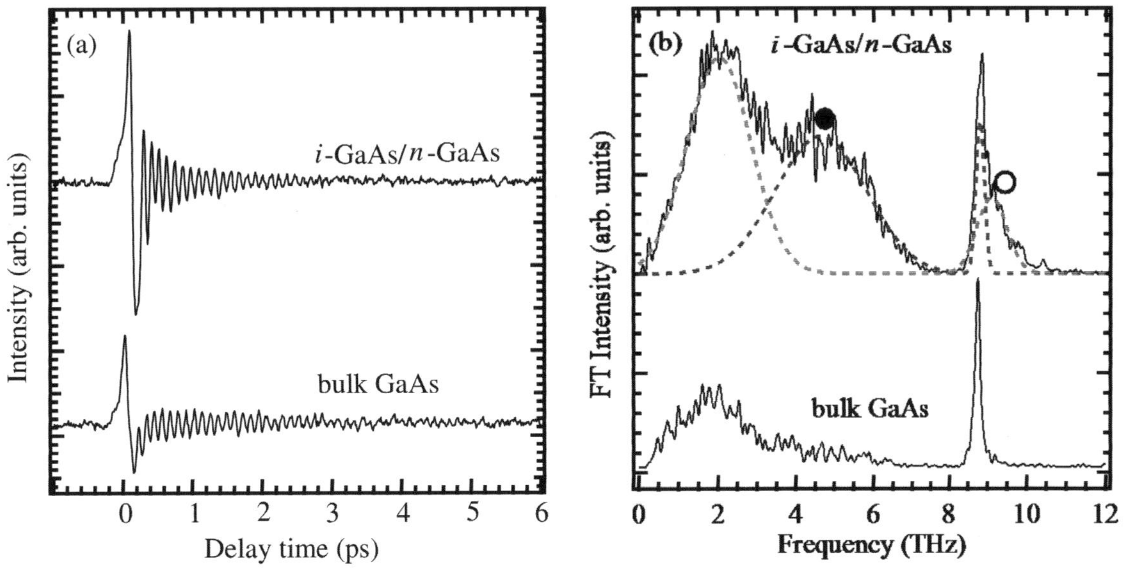

Figure 1. (a) Terahertz waveforms at the pump power of 100 mW, which is the highest power in the present work, in the *i*-GaAs/*n*-GaAs sample and a bulk (001) GaAs crystal as a reference. (b) Fourier power spectra of the terahertz waveforms.

coefficient and refractive index for GaAs of the present pump beam are 1.4×10^4 cm^{-1} and 3.69, respectively [13]. Taking account of these values, for example, in the illumination of the pump beam with 50 mW, the density of the photogenerated carriers is estimated to be 1.2×10^{17} cm^{-3}. The emitted terahertz wave was detected by an optically gated dipole antenna with a gap of 6 μm formed on a low-temperature-grown GaAs layer. The power of the gate beam was fixed to 10 mW. All the measurements were performed at room temperature. The humidity was controlled to be 5% during the measurement under a nitrogen-gas-purge condition.

3. Results and discussion

Figure 1(a) shows the terahertz waveforms at the pump power of 100 mW, which is the highest power in the present work, in the *i*-GaAs/*n*-GaAs sample and a bulk (001) semi-insulating GaAs crystal as a reference. The monocycle signal, which is attributed to the surge current component, is observed around the delay time of 0 ps. Furthermore, the terahertz waveforms exhibit oscillatory patterns. In order to clarify the components of the terahertz waveform, we treat the Fourier power spectrum in frequency domain, which is shown in Fig. 1(b). It is obvious that the profile of the Fourier power spectrum in the *i*-GaAs/*n*-GaAs sample is much different from that in the bulk GaAs crystal. The Fourier power spectrum in the bulk GaAs crystal consists of two bands. One is a broad band peaking at ~2 THz due to the monocycle signal. The other is a sharp band at 8.8 THz due to the coherent LO phonon because 8.8 THz just agrees with the frequency of the GaAs LO phonon. This is an ordinary Fourier power spectrum of the terahertz waveform emitted from a semiconductor surface. In the *i*-GaAs/*n*-GaAs sample, the Fourier power spectrum exhibits a complicated profile. We phenomenologically decompose the Fourier power spectrum with four Gaussian functions. The dashed curves in Fig. 1(b) show the fitted results, which reasonably explain the spectrum profile. The two bands indicated by the solid and open circles newly appear in addition to the two bands that are observed in the bulk GaAs crystal. Thus, the newly appearing two bands are peculiar to the *i*-GaAs/*n*-GaAs sample and to the high density excitation condition.

In order to reveal the origin of the two bands, we focus on the pump-power dependence of the Fourier power spectrum of the terahertz waveform in the *i*-GaAs/*n*-GaAs sample, which is shown in Fig. 2. The broad band shows an asymmetric shape with a tail extending to the high frequency side at the pump powers higher than 20 mW, whereas the peak frequency is almost fixed at about 2 THz. At the pump powers of 70 and 100 mW, the broad band clearly splits into two components. The higher

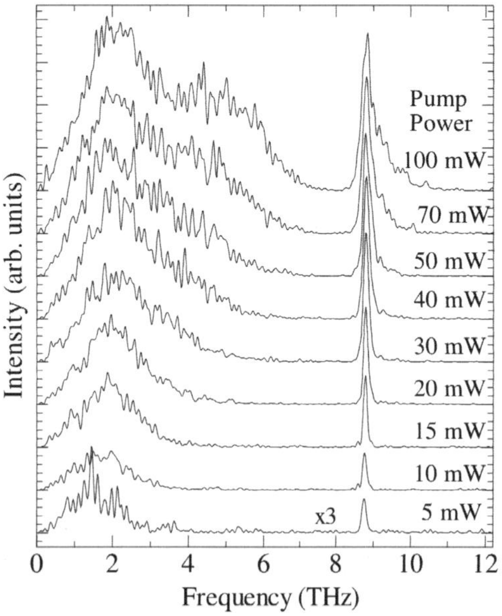

Figure 2. Pump-power dependence of the Fourier power spectrum of the terahertz waveform in the *i*-GaAs/*n*-GaAs sample.

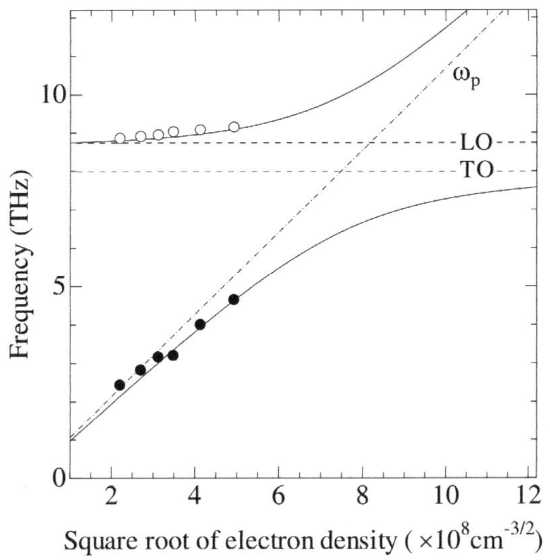

Figure 3. Calculated dispersion relations (solid curves) of the lower and upper branches of the LOPC mode using Eq.(1) and the observed frequencies (solid and open circles) as a function of square root of electron density.

frequency component corresponds to the terahertz band indicated by the solid circle in Fig. 1(b). In addition, the band of the coherent GaAs LO phonon has an asymmetric tail toward the higher frequency side at the pump powers higher than 20 mW. The asymmetry becomes remarkable with an increase in pump power. The pump-power dependent phenomena described above suggest that photogenerated carriers modify the terahertz waveform; namely, the formation of the LOPC mode is expected.

The frequencies of the lower and upper branches of the LOPC mode, ω_{L-} and ω_{L+}, are given by [14]

$$\omega_{L\pm}^2 = \frac{1}{2}\left[\left(\omega_p^2 + \omega_{LO}^2\right) \pm \sqrt{\left(\omega_p^2 + \omega_{LO}^2\right)^2 - 4\omega_p^2\omega_{TO}^2}\right], \tag{1}$$

where ω_{LO} and ω_{TO} are the LO and transverse optical (TO) phonon frequencies, respectively. The plasmon frequency is ω_p written by

$$\omega_p = \sqrt{\frac{ne^2}{\varepsilon_\infty m^*}}, \tag{2}$$

where n, m^*, and ε_∞ are the electron density, effective mass, and dielectric constant, respectively. We note that the plasmon of holes does not produce the well-defined lower and upper branches of the LOPC mode described by Eq.(1) because of large damping of the hole plasmon mainly due to the low mobility [15]. Thus, we neglect the contribution of the hole plasmon in this case. In the calculation, the values of the parameters are taken from those of the bulk GaAs crystal [16]. Figure 3 shows the peak frequencies of the two pump-power dependent terahertz bands (solid and open circles), which are observed at pump powers from 20 to 100 mW in Fig. 2, as a function of square root of electron density, where four Gaussian functions were used to resolve the Fourier power spectrum as shown in Fig. 1(b). The electron density was evaluated from the pump power as described above. The solid curves in Fig. 3 depict the calculated dispersion relations of the lower and upper branches of the LOPC mode using Eq.(1). From Fig. 3, it is evident that we detect the terahertz radiation from the LOPC mode. Furthermore, the present result indicates that the peak frequency of the observed LOPC mode is determined only by the photogenerated carrier density. Thus, we conclude that the LOPC mode is

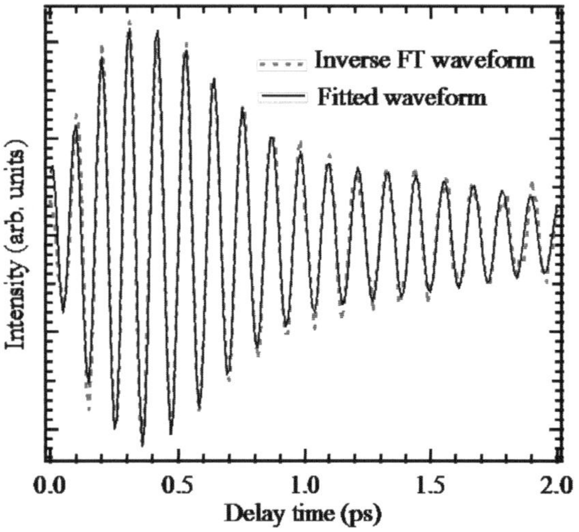

Figure 4. Inverse FT terahertz waveform (dashed curve) at the pump power of 100 mW and the fitted result (solid curve) using Eq. (3).

formed in the *i*-GaAs layer; namely, the surge current through the *i*-GaAs layer involves the plasmon component that couples with the coherent LO phonon.

Finally, we discuss the dynamics of the LOPC mode from the viewpoint of the terahertz waveform. In order to eliminate the monocycle signal in the terahertz waveform, we performed the inverse FT of the FT spectrum in the frequency region from 8 to 12 THz at the pump power of 100 mW. The dashed curve in Fig. 4 shows the inverse FT terahertz waveform. We use the following equation, which consists of the oscillation terms of the coherent LO phonon and the upper branch of the LOPC mode, to fit the terahertz waveform:

$$A(t) = A_{\mathrm{LO}} \exp(-t/\tau_{\mathrm{LO}}) \cos(\omega_{\mathrm{LO}} + \phi_{\mathrm{LO}}) + A_{\mathrm{L+}} \exp(-t/\tau_{\mathrm{L+}}) \cos(\omega_{\mathrm{L+}} + \phi_{\mathrm{L+}}) , \tag{3}$$

where A, τ, and ϕ are the amplitude, lifetime, and initial phase, respectively. The fitted terahertz waveform using Eq. (3) is depicted by the solid curve in Fig. 4. In the fitting process, we fix the values of the initial phases (ϕ_{LO} and $\phi_{\mathrm{L+}}$) to zero, assuming the generation mechanism that is instantaneous screening of the surface electric field [17], which is the major generation mechanism of the coherent LO phonon in polar semiconductors such as GaAs. It is evident that the terahertz waveform is well reproduced by the fitting. The ratio of A_{LO} to $A_{\mathrm{L+}}$ is estimated to be 1:1.3. The important parameters are the lifetimes of the LOPC mode and coherent LO phonon. The evaluated values of $\tau_{\mathrm{L+}}$ and τ_{LO} are 0.3 and 1.1 ps, respectively. The ultrashort lifetime of the LOPC mode indicates that the photogenerated carriers including the plasmon component in the *i*-GaAs layer are very efficiently swept out by the built-in electric field. In contrast, the coherent LO phonon continues to oscillate after the disappearance of the LOPC mode. This suggests that the LOPC mode and coherent LO phonon are spatially separated.

4. Conclusions

We have detected the time-domain terahertz radiation from the LOPC mode in the *i*-GaAs (200 nm)/*n*-GaAs (3 μm and 3×10^{18} cm^{-3}) epitaxial structure using the optical gating method. The observed frequencies of the lower and upper branches of the LOPC mode are reasonably explained by the calculated dispersion relations of the LOPC mode. In addition, the frequencies of the LOPC mode depend only on the pump power (photogenerated carrier density). These facts demonstrate that the LOPC mode is instantaneously formed in the *i*-GaAs layer; namely, the surge current flowing through the *i*-GaAs layer has a plasmon component that couples with the coherent LO phonon. Furthermore, the analysis of the terahertz waveform indicates that the lifetime of the LOPC mode is ultrashort: 0.3

ps. Such an ultrashort lifetime results from the fact that the sweep-out of photogenerated carriers in the *i*-GaAs layer is ultrafast due to the built-in electric field peculiar to the *i*-GaAs/*n*-GaAs structure.

Acknowledgement
One of the authors, H.T., acknowledges Grant-in-Aid for Young Scientists B (No. 22760010) from the Japan Society for the Promotion of Science.

References

[1] For a review, *Terahertz Optoelectronics*, edited by Sakai K 2005 (Springer, Berlin)
[2] Zhang X -C and Auston D H 1992 *J. Appl. Phys.* **71** 326
[3] Sarukura N, Ohtake H, Izumida S and Liu Z 1998 *J. Appl. Phys.* **84** 654
[4] Gu P, Tani M, Kono S, Sakai K and Zhang X -C 2002 *J. Appl. Phys.* **91** 5533
[5] Kuznetsov A V and Stanton C J 1995 *Phys. Rev. B* **51** 7555
[6] Dekorsy T, Auer H, Waschke C, Bakker H J, Roskos H G, Kurz H, Wagner V and Grosse P 1995 *Phys. Rev. Lett.* **74** 738
[7] Tani M, Fukasawa R, Abe H, Matsuura S, Sakai K and Nakashima S 1998 *J. Appl. Phys.* **83** 2473
[8] Mizoguchi K, Furuichi T, Kojima O, Nakayama M, Saito S, Syouji A and Sakai K 2005 *Appl. Phys. Lett.* **87** 093102
[9] Nakayama M, Ito S, Mizoguchi K, Saito S and Sakai K 2008 *Appl. Phys. Express* **1** 012004
[10] Kersting R, Heyman J N, Strasser G and Unterrainer K 1998 *Phys. Rev. B* **58** 4553
[11] Hasselbeck M P, Stalnaker D, Schlie L A, Rotter T J, Stinz A and Sheik-Bahae M, 2002 *Phys Rev. B* **65** 233203
[12] Takeuchi H, Yanagisawa J, Hasegawa T and Nakayama M, 2008 *Appl. Phys. Lett.* **93** 081916
[13] Aspnes D E and Studna A A 1983 *Phys. Rev. B* **27** 985
[14] Yokota I 1961 *J. Phys. Soc. Jpn.* **16** 2075
[15] Fukasawa R and Perkowitz S 1994 *Phys. Rev. B* **50** 14119
[16] Adachi S 1985 *J. Appl. Phys.* **58** R1
[17] Pfeifer T, Dekorsy T, Kütt W and Kurz H, 1992 *Appl. Phys. A* **55** 482

Journal of Physics: Conference Series **417** (2013) 012052 doi:10.1088/1742-6596/417/1/012052

Study on the Growth of Microcrystalline Silicon Films in Atmospheric-Pressure VHF Plasma Using Porous Carbon Electrode

H. Kakiuchi, H. Ohmi, T. Yamada, A. Hirano, T. Tsushima, and K. Yasutake

Department of Precision Science and Technology, Graduate School of Engineering, Osaka University, 2–1 Yamada-Oka, Suita, Osaka 565–0871, Japan

E-mail: kakiuchi@prec.eng.osaka-u.ac.jp

Abstract. Using atmospheric pressure (AP) plasma excited by a 150-MHz very high-frequency (VHF) power, the growth process of hydrogenated microcrystalline silicon (*μc*-Si:H) films was studied. A porous carbon electrode, which had originally been developed for the high-quality Si epitaxy at low temperatures, was used for the plasma generation in the area of 30×100 mm^2. VHF power density (P_{VHF}) and H_2/SiH_4 ratio were varied as parameters under a fixed substrate temperature of 220 °C. It was shown that *μc*-Si:H films having good uniformities of thickness and crystallinity in the area of 20×80 mm^2 were obtained without being contaminated by dusty particles. Although increasing H_2/SiH_4 ratio led to the improvement of both deposition rate and film crystallinity, a crystallized Si film with a Raman crystalline volume fraction of ~72% was obtained with a high deposition rate of 5 nm s^{-1} at P_{VHF} = 24 W cm^{-2}, even if hydrogen was not added to the process gas mixture (H_2/SiH_4 = 0). Numerical simulations of the gas flow revealed that the process gas mixture was uniformly injected into the narrow plasma gap region from the opposite side of the substrate surface and suggested that the gas residence time in the plasma at the central part of the plasma region was long enough (>5 ms) to sufficiently deplete the source SiH_4, promoting the crystallization of the growing film without the necessity of H_2 addition.

1. Introduction

Low-temperature hydrogenated microcrystalline silicon (*μc*-Si:H) is a promising material for the use in large-area electronic devices. In particular, *μc*-Si:H films deposited on plastic sheet substrates can realize flexible and low-cost thin film transistor-based devices and solar cells. Thus, there is a strong demand for the high-rate deposition of good-quality *μc*-Si:H films at low temperatures. Generally, *μc*-Si:H films are prepared by plasma-enhanced chemical vapor deposition (PECVD) using SiH_4 and H_2 as the source gases. In conventional radio-frequency PECVD processes, high H_2/SiH_4 ratio conditions are utilized to promote the crystallization of Si films at low temperatures. However, the depositions at substrate temperatures lower than 100 °C result in the decrease in deposition rate and the deterioration of film quality as far as conventional PECVD processes are used. This limits the practical application of those processes as the key tools for the deposition of good-quality *μc*-Si:H with high rates on plastic materials, such as polyethylene terephthalate and polyethylene naphthalate (PEN). Probable methods to overcome this technical issue are the use of very high-frequency (VHF) excitation of plasma and/or high working pressure [1–3].

We have been developing a thin film growth technology using atmospheric-pressure (AP) plasma excited by a 150-MHz VHF power. Note that the VHF excitation of plasma and the very high collision frequency between gas-phase species at AP lead to the trapping of charged particles in a narrow (≤1 mm) gap region between electrodes and also to the small kinetic energies of ions [4]. This can avoid the ion damage of the film-growing surface, while keeping high power density in the plasma, which enables us to deposit Si films with high rate and at low temperatures [4–9].

The aim of the present study is to develop an efficient formation process of good-quality functional thin films on polymers using the AP plasma technology. By using a porous carbon electrode that has originally been developed for the high-quality Si epitaxy at low temperatures [7], the formation of μc-Si:H on a PEN sheet of 0.125 mm thickness has been achieved with a deposition rate of 2.5 nm s^{-1} [9].

In this paper, we discuss the growth process of μc-Si:H in AP-VHF plasma excited using a porous carbon electrode, focusing especially on the origin of the uniformities of thickness and crystallinity of the μc-Si:H films in the plasma region.

2. Experiments

The experiments were conducted in AP plasma CVD system, which is the same one as that used in our previous studies [8,9]. Figure 1 shows a schematic diagram of a porous carbon electrode used for the plasma generation. By using this type of electrode, fresh process gas mixtures can be directly supplied into the narrow plasma gap (0.5 mm) through the porous carbon plate of 3 mm thickness [7]. The size of electrode is 30×100 mm^2. By supplying a 150 MHz VHF power, AP plasma was excited stably in the gap region between the electrode and a substrate as shown in Figure 1. The substrate was fixed on a TiN-coated copper susceptor of the substrate heating stage by a vacuum chuck system that ensured a sufficiently high efficiency of heat transfer across the substrate/susceptor interface.

In a series of experiments, Corning EAGLE2000 glass plates of 0.7 mm thickness (100×100 mm^2) were used as substrates. The films were prepared with a He flow rate of 50 SLM (SLM denotes liter per minute at STP) and a SiH_4 flow rate of 50 SCCM (SCCM denotes cubic centimeter per minute at STP) under a constant process pressure of 1×10^5 Pa. The H_2 flow rate and VHF power density (P_{VHF}) were varied as parameters. The temperature of the back surface of substrate (T_{sub}) was monitored by a Chromel-Alumel thermocouple embedded in the position 3 mm apart from the susceptor surface and kept at 220 °C during deposition.

The film thickness was measured by a step profiler. The film structure was characterized by Raman scattering spectroscopy. Raman spectra were recorded with a 514.5-nm Ar^+ laser radiation in the back-scattering configuration using a Raman spectrometer, which comprised a double monochrometer with a resolution of 1 cm^{-1}, a standard cooled photomultiplier and a photon counting system.

To discuss the growth process of Si in AP-VHF plasma, the gas flow around the plasma region was simulated by solving coupled sets of partial differential equations governing the mass, momentum, and energy transports, where the transport parameters, such as viscosity and thermal conductivity, were derived from gas kinetic considerations. The simulation model representing the actual experimental setup and conditions was calculated using the commercially available simulator PHOENICS-CVD that allowed three-dimensional visualizations of the gas flow and temperature profiles.

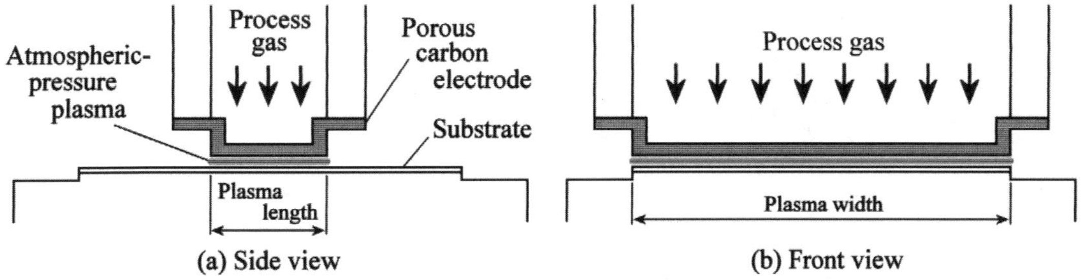

Figure 1. Schematic diagram of the experimental setup.

3. Results and discussion

3.1. Deposition characteristics of Si films

Figure 2 shows the photographs of the Si films under a fixed H_2 flow rate of 100 SCCM ($H_2/SiH_4 = 2$) as a function of P_{VHF}. It is clearly observed in Figure 2 that increasing P_{VHF} results in the generation of dusts at the film edges downstream of the gas flow and also appears that dusty particles are not formed in the plasma region. Indeed, the observation by a scanning electron microscope revealed that no dusts were seen on the film surface. This indicates that the film-forming precursors that do not contribute to the film growth immediately condense into Si powders after being swept out from the plasma. These are consistent with the results obtained in our previous studies on the low-temperature Si epitaxy [6,7], suggesting that the VHF excitation of AP plasma is effective for the film-forming precursors to not induce the gas-phase condensation, although the possibility of the formation of Si nano clusters must be considered. As a result, by sucking the gas flow from the electrode edge zone, sticking of dusts on the substrate surface can be avoided as demonstrated in Figure 2e, which enables us to realize dust-free and continuous depositions by scanning substrate. However, all the Si films shown in this paper have been prepared without substrate scanning.

Figure 3 shows the thickness profiles of the Si film after 4 min deposition with 100 SCCM H_2 at $P_{VHF} = 24$ W cm^{-2}. The position at 0 mm corresponds to the center of the deposition area. It has been confirmed that, in this condition, the film has extremely thin or even negligible amorphous incubation layer [8,9]. The profiles in Figure 3 indicate that the film thickness is almost uniform at around 1.5 μm in the area of ~20×80 mm^2. The distribution of crystallinity of the same film has been also evaluated. The Raman spectra are shown in Figure 4. Since no noticeable differences are observed in the spectra, it can be noted that the film crystallinity has uniformity similar to the film thickness. These suggest the almost uniform distribution of reactive species in the area of 20×80 mm^2, which is discussed later.

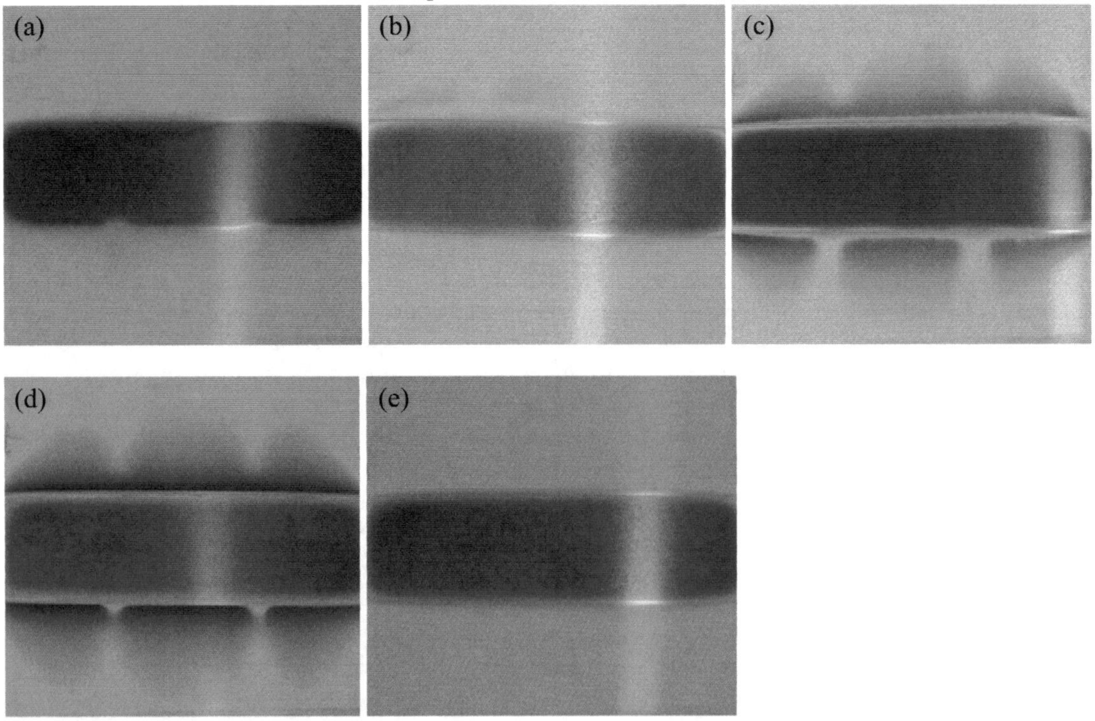

Figure 2. Photographs of the Si films under the room light. The films were deposited with P_{VHF} = (a) 6, (b) 12, (c) 18, and (d,e) 24 W cm^{-2} under a fixed H_2 flow rate of 100 SCCM ($H_2/SiH_4 = 2$). The film thickness is (a) 1000, (b) 640, (c) 780, and (d,e) 770 nm. The sample (e) was prepared by sucking the gas flow from the electrode edge zone to remove the dusty particles.

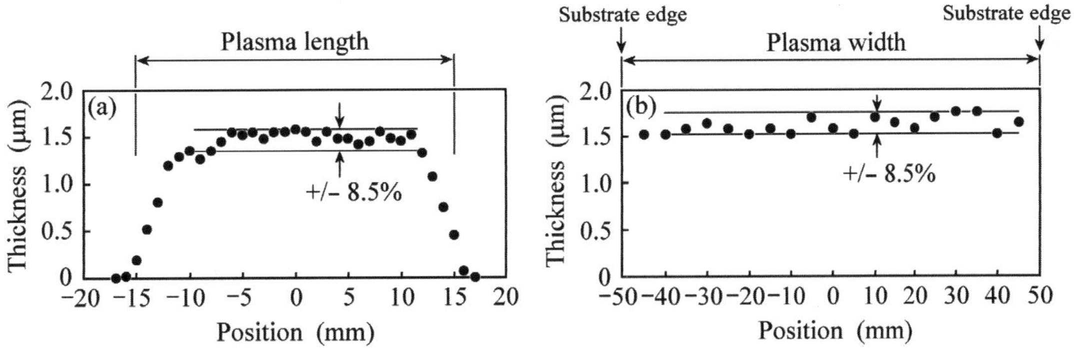

Figure 3. Thickness profiles of the Si films in the (a) vertical and (b) transverse directions after 4 min deposition with P_{VHF} of 24 W cm^{-2} and H_2 flow rate of 100 SCCM. The plasma width in the transverse direction is equal to the substrate size.

3.2. H_2/SiH_4 ratio dependences of deposition rate and crystallinity

The Raman spectra for the μc-Si:H films prepared in this study had a sharp peak at around 520 cm^{-1} with a shoulder on the lower wavenumber side as shown in Figure 4, which could be deconvoluted by considering the presence of two additional Gaussian peaks in the vicinity of 510 and 480 cm^{-1}. Raman crystalline volume fraction, I_C^{RS}, was calculated as $I_C^{RS} = (I_{520}+I_{510}) / (I_{520}+I_{510}+\gamma I_{480})$, where I_{520}, I_{510} and I_{480} are the integrated intensities of the peaks. γ is the ratio of the back scattering cross sections and has a value of 0.8 [8,9].

Figure 5 shows the H_2/SiH_4 ratio dependences of deposition rate and I_C^{RS} of the Si films deposited at P_{VHF} = 12 and 24 W cm^{-2}. Both the deposition rate and I_C^{RS} increase with increasing H_2/SiH_4 ratio, indicating that the chemical reactions in gas phase and on the film-growing surface are more enhanced at the higher H_2/SiH_4 ratio. Increasing P_{VHF} gives rise to the further enhancement of the film-forming

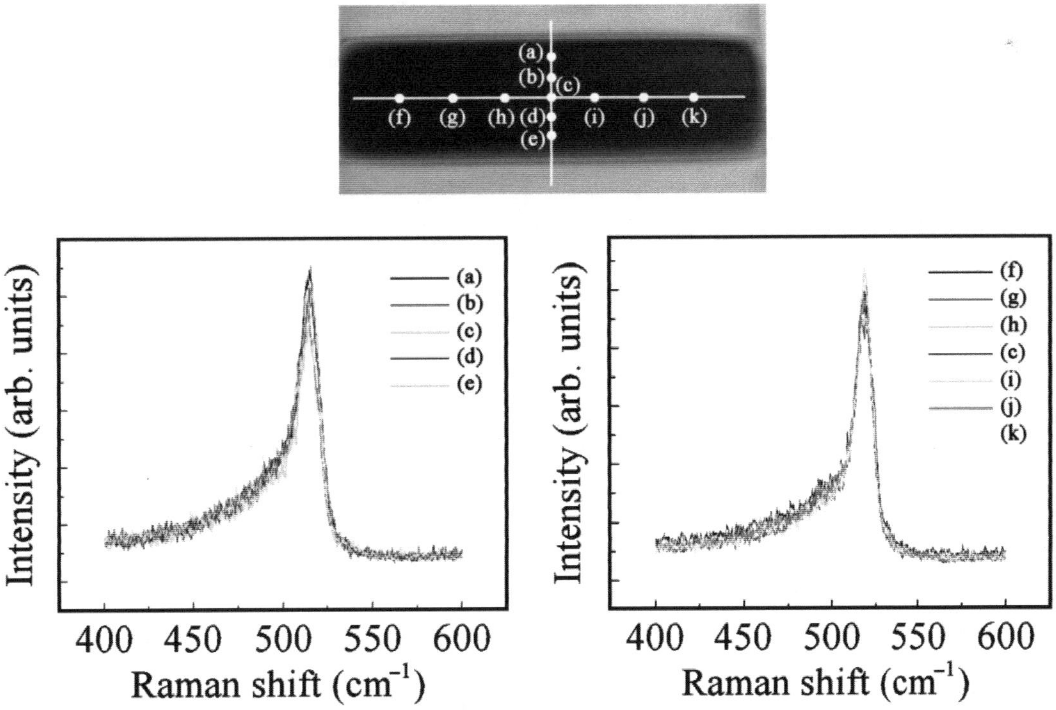

Figure 4. Raman spectra for the Si films deposited in the same condition as that in Figure 3 as a function of position in the vertical and transverse directions.

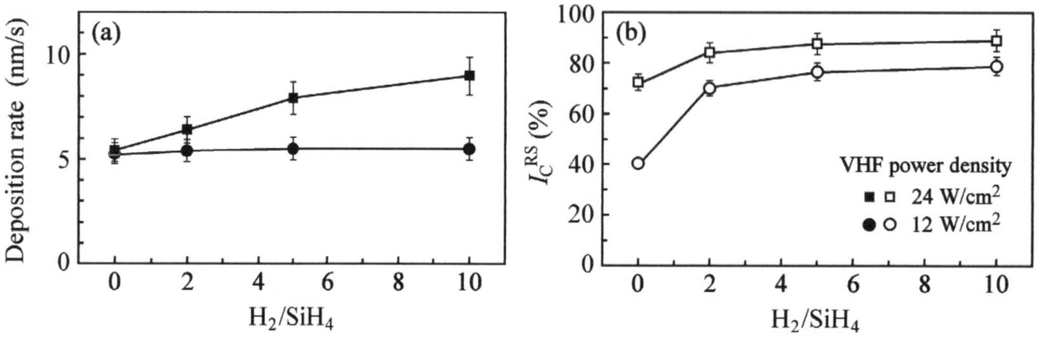

Figure 5. H_2/SiH_4 ratio dependences of (a) deposition rate and (b) I_C^{RS} of the Si films.

reactions. Note that crystallized films are obtained even if H_2 is not added to the process gas mixtures ($H_2/SiH_4 = 0$). This suggests the saturation of the film-forming reactions both in gas phase and on the growing film surface and is considered to be linked to the uniformities of thickness and crystallinity of the resultant films described in the former section.

3.3. Simulations of the gas flow around the plasma region

As described above, a crystallized Si film with good uniformities of film thickness and crystallinity in the plasma region are obtained with a high rate by using a porous carbon electrode even at $H_2/SiH_4 = 0$. Such a unique feature is considered to be closely related to the gas flow pattern in the plasma, because the process of both the gas-phase and surface reactions is strongly affected by the gas residence time in the plasma. Thus, we performed numerical simulations of gas flow around the plasma region. In the simulations, AP-VHF plasma was simply treated as a heat source radiating approximately 50% of the input VHF power as heat, which was determined based on the experimental data [9].

Figure 6 illustrates the contours of pressure and flow velocity on a vertical midplane through the plasma region. The flow vectors are also shown by arrows in the figure. It is observed that the pressure inside the electrode box is higher than the outside and that the pressure drop is generated in the inside of the porous carbon plate. As a result, the process gas is introduced uniformly into the plasma region from the opposite side of the substrate surface, as shown by the flow vectors. In the plasma region, the velocity in the horizontal direction is nearly zero at the center of the electrode and increases toward the electrode edge. Thus, the gas residence time in the plasma varies depending on the horizontal position. At the central part of the plasma area, the gas residence time is estimated to be longer than 5 ms.

On the basis of these results, the reaction process in the plasma can be discussed as follows. Taking into account the gas residence time (>5 ms) in the central area of the plasma region, the source SiH_4 molecules injected into the plasma have enough time to reach the substrate surface by diffusion. In such circumstances, atomic hydrogen generated by the decomposition of the source gas molecules can promote further decomposition of the available SiH_4 molecules through the repeated interactions in the plasma. As a result of such enhanced interactions between atomic hydrogen and SiH_4 molecules, the source SiH_4 might be depleted, ensuring a large flux of atomic hydrogen onto the substrate surface. By virtue of the frequent interactions of abundant atomic hydrogen with the film-growing surface together with the moderate surface heating by the plasma, the formation of crystalline Si network is greatly enhanced even at the high deposition rates. In this context, it is reasonable that the uniform supply of the source SiH_4 from the electrode leads to the film growth with uniform thickness and crystallinity.

4. Conclusion

We have studied the growth process of μc-Si:H prepared with very high rates (>5 nm s^{-1}) in AP-VHF plasma excited by a porous carbon electrode of 30×100 mm^2 in size at a substrate temperature of 220 °C. It is demonstrated that μc-Si:H films showing good uniformities of thickness and crystallinity are obtained in the area of 20×80 mm^2 and that a μc-Si:H film having a crystalline volume fraction of 72% can be deposited even if hydrogen is not added to the process gas mixture ($H_2/SiH_4 = 0$). We conclude

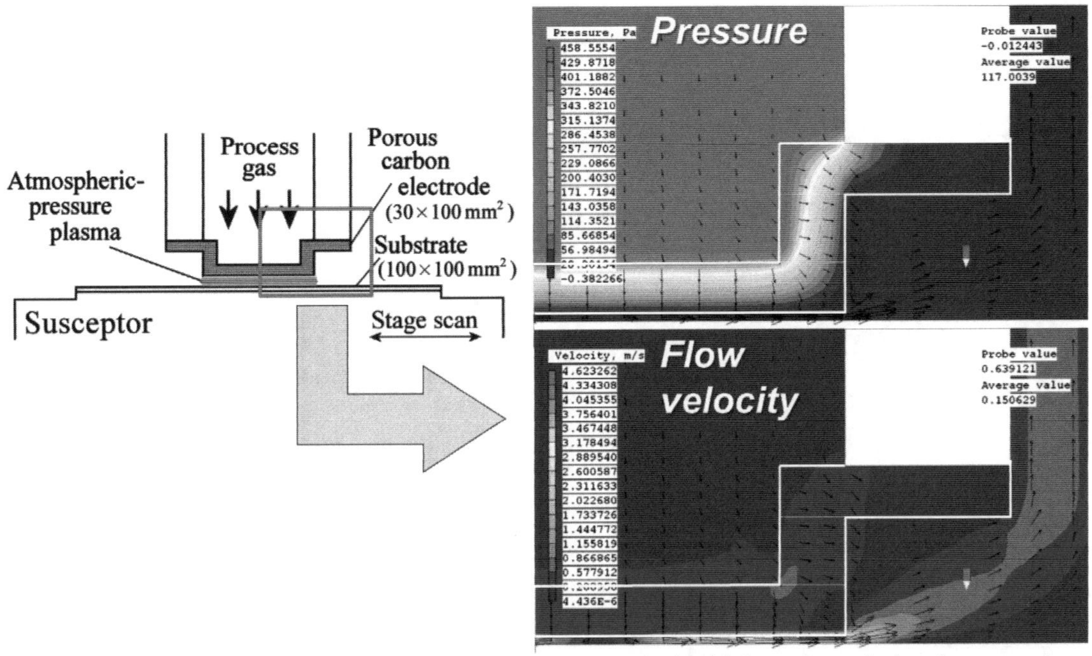

Figure 6. Contours of pressure and flow velocity on a vertical midplane through the plasma region. The simulation was done in an actual deposition conditions: He, H_2 and SiH_4 flow rates of 50 SLM, 100 SCCM and 50 SCCM, respectively, $P_{VHF} = 24$ W cm^{-2}, and T_{sub} of 220 °C.

that the origin of such unique deposition characteristics result from the long gas residence time in the plasma, which greatly enhances the chemical reactions both in gas the phase and on the film-growing surface. The realization of actual thin film devices, particularly using plastic materials as substrates, awaits further empirical and theoretical investigations.

Acknowledgements
This work was carried out in Ultra Clean Room of Department of Precision Science and Technology, Osaka University. This work was funded by KAKENHI (20676003) and a Grant-in-Aid for the Global COE Program from the Ministry of Education, Culture, Sports, Science and Technology. The authors would like to thank Mr. A. Takeuchi of Osaka University for his technical assistance.

References

[1] Fukawa M Suzuki S Guo L Kondo M and Matsuda A 2001 *Sol. Energy Mater. Sol. Cells* **66** 217
[2] Graf U Meier J Kroll U Bailat J Droz C Vallat-Sauvain E and Shah A 2003 *Thin Solid Films* **427** 37
[3] Mai Y Klein S Carius R Wolff J Lambertz A Finger F and Geng X 2005 *J. Appl. Phys.* **97** 114913
[4] Kakiuchi H Ohmi H and Yasutake K 2007 *Trends in Thin Solid Films Research* ed A R Jost (New York: Nova Science) Chapter 1 pp 1–50.
[5] Kakiuchi H Matsumoto M Ebata Y Ohmi H Yasutake K Yoshii K and Mori Y 2005 *J. Non-Cryst. Solids* **351** 741
[6] Yasutake K Ohmi H Kakiuchi H Wakamiya T Tawara N and Watanabe H 2006 *Jpn. J. Appl. Phys.* **45** 3592
[7] Ohmi H Kakiuchi H Tawara H Wakamiya T Shimura T Watanabe H and Yasutake K 2006 *Jpn. J. Appl. Phys.* **45** 8424
[8] Kakiuchi H Ohmi H Inudzuka R Ouchi K and Yasutake K 2008 *J. Appl. Phys.* **104** 053522
[9] Kakiuchi H Ohmi H Ouchi K Tabuchi K and Yasutake K 2009 *J. Appl. Phys.* **106** 013521

Magneto-optical effect in GdN epitaxial thin film

R.Vidya Sagar[1], H. Yoshitomi[1], S. Kitayama[1], T. Kita[1], H. Ohta[2], T. Sakurai[3]
[1]Department of Electrical and Electronic Engineering, Graduate School of Engineering, Kobe University.
[2]Molecular Photoscience Research Center and Graduate School of Science, Kobe University.
[3]Center for Support to Research and Education Activities, Kobe University.

E-mail: drsagar@sapphire.kobe-u.ac.jp

Abstract. We report the cooperative phenomena of magneto-optical effect in AlN/GdN/AlN heterostructure grown by reactive rf magnetron sputtering technique performed under a base pressure less than 5×10^{-6} Pa. The ultra-high vacuum conditions during growth of GdN enable to curtail the oxophilicity character. The temperature dependent optical absorbance measurements have been carried out to evaluate the direct optical band gaps at the X point through the Tauc plots for the heterostructure. We observed for the first time that the optical absorbance studies demonstrate an exceptional splitting in valance and conduction bands in GdN below the critical temperature. The splitting energy is slightly increased with increase of temperature range of 30-55 K and the difference is merely ΔE=28 meV. Furthermore, the splitting energy drastically increases from 30 K to 3 K and the difference is ΔE=102 meV. This startling increase convincing that there is rapid increase of magnetic moments caused by long range correlation of spins.

1. Introduction

For the past few decades, the researchers are drive to exploit the quantum behaviour of spins in standard semiconductor, which can offer a new frontier in design and function towards next generation of semiconductor based spin transport devices [1-2]. For instance, the rare-earth nitride compounds have found to be contemporary ferromagnetic semiconductors since they have partially filled $4f$ orbital. Amongst, Gd has both partially filled $4f$ and $5d$ orbitals which can take part in a new coupling mechanism proceeding via intra-ion $4f$-$5d$ exchange interaction followed by inter-ion $5d$-$5d$ coupling medicated by charge carriers. The crucial spin effects can often be accessed through the partially filled $4f$ & $5d$ orbital properties of the electrons in semiconductor. Thus, GdN has become one of the promising ferromagnetic semiconductors due to its exceptional optical, electronic and magnetic properties [3]. At this exciting stage, the fundamental problems are still being contradictory on its oxophilicity, stoichiometry, electronic and ferromagnetic characteristics. According to the earlier reports, the theoretical and experimental results have showed different properties and also sometimes contradicting [4-5]. However, the highly stoichiometric GdN films are required to understand the relevant electronic and ferromagnetic properties. In the present study, we have made an effort on the deposition of AlN/GdN/AlN epitaxial heterostructure grown by reactive r-f sputtering at ultra-pure conditions; and the film is selected for temperature dependent optical spectroscopy to analyse the band structure.

2. Experimental Details

Double heterostructure of AlN/GdN/AlN grown on *c*-sapphire (0001) substrate at 500 °C using reactive rf magnetron sputtering in ultra-high vacuum conditions. The *c*-sapphire substrate (10×10 mm^2) was cleaned with acetone, methanol, and deionised (DI) water in an ultrasonic bath to remove residual organic contamination on the surface. The substrate is dipped in HF solution to remove surface native oxides then the substrate was again cleaned with DI water. After the chemical cleaning process, the substrate was mounted onto sample holder; loaded into the buffer chamber and then transferred into the growth chamber with the transfer arm. The substrate temperature is fixed at 500 °C. Initially, we deposited 100 nm AlN-buffer layer on the *c*-sapphire then the GdN layer is grown on the buffer layer. The thickness of the GdN thin layer is 68 nm. Oxophilicity character is curtailed by high vacuum depositing conditions (less then 5×10^{-6} Pa) and AlN capping layers. *Our earlier investigations showed that the GdN has (111) preferential orientation of the rocksalt structure on the c-axis oriented AlN with the wurtzite structure; the secondary ion mass spectroscopic analysis (SIMS) showed the oxygen content is about* 1.0×10^{-19} cm^{-3}[3]. The heterostructure was characterized by temperature dependent spectroscopy to understand the band structure of GdN.

3. Results &Discussion

Optical absorbance measurements have been carried out by temperature dependent spectrophotometer for 68 nm thin GdN layer grown as sandwich layer between AlN buffer layer and capping layer, and the results are displayed in Figure.1. Inhomogeneous background in the optical absorbance spectra can be negligible because the background is cancelled by subtracting the transmitted signal passing through reference sapphire substrate. From the Figure.1, it is obvious that the GdN film exhibited the broad optical absorption signal. Interestingly, an anomalous change in the absorbance spectra has been identified in the region of 725 to 810 nm explicitly below 30 K. This remarkable result leads to two absorbance edges, prior to that only one absorbance edge has been observed for above 50 K. According to the literature, such kind of phenomena has not been presented experimentally. Notably, the absorbance values are decreased from 300-100 K then the absorbance value increases slightly from 50 K to 3 K. The anomalous change in optical absorbance spectra lead us to compute two optical direct band gaps for GdN thin film below 30 K.

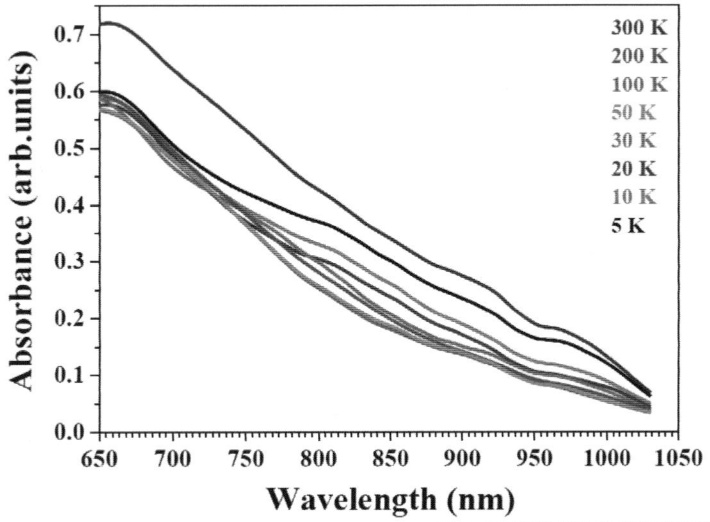

Figure.1: Temperature dependent optical absorbance spectra for AlN/GdN/AlN heterostructure.

The optical band gaps of the heterostructure have been computed based on the optically induced transitions from Visible-NIR absorbance spectra. The principle behind in this technique is that a photon with an energy greater than the band gap energy will be absorbed. There are two kinds of optical transitions at the fundamental absorption edge: direct and indirect transitions, both of which involve the interaction of an electromagnetic wave with an electron in the valance band. According to the literature [6], the optical direct band gap of the heterostructure can be obtained by Tauc plots which are shown in Figure. 2. From the Tauc plots, the respective values of optical band gaps are obtained by extrapolating the linear portion of the curve onto the $(\alpha h\nu)^2=0$ for the heterostructure.

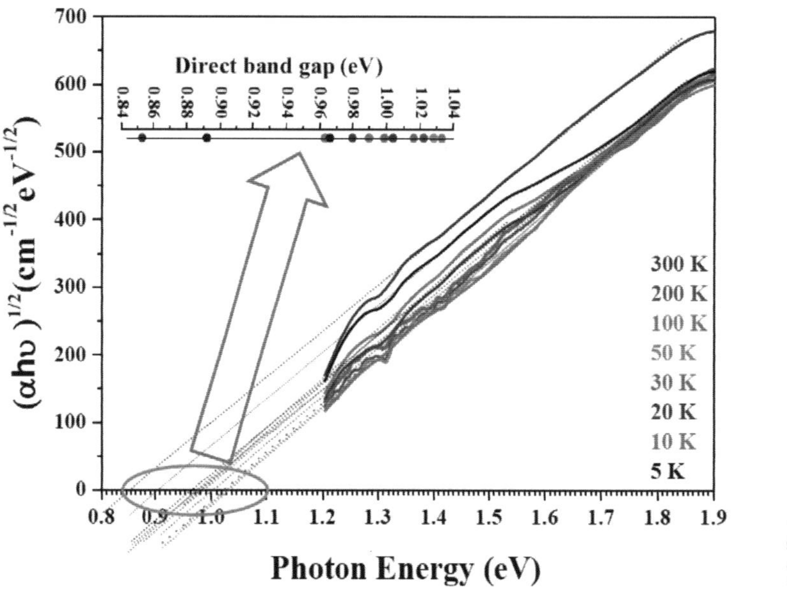

Figure. 2: Tauc plots for AlN/GdN/AlN heterostructure

Figure. 3 shows the temperature dependent (3-300 K) optical direct band gaps for the AlN/GdN/AlN double heterostructure. A very interesting magneto-optical effect has been observed for the AlN/GdN/AlN double heterostructure. From the plot, it is obvious that the band gap decrease (red shift) with the temperature below 35 K and the splitting of band gap in the conduction and valance band at X point from 55 K to 3 K have clearly been identified. The dramatic reduction in optical band gap *(blue data)* below 35 K is about 85 meV; the red shift is strongly contemplated due to the long range ferromagnetic ordering of spins. The optical direct energy *(green data)* drastically increases from 35 K to 3 K and is about 102 meV. With increasing the temperature from 35 K to 300 K, the optical band gap reduced (blue-shift) unto 55 K then gradually decreases till reached to 300 K; such trend is somewhat sharply referred to semiconducting nature. The splitting energy difference (ΔE) *and the magnetization* as function of temperature are shown in inset of Figure. 3. The splitting energy difference is *drastically increases from 35 K to 3 K and the difference is ΔE=131 meV. According to the LSDA+U calculations, the splitting energy difference could be attributed to the difference between minority and majority spin band energies [7]. The valence band maximum has full with majority spin character, but the conduction band minimum as a mixed characters thus the lower one is lightly more majority while the next one is lightly more minority spin in character. Thus, it could be presume that the splitting energy difference is intensively dependent on the alignment of magnetic spins via long range correlation of spins below its critical temperature for ferromagnetic materials. The splitting energy difference is slightly decreased from 35 K to 55 K; the difference is merely ΔE=28 meV.* This slight difference pertaining to the short range correlation of spins lead to paramagnetic behaviour. The red-shift and splitting energy difference in the GdN are the important features of the intrinsic ferromagnetic semiconductors and such phenomenon never been observed in diluted magnetic semiconductors until now. The obtained results have significantly important, because of a significant contribution to the development of magneto-optical and spintronic elements and devices.

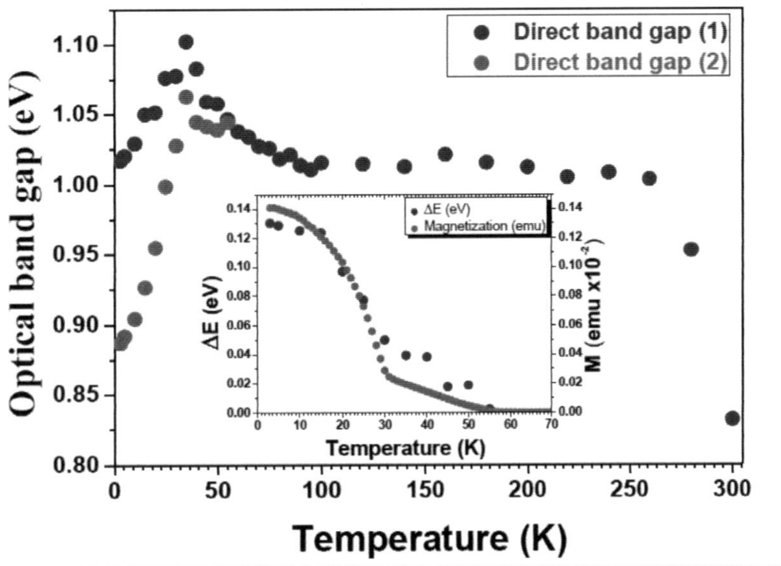

Fig. 3: Direct transition energies of AlN/GdN/AlN heterostructure

4. Conclusion

In summary, it is concluded that the AlN/GdN/AlN double heterostructure has successively grown on *c*-sapphire by reactive rf sputtering at ultrapure conditions were investigated. The temperature dependent optical absorbance measurements for the heterostucture have been carried out to evaluate the optical induced transitions. The optical band gaps at X point have been computed using Tauc plots from 3 K to 300 K. Red shift in the optical band gap of GdN observed below 35 K and the blue shift above 35 K reveals that the GdN is intrinsic ferromagnetic semiconductor. The direct band gaps energy difference (ΔE) is due to the splitting energy difference between the conduction and valance band of GdN below its critical temperature. This remarkable investigation on magneto-optical effect though of red-shift and splitting phenomena is open for the new frontier in development of low temperature semiconductor spintronics devices.

References:

[1] Wolf SA, Awschalom DD, Buhrman RA, Daughton JM, von Molnar S, Roukes ML, Chtchelkanova AY, Treger DM 2001 *Science* **294** 1488
[2] Senapati K, Blamire MG, Barber ZH 2011 *Nat. Mater* **10** 849
[3] Yoshitomi Y, Kitayama S, Kita T, Wada O, Fujisawa M, Ohta H, Sakurai T 2011 *Phy. Rev B* **83** 155202
[4] Mitra C and Lambrecht WRL 2008 *Phy. Rev B* **78** 134421
[5] Sharma A and Nolting W 2010 *Phy. Rev B* **81** 125303
[6] Tauc J 1974 *Amorphous and Liquid Semiconductors* (Plenum, London).
[7] Trodahl HJ, Preston AR, Zhong J, Ruck BJ, Strickland NM, Mitra C, Lambrecht 2007 *Phy. Rev B* **76** 085211

Effect of Er^{3+} Concentration on Upconversion in Hexagonal-Phase $NaYF_4:Er^{3+}$ Nanocrystals

X J Luo, R Yuminami, T Sakurai and K Akimoto

Institute of Applied Physics, University of Tsukuba, Tsukuba, Ibaraki 305-8573, Japan

corresponding author:luoxj2006@gmail.com

Abstract. A facile synthesis method was developed to produce hexagonal-phase of $NaYF_4$ nanocrystals (NCs) doped with Er^{3+} in different concentration, which showed upconversion (UC) emission from infrared to visible spectral region. This proposed method is simple and less toxic compared with generally used method so far. It was found that up-conversion emission spectra of $NaYF_4:Er^{3+}$ NCs, excited at 1550 nm, included four peaks at about 980 nm, 800 nm, 660 nm and 540 nm. The effect of Er^{3+} concentration on UC in β-phase $NaYF_4:Er^{3+}$ NCs were discussed based on the excitation power dependence. The optimum Er^{3+} concentration for 2-step and 3-step UC was found to be around 10~30%.

1. Introduction

Upconversion (UC) nanocrystals (NCs) have recently attracted significant attentions, due to their potential applications on solar cells, light emitting devices, flat-panel displays, biological assays and medical imaging [1]. One of the promising UC nanomaterial is lanthanide ion doped $NaYF_4$, since the phonon energy of the host is relatively low which suppresses non-radiative relaxation. The phase control of $NaYF_4$ is very important, because the hexagonal (β)-phase $NaYF_4$ has higher UC efficiency than the cubic (α)-phase [2]. However, the synthesis method generally used is relatively complex, sensitive to the condition and with discharging toxic gas like COF_2 [2]. Moreover, the chemical product is generally mixture of α- and β-phase of $NaYF_4$ which is sensitive to the synthesis conditions. Therefore, the UC efficiency is not stable and depends on the lot of the production.

In this work, a facile synthesis method for hexagonal $NaYF_4:Er^{3+}$ NCs with less toxicity was presented. The $NaYF_4:Er^{3+}$ NCs were synthesized in oleic acid (OA) and 1-octadecene (ODE) with $Y_2(CO_3)_3{\cdot}xH_2O$, $Er_2(CO_3)_3{\cdot}xH_2O$, Na_2CO_3 and NH_4F as precursors [3]. During the reaction process, small $NaYF_4:Er^{3+}$ crystal nucleation was progressed at low temperature and the discharged gas was mainly ammonia. The optimum synthesis conditions were obtained and the properties of the UC were studied.

2. Experimental Details

$Er_2(CO_3)_3{\cdot}xH_2O$ (99.9%) was purchased from ChemPur, $Y_2(CO_3)_3{\cdot}xH_2O$ (99.9%), Na_2CO_3 (99.5%), NH_4F (98%), OA, and ODE were purchased from Aldrich. The $NaYF_4$:10%Er^{3+} NCs were synthesized as follows: $Y_2(CO_3)_3{\cdot}xH_2O$ (0.4 mmol) , $Er_2(CO_3)_3{\cdot}xH_2O$ (0.1 mmol), Na_2CO_3 (0.5 mmol) and NH_4F (4 mmol) were mixed with OA (10 mmol) and ODE (30 mmol) and put into 100 ml flask. Then the mixture solution was degassed at 100 ℃ for 30 min. The temperature was increased to around 310 ℃ for 0.5 h followed by 330 ℃ for 0.5 h heating. After cooling down, the NCs were collected by centrifugation and washed with ethanol for three times. Finally the products were dried at 80℃ for 3 h. The chemical reaction can be expressed as below:

$$Na_2CO_3 + RE_2(CO_3)_3 + NH_4F \rightarrow NaREF_4 + CO_2\uparrow + NH_3\uparrow + H_2O \qquad (1)$$

In fact, Na_2CO_3 and rare earth carbonates were mixed with NH_4F in OA and ODE, immediately resulting in occurrence of ammonia gas and the formation of nanoscale α–phase $NaREF_4$. When the reaction temperature was increased to 100 ℃, the formation rate of nanoscale α–phase $NaREF_4$ was also increased. It was found that the hexagonal-phase of $NaYF_4$:Er^{3+} NCs were controlled by the reaction temperature during synthesis.

The prepared NCs powders were characterized by X-ray diffraction (XRD) on a Panalytical X'Pert diffractometer. The UC spectra were measured by using a Seki Technotron spectrometer with the excitation wavelength of 1550 nm with the density of ranging from 0.22 W/cm^2 to 45.19 W/cm^2.

3. Results and Discussion

Figure 1 showed the XRD pattern of $NaYF_4$ NCs doped with 10%, 20%, 30%, 40% and 60% Er^{3+}, prepared separately at 310 ℃ for 1 h, or at 310 ℃ for 0.5 h followed by 330 ℃ for 0.5 h heating. The XRD pattern demonstrated that the pure β-phase $NaYF_4$ (PDF card No. 160334) NCs doped with 10~60% were successfully obtained under the reaction at 310 ℃ for 0.5 h followed by 330℃ for 0.5 h heating. In case of $NaYF_4$:10%Er^{3+} prepared at 310 ℃ for 1 h, both β- and α- phase (PDF card No. 060342) were observed by XRD with comparable intensity. Therefore it was found that the temperature was key factor for the formation of pure β-phase of $NaYF_4$ NCs. The mean size of $NaYF_4$:30%Er^{3+} NCs calculated by XRD was about 27 nm and its size obtained from TEM image in figure 2 was about 32 nm.

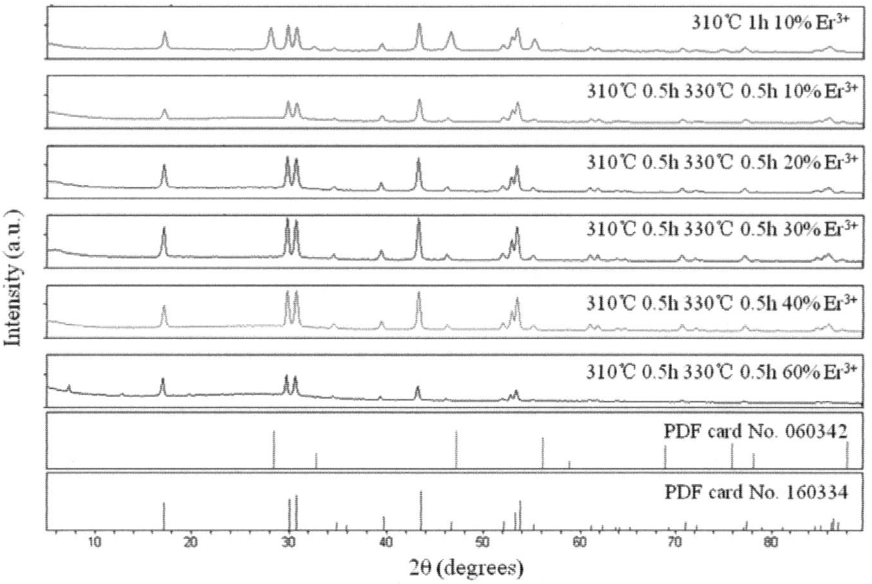

Figure 1. XRD patterns of the $NaYF_4$:10%, 20%, 30%, 40%, 60% Er^{3+} NCs obtained after heating separately at 310 ℃ for 1 h, or at 310 ℃ for 0.5 h followed by 330 ℃ for 0.5 h heating and the line pattern of cubic (PDF card No. 060342) and hexagonal (PDF card No. 160334) phase of $NaYF_4$ were presented.

The UC emission spectra of $NaYF_4$:30%Er^{3+} NCs, excited by the light with 1550 nm, were dominated by four peaks at around 980 nm, 800 nm, 660 nm and 540 nm as shown in figure 3. The

Energy levels diagram and corresponding transitions were demonstrated in figure 4. Two major up-conversion mechanisms, energy transfer up-conversion (ETU) and ground state absorption followed by excited state absorption (GSA+ESA), are commonly accepted [4], and following the accepted model, the optical process for the luminescence observed in figure 3 may be considered as below. The ground state of Er^{3+} ion absorbs the excitation light with 1550 nm, resulting in a generation of excited $Er^{3+}(^4I_{13/2})$ ion. A subsequent absorption of excitation light gives rise to high excited $Er^{3+}(^4I_{9/2})$ ion, from which 800 nm (Er^{3+}: $^4I_{9/2}\rightarrow{}^4I_{15/2}$) peak emission occurs. Green emission, its peak at 540 nm (Er^{3+}: $^4S_{3/2}\rightarrow{}^4I_{15/2}$), occurs via $Er^{3+}(^4I_{9/2})\rightarrow Er^{3+}(^4S_{3/2})$ transition by another subsequent absorption of excitation light. From $Er^{3+}(^4I_{9/2})\rightarrow Er^{3+}(^4I_{11/2})$ relaxation, 980 nm (Er^{3+}: $^4I_{11/2}\rightarrow{}^4I_{15/2}$) emission can take place. Red emission, its peak at 660 nm (Er^{3+}:$^4F_{9/2}\rightarrow{}^4I_{15/2}$), occurs via $Er^{3+}(^4S_{3/2})\rightarrow Er^{3+}(^4F_{9/2})$ relaxation or $Er^{3+}(^4I_{9/2})\rightarrow Er^{3+}(^4I_{11/2})$ relaxation followed by $Er^{3+}(^4I_{11/2})\rightarrow Er^{3+}(^4F_{9/2})$ transition. From the energy transitions process, it was found that emission at 980 nm and 800 nm were generated by 2-step UC and at 660 nm and 540 nm were by 3-step UC.

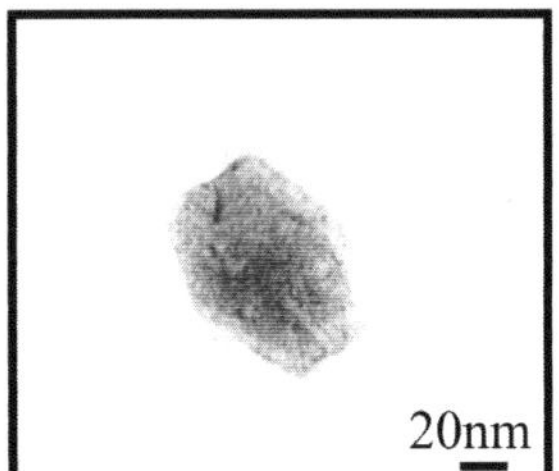

Figure 2. TEM image of $NaYF_4$:30%Er^{3+} NCs.

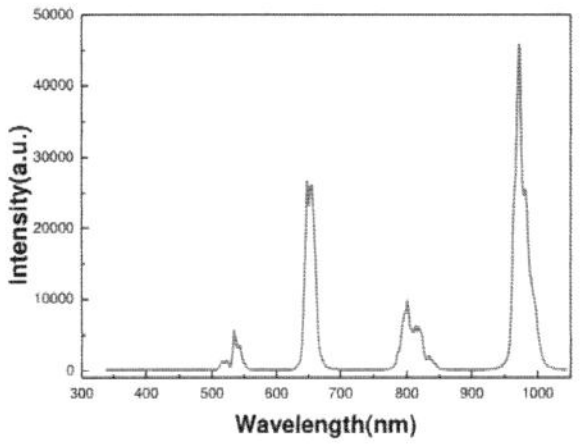

Figure 3. Emission spectra of $NaYF_4$:30%Er^{3+} NCs prepared at 310 ℃ for 0.5 h followed by 330 ℃ for 0.5 h heating, excited by the light with 1550 nm.

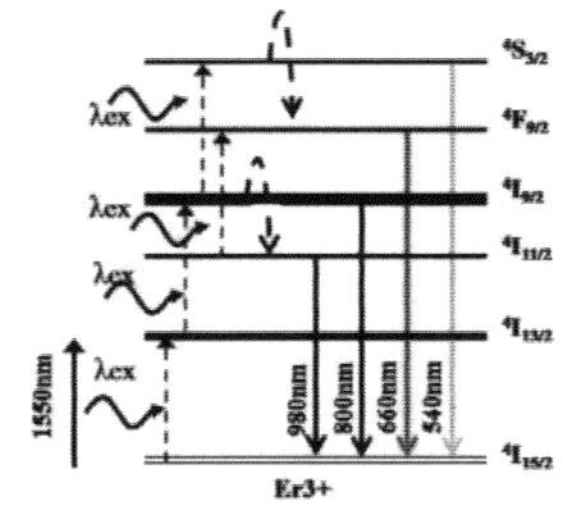

Figure 4. Energy level diagram showing the optical process of UC in $NaYF_4$:Er^{3+}.

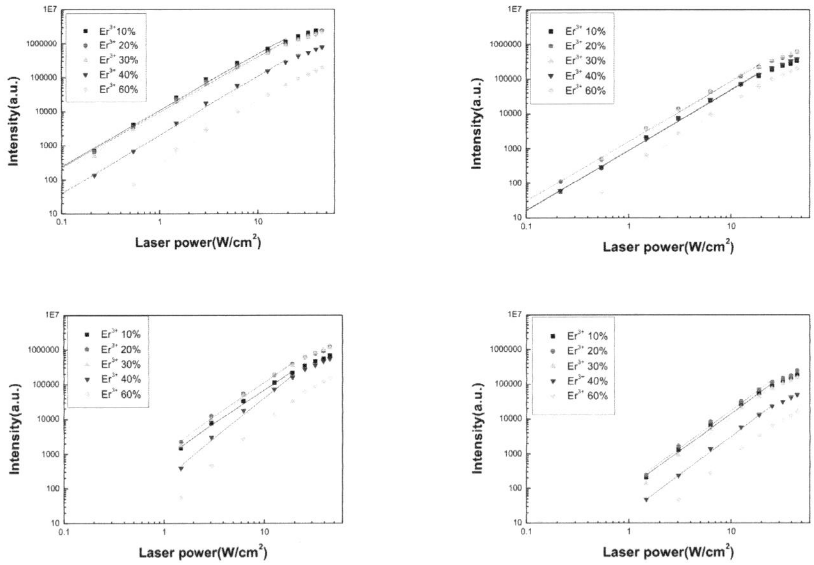

Figure 5. Pump-power dependence for (a) 980 nm, (b) 800 nm, (c) 660 nm and (d) 540 nm UC emission intensity of $NaYF_4$:10%, 20%, 30%, 40%, 60% Er^{3+} NCs with 1550 nm laser excitation.

The excitation power dependence of UC plotted by a double-logarithmic representation was illustrated in figure 5. The slope n shown in figure 5 decreased at high laser powers, which indicated a saturation effect in high excitation energy [5]. The $NaYF_4$ NCs with the Er^{3+}concentration of around 10~30% had relatively high intensity of UC emission. In higher doped Er^{3+} concentration, the loss process, such as cross relaxation, may be increased [6].

Table 1 showed the values of the slope n for 980 nm, 800 nm, 660 nm and 540 nm emissions of $NaYF_4$:10%, 20%, 30%, 40%, 60% Er^{3+} NCs. The n values for 980 nm and 800 nm emissions were about 1.7, and for 660 nm and 540 nm emissions were about 2.2, respectively. Such n values showed that the emission at 980 nm and 800 nm were generated by 2-step UC ($n\approx2$) and at 660 nm and 540 nm were 3-step UC ($n\approx3$). Compared to the n value of the 2-step UC and 3-step UC, it was found that higher excitation density preferred to increase 3-step UC rather than 2-step UC. Table 1 also demonstrated that the n value was strongly affected by the Er^{3+} concentration. The n value in higher Er^{3+} concentration was basically larger value than those in lower Er^{3+} concentration.

Table 1. The slope n for 980nm, 800nm,660nm and 540nm emission.

	980nm	800nm	660nm	540nm
10%Er^{3+}	1.7	1.7	2.0	2.1
20%Er^{3+}	1.6	1.7	2.0	2.1
30%Er^{3+}	1.7	1.8	2.1	2.2
40%Er^{3+}	1.7	1.7	2.3	2.2
60%Er^{3+}	1.8	1.9	2.5	2.3

4. Conclusion

A simple and less toxicity synthesis method was developed for pure hexagonal-phase of $NaYF_4$ NCs doped with 10%, 20%, 30%, 40% and 60% Er^{3+}. The hexagonal of $NaYF_4:Er^{3+}$ can be fabricated by controlling the temperature during synthesis. UC emission spectra of $NaYF_4:Er^{3+}$ NCs, excited at 1550 nm, showed four peaks at about 980 nm, 800nm, 660 nm and 540 nm. The values of the slope n in pump-power dependence of UC emission intensity confirmed that 980 nm, 800 nm emission were 2-step UC, and 660 nm, 540 nm emission were 3-step UC. It is found the optimum Er^{3+} concentration for 2-step and 3-step UC was around 10~30%.

References

[1] Wang F, Han Y, Lim C S, Lu Y H, Wang J, Xu J, Chen H Y, Zhang C, Hong M H and Liu X G 2010 *Nature* **463** 1061

[2] Mai H X, Zhang Y W, Si R, Yan Z G, Sun L D, You L P and Yan C H 2006 *Journal of the American Chemical Society* **128** 6426

[3] Schafer H, Ptacek P, Eickmeier H and Haase M 2009 *Advanced Functional Materials* **19** 3091

[4] Auzel F 2004 *Chemical Reviews* **104** 139

[5] Pollnau M, Gamelin D R, Luthi S R, Gudel H U and Hehlen M P 2000 *Physical Review B* **61** 3337

[6] Kozhan T M, Kuznetsova V V, Sergeev I I, Timofeev Y P, Khomenko V S and Chau M 1994 *Journal of Applied Spectroscopy* **61** 500

Dislocation generation in GaN by dicing process

Hideyuki Taguchi[1, 2], Amane Kitahara[1], Syugo Miyake[1], Akimitu Nakaue[1], Atsushi Nishikawa[2] and Yasufumi Fujiwara[2]
[1]Electronics Division, Technology Department, KOBELOCO Research Institute, inc., 1-5-5 Takatsukadai, Nishi-ku, Kobe, Hyougo 651-2271, Japan
[2]Division of Materials and Manufacturing Science, Graduate School of Engineering, Osaka University, 2-1 Yamadaoka, Suita, Osaka 565-0871, Japan

E-mail:taguchi.hideyuki@kki.kobelco.com

Abstract. In order to analyze effect of the dicing process on the GaN epitaxial layer, the GaN-wafer is cut in sizes of the 0.7 mm square and the 1.7 mm square. The crystal characteristics of the GaN-chips have been measured using X-ray measurements and Raman spectra measurements. The full-width half maximum (FWHM) values of the X-ray rocking curves of (0002), (10-13) and (10-12) of the 0.7 mm square GaN-chip become wider than that of before the dicing process. The E_2 (high) peak of Raman spectra at the edge in the 0.7mm square GaN-chip is shifted to lower wave number. In consideration of crystallography, we infer from these results that both the crystal strains and the screw dislocations have been generated during the dicing process.

1. Introduction

Light emitting diodes (LEDs) have been expected to be one of the next generation key devices because of their excellent properties such as low power consumption and high efficiency. For instance, InGaN/GaN multiple quantum wells (MQW) LEDs over sapphire substrates are well-known as full color lighting devices and are commonly used. One of the key issues for performance of the GaN based LEDs is the control of dopant elements distribution, optimization of the structure of MQW and reduction of crystal defects in the GaN epitaxial layer [1]. Low density of structural defects has been achieved, in spite of a large lattice mismatch, due to a buffer layer between the GaN epitaxial layer and the sapphire substrate. However, structural defects in the GaN epitaxial layer are still high density. Therefore, much effort has been devoted to reducing the structural defects by epitaxial growth techniques [2-4].

Even though the wafer dicing process is one of the most considerable processes for assembling LEDs modules, performance degradations as a LEDs device, have been occurred in the GaN epitaxial layer such as structural defects and crystal strains. Effect of the dicing process on fabricated chips was read many papers and patented for semiconductor devices manufacturing method, but they were studied about the shape of the diced surface, chip strength and electrical characteristics [5-8]. Drop in the chip properties is caused by degradations of the crystal structure. As a first step of the analysis of structural defects and crystal strains generated in the GaN epitaxial layer by the dicing process, we report on the relationship between the variation of the crystal structures and the dicing process.

2. Experiments

The GaN wafer samples, which were grown in the <0002>-oriented GaN epitaxial layer on sapphire substrate by organometallic vapor phase epitaxy, were prepared for subsection by the dicing process. The structure of the GaN epitaxial layer as a sample is shown in Figure 1. The buffer layer is 1.1 μm, the Si doped n-GaN layer is 2.5 μm, the MQW is 0.1 μm and the p-GaN layer is 0.1 μm in thickness. In order to dice the GaN-wafer, that process was performed using a dicing machine (DAD3350 (1.8Kw) Automatic Dicing Saw attached with a VT07-SD400-VC100-75 dicing blade, DISCO). The GaN-chips were cut in sizes of 0.7 × 0.7 mm and 1.7 × 1.7 mm as shown in Figure 2.

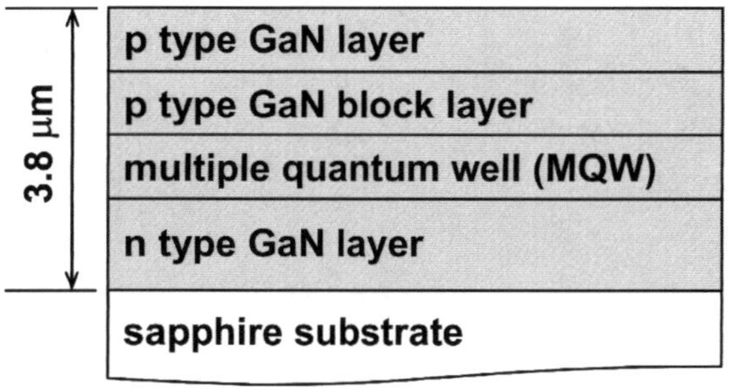

Figure 1. Device structure of the GaN epitaxial layer on sapphire substrate.

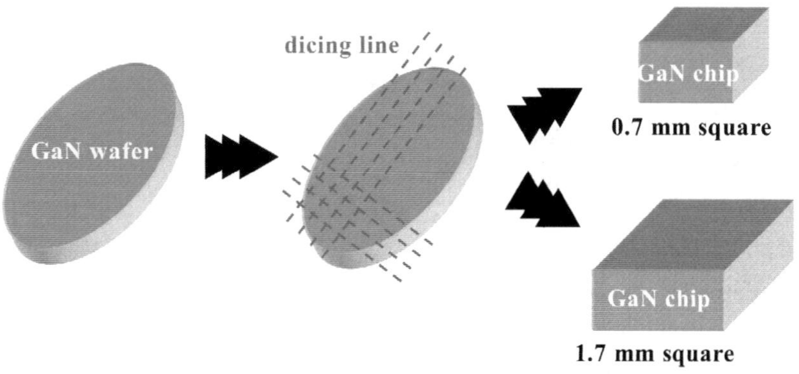

Figure 2. Schematic of the GaN-wafer dicing process.

In order to characterize the crystal structures of the GaN-chip, X-ray measurements and Raman spectra measurements were carried out. X-ray measurements were performed using a high resolution x-ray diffractometer (Smart-Lab, Rigaku) with a Cu rotation anode target and a four-crystal monochromator in Ge(220) configuration. Double-axis rocking curves of (002) and (*hkl*) reflections with $h,k \neq 0$ were recorded, the latter one in skew symmetric diffraction geometry by using an Eulerian cradle. Raman spectra measurements were performed using a Czerney-Turner type micro-Raman spectrometer (JOBIN YVON LabRAM HR-800, HORIBA) equipped with a thinned back-illuminated charge-coupled device (CCD) detector. Raman scattering was excited with the 514.5 nm line of Ar^+ ion laser at a power of 5 mW. An infinity corrected microscope objective focused the laser beam to a spot size of about 1 μm in diameter.

3. X-ray measurement results

Figure 3 presents the full-width half maximum (FWHM) values of X-ray rocking curves for the GaN-wafer, the 0.7 mm square and the 1.7 mm square GaN-chips. The FWHM values of the each lattice plane of the GaN-wafer increased simply with the increasing angle of inclination. The FWHM values of (0002), (10-13), and (10-12) of the 0.7 mm square GaN-chip were become wider than those of the GaN-wafer. V. Srikant *et al.* had reported that (0002), (10-13), and (10-12) measured by the tilt angle were correlated with the density of the screw dislocations [9, 10]. Thus, the results in Figure 3 indicate that the screw dislocations might be generated in the GaN-chip by the dicing process. Meanwhile, the FWHM values of (0002), (10-13), and (10-12) of the 1.7 mm square GaN-chip were shown as wide as the GaN-wafer. Due to X-ray spot size of about 1mm in diameter, the FWHM values of the 1.7mm square GaN-chip were measured only about 1 mm in diameter of the GaN-chip center. These results suggest that the screw dislocations were generated just on the edge of the GaN-chip by the dicing process.

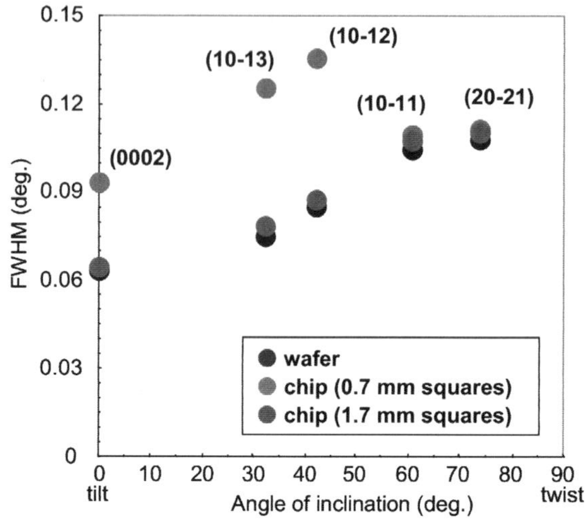

Figure 3. The FWHM values of X-ray rocking curves for (*hkl*) reflections as function of the inclination angle ϕ of the reflecting lattice planes with the GaN-wafer, the 0.7 mm square and the 1.7 mm square GaN-chips.

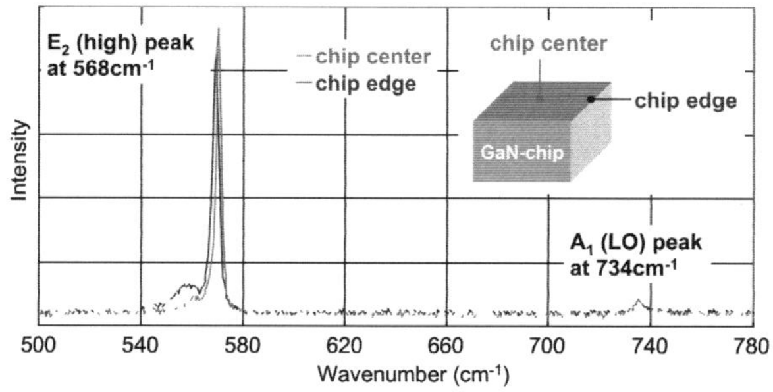

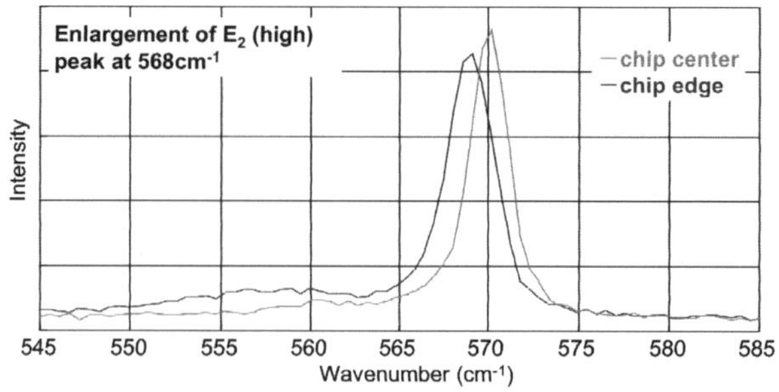

Figure 4. The Raman spectra of the GaN epitaxial layer at the center and edge points in the 0.7 mm square GaN-chip measured with 514.5 nm excitation.

4. Micro-Raman spectra measurement

Figure 4 shows the Raman spectra at the center and the edge points in the 0.7 mm square GaN-chip. The direction of the laser beam irradiation was parallel to the <0002>-oriented GaN epitaxial layer. The E_2 (high) peak and the A_1 (LO) broad peak are observed at 568 cm^{-1} and 734 cm^{-1}, respectively. This wave number accords with literature date [11]. F. A. Ponce *et al.* had reported that in a highly Si doped GaN material the quenching of the A_1 (LO) peak by donor impurities had been used to image spatially varying carrier concentrations in faceted GaN crystallites, therefore the A_1 (LO) peak was broadened [12, 13]. The results suggest that the Raman spectra were measured from the p-GaN layer to the n-GaN layer. The E_2 (high) peak at the edge point in the 0.7 mm square GaN-chip was shifted to lower wave number and the FWHM was broadened, as compared with the 0.7 mm square GaN-chip center point. The Raman peak shift was influenced not only the crystal structures but also the atmosphere temperatures. Thus, the Raman peak shift can not be determined to be caused by the variation of the crystal structures of the 0.7 mm square GaN-chip edge point. The FWHM of the Raman peak is depended only on the crystal structures. The results indicate that the crystal strains occurred just on the edge of the GaN-chip by the dicing process. We are considering that the phenomenon occurred by the dicing process was both the crystal strains and the screw dislocations.

5. Conclusions

In order to characterize the variation of the crystal structures in the GaN epitaxial layer by the dicing process, X-ray measurements and Raman spectra measurements were carried out. In the results of X-ray measurements, the FWHM values of (0002), (10-13), and (10-12) of the GaN-chip became wider than those of the GaN-wafer. In the results of Raman spectra measurements, the E_2 (high) peak at the edge point in the GaN-chip was shifted to lower wave number and broadened. These results infer that both the crystal strains and the screw dislocations were generated just on the edge of the GaN-chip by the dicing process.

Acknowledgements

The work described in this paper is under the supervision of T. Okano, K. Ikeda, H. Takada, R. Adachi, K. Isoo, and K. Ikeda.

References

[1] C. Mion J. F. Muth E. A. Preble and D. Hanser 2006 *Appl. Phys. Lett.* **89** 92123

[2] A. Sakai H. Sunakawa and A. Usui 1997 *Appl. Phys. Lett.* **71** 2259

[3] M. A. Moram Y.Zhang M. J. Kappers Z. H. Barber and C.J. Humphreys 2007 *Appl. Phys. Lett.* **91** 152101

[4] M. H. Oliver J. L. Schroeder D. A. Ewoldt I. H. Wildeson V. Rawat R. Colby P. R. Cantwell E. A. Stach and T. D. Sands 2008 *Appl. Phys. Lett.* **93** 23109

[5] E. Gu C.W. Jeon H.W. Choi G. Rice M.D. Dawson E.K. Illy and M.R.H. Knowles 2004 *Thin Solid Film* **453-454** 462

[6] C. Xiong F. Jiang W. Fang L. Wang C. Mo and H. Liu 2007 *Journal of Luminescence* **122-123** 185

[7] J. Li H. Hwang E.C. Ahn Q. Chen P. Kim T. Lee M. Chung and T. Chung 2007 *Electronic*

Components and Technology Conference 761

[8] S. Takyu T. Kurosawa N. Shimizu and S. Harada 2006 *Electronic Components and Technology Conference* 1623

[9] V. Srikant J. S. Speck and D. R. Clarke 1997 *J. Appl. Phys.* **82** 4286

[10] H. Heinke V. Kirchner S. Einfeldt and D. Hommel 2000 *Appl. Phys. Lett.* **77** 2145

[11] Hiroshi Harima 2002 *J. Phys.: Condens. Matter.* **14** R967

[12] F. A. Ponce J. W. Steeds C. D. Dyer and G. D. Pitt 1996 *Appl. Phys. Lett.* **69** 2650

[13] G. Nootz A. Schulte L. Chernyak A. Osinsky J. Jasinski M. Benamara and Z. Liliental-Weber 2002 *Appl. Phys. Leet.* **80** 1355

Development of new microscope unit for single molecule spectroscopy under various ambient conditions

T Yamada*, T Kaji*, R Ueda, and A Otomo

Advanced ICT Research Institute, National Institute of Information and Communications Technology, 588-2 Iwaoka, Kobe 651-2492, Japan

E-mail: toshiki@nict.go.jp, kaji@nict.go.jp

Abstract. This paper introduces techniques we previously developed for single molecule spectroscopy and continues on to describe our studies on dipole orientation imaging of single molecules under various ambient conditions. In these studies, we successfully obtained defocused images of single perylene diimide (PDI) molecules under air, high-vacuum, and pure N_2 gas conditions by utilizing the advantages of our new microscope unit. The studies are positioned as one of the important applications of our microscope unit for single molecule spectroscopy. We expect a wide range of applications for this unit for various microscope measurements for many types of materials.

1. Introduction

Single molecule spectroscopy, along with use of the time-correlated single photon counting (TCSPC) system, has been widely applied in studying the emission characteristics of individual organic fluorescent molecules, colloidal semiconductor nanocrystals, and biomolecules labeled with fluorescent moieties [1,2].

We have developed new techniques for efficiently performing single molecule spectroscopy. First, we demonstrated the usefulness of the TCSPC system with modified photomultiplier tubes as photon detectors, which have high quantum efficiency, a relatively large active area, relatively high response speed, and extremely small dark counts [3,4]. Our system has sufficiently high sensitivity for performing single molecule spectroscopy by utilizing the above advantages. We then reported the fabrication of a two-dimensional photonic crystal (PC) slab using tantalum pentoxide (Ta_2O_5) that has low background emission [5]. Enhancement of fluorescence from single molecules of N,N'-bis(2,6-dimethy-phenyl)perylene-3,4,9,10-tetracarboxylic diimde (PDI) on a Ta_2O_5- PC slab with low background emission was observed due to efficient coupling of excitation and fluorescence light to PCs' modes [6]. This was the first demonstration of enhancement of fluorescence from single fluorescence dyes via the effect of two-dimensional PCs. Then we developed a new microscopy unit that has an immersion objective with a high numerical aperture (NA) (1.3-1.4) and an ionic liquid as a non-volatile refractive index matching medium. The unit enabled us to perform single molecule spectroscopy in a high vacuum (10^{-5}-10^{-6} Torr) and ambient gas with an appropriate pressure [7]. The microscopy unit was applied to single molecule spectroscopy for single colloidal quantum dots (CdSe/ZnS core shell nanocrystal), and the differences of the optical characteristics between in air, high vacuum, and pure N_2 gas were elucidated [8]. Moreover, we performed studies toward optimization of the microscope unit [9]. Thus, we have developed three techniques toward a wide range of applications of single molecule spectroscopy as well as an efficient form of this spectroscopy.

In this article, we focus on the studies on dipole orientation imaging of single PDI molecules under air, high vacuum, and pure N_2 gas conditions. These studies are positioned as another important application of our microscope unit for single molecule spectroscopy. We would like to emphasize a wide range of applications of this unit for various kinds of microscope measurements for numerous types of materials.

2. Experimental procedures

Figure 1 shows the schematic of our laser system and new microscope unit. The main differences compared with our previous studies are the excitation light source and detection system. In the present study, we used a continuous wave (cw) laser with a wavelength of 488 nm as an excitation light source and a charge coupled device (CCD) camera as a detector, while in the previous studies, we used a laser with a wavelength of 446 nm or 488 nm, pulse width of 2 ps, and repetition frequency of 8 MHz as an excitation source and a modified PMT (H7422P-40-MOD, Hamamatsu) as a detector. We thus performed epifluorescence microscopy with wide-field illumination for dipole orientation imaging rather than the confocal microscopy in previous studies.

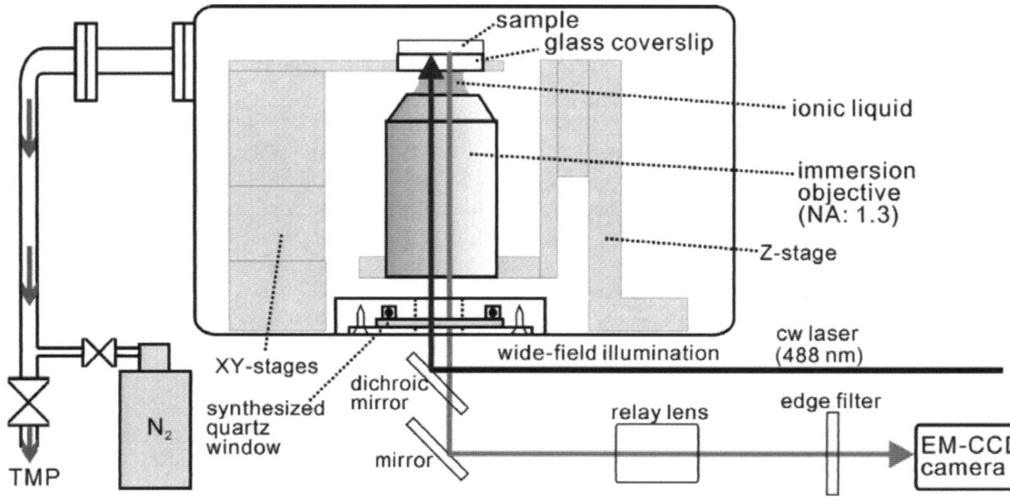

Figure 1. Schematic of laser system and new microscope unit

Our microscope unit is essentially the same as in previous studies [8,9]. It consists of a turbo molecular pump (TMP), gas purge line, small vacuum chamber with a synthesized quartz window, sample substrate, XY- and Z-translation stages, immersion objective with an NA of 1.3 and magnification of 100×, and an ionic liquid. As previously shown, the inside of our microscope unit also offers ambient gas conditions at an appropriate pressure, as well as high-vacuum conditions. The phosphonium-based ionic liquid (IL-AP3-3, Koei) as shown in figure 2(b) with a refractive index of 1.5 was used as non-volatile refractive index matching medium [9]. This ionic liquid has low background fluorescence comparable to that of non-fluorescence (NF) immersion Oil (Nikon) and a good refractive index matching with a general glass coverslip [9].

We prepared a toluene solution containing PDI as shown in figure 2(a) and 0.5 wt% poly(methylmethacrylate) (PMMA). The PDI/PMMA/toluene solution was spin cast onto a pre-cleaned glass coverslip substrate at 3000 rpm. This sample substrate was used in our experiments.

We performed epifluorescence microscopy of the sample with wide-field illumination by using a 488 nm cw laser. The fluorescence from the sample was collected by the same microscope objective and detected by an electron multiplying (EM) CCD camera (iXon3, Andor), after passing through a dichroic mirror, relay lens, and edge filter to eliminate the excitation light. The dipole orientation images were obtained by defocusing about 1 μm displacement of the immersion objective toward the sample substrate.

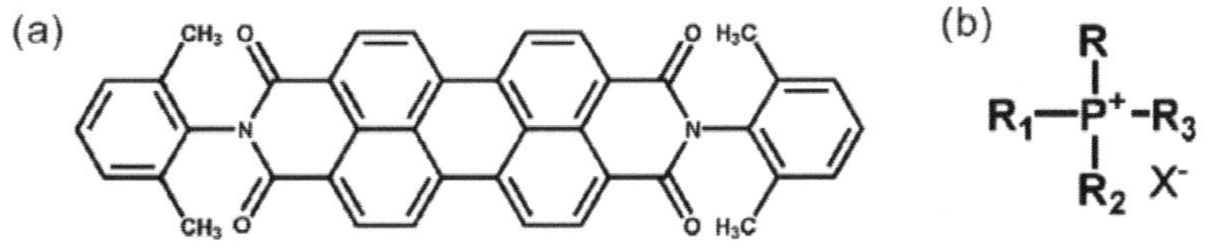

Figure 2. Chemical structures of (a) PDI and (b) ionic liquid, IL-AP3-3

3. Results and discussion

First, we confirmed the usefulness of the immersion objective with the ionic liquid, IL-AP3-3. The microscope unit shown in figure 1 was used without vacuum evacuation, that is, in air. Figures 3(a) and 3(b) show the detected images at the focal plane and by defocusing, respectively, obtained by using the ionic liquid and immersion objective. The detected images were essentially the same as those by using the immersion objective with Nikon NF immersion oil. These results indicate that the ionic liquid works as a good refractive index matching medium with low background fluorescence. We also tried to obtain defocused images by using a general high NA (0.95) microscope objective that is used without a refractive index matching medium. However, we failed to obtain defocused images similar to figure 3(b). Thus, the use of an immersion objective and refractive index matching medium is necessary for obtaining defocused images, as found a number of studies [10-12].

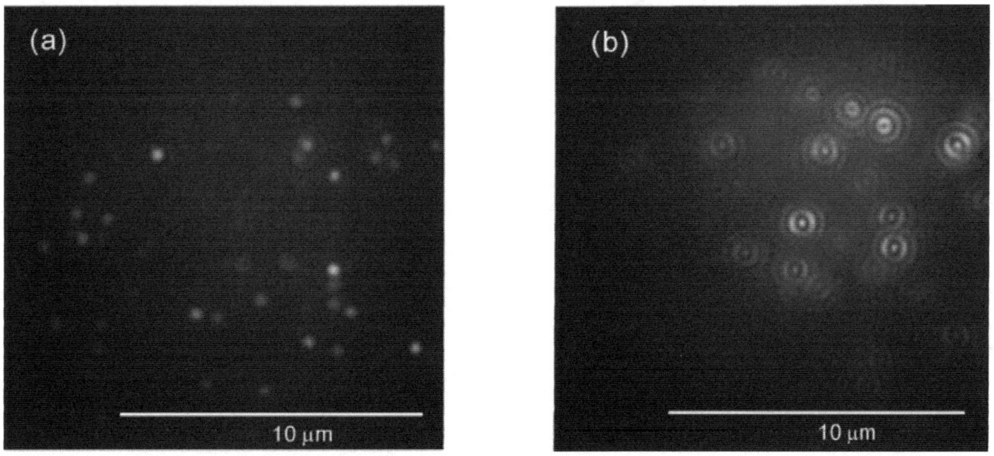

Figure 3. Detected images (a) at focal plane and (b) by defocusing, obtained by using the ionic liquid and the immersion objective in air. Bright spots in (a) correspond to single PDI molecules.

Figure 4(a) shows a typical defocused image obtained by using the microscope unit under a high vacuum condition; thus, we successfully obtained the defocused image. Figures 4(b) and 4(c) are selected magnified images. In Fig. 4(b) are found concentric rings in which the center is dark, and the corresponding dipole orientation has vertical orientation with respect to the surface plane. Figure 4(c) is found a typical dipole radiation pattern and the corresponding dipole orientation has parallel orientation with respect to the surface plane. Böhmer and Enderlein [10] investigated the relationship between the defocused images and dipole orientation. During the measurements in high vacuum, we observed a relative temporal stability of defocused images that would originate from inhibition of the photobleaching in the high vacuum. Figure 4(d) shows a typical defocused image obtained by using the microscope unit in pure N_2 gas with an appropriate pressure (0.12 MPa); thus, we successfully obtained the defocused image in ambient gas conditions. We for the first time succeeded in obtaining defocused images of single PDI molecules under air, high vacuum, and pure N_2 gas conditions, by utilizing the advantages of our microscope unit.

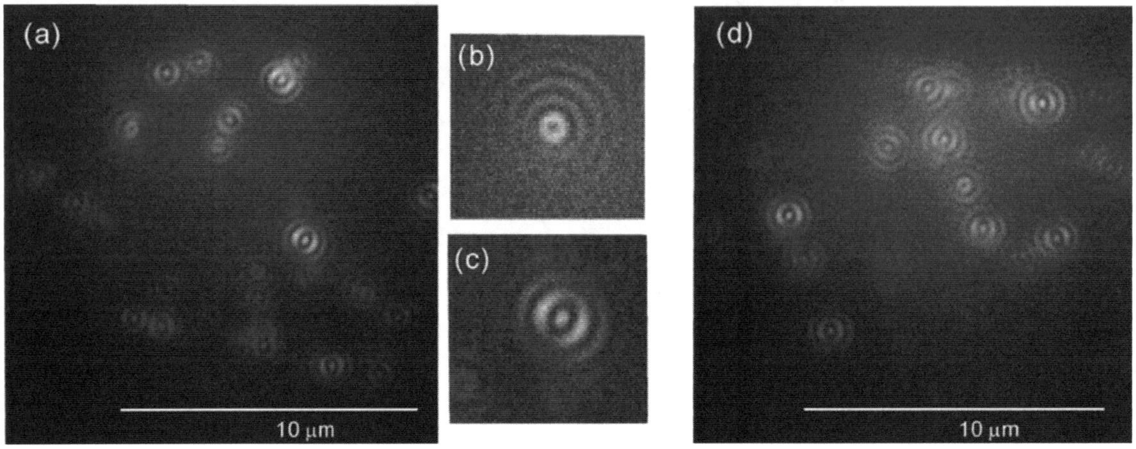

Figure 4. (a) Typical defocused image obtained by using the microscope unit under a high vacuum condition. (b) and (c) are selected magnified images. (d) Typical defocused image obtained by using the microscope unit in pure N_2 gas with an appropriate pressure (0.12 MPa)

4. Conclusions

We introduced the techniques that we previously developed for single molecule spectroscopy and then described the recent studies on dipole orientation imaging under various ambient conditions. In these studies, we successfully obtained defocused images of single PDI molecules under air, high-vacuum, and pure N_2 gas conditions by utilizing the advantages of our microscope unit. The application of our microscope unit is not restricted to single molecule fluorescence spectroscopy under various ambient conditions. It is expected to have a wide range of applications, such as surface plasmon, and Raman scattering, and nonlinear optical microscopy, which require high resolution and high brightness under various ambient conditions. We expect many applications for the unit in the fields of analysis of organic and inorganic thin films, and in nano-technology and bio-technology.

References

[1] Lakowicz R 2006 *Principles of Fluorescence Spectroscopy (Third Edition)* (Springer)
[2] Becker W 2008 *The bh TCSPC Handbook (Third Edition)* (Becker & Hickl GmbH)
[3] Yamada T, Goushi K, Xingsheng X and Otomo A 2008 *Thin Solid Films* **517** 1507
[4] Goushi K, Yamada T and Otomo A 2009 *J. Phys. Chem. C* **113** 20161
[5] Kaji T, Yamada T, Ueda R, Xingsheng X and Otomo A 2011 *Opt. Express* **19** 1422
[6] Kaji T, Yamada T, Ueda R and Otomo A 2011 *J. Phys.Chem. Lett.* **2** 1651
[7] Yamada T, Goushi K and Otomo A 2009 *Thin Solid Films* **518** 432
[8] Yamada T and Otomo A 2010 *Jpn. J. Appl. Phys.* **49** 01AE04
[9] Yamada T and Otomo A 2011 *Physics Procedia* **14** 7
[10] Böhmer M and Enderlein J 2003 *J Opt. Soc. Am. B* **20** 554
[11] Patra D, Gregor I, and Enderlein J 2004 *J. Phys. Chem. A* **108** 6836
[12] Chen X, Xu Y, Lan S, Dai Q, Lin X, Guo Q, and Wu L 2009 *Phys. Rev. B* **79** 115312

Electronic structure of the L-cysteine films on dental alloys studied by ultraviolet photoelectron spectroscopy

K Ogawa[1], T Tsujibayashi[2], K Takahashi[1], J Azuma[1], M Ichimiya[2], and M Kamada[1]

[1] Synchrotron Light Application Center, Saga University, Saga 840-8502, Japan
[2] Department of Physics, Osaka Dental University, Osaka 573-1121, Japan

E-mail: e7141@cc.saga-u.ac.jp

Abstract. The valence electronic structures of the dental alloys, type 1, type 3, K14, and MC12 and their interaction with L-cysteine have been studied by ultraviolet photoelectron spectroscopy with synchrotron radiation. It was found that the electronic structures of the type-1 and type-3 dental alloys are similar to that of polycrystalline Au, while that of the K14 dental alloy is much affected by Cu. The electronic states of the MC12 dental alloy originate dominantly from Cu 3*d* states and Pd 4*d* states around the top of the valence bands, while the 4~7-eV electronic structure of MC12 originates from the Ag 4*d* states. The peak shift and the change in shape due to alloying are observed in all the dental alloys. For the L-cysteine thin films, new peak or structure observed around 2 eV on all the dental alloys is suggested to be due to the bonding of S 3*sp* orbitals with the dental alloy surfaces. The Cu-S bond as well as the Au-S and Au-O bonds may cause the change in the electronic structure of the L-cysteine on type 1, type 3 and K14. For MC12, the interaction with L-cysteine may be dominantly due to the Pd-S, Cu-S, and Ag-O bonds, while the contribution of the Ag-S bond is small.

1. Introduction

Dental casting alloys are categorized into noble and base metal alloys. Noble dental alloys consist of Au, Ag, Cu, Pd, Pt, and other elements[1]. Anti-corrosive nature of the noble metals is the reason of their use with a long history. However, they are not completely immune to corrosion. When these alloys are used for dental restoration, they are repeatedly exposed to proteins contained in foods. It is therefore interesting to investigate the valence electronic structures of the dental alloys and their interaction with amino acids.

Amino acids are building blocks of protein, which is the dominant component of the living things. They are normally restricted to twenty kinds of L-form α-amino acids and only two of them contain sulfur[2]. Cysteine [HS-CH_2-CH(NH_2)-COOH] is one of the two such amino acids, and contains a sulfanyl group in its -CH_2SH side chain. The sulfanyl group (-SH) plays essential and ubiquitous roles in biology and bioelectronics[3]. For examples, the proteins involving an amino-acid cysteine are related to the movement of living cells and ion channel cellar[4]. The formation of a disulphide bond between two cysteine residues results in the unraveling of the α-helix and the distortion of the ligand-binding groove[5]. In order to investigate the functionality of cysteine, we measured the electronic structures of L-cysteine using ultraviolet photoelectron spectroscopy (UPS) with synchrotron radiation[6]. It was revealed that the highest occupied molecular orbital (HOMO) of L-cysteine consists mainly of the S $3sp$ orbitals.

Recently, much interest from the basic bioscience and bioelectronics focuses on the amino-acid cysteine on well-defined metallic surfaces and its adsorption geometry, chemical reaction, electrical function, and so on. Experimental and theoretical investigations on the cysteine/Au system showed interesting findings[7–11]. Beerbom *et al.* conducted a UPS measurement on L-cysteine film prepared by dipping on polycrystalline Au[7]. Although they succeeded to observe the photoelectron spectrum of L-cysteine thick film which fairly agrees with that in the present study, they had a difficulty in observing a weak signal from the interface between L-cysteine and the substrate. Canepa *et al.* performed metastable deexcitation spectroscopy (MDS) for L-cysteine thin films on the surfaces of Au, Ag, and Cu single crystals[8, 9]. Since MDS is very sensitive to the molecular orbitals protruding on the vacuum side from the film surface only, they could not detect the interfacial state between L-cysteine and the substrates. Therefore, there is still a lack of thorough investigation on the valence electronic structures of L-cysteine thin films on metallic surfaces so far. Previously, we measured the photoelectron spectra of the thick and thin L-cysteine films on the dental alloy, type 1, and found the difference between them due to the interaction of L-cysteine with the dental alloy surface[12]. In order to study the interaction of L-cysteine with the dental alloy surfaces in more details, the UPS spectra of the L-cysteine films with the thickness of 3 Å on four dental alloys including type 1 have been carefully measured under the same conditions in the present study[13].

2. Experimental

UPS experiments were carried out at the Saga-University beamline BL13 in the Saga Light Source[14]. Synchrotron radiation was provided from the 1.4-GeV storage ring and monochromatized using a plane-grating monochromator[15]. For photoelectron spectroscopy, a hemi-spherical electron analyzer, MBS A-1, was used with a pass energy of 10 eV, corresponding to a resolution of about 40 meV. UPS spectra were obtained with the photon energy ($h\nu$) of 40 eV, where the overall resolution involving the monochromator and photoelectron spectrometer was checked by measuring the Fermi edge of Au to be about 100 meV.

Four types of dental casting alloys, type 1, type 3, K14 and MC12[1], were studied in the present study. type 1 (Casting Gold M.C. Type I), K14 (M.C. Gold Alloy) and MC12 (Castwell M.C. Gold 12 %) were purchased from GC Corporation (Japan), while type 3 (Casting Gold M.C. Type III) was obtained from ISHIFUKU Metal Industry Co., Ltd. (Japan). The compositions of the alloys supplied from the manufacturers are shown in Tab. 1. Four polycrystalline metallic plates, Ag, Cu, Pd, and Au were purchased from Nilaco Corporation (Japan). Before measurements, the surface of these alloys and metals was cleaned up by Ar^+ ion sputtering with 0.5 kV in the sample-preparation chamber whose base pressure was about 8×10^{-8} Pa. The samples were transferred from the sample-preparation chamber to the analyzing chamber without exposing to air. UPS measurements were carried out in the analyzing chamber whose base pressure was about 4×10^{-8} Pa. The UPS spectra of polycrystalline Au, Ag, Pd and Cu plates were measured under the same conditions for comparison.

The L-cysteine films were prepared on the dental alloys by means of vacuum deposition. Powder samples of L-cysteine were purchased from Wako Pure Chemical Industries, Ltd. (Japan). The evaporation chamber was evacuated with a turbo molecular pump to about 4×10^{-5} Pa. A quartz thickness gauge was used to monitor the thickness of the deposited films. The deposition rate was set to be 0.3 Å/s. The L-cysteine films prepared on the alloys were transferred in situ to the analyzing chamber. It has already been confirmed with Raman scattering and atomic force microscopy that smooth L-cysteine films can be formed without decomposition in the present vacuum deposition method[6].

Table 1. The composition of the type-1, type-3, K14, and MC12 dental alloys.

Alloy	Composition (atomic%)				
	Au	Ag	Pt	Pd	Cu
type 1	69	18	0	0	13
type 3	54	14	2	3	24
K14	37	19	0	4	41
MC12	6	43	0	19	32

3. Results and discussion

Figure 1 shows the UPS spectra of the four dental alloys taken at $h\nu$=40 eV. The UPS spectra of polycrystalline Au, Ag, Pd and Cu are also shown. Each UPS spectrum of the dental alloys and metals shows characteristic structures reflecting different valence electronic states. In the spectrum of polycrystalline Au, there are Au $5d_{5/2}$ and Au $5d_{3/2}$ peaks about 3.0 and 6.0 eV, respectively[16, 17]. A peak about 4.2 eV and a shoulder about 6.6 eV observed on polycrystalline Ag are attributed to Ag $4d_{5/2}$ and Ag $4d_{3/2}$, respectively [16, 18]. The Pd $4d$ states have the large density of states in the range of 0∼4 eV. For polycrystalline Cu, a prominent peak from $3d$ states locates at 2.7 eV [17, 19]. We can understand the electronic structures of the dental alloys by comparing the UPS spectra in Fig. 1 and also by referring the existing reports on the electronic structures of metals and alloys[16–19].

The spectral features of type 1 and type 3 in the range of 2.0∼8.0 eV are very close to each other, and also they are in good agreement with that of polycrystalline Au because of the large content of Au, but the details are slightly different between them. In the present study, we synthesized the spectra for each alloy by simply summing the observed spectra of polycrystalline metals weighted according to the composition listed in Tab. 1. The synthesized spectrum for type 1 fairly agrees the features of the observed one with some exceptions (not shown). The peak at ∼6 eV in the synthesized spectrum was much smaller than that in the observed one. The difference may be due to the peak shifts of Au $5d$ and Ag $4d$ states upon alloying[16]. The discrepancy between the observed and the synthesized spectra for type 3 was much bigger than for type 1. One of the differences is the shoulder around 1.0∼2.0 eV which may be due to peak shift upon alloying of the Cu $3d$ states increased in type 3[17]. On the other hand, the peaks and shoulders are observed in the range of 1.0∼8.0 eV for the UPS spectra of K14 and MC12 which are not similar to that of polycrystalline Au. Since the shoulder around 1.9 eV in the spectrum of K14 is close to that of polycrystalline Cu, it may be due to the Cu $3d$ states with a small energy shift just as observed for type 3. Since the Au $5d$ states appear in the range of 2.0∼8.0 eV and also the peaks due to the Ag $4d$ states exist at 3.8∼7.7 eV, the peaks in the range of 2.3∼8.0 eV in the K14 are attributed to the mixture of Au $5d$ and Ag $4d$ states with small shifts and changes in shape. Since the Pd $4d$ and Cu $3d$ states have the large density of states in the range of 0∼4 eV and 2∼4 eV, the structure of MC12 in the range of 1∼4 eV may be due to the Pd $4d$ and Cu $3d$ states. The peaks at ∼5.0 and ∼6.3 eV are due to these Ag $4d$ states. In summary, the electronic states of type 1 originate dominantly from Au and additionally from Ag and Cu as constituents, and those of type 3 originate dominantly from Au and largely from Cu with additional contribution from Ag. The electronic states of K14 originate from Au, Ag, and Cu, while those of MC12 originate dominantly from the Cu $3d$ and Pd $4d$ states around the top of the valence bands and the 4∼7-eV structure originates from the Ag $4d$ states. The electronic structures of the dental alloys are not a simple sum of those of the constituents, showing energy shifts and changes in shapes upon alloying.

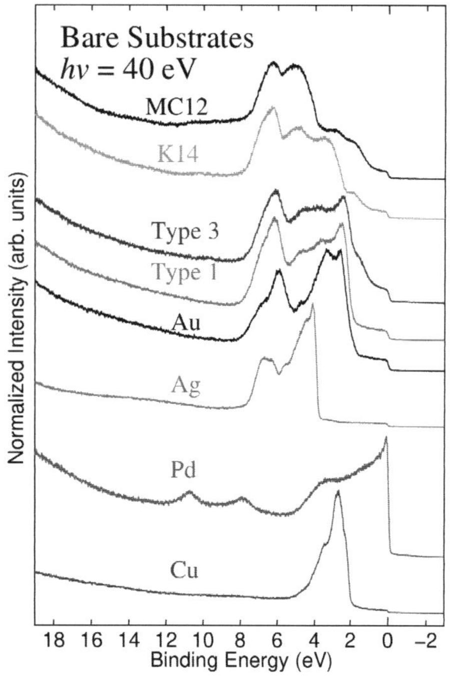

Figure 1. Photoelectron spectra of the dental alloys, type 1, type 3, K14, and MC12 taken at $h\nu$=40 eV. Photoelectron spectra of polycrystalline Au, Ag, Pd and Cu measured under the same conditions are also shown for comparison.

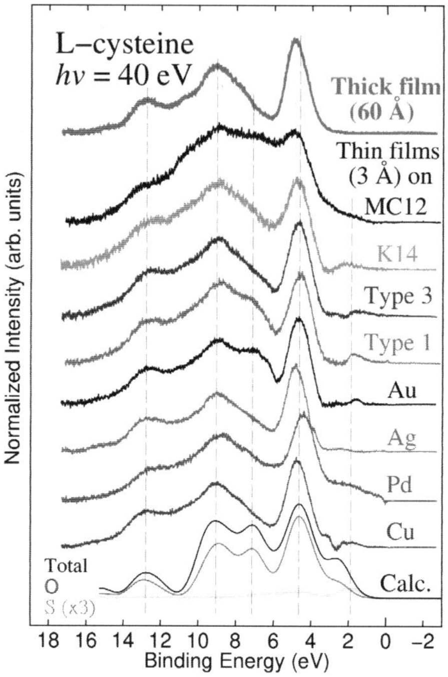

Figure 2. Photoelectron spectra of the L-cysteine thick and thin films on the dental alloys, type 1, type 3, K14, and MC12 taken at $h\nu$=40 eV. Photoelectron spectra of the L-cysteine thin films on polycrystalline Ag, Cu, Pd, and Au taken under the same conditions are also shown for comparison. At the bottom is the total and partial photoelectron spectra of L-cysteine molecule which are calculated by using the molecular orbital calculation and photoionization cross sections. Vertical lines are drawn as a mere guide to eyes.

Figure 2 shows the UPS spectra of L-cysteine thin films on the dental alloys with the thickness of 3 Å. The contribution of the substrates and the background due to inelastic scattering were subtracted on the assumption that the intensity just below the Fermi edge is due to the dental alloys and the background due to inelastic scattering is sufficiently smooth with no structure. In Fig. 2, the spectra of the L-cysteine thin film on polycrystalline Au, Ag, Pd, Cu and of the L-cysteine thick film (60 Å), which were measured under the same conditions, are also shown for comparison. The spectra of the L-cysteine thin films are apparently different from that of the thick film, which clearly indicates that they exhibit the electronic structures of L-cysteine interacting with the substrate dental alloys at interface. An additional peak appears around 2 eV for type 1, type 3, K14 and an additional structure appears for MC12 similar to polycrystalline Pd. The peak around 5 eV shifts to the lower binding energy side for type 1, type 3, and by 0.2 eV, while the peak for K14 shifts only by 0.1 eV. There is no appreciable shift for MC12. The structure around 7 eV changes for type 1 and MC12, but not for type 3 and K14. The peaks at 9 and 13 eV are relatively insensitive to the substrates. These discrepancies are due to the different interaction between the L-cysteine thin films and the substrate surfaces as explained below.

In order to understand the interaction, the electronic structure of L-cysteine molecule is required. In the previous study, the photoelectron spectra and HOMO and lowest unoccupied molecular orbitals of L-cysteine molecule were calculated using a molecular-orbital (MO) calculation and the photo-ionization cross section[6]. The MO calculation was carried out using the Gaussian03 program with the DFT B3LYP/6-311+G(d, p) method. The values of the photo-ionization cross section were taken from the well-known data[20]. The good agreement between the calculated and observed spectra of L-cysteine showed that the C3 geometry may be the most probable geometry among the seven possible geometries of cysteine molecule (C1∼C7). The calculated UPS spectra of L-cysteine molecule is shown in Fig. 2. The partial photoelectron spectra of constituent elements of O and S are also shown for convenience. Figure 2 clearly shows that the HOMO originates from the S $3sp$ orbitals, and the peaks around 5.0 and 7.4 eV are due to the O $2sp$ orbitals.

The additional peak at ∼2 eV on the L-cysteine thin films in Fig. 2 locates at the position very close to the valence states derived from the Au $5d$ and Cu $3d$ states in Fig. 1 and the S $3sp$ orbitals of L-cysteine in Fig. 2. Therefore, the formation of orbital bonding may occur between the Au $5d$ and Cu $3d$ states and S $3sp$ orbitals. The peak at 5 eV on the L-cysteine thick film shows some shifts to 4.9 eV and 4.8 eV on the thin films because of the interaction with the substrates. The O $2sp$ orbitals of L-cysteine exist around 5.0 eV as shown in Fig. 2. The large density of states (DOS) originating from the Au $5d$ and Ag $4d$ states exist around 4 eV. So, the interaction between the Au $5d$ and/or Ag $4d$ states and O $2sp$ orbitals may cause the shift of the 5-eV peak. The structure around 7 eV changes for type 1, MC12 and polycrystalline Au, but not for type 3, K14 and polycrystalline Ag. The DOS originating from the Au $5d$ and Ag $4d$ states extend in this energy as in Fig. 1. The O $2sp$ orbitals are dominant in the energy range in Fig. 2. Therefore, the interaction of the Au $5d$ states with O $2sp$ orbitals may change the 7-eV structure in type 1. The interaction of the Ag $4d$ states with the O $2sp$ orbitals may also change the 6∼8 eV structure because MC12 has a strong peak mainly due to the Ag $4d_{3/2}$ states at 6.3 eV upon alloying with Cu and Pd while the interaction between the Ag $4d$ states and O $2sp$ orbitals may be weak on polycrystalline Ag.

In conclusion, for the L-cysteine thin film on type 1, the new peak around 2 eV, the energy shift of the 5-eV peak, and the 7-eV shoulder structure are due to the Au-S, Au-O, and Au-O bonds, respectively according to the discussion above. Similar Au-S and Au-O bonds may be formed for the L-cysteine thin film on type 3, resulting in the 2-eV new peak and the shift of the 5-eV peak. The Cu-S bond may also contribute to the 2-eV peak, and a little more Cu atoms on the surface may hinder the formation of Au-O bond, resulting in the smaller 7-eV shoulder. For the L-cysteine thin film on K14, the energy position of the 2-eV peak is slightly different with others. The energy shift of the 5-eV peak is smaller for K14, even though the amount of Ag in K14 is not so different from those in type 1 and type 3. The 7-eV shoulder is not appreciable on K14. These differences may be due to the contributions from Cu since its composition is much higher. So, the Cu-S bond may contribute to the 2-eV peak, and Cu atoms on the surface may hinder the formation of Au-O bond, resulting in the small shift of the 5-eV peak and the smaller 7-eV shoulder just as for type 3. For the L-cysteine thin film on MC12, the spectrum shows only a small and broad structure around 2 eV just as the thin film on polycrystalline Ag. Therefore, it is supposed that the Ag-S bond is not strong, although Ag is the most dominant component in MC12. This may be due to the energy difference between the Ag $4d$ states and S $3sp$ orbitals and the shielding effect by conducting electrons from Cu and Pd components in MC12. The interaction of the S $3sp$ orbitals with Pd $4d$ and Cu $3d$ states may result in the broad structure around the top of the valence band. The energy shift of the 5-eV peak is not observed in Fig. 2. This may be due to the strong interaction of the Ag $4d$ states with the O $2sp$ orbitals. The interaction of the Ag $4d$ states with the O $2sp$ orbitals may also change the 6∼8 eV structure.

4. Summary

The electronic structures of the dental alloys, type 1, type 3, K14, and MC12 and their interaction with L-cysteine have been studied by UPS using synchrotron radiation. It was found that the electronic structures of type 1 and type 3 are similar to that of polycrystalline Au, while that of K14 and MC12 are much affected by Cu and Ag, respectively. The peak shifts and the changes in shape upon alloying are also observed in all the dental alloys. It should be noted that these electronic structures play an important role for the chemical interaction of the dental alloys with L-cysteine.

For the L-cysteine thin film on type 1, it was found that the electronic structure is almost identical with that on polycrystalline Au and is close to those on type 3 and K14 with some differences. A new peak or structure was observed around 2 eV for the L-cysteine thin films on all the dental alloys and elemental metals, while the 5-eV peak shifted to the lower binding energy side on type 1 and type 3 by 0.2 eV and on K14 by 0.1 V, and there was no appreciable shift on MC12. The 7-eV shoulder was observed on type 1 and MC12. It is suggested that the new peak or structure observed around 2 eV for the L-cysteine thin films on all the dental alloys may be due to the bonding of S 3*sp* orbitals with the dental alloy surfaces forming Au-S, Cu-S and Pd-S bonds, while the contribution of the Ag-S bond is small. The Au-O bond may be responsible for the shift of the 5-eV peak and the 7-eV shoulder. We believe that the deeper understanding of the electronic structure of L-cysteine films on the dental alloys leads to the improvement of dental alloy toward perfect anti-corrosive nature against amino acids.

Acknowledgments

A part of this work was supported by Nanotechnology Network Project (Kyushu-area Nanotechnology Network) of the Ministry of Education, Culture, Sports, Science and Technology (MEXT), Japan.

References

[1] Powers J M and Sakaguchi R L 2006 *Craig's Restorative Dental Materials* 12th ed (St. Louise: Mosby)
[2] Watson J D, Baker T A, Bell S P, Gann A, Levine M and Losick R 2007 *Molecular Biology of the Gene* 6th ed (San Francisco: Benjamin Cummings)
[3] Willner L and Katz E 2005 *Bioelectronics: From Theory to Applications* (Weinheim: Wiley-VCH)
[4] Montell C 2007 *Cell* **131** 19
[5] Hardie R C 2007 *Nature* **450** 37
[6] Kamada M, Sugiyama H, Takahashi K, Azuma J, Kitajima S, Ogawa K, Sumimoto M, Hori K and Fujimoto H 2010 *J. Phys. Soc. Jpn.* **79** 034709
[7] Beerbom M M, Gargagliano R and Schlaf R 2005 *Langmuir* **21** 3551
[8] Canepa M, Pelori P, Lavagnino L, Bisio F, Moroni R, Terreni S and Mattera L 2007 *Nucl. Instr. Meth. Phys. Res. B* **256** 324
[9] Canepa M, Lavagnino L, Pasquali L, Moroni R, Bisio F, De Renzi V, Terreni S and Mattera L 2009 *J. Phys.: Condens. Matter* **21** 264005
[10] Höffling B, Ortmann F, Hannewald K and Bechstedt F 2010 *Phys. Rev. B* **81** 045407
[11] Höffling B, Ortmann F, Hannewald K and Bechstedt F 2010 *Phys. Status Solidi C* **7** 149
[12] Ogawa K, Tsujibayashi T, KTakahashi, Azuma J, Kakimoto K and Kamada M 2011 *J. Electron Spectrosc. Relat. Phenom.* **184** 327
[13] Ogawa K, Tsujibayashi T, KTakahashi, Azuma J, Kakimoto K and Kamada M 2011 *J. Appl. Phys., doi:10.1063/1.3662146*
[14] Takahashi K, Kondo Y, Azuma J and Kamada M 2005 *J. Electron Spectrosc. Relat. Phenom* **144-147** 1093
[15] Sugiyama H, Takahashi K, Ogawa K, Azuma J and Kamada M 2010 *AIP Conf. Proc.* **1234** 685
[16] Wu S C, QWang Z, Sondericker D, Sokolov J, Garrett R F and Jona F 1986 *Phys. Rev. B* **34** 8418
[17] Nahm T U, Park K H, Oh S J, Chung S M and Wertheim G K 1995 *Phys. Rev. B* **52** 16 466
[18] Coulthard I and Sham T K 1996 *Phys. Rev. Lett.* **77** 4824
[19] Stöhr J, McFeely F R, Apai G, Wehner P S and Shirley D A 1976 *Phys. Rev. B* **14** 4431
[20] Yeh J J and Lindau I 1985 *Atomic Data Nuclear Data Tables* **32** 1

Local Structural Ordering in Cluster-Glass RE_2CuSi_3 (RE = Ce and Nd) compounds from HRTEM image

K. Yubuta, T. Yamamura and D. X. Li

Institute for Materials Research, Tohoku University, Sendai 980-8577, Japan

E-mail: yubuta@imr.tohoku.ac.jp

Abstract. Direct mapping of nano-domains with short-range order was presented via high-resolution transmission electron microscopy (HRTEM) on cluster-glass (CG) magnetic compounds RE_2CuSi_3 (RE=Ce and Nd) related to a hexagonal AlB_2-type structure. A new image processing technique using scanning a region of interest can be used in order to obtain the information of the structural ordering with nano-meter range. The degrees of CG on the short-range order, which shows $\frac{1}{2}\frac{1}{2}\frac{1}{2}$ -typed superstructure spots, are consistent with those of magnetic properties. The present observations strongly support that local structures with clustering relate directly to the origin of the CG magnetic behavior in RE_2CuSi_3.

1. Introduction

Recently ternary intermetallic R_2TX_3-type (R : U or rare earth metal, T : d-electron transition metal and X : Si, Ge or Ga) compounds have accumulated many researchers' attention as a representative of non-magnetic atom disordered compounds. R_2TX_3 compounds with disordered AlB_2-type structure [1-3], fully ordered U_2RuSi_3-type structure [1,4] and partially ordered U_2RhSi_3-type structure [1,5] show spin-glass (SG), paramagnetic and ferromagnetic cluster-glass (CG) behaviors, respectively. In previous studies, magnetic properties of Ce_2CuSi_3 [6,7] and Nd_2CuSi_3 [8,9] were measured. Ce_2CuSi_3 exhibits the CG behavior below the spin freezing temperature T_f = 2.7 K. In contrast the CG feature was also observed for the almost long-range ferromagnetic ordered compound Nd_2CuSi_3 below the Curie temperature in spite of its weak effect.

It is significantly important to investigate the crystallographic features of these compounds, because of the correlation between the magnetic properties and crystal structures in order to understand the observed CG behavior. Microscopic measurements such as small angle neutron scattering (SANS), μSR and TEM are very useful for the local crystal structure investigation. Recently, Marcano *et al.* investigated a percolative process of magnetic clusters for $CeNi_{1-x}Cu_x$ compounds by the SANS and μSR experiments [10-12]. From results obtained by many techniques, a phenomenological model with magnetic clusters was proposed; magnetic correlation length is about 2 nm at the cluster-glass state below T_f.

In addition, HRTEM is also a powerful tool for investigating structural features in the real space by the direct observation of microstructure, especially for locally ordered arrangements of atoms. Using a filtered FFT technique for HRTEM images, we have successfully presented the domain structure of

Published under licence by IOP Publishing Ltd

CG Ce_2CuSi_3 and Nd_2CuSi_3 compounds showing the $\frac{1}{2}\frac{1}{2}\frac{1}{2}$-typed superstructure spots [13]. However, it was hard to reveal features of small domain clusters in details because of a low resolution and quantitativity of filtered HRTEM images. In this paper, we re-present the domain structure of Ce_2CuSi_3 and Nd_2CuSi_3 by using a new image processing for observed HRTEM images.

2. Experimental

Polycrystalline samples of RE_2CuSi_3 (RE = Ce and Nd) were synthesized by melting stoichiometric amounts of constituent elements using an arc furnace in an argon atmosphere. The samples were then annealed at 800 °C for a week. X-ray powder diffraction patterns collected at room temperature using a diffractometer can confirm that the samples are single phases. The crystalline phases were identified by X-ray and electron diffractions. Thin samples for transmission electron microscopy (TEM) were prepared by dispersing crushed materials on holey carbon films. Electron diffraction (ED) patterns and HRTEM images were taken on TOPCON EM-002B, using an accelerating voltage of 200 kV at a resolution of 0.14 nm, and recorded using a charge coupled device (CCD) camera attached to the TEM. Image processing was performed using the *ImageJ* software [14].

3. Results and Discussion

No superstructure reflection cannot be detected by the X-ray diffraction measurement as reported in the previous study [13]. On the other hand, one can see clearly superstructure reflections in ED patterns of Ce_2CuSi_3 (Fig. 1(a)) and Nd_2CuSi_3 (Fig. 1(b)) compounds. ED patterns taken with incident beams parallel to the $[\bar{1}\bar{1}0]$ direction show that superstructure reflections are located at $\frac{1}{2}\frac{1}{2}\frac{1}{2}$-typed positions, as indicated by arrowheads with letters (s_1 and s_2). No diffuse scattering intensity indicates that the ordered phase exist in coherent with the matrix of AlB_2-type structure (the disordered phase). As reported in the previous studies [13,15], superstructure reflections originated from the new ordered structure, which is different from the U_2RuSi_3-type structure [4] and has a double stacking sequence of *ABAB*', *ABAB*'' and *AB*'*AB*'' along the *c*-axis. Figures 1(c) and 1(d) show HRTEM images of the Ce_2CuSi_3 and Nd_2CuSi_3 compounds, taken with the incident beam parallel to the $[\bar{1}\bar{1}0]$ direction. It can be noticed at first glance that image contrasts are homogeneous in the both observed images. Although it is not be recognized to exit the ordered structure producing superstructure reflections, the Fourier diffractograms of HRTEM in Figs. 1(e) and 1(f) show weak reflection spots at $\frac{1}{2}\frac{1}{2}\frac{1}{2}$-type positions as indicated by arrows. To make clear the effect of the superstructure reflections, an image processing was carried out by the Fourier filtering technique using only $\frac{1}{2}\frac{1}{2}\frac{1}{2}$-type superstructure reflections. In the image processing, fundamental reflections were not used in order to enhance the effect of superstructure reflections. As obliquely viewing of Figs. 1(g) and 1(h), the region with symmetry of a base-centered atomic arrangement can be noticed as strong fringes. Enhanced regions correspond to those having the ordered structure. The clusters in the Ce_2CuSi_3 compound exist with sizes of about 5 nm (this length corresponds to about 7 seven unit cell of the disordered AlB_2-type structure). The cluster size of Nd_2CuSi_3 is much larger than that of Ce_2CuSi_3 and the region of the ordered phase is extended on a right and bottom region in Fig. 1(h). That is, domains with the ordered structure in the Nd-system were well developed rather than those of the Ce-system. This feature is consistent with the magnetic properties obtained by measurements of ac magnetic susceptibility.

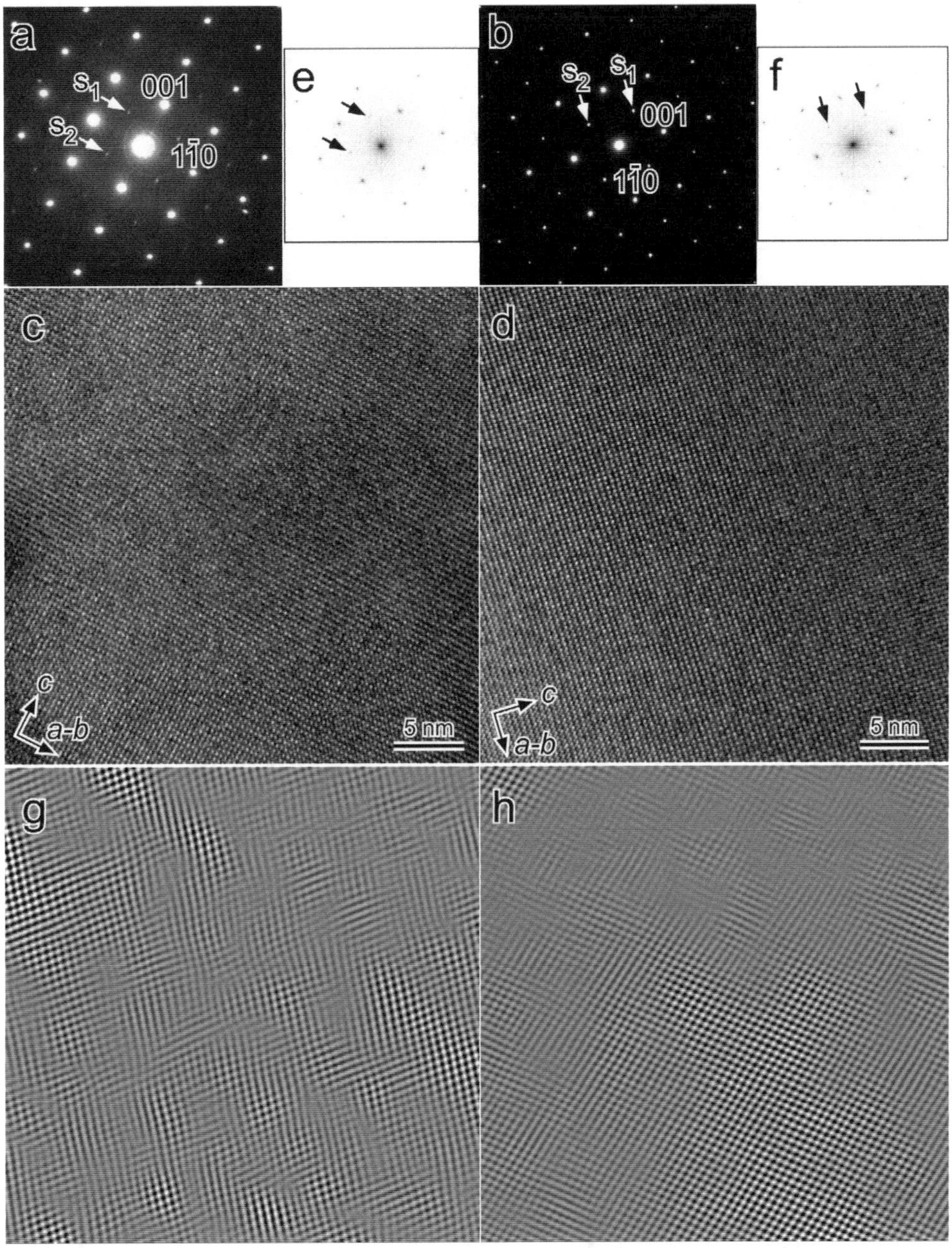

Figure 1. ED patterns (a, b), HRTEM images (c, d), power spectra (e, f) and Fourier filtered images (g, h) of the Ce_2CuSi_3, (left) and Nd_2CuSi_3 (right) compounds, taken with the incident beams parallel to the $[\bar{1}\bar{1}0]$ direction. The image processing was carried out by Fourier filtering technique, using superstructure $\frac{1}{2}\frac{1}{2}\frac{1}{2}$-typed reflections.

Although the Fourier filtered images give us significantly important information for the clustering, a spatial resolution and quantitativity of the ordering in the processed images were too low to discuss features of the clustering in details. Thus, in order to make the distribution of ordered domains clarify furthermore, new image analysis was carried out. Figure 2 illustrates a scheme of the image processing in this study. Each power spectrum was calculated from a region of interest that moves scanning on the image. Subsequently the intensity of an area which corresponds to the diffraction spot in the reciprocal space was measured. A size and a gradation of observed HRTEM images were 2048 × 2048 pixel^2 (~ 45.1 × 45.1 nm^2) and 16 bit gray scale, respectively. In the narrow regions, it can be postulated that thickness of samples is constant in the each sample. Power spectra were obtained from a square region having a size of 512 × 512 pixel^2 (~ 11.3 × 11.3 nm^2). A step width in the scanning of the square region was 4 pixel (~ 0.1 nm). Intensity of satellite and fundamental spots, I_s and I_f, was estimated as an integrated density of oval regions after background subtraction. Diffraction spots, $\frac{\bar{1}\,1\,1}{2\,2\,2}$, $\frac{\bar{1}\,1\,\bar{1}}{2\,2\,2}$ and 001, were chosen as satellite ones s_1, s_2 and fundamental one f, respectively, in Fig. 2(b). The degree of the structural ordering was defined as I_s / I_f in order to minimize dynamical effect in the real imaging. The present imaging procedure is considered as a kind of the *scanning dark-field imaging method*. It is believed that obtained maps were relatively credible to compare results between Ce- and Nd-compounds.

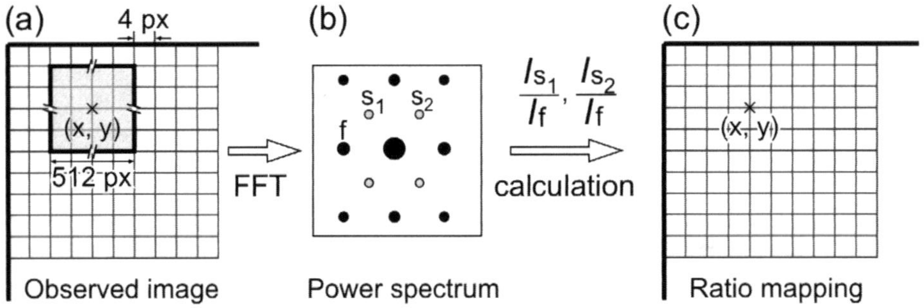

Figure 2. Schematic illustration of image analysis. A power spectrum (b) is performed from a gray square in (a) having a size of 512 × 512 pixels^2.

Figure 3 shows maps of the degree of the structural ordering, (a, b) I_{s1} / I_f, (c, d) I_{s2} / I_f and (e, f) (I_{s1} + I_{s2}) / $2I_f$, for the Ce_2CuSi_3 (left) and Nd_2CuSi_3 (right) compounds from Figs. 1(c) and 1(d). The high ratio of I_s / I, indicates that the high structural ordering occurs in the corresponding region. It can be recognized that a fine distribution of the clustering, which does not appear in Figs. 1(g) and 1(h), exist. The dispersed domain structure can be noticed more directly by the scanning-microscopic method compared to the Fourier filtering technique. From a detailed analysis, it was found in both systems that enhanced regions for s1 and s2 diffraction spots are slightly different, i.e., the mirror symmetry was broken. It is difficult for conventional enhanced FFT images as shown in Figs. 1(g) and 1(h) to find out the difference. The reason of the characteristic was considered that complex short-range order, rather than long-range order, originated from. In the other words, this result is due to three types of stacking sequences along the *c*-axis, namely, *ABAB*', *ABAB*'' and *AB*'*AB*'' for the ordered structure [13,15]. A size of clusters in the Ce-compound is estimated as 2 ~ 3 nm, on the other hand, that is more than 30 nm in the Nd-one. Apparently, larger ferromagnetic clusters are developed in the Nd-system. The topological feature is consistent with the magnetic properties reported in the previous studies [7,9].

It is demonstrated that the scanning dark-field imaging method is useful for investigating sscale ordered domain structures. A HAADF-STEM observation, which is generally interpreted as chemical atomic number Z-contrast imaging, is an effective method especially for a compositional modulation. On the other hand, because the frameworks were constructed by heavy *RE* atoms in the present compounds, it is difficult to detect the chemical ordering between relative light Cu and Si atoms by the HAADF-STEM. We consider that the present image processing is a suitable method for clearing the existence of the coherently short-range ordered domain structure in nano-meter order.

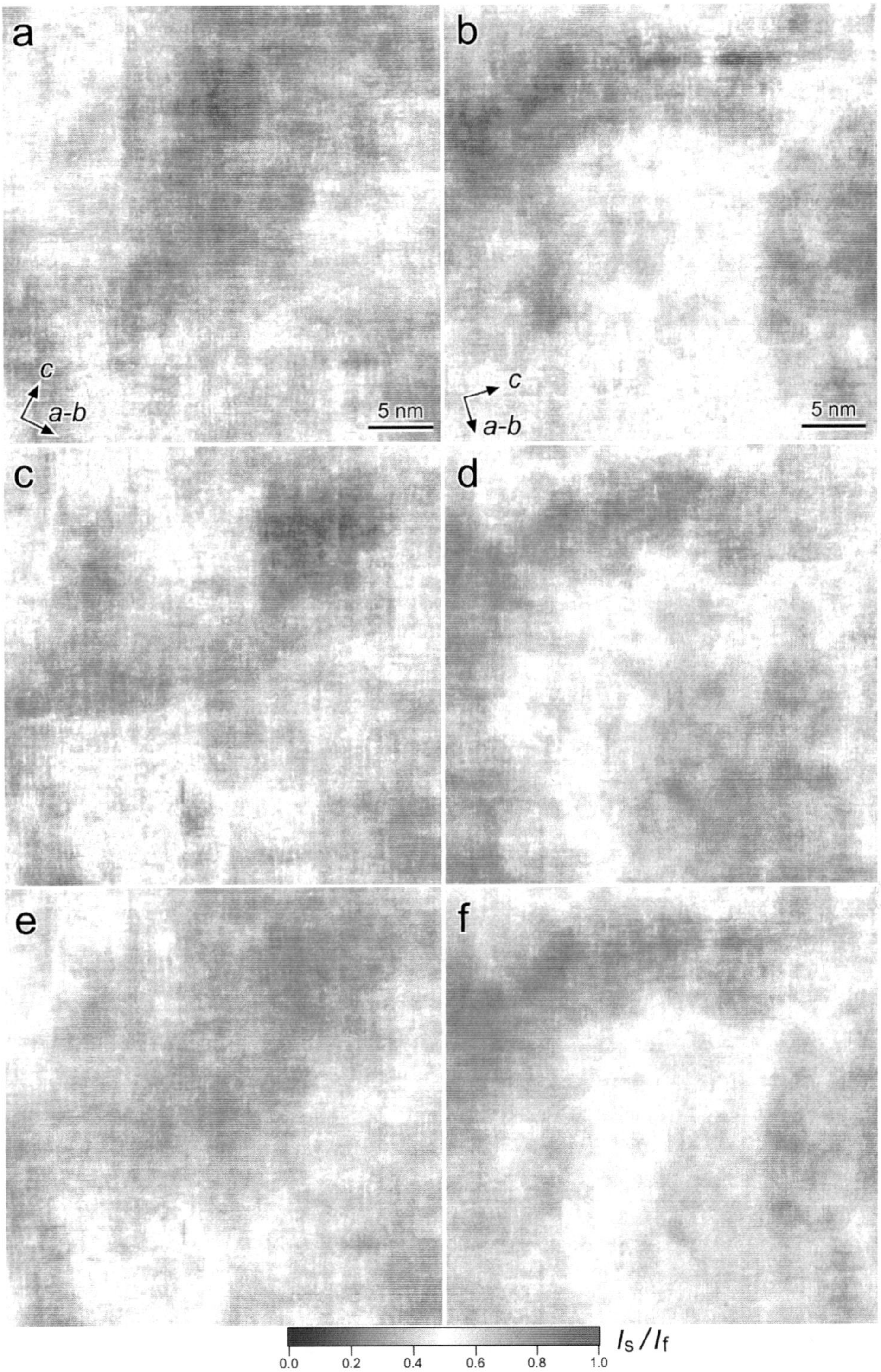

Figure 3. Maps of the degree of the structural ordering from observed HRTEM images (Figs. 1(c) and (d)) of the Ce_2CuSi_3, (left) and Nd_2CuSi_3 (right) compounds. Top, middle and bottom images are obtained using s_1, s_2 and s_1+s_2 satellite spots of Figs. 1(a) and (b), respectively. Red and blue pixels indicate strong and weak ordered regions.

4. Conclusions

We present the inhomogeneous distribution of ordered domains of the RE_2CuSi_3 (RE=Ce and Nd) compounds by using scanning microscopic processing for HRTEM images. Obtained mappings show more fine distributions of the ordered structure rather than those of the conventional FFT processing. In addition, mappings can exhibit the degree of the structural ordering more precisely. On the basis of the new image processing, it could be concluded that the observed domain structures are strongly related to an origin of the CG behavior of the RE_2CuSi_3 compounds.

References

[1] Chevalier B, Pöttgen R, Darriet B, Gravereau P and Etourneau J 1996 *J. Alloys Compd.* **233** 150
[2] Pöttgen R and Kaczorowski D 1993 *J. Alloys Compd.* **201** 157
[3] Kaczorowski D and Noël H 1993 *J. Phys.: Condens. Matter* **5** 9185
[4] Pöttgen R, Gravereau P, Darriet B, Chevalier B, Hickey E and Etourneau J 1994 *J. Mater. Chem.* **4** 463
[5] Li DX, Dönni A, Kimura Y, Shiokawa Y, Homma Y, Haga Y, Yamamoto E, Honma T and Onuki Y 1999 *J. Phys.: Condens. Matter.* **11** 8263
[6] Hwang JS, Lin KJ and Tien C 1996 *Solid State Commun.* **100** 169
[7] Li DX, Shiokawa Y, Nimori S, Haga Y, Yamamoto E, Matsuda TD and Onuki Y 2003 *Physica* B **329-333** 506
[8] Tien C, Luo L and Hwang JS 1997 *Phys. Rev.* B **56** 11710
[9] Li DX, Zhao X, Nimori S, Koyama K and Shiokawa Y 2009 *J. Phys.: Condens. Matter* **21** 026006
[10] Marcano N, Espeso JI, Gómez Sal JC, Rodríguez Fernández J, Herrero-Albillos J and Bartolomé F 2005 *Phys. Rev. Lett.* **71** 134401
[11] Marcano N, Gómez Sal JC, Espeso JI, De Teresa JM, Algarabel PA, Paulsen C and Iglesias JR 2007 *Phys. Rev. Lett.* **98** 166406
[12] Marcano N, Gómez Sal JC, Espeso JI, Fernández Barquín L and Paulsen C 2007 *Phys. Rev.* B **76** 224419
[13] Yubuta K, Yamamura T and Li DX 2009 *Solid State Commun.* **149** 286
[14] http://rsb.info.nih.gov/ij
[15] Yubuta K, Yamamura T and Shiokawa Y 2006 *J. Phys.: Condens. Matter* **18** 6109

Preparation of ZnO thin films by MO-CVD using fibrous bis (acetylacetonato) zinc (II) and ozone

K Haga[1], S Abe[1], Y Takizawa[1], K Yubuta[2] and T Shishido[2]

[1]RIC, Sendai National College of Tech. 4-16-1 Ayashi-chuo, Aobaku, Sendai, Miyagi 989-3128, Japan
[2]IMR, Tohoku University, 2-1-1 Katahira, Aobaku, Sendai, Miyagi 980-8577, Japan

haga@sendai-nct.ac.jp

Abstract. The fibrous bis (acetylacetonato) zinc (II) ($Zn(C_5H_7O_2)_2{\cdot}xH_2O$) was prepared with novel apparatus involved the vaporization and the recrystallization, and was investigated by differential thermal analysis (DTA), flourier transform infrared spectroscopy (FT-IR) with attenuated total reflection (ATR), and the inductively coupled plasma optical emission spectroscopy (ICP-OES). The zinc oxide (ZnO) films were prepared by the low-pressure CVD (LP-CVD) using the fibrous $Zn(C_5H_7O_2)_2{\cdot}xH_2O$ and ozone(O_3). The DTA curve of the fibrous material with a wide-ranging endothermic peak under 100°C according to desorption of the hydrated water separated to two peaks decreasing in the evaporation temperature. The resistivity of the ZnO films has changed widely from 10^4 to 10^8 Ωcm by the fibrous material with a different preparation condition.

1. Introduction

Zinc oxide (ZnO) including the basis metal zinc with a wide bandgap of 3.37eV is transparent in the visible region, and are a luminescence property and a photoelectric transfer characteristic in the near-ultraviolet region. The ZnO film with these properties is applied to transparent conductive film [1], near-ultraviolet light emitting device [2], and transparent thin film transistor (TFT) [3]. A high-quality ZnO thin film is prepared by molecular beam epitaxy (MBE) [4], pulse laser deposition (PLD) [5], and metalorganic chemistry vapor deposition (MO-CVD) [6-7]. In these preparation methods, MO-CVD is advanced the industrial production because the high vacuum is not needed, and the continuous feed of the material is possible, and has been introduced into a lot of semiconductor factories. However, the crystallinity and the surface flatness of oxide films prepared with MO-CVD decrease dramatically by an intense gas phase reaction of an organometallic material and the oxygen.

Acetylacetonato zinc with the hydrated water ($Zn(C_5H_7O_2)_2{\cdot}xH_2O$) that doesn't react intensely to the oxygen has been used as a zinc material in the present study [8]. A high-quality $Zn(C_5H_7O_2)_2{\cdot}xH_2O$ of the semiconductor grade was not sold in the market, and we had to use the material of the industrial use of the fine powder. Moreover, these materials solidified as a residue, and the evaporation of $Zn(C_5H_7O_2)_2{\cdot}xH_2O$ has not steadied. We recrystallized commercially available $Zn(C_5H_7O_2)_2{\cdot}xH_2O$ powder to vaporize in the vacuum chamber, and prepared the MO material of the fibrous shape not solidified easily. The fibrous material has the excellent carrier gas transmission characteristic that can be stably supplied without solidifying the MO material. The ZnO-TFT with the electron field-effect mobility of 5.2 cm^2/V.sec is obtained by using this material [9]. Moreover, homo-epitaxially growth ZnO films with excellent flatness and the crystallinity was deposited on the ZnO single-crystal substrate [10].

In this paper, we investigate the difference of the $Zn(C_5H_7O_2)_2{\cdot}xH_2O$ with the fibrous and the powder shape, and the property of the fibrous materials by using DTA, FT-IR with ATR, and ICP-OES. The resistivity of the ZnO films prepared by MO-CVD using the fibrous $Zn(C_5H_7O_2)_2{\cdot}xH_2O$ and ozone(O_3) were measured by the co-planer electrode of evaporated Al film. The structure and the composition of the ZnO films were analyzed with XRD, XPS and AFM.

2. Experimental Procedures

The fibrous $Zn(C_5H_7O_2)_2{\cdot}xH_2O$ produced by the unique method which has the process of vaporization and recrystallization $Zn(C_5H_7O_2)_2$ powder of 99.95% purity that is inexpensive, less toxic and less reactive to oxygen. The fibrous specimen was prepared by the apparatus shown in Figure 1. The preparation condition were the vaporized temperature of 160-180°C, and pressure of 500-1000hPa. The fibrous specimen recrystallized on the wire located in the upper part of vaporized container as shown in the insertion figure.

The ZnO films were prepared by the low-pressure CVD (LP-CVD) using the fibrous $Zn(C_5H_7O_2)_2{\cdot}xH_2O$ and ozone (O_3) [9]. MO gas carried with Ar flows in the reaction chamber. The substrate temperature was maintained at 400-530°C. O_3 gas was generated in the range of 0.60-2.04 g/h by adjusting both the flow rate of O_2 and the discharge voltage in the ozonizer.

The fibrous $Zn(C_5H_7O_2)_2{\cdot}xH_2O$ was analyzed by the differential thermal analysis 4(DTA), inductively coupled plasma optical emission spectroscopy (ICP-OES) and fourier transform infrared spectroscopy (FT-IR) with attenuated total reflection (ATR). The film thickness was measured by stylus instrument. Surface profile of the film was observed by atomic force microscope (AFM). The crystal structure of the film was evaluated by X-ray diffraction analysis (XRD). Chemical composition analysis was examined by X-ray photoelectron spectroscopy (XPS).The resistivity was determined by pA meter/DC voltage source (HP 4140B) with the evaporated aluminum (Al) electrodes.

Figure 1. Vaporization and recrystallization apparatus used to prepare fibrous $Zn(C_5H_7O_2)_2{\cdot}xH_2O$

3. Results and Discussion

DTA analysis for the fibrous $Zn(C_5H_7O_2)_2{\cdot}xH_2O$ were carried under vacuum of 500hPa similar to preparation condition of MO-CVD. The specimen was heated from room temperature to 220°C at the heating rate of 10°C/min. Fig. 2 shows the DTA curves of the fibrous specimen with the different preparation condition and commercially available $Zn(C_5H_7O_2)_2$ powder. Figure 2(a) is DTA curve of the specimen prepared by different vaporized temperature (V.T.) and pressure (V.P.). The result of the powder sample is shown as a comparison. It is thought that the endothermic peak within the range from 70 to 100 °C depends on the vaporization of the several molecules coupling hydrated water or the organic substance of low molecular weight. An endothermic peak of the temperature range of 120-150 °C depends on the melting point of the fibrous $Zn(C_5H_7O_2)_2{\cdot}xH_2O$. The endothermic within the range from 70 to 100 °C separated to two for the sample of V.P. =500hPa, and it was not able to be identified to the powder sample. The endothermic peak by the melting point almost becomes a constant without depending on the preparation condition, and it is assumed that all specimens maintain the acetylacetonato zinc structure.

Fig. 2(b) is DTA curves of the fibrous specimen prepared by V.T.=180°C and V.P.=500-900hPa. A large endothermic peak within the range of 70-100°C moves to the low temperature side as VP increases, and a small endothermic peak move to the high temperature side and disappear. It is reported that the hydrated water contained in the $Zn(C_5H_7O_2)_2{\cdot}xH_2O$ contaminates the substrate surface, and brings a remarkable degradation of the surface quality of the ZnO film [10]. Therefore, it is very important to reveal the relation between the amount in the hydrated water included in MO-material and the physical property of the ZnO film.

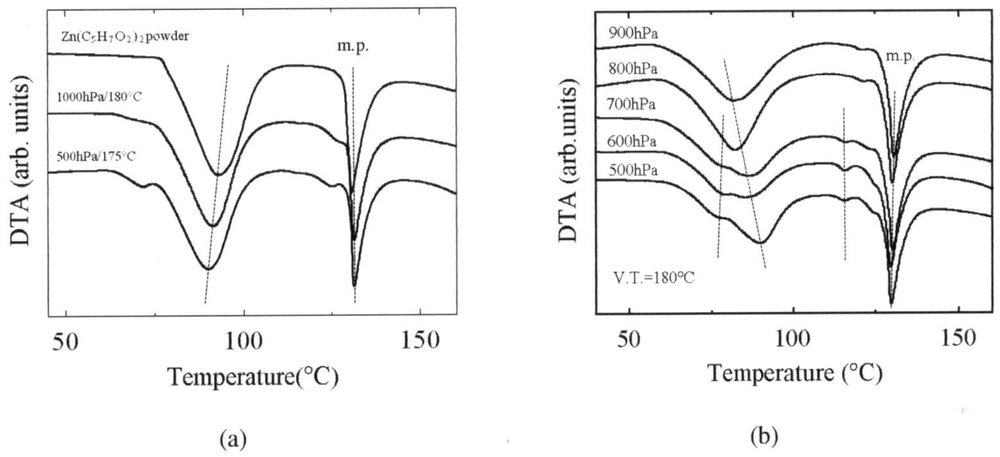

Figure 2. DTA curves of the fibrous $Zn(C_5H_7O_2)_2{\cdot}xH_2O$ prepared by different preparation condition

Table 1 is a composition ratio of the fibrous materials analyzed by ICP-OES. The powder material is shown as a comparison. The powder and the fibrous specimen prepared with V.P.=500hPa, V.T.=165°C and V.P.=800hPa, V.T.=180°C were analysed respectively. The composition ratio for four elements of Zn, C, H, and O was determined from the average of the measurement data. The difference of each element was not verified to all specimens including the powder material.

Table 1. Composition ratio analyzed with ICP-OES of fibrous specimen obtained by different preparation condition

Sample	Zn (at.%)	C (at.%)	H (at.%)	O (at.%)
Powder	22.7	41.1	6.4	29.9
500hPa/165°C	22.7	45.9	6.51	29.6
800hPa/180°C	22.5	46.6	5.95	29.7

Fig. 3 shows FT-IR spectrum with ATR of the same specimen as Fig. 2. The measurement result of the residue that remains in the vaporization container is shown in Fig. 3(a) as a comparison. All the characteristic IR vibrational frequencies have been assigned is in agreement with the proposed molecular structure. In Fig. 3(a) and (b), the O-H stretching frequencies were broadly found at 3650-

2750 cm^{-1}, and the symmetric and antisymmetric CH_3 vibrations observed around 2965 cm^{-1} and 2833cm^{-1} . The absorption band at 1518 cm^{-1} is due to the in plain bending vibration of CH_2 group and bands at 1138 cm^{-1} indicates the symmetric vibration of CH_3 group attached to carbon atom. From FT-IR spectrum of Fig. 3(a), the O-H vibration remarkably observed in the powder sample decreased in the fibrous specimen, and was hardly verified in the residue. Moreover, the symmetric vibration of CH_3 disappeared in the powder and the residue specimen. A remarkable difference of symmetric vibration of CH_3 was seen though FT-IR spectrum of almost every specimen was the same approximately in Fig. 3(b).

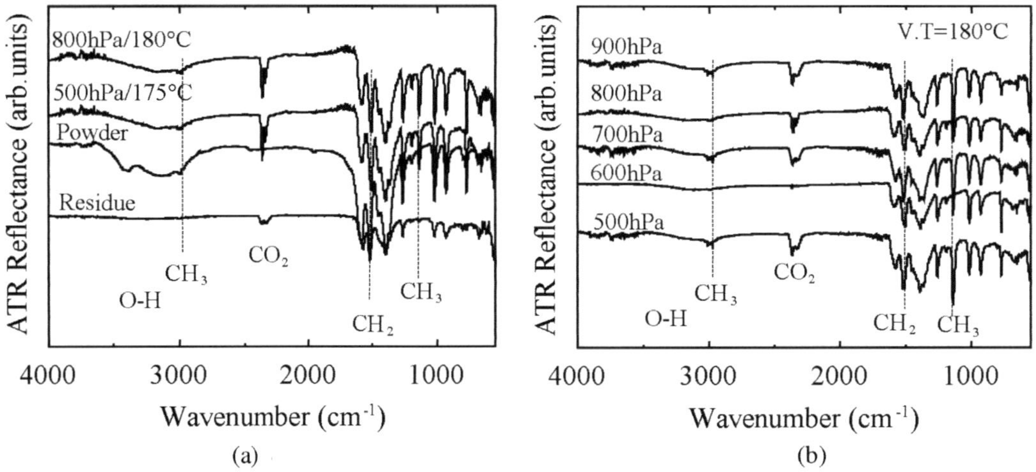

Figure 3. FT-IR spectrum of the fibrous $Zn(C_5H_7O_2)_2 \cdot xH_2O$ prepared by different preparation condition

Fig. 4 shows the effect of various CVD pressure and substrate temperature on the resistivity of the ZnO films. The fibrous $Zn(C_5H_7O_2)_2 \cdot xH_2O$ were prepared with V.T.=175°C and V.P.=500hPa. The ZnO films were prepared on the condition of deposition pressure 600-800hPa and the substrate temperature 500-530°C. As for the ZnO film deposited by the pressure of 600-800hPa, the resistivity has increased according to an increase of the substrate temperature. The results in Fig. 4 are different from the electric characteristics of the ZnO film deposited by the powder material. Generally, the crystallinity of the ZnO film improves according to the rise of the substrate temperature, and the resistivity of the ZnO film decreases with increasing the crystallinity of the film. The improvement of the crystallinity increases the grain size of the polycrystalline ZnO film, and increases the mobility of the film that brings a decrease in the resistivity. Fig. 4 shows a result quite opposite to the paper reported before [8]. As for a decrease of the resistivity of the ZnO film, the following reasons are supposed. The first reason is a bilayer theory that the ZnO film becomes a layer structure due to the decrease in the crystallinity of the ZnO film in the substrate interface. The following reason is a mobility improved model that the resistivity decreases because of an increase in the mobility of the ZnO film. However, the result of Fig. 4 doesn't coincide to both of the above-mentioned model. It is thought that an increase in the resistivity of the ZnO film according to the rise of the substrate temperature was obtained by the synergistic effect of the optimization of the stoichiometry by the introduction of active ozone and the improvement of the crystallinity.

Fig. 5 shows the relation between resistivity of the ZnO film and vaporized pressure (V.P.) of the powder material. Vaporized Temperature (V.T.) in the vessel including the powder material was set to 180°C, and the preparation condition of the ZnO films were the substrate temperature of 530°C and the deposition pressure of 650hPa, respectively. The resistivity of the ZnO film indicates the highest value at VP=800hPa in figure. It is very difficult to explain the reason why the resistivity of the ZnO film changes remarkably by different V.P. and V.T. value by using present data. The further analysis is needed to clarify these phenomena though it is considered that the structure of the material and the existence of the hydrated water greatly influence the electric characteristics of the ZnO film.

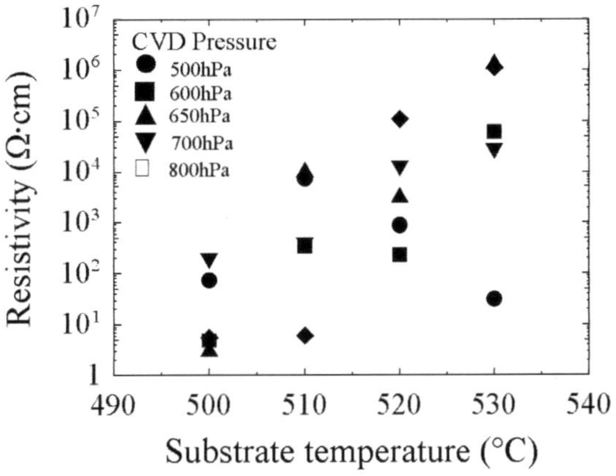

Figure 4. Dependence of the resistivity of the ZnO films as a function of substrate temperature at various CVD pressure.

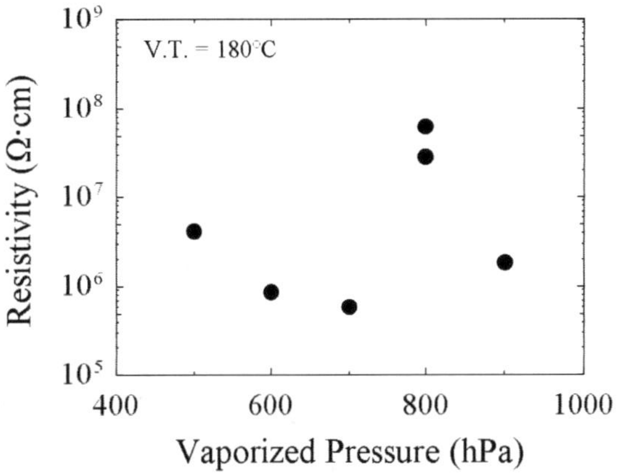

Figure 5. The resistivity of the ZnO films versus vaporized pressure of the fibrous $Zn(C_5H_7O_2)_2{\cdot}xH_2O$.

Fig. 6 shows ozone (O_3) density (g/h) dependence of the X-ray diffraction (XRD) spectrum for the ZnO films prepared on the Pyrex glass substrate at the substrate temperature of 500°C by using the O_2-O_3 mixed gas as an oxygen material. The fibrous $Zn(C_5H_7O_2)_2{\cdot}xH_2O$ materials prepared in V.T.=175°C, V.P.=500hPa were used for deposition of the ZnO film. As for all samples, the diffraction peak of the ZnO (0002) plane indicating the c-axis orientation was observed near 34.2 degrees. An increase in the diffraction peak intensity corresponding to the crystallinity of the ZnO film was observed according to an increase in the O_3 density. An increase in the O_3 density is assumed to improve activation energy to promote the ZnO crystal growth.

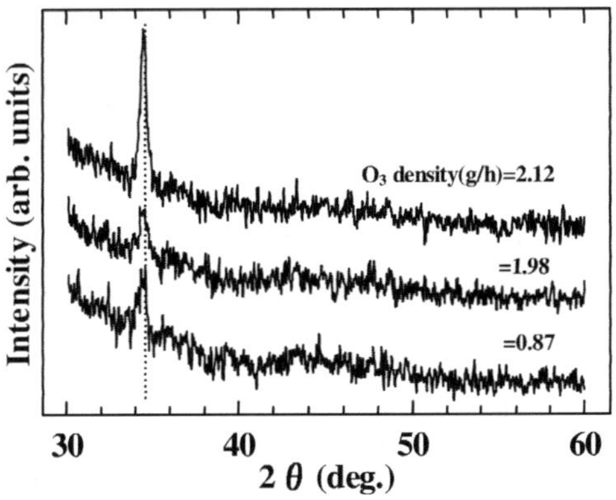

Figure 6. The ozone density dependence of the XRD spectrum for the ZnO films prepared on the Pyrex glass substrate.

Surface composition of these films were analysed by X-ray photoelectron spectroscopy (XPS). The Mg-Kα excitation was used for this study. The film was etched by Ar^+ ion beam in order to remove impurity on the surface. Zn-$2p_{3/2}$, Zn-$2p_{1/2}$ and O-1s can be observed easily. O-1s peak are observed at about 530.8 eV as normal shown in same binding energy, and Zn-$2p_{3/2}$, Zn-$2p_{1/2}$ in 1022 eV and 1045 eV, respectively. The C-1s peak that decided the amount of the carbon contained in the ZnO films was negligibly small.

Fig. 7 is an atomic force microscope (AFM) image to the same samples measured by XRD. The ZnO films in (a) - (c) are polycrystalline, and the crystal shape of (a) and (b) can identify a diameter as a sphere of around 0.05 μm from an AFM image. As for the AFM image of sample (c), the grain size increased to about 0.1 μm, and a dense, flat surface morphology was observed. The results of XRD spectrum and the AFM image show a similar tendency, and it is concluded that the increase of the ozone density contributes to expansion of the grain size and the planarization of the film surface.

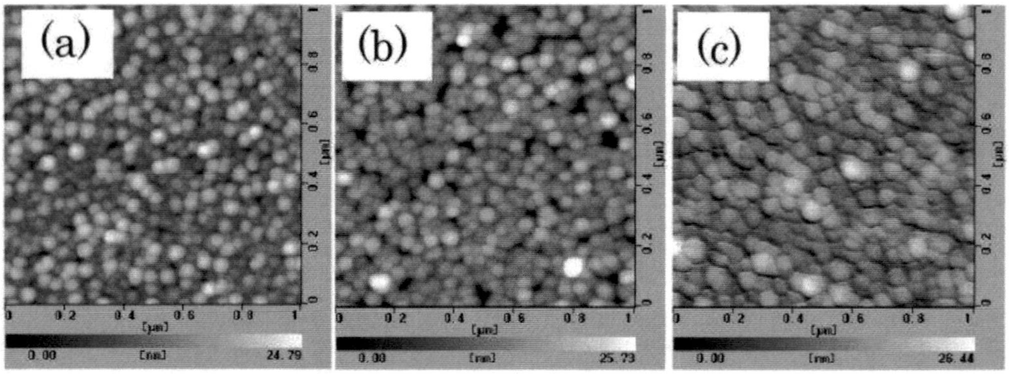

Figure 7. The AFM images for the surface morphology of the ZnO films on Pyrex glass substrate. The ozone density is (a) 0.87 (g/h), (b) 1.98 (g/h), (c) 2.12 (g/h).

4. Conclusion

The fibrous $Zn(C_5H_7O_2)_2 \cdot xH_2O$ was obtained by the unique method which has the process of vaporization and recrystallization the commercially available acetylacetonato zinc powder. Fibrous MO material to which the vaporization of the material gradually progressed from the interstice of the fiber was able to supply the material of Zn stably until the growth of the ZnO film finished. As for the DTA curve of the acetylacetonato zinc material, the temperature of an endothermic peak by the vaporization of the hydrated water in the fiber shape specimen is lower than that of the commercially available powder sample, and a small endothermic peak appears in a further low temperature regime according to the preparation condition of the fiber specimen. The resistivity of the ZnO film prepared by using the fibrous $Zn(C_5H_7O_2)_2 \cdot xH_2O$ was a very high value of 1.92×10^8 Ω cm at V.T.=175°C.

References

[1] Haga K, Katahira F and Watanabe H 1999 *Thin Solid Films* **343** 145.
[2] Hayamizu M, Tabata H, Tanaka H and Kawai T 1996 *J. Appl. Phys.* **80** 787.
[3] Hoffman R L, Norris B J and Wager J F 2003 *Appl. Phys. Lett* **82** 733.
[4] Vasco E, Rubio-Zuazo J, Vazquez L, Prieto C and Zaldo C 2001 *J. Vacuum Soc. Tech.* **B19** 224.
[5] Takai O, Futshara M, Simizu G, Lungu C P 1998 *Thin Solid Films* **318** 117.
[6] Chen Y, Bagnall D M, Zhu Z, Sekiguchi T, Park K, Hiraga K, Yao T, Koyama S, Shen M Y and Goto T 1997 *J. Cryst. Growth* **181** 165.
[7] Roth A P and Williams D F 1981 *J. Electrochem. Soc.* **128** 2684.
[8] Haga K, Wijesena P S and Watanabe H 2001 *Applied Surface Science* **504** 169.
[9] Haga K, Sakuma M, Takizawa Y and Seki S 2010 *Phys. Status Solidi* **C7** 1715.
[10] Seki S, Onodera H, Sekizawa T, Sakuma M, Haga K, Seki Y, Sawada Y, and Shishido T 2010 *Phys. Status Solidi* **C7** 1565.

Epifluorescence microscopic detection of photolithographically micropatterned aldehyde- and carboxy-terminated self-assembled monolayer

T Saito[1], Y Kyoda[2], M Sakurai[2] and A Hozumi[1]
[1] National Institute of Advanced Industrial Science and Technology (AIST), 2268-98 Anagahora, Shimoshidami, Moriyama-ku, Nagoya 463-8560, JPN
[2] Department of Applied Chemistry, Chubu University, 1200 Matsumoto-cho, Kasugai 487-8501, Japan

E-mail: t-saito@aist.go.jp

Abstract. An aldehyde (CHO)-terminated self-assembled monolayer (SAM) was prepared on silicon substrates covered with native oxide (SiO_2/Si) by chemical vapor deposition of triethoxysilylundecanal (TESUD). The CHO-terminated SAM was subsequently converted to the COOH-terminated SAM by photooxidation using 172-nm vacuum UV light through photomasks with square windows of 10×10 and 2.5×2.5 μm^2. The chemical reactivity of each microspot composed of the COOH-terminated SAM was detected with high resolution by epifluorescence microscopy using Qdot® 655. These results demonstrated that the microspots (spot size: 2.5×2.5 μm^2) composed of active COOH-terminated SAM were successfully fabricated on the SiO_2/Si substrate at a high density of 8×10^4 spots/cm^2, which was four times higher than that of previously reported miscrospots.

1. Introduction

Microarray technology has become a crucial tool for large-scale and high-throughput biological science and technology, facilitating fast, easy, and simultaneous detection of thousands of addressable elements within a single experiment under identical conditions [1]. In recent years, microarrays of DNA, proteins, and cells are becoming indispensable for studies on genomics, proteomics, and cellomics. Spatially defined immobilization of DNA, proteins, and cells [2] on the surfaces of various microarray platforms is an important technology. Many strategies for constructing protein microarrays on a wide range of substrates have been proposed in the past decades. The method for immobilizing proteins on surfaces will determine the functional properties of protein microarrays. There are two approaches for developing protein microarrays: tailoring the surfaces to functionalize natural target proteins and modifying target proteins with tags such as histidine, streptavidin, and biotin in order to effectively bind them to surfaces [3]. The former approach is considered to be more practical because it can tailor microarray surfaces to be used for wide varieties of proteins without compromising their functional properties. Almost all types of proteins have reactive functional groups such as carboxy (COOH), amino (NH_2), thiol (SH), and hydroxy (OH) groups in their side chains. These functional side groups can be used as a starting point for immobilizing proteins on tailored microarray surfaces. One of the most promising approaches to regulate site-selective attachment of proteins is employment of patterned self-assembled monolayers (SAMs) terminated with chemically reactive groups that bind to these functional side groups in proteins. Various constructing strategies and a wide range of substrates, such as glass, gold, polymer and other special formats, have been examined for constructing protein microarrays [4]. In a previous study [5], we reported the fabrication of well-defined microspots (5 × 5 μm^2) composed of the COOH-terminated SAM by photooxidation of triethoxysilylundecanal (TESUD, $[CHO]–[CH_2]_{10}Si[OC_2H_5]_3$) SAM on silicon substrates by masked 172-nm vacuum UV (VUV) light. The chemical reactivity was confirmed with a novel method using epifluorescence microscopy of the Qdot® 655 streptavidin conjugate tagged to biotin hydrazide bound to the activated COOH groups. In this study, we report the fabrication of microspots (2.5 × 2.5 μm^2) composed of the COOH-terminated SAM; these microspots were smaller than those previously reported using masked photooxidation of TESUD SAM [5]. Moreover, it was demonstrated these microspots could be detected with high resolution by epifluorescence microscopic detection using Qdot® 655.

2. Materials and Methods

2.1. Preparation of CHO-terminated SAM

The CHO-terminated SAM was prepared on silicon substrates covered with native oxide (SiO_2/Si) by chemical vapor deposition (CVD) of TESUD (Gelest Inc.), as previously described [3]. In brief, Si (100) wafers (1.5 × 1.5 cm^2) were exposed for 30 min at 10^3 Pa to 172-nm VUV light (λ= 172 nm, 10 mW/cm^2) radiated from an excimer lamp (UER20-172V; Ushio Inc.); the distance of between the lamp window and the sample surface was 20 mm. The VUV/ozone-cleaned Si wafers were placed with 0.1 cm^3 of TESUD diluted with 0.7 cm^3 of absolute toluene in a Teflon container having a volume of 65 cm^3 in a dry N_2 atmosphere with less than 10% relative humidity. The container was sealed and heated in an oven maintained at 130°C for 6 h. Each sample exposed to TESUD vapors was sonicated for 20 min successively in absolute toluene, absolute hexane, and acetone. Finally, the sample was rinsed with deionized water and blow dried with a N_2 gas stream.

2.2. Photooxidation of CHO-terminated SAM

Next, the CHO-terminated SAM was irradiated for 25 min at 10^5 Pa by the masked 172-nm VUV light radiated from a distance of 90 mm between the window and the sample surface [5]. As shown in Figure 1, the CHO-terminated SAM on SiO_2/Si substrates was fabricated by site-selective photooxidation with activated oxygen species ($O(^1D)$, $O(^3P)$) through photomasks with square windows of 10 × 10 and 2.5 × 2.5 μm^2.

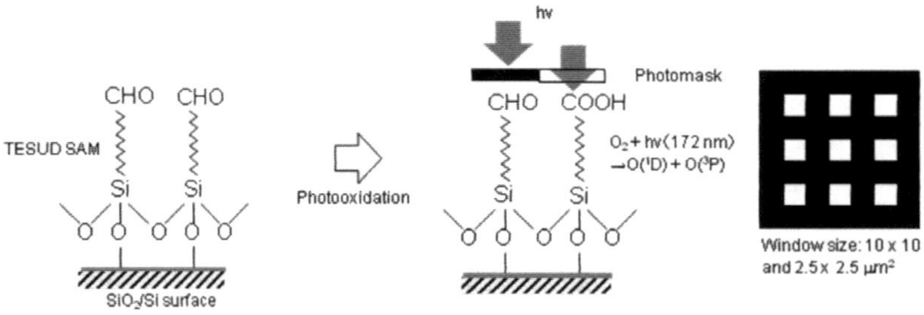

Figure 1. Fabrication of micropatterned CHO- and COOH-terminated SAM on the SiO_2/Si surface by photooxidation using masked 172-nm VUV light.

2.3. Chemical Reactivity of 172-nm VUV irradiated microspots and unirradiated regions.

To investigate the chemical reactivity in the 172-nm VUV unirradiated regions composed of the CHO-terminated SAM (Figure 2A), the micropatterned sample was immersed overnight in 1 ml of 50 mM phosphate buffer (pH 5.8) containing 5μg/ml biotin hydrazide at 4°C. Excess physisorbed biotin hydrazide was removed by washing with 50 mM phosphate buffer (pH 7.4) containing 1 M NaCl and

0.05% Tween 20. Next, the sample was incubated at 37°C for 2 h in a 2% phosphate buffer (pH 7.4) solution of the Qdot® 655 streptavidin conjugate (average diameter of ~15-20 nm, Invitrogen). On the other hand, to investigate the chemical reactivity of the 172nm-VUV irradiated microspots (Figure 2B), another micropatterned sample was immersed overnight in 1 ml of 50 mM phosphate buffer (pH 7.4) containing 1μg/ml bovine serum albumin (BSA) at 4°C to block the CHO-terminated regions. Next, the sample was washed with 50 mM phosphate buffer (pH 7.4) containing 1 M NaCl and 0.05% Tween 20. The sample was then incubated at room temperature for 2 h with 25 mM N-hydroxysuccinimide (NHS) and 20 mM 1-ethyl-3-(3-dimethylaminopropyl)-carbodiimide hydrochloride (EDC) in ethanol to activate the 172-nm VUV irradiated microspots. Immobilization of biotin hydrazide and subsequent labeling with the Qdot® 655 streptavidin conjugate were conducted in a manner similar to that performed in the 172-nm VUV unirradiated regions.

2.4. Epifluorescence microscopy
Fluorescence (Em: 655 nm) from Qdot® 655 tagged to biotin hydrazide immobilized on each sample surface (Figure 2A and 2B) was detected with an epifluorescence microscope (Olympus, BX51) equipped with a CCD camera (Olympus, DP70) and a U-MWG2 filter set (Olympus, excitation filter: 510–550 nm; emission filter: 590 nm)

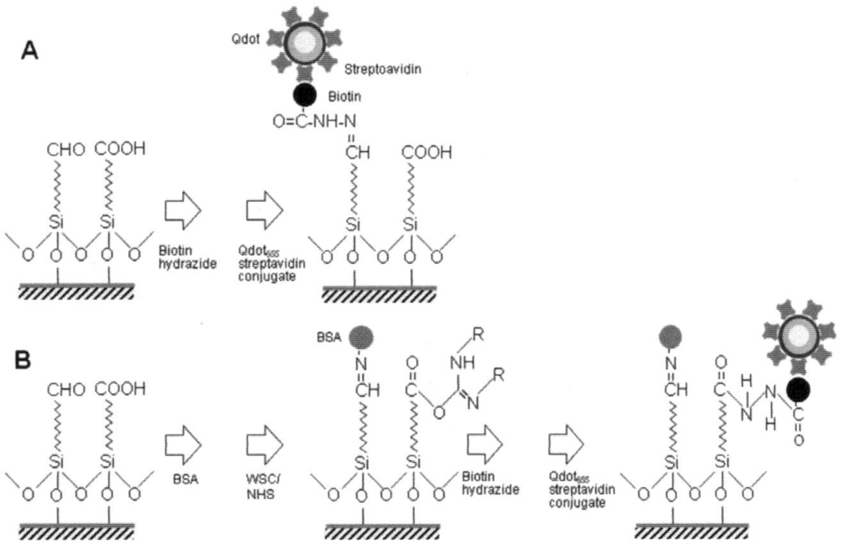

Figure 2. Labeling micropatterned CHO- (A) and COOH- (B) terminated SAM on the SiO_2/Si surface with the Qdot® 655 streptavidin conjugate.

3. Results and Discussion
As shown in previous studies [5, 6], the hydrophobic TESUD SAM surface on a silicon substrate became hydrophilic through the 172-nm VUV light irradiation and the water-contact angle decreased from approximately 80° to approximately 33° during irradiation for the first 25 min, while the thickness remained unchanged at approximately 1.1 nm, suggesting no alkyl chains were reacted during the VUV irradiation. To investigate the chemical reactivity in the 172-nm VUV irradiated and unirradiated regions, the photolithographically micropatterned samples using photomasks with the square windows of 10 × 10 and 2.5 × 2.5 μm^2 were treated with biotin hydrazide, which was subsequently tagged with the Qdot® 655 streptavidin conjugate and fluorescence from Qdot® 655 was detected with epifluorecence microscopy. As shown in Figure 3A, Qdot® 655 (red fluorescence) was selectively immobilized in the unirradiated regions but did not adsorb on the irradiated microspots (no fluorescence). Because NH_2 groups in biotin hydrazide react with CHO groups in TESUD SAM to

form a Schiff base, we concluded that the area-selective adsorption of biotin hydrazide in the unirradiated regions is based on the chemical reaction between NH_2 and CHO groups. Moreover, the fact that Qdot® 655 did not adsorb on the irradiated microspots indicated the area-selective photooxidation of CHO groups in TESUD SAM. If the photooxidation condition is precisely controlled for conversion of CHO groups to COOH groups, these irradiated microspots are expected to again become reactive to biotin hydrazide by activating COOH groups with NHS/EDC treatment. To confirm this, blocking the CHO groups in the unirradiated regions was carried out by BSA adsorption and subsequent activation of the irradiated microspots of 10×10 and 2.5×2.5 μm^2 with NHS/EDC treatment. As shown in Figure 3(B), well-resolved fluorescence from Qdot® 655 (see insets in Figure 3(B)) was detected only in the irradiated microspots, indicating that these microspots became reactive to biotin hydrazide, while no fluorescence was detected from the unirradiated regions where the reactivity of the CHO groups was blocked by BSA adsorption. These results show that the CHO groups in the 172-nm VUV-irradiated microspots were photochemically converted to COOH groups, which were further converted to active N-hydroxysuccinimidyl esters by the activation treatment with NHS/EDC. Therefore, stable amide linkages were formed between the activated COOH groups and biotin hydrazide.

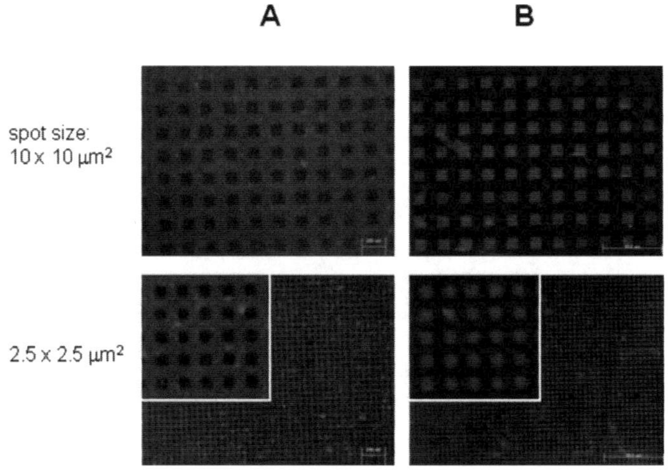

Figure 3. Epifluorescence microscopic images of micropatterned CHO- (A) and COOH- (B) terminated SAM on the SiO_2/Si surface. Each functional group was labeled with the Qdot® 655 (Em: 655 nm) streptavidin conjugate.

4. Conclusions

In this study, the COOH groups photochemically converted from the CHO groups of TESUD SAM were found to show excellent chemical reactivity to biotin hydrazide even in significantly small microspots (spot size: 2.5×2.5 μm^2). Therefore, such patterned microspots with chemically active COOH groups arrayed in high density (8×10^4 spots/cm^2) as shown in this study are expected to serve as protein microarrays for high-throughput protein analysis.

References

[1] Ito Y, Nogawa M, Takeda M and Shibuya T 2005 *Biomaterials* **26**, 211
[2] Saito T, Goto H, Yagihashi M and Hozumi A 2010 *Chem. Lett.* **39**, 1048
[3] Templin MF, Stoll D, Schrenk M, Traub PC, Vöhringer CF and Joos TO 2002 *Drug Discov. Today* **7**, 815
[4] Schäferling M and Kambhampati D 2008 *Protein Sci. Encyclopedia* (New York: Wiley-VCH)
[5] Hozumi A, Taoda H, Saito T and Shirahata N 2009 *J. Nanosci. Nanotechnol.* **9**, 455
[6] Hozumi A, Taoda H, Saito T and Shirahata N 2008 *Surf. Interface Anal.* **40**, 408

Au Coating of Carbon Nanofiber-Tipped SPM Probes for Immobilization of Thiolated Biomolecules

S Park[1], H Goto[1], T Saito[2], T Sonoda[2], M Kitazawa[3] and M Tanemura[1]
[1] Department of Frontier Materials, Nagoya Institute of Technology, Gokiso-cho, Syowa-ku, Nagoya 466-855, Japan
[2] National Institute of Advanced Industrial Science and Technology (AIST), 2268-98 Anagahora, Shimoshidami, Moriyama-ku, Nagoya 463-8560, Japan
[3] Olympus Co. Ltd., 6666 Inatomi, Tatsuno, Kami-Ina-Gun, Nagano 399-0495, Japan

E-mail: t-saito@aist.go.jp

Abstract. An argon-ion-induced carbon nanofiber (CNF) is expected to be an ideal nanocarbon for developing probes for scanning probe microscopes (SPM) owing to its superior physical properties and structures. To use this pristine CNF as a biofunctionalized SPM probe, its surface needs to be further modified with probing molecules that have specific biochemical affinities toward biological targets. In this study, using a carbon-coated Si wafer as a substitute for a CNF-tipped SPM probe, we investigated ion sputtering for the formation of an Au film on a carbon surface, followed by thermal annealing for making the Au-crystal orientation preferable for the chemisorption of thiolated biomolecules. Finally, we confirmed the biofunctionality of the immobilized thiolated oligonucleotide probe on the annealed Au film in order to hybridize it with the complementary oligonucleotide target. The hybrids were detected by fluorescence emitted from streptavidin-conjugated quantum dots tagged to the biotin molecules. From these results, it was shown that Au coating, thermal annealing, and the subsequent immobilization of thiolated biomolecules can be used for the development of new biofunctional CNF-tipped SPM probes.

1. Introduction

Tanemura *et al.* demonstrated that argon ion (Ar^+) bombardment of both bulk carbon [1] and carbon-coated substrates [2] induced growth of conical protrusions, and carbon nanofibers (CNFs) grew on the cone tips without the presence of any catalysts, even at room temperature. The CNFs created by Ar^+ bombardment were one-dimensional and amorphous nanocarbon materials without a hollow structure, having a high aspect ratio and nanoscale tip radius of curvature [2]. Considering their physical properties and structures, the CNFs, along with carbon nanotubes (CNTs), were expected to be an ideal material for use in the imaging probe of a scanning probe microscope (SPM) [3]. Several methods have been developed for fixing a single CNT on the apex of an SPM tip, which require manual or electrophoretical manipulation and additional post-processing [4-6]. These methods are considered to be time consuming and costly, and are not suitable for mass production of commercial CNT-tipped SPM probes. On the other hand, by bombarding Ar^+ to carbon-coated silicon (Si) cantilevers, batch growth of a linear-shaped single CNF with a typical diameter of 16-30 nm on each tip of the Si cantilevers has been achieved on a wafer scale [7, 8]. The CNF-tipped SPM probes have been mass-produced and supplied by Olympus Co. for practical use in nanotechnology. Recently, researches in nanobiotechnology need sensing and manipulating techniques using SPMs equipped with cantilevers containing biofunctionalized needlelike tips (nanoneedle) [9]. CNFs are considered to be superior to CNTs as a basal nanoneedle for developing biofunctionalized SPM probes because of their superior mass productivity compared with that of CNTs, as mentioned above. In the present study, using a carbon-coated Si wafer as a substitute for a CNF-tipped SPM probe, we examined ion sputtering to form an Au film on carbon-coated surfaces, thermal annealing to make the Au-crystal orientation preferable for chemisorption of thiolated biomolecules, and the immobilization of thiolated oligonucleotide probes on the annealed Au film.

2. Materials and Methods

2.1. Preparation of carbon-coated Si wafers

Prior to carbon vapor deposition, Si (100) wafers (5×5 mm^2; Shin-Etsu Chemical Co.) were cleaned by exposing them for 30 min at 10^3 Pa to 172-nm vacuum UV (VUV) radiation ($\lambda = 172$ nm, 10 mW/cm^2) from an excimer lamp (UER20-172V; Ushio Inc.) with a distance of 20 mm between the lamp window and the sample surfaces. A carbon layer of about 40 nm thickness was formed on the surface of the Si wafers by carbon vapour deposition using a carbon coating unit (Hitachi High-Technologies Co.). During the deposition process, the vacuum of the chamber was maintained below 7 Pa and the current was held at 20 A.

2.2. Au coating on carbon-coated Si wafers

A carbon-coated Si wafer was then coated with an Au film at a rate of about 6 nm s^{-1} by ion sputtering using an ion sputter coater (E-1020; Hitachi High-Technologies Co.) at a discharge current of 15 mA.

2.3. Thermal annealing of the Au film on the carbon-coated Si wafer

The Au film deposited on the carbon-coated Si wafer was annealed in air at 350°C for 1 h with a lamp heater (MILA-3000; ULVAC-RIKO Co., Japan). The crystal orientations of the Au films as-deposited and after annealing on the carbon-coated Si wafer were examined by X-ray diffraction (XRD) (MXP-1; Bruker AXS K.K.) with CuKα anode radiation generated at 40 kV and 20 mA. The crystal orientations were determined by comparing the relative intensities of Au (111) and Au (200) peaks. The surfaces of the Au films as-deposited and after annealing were observed by atomic force microscopy (AFM) (SPA400; SII Nanotechnology) and field emission scanning electron microscopy (FE-SEM) (S-4300; Hitachi High-Technologies Co.).

2.4. Immobilization of thiolated biomolecules

The annealed samples were incubated in 1 μM HS-$(CH_2)_6$-$(dT)_{30}$ for 2 h, washed with TE-1 M NaCl buffer three times, incubated in 1 mM 6-mercapto-1-hexanol for 1 h, washed with TE-1 M NaCl buffer thrice, rinsed with milli-Q water, and then dried in air. To confirm the biofunctionality of the immobilized HS-$(CH_2)_6$-$(dT)_{30}$ on the annealed Au film deposited on the carbon-coated Si wafer, samples irradiation by the masked 172-nm vacuum UV (VUV) were incubated in 0.1 μM 5'-biotin-labeled A_{30} at 0°C for 1 h, and then washed with TE-0.5 M NaCl buffer three times. The hybrids formed between $(dT)_{30}$ and 5'-biotin-labeled A_{30} were detected by tagging streptavidin-conjugated quantum dots (Q-dot$_{655}$$^{®}$) to the biotin. Fluorescence (Em: 655 nm) from Q-dot$_{655}$$^{®}$ coupled to the biotin hydrazide immobilized on the sample surface was detected using an epifluorescence microscope (BX51; Olympus Co.) equipped with a charge-coupled-device (CCD) camera detector (DP70; Olympus Co.), and a U-MWG2 filter set (excitation filter: 510-550 nm, emission filter: 590 nm; Olympus Co.).

3. Results and Discussion

Kitazawa *et al.* demonstrated that Ar^+-induced CNFs possess excellent mechanical properties similar to those of high quality CNT probes [10] and electrical properties superior to conventional electroconductive metal-coated Si probes [11]. Because of these excellent physical properties and mass productivity, the CNF-tipped SPM probe can be practically used with high resolution in nanotechnology. To use the CNF-tipped SPM probe as a biofunctionalized nanoneedle equipped in SPMs for sensing and manipulating living cells and biomolecules in life science and medical fields, the surface of the pristine CNF must be further modified with probe molecules having specific biochemical affinities toward these targets. One of the most promising approaches for immobilizing probing molecules is to coat CNFs with a thin layer of Au, and then subsequently functionalize the Au-coated CNFs with probing molecules having thiol terminal groups by forming self-assembled monolayers (SAMs) through the chemisorption of thiols on the Au surfaces. In the present study, conditions for Au coating and thermal annealing were examined initially using carbon-coated Si wafers as a substitute for the CNF-tipped SPM probe. An Au film was deposited by ion sputtering on the surface of a carbon-coated Si wafer with a carbon film of about 40 nm in thickness.

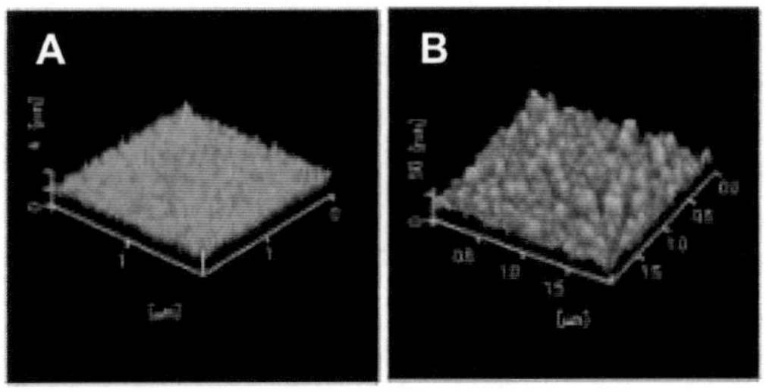

Figure 1. AFM images of Au films (A) as-deposited (thickness of 15 nm) and (B) after annealing at 350°C for 1 h.

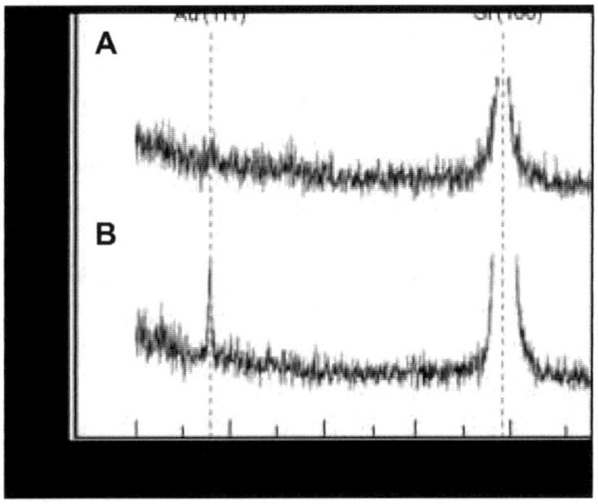

Figure 2. XRD patterns of Au films (A) as-deposited (thickness of 15 nm) and (B) after annealing at 350°C for 1 h.

The Au film deposited for 2.5 min had a thickness of about 15 nm and a surface roughness of about 1.2 nm (Fig. 1A). The deposited Au films with thicknesses ranging from 15 nm to 120 nm on the carbon-coated Si wafers were then annealed in air at 350°C for 1 h. As a result of the thermal annealing, growth of Au polycrystals was observed in the Au film by AFM, with more polycrystals seen in the thicker samples (Fig. 1B). The crystal orientations of the Au films as-deposited and after

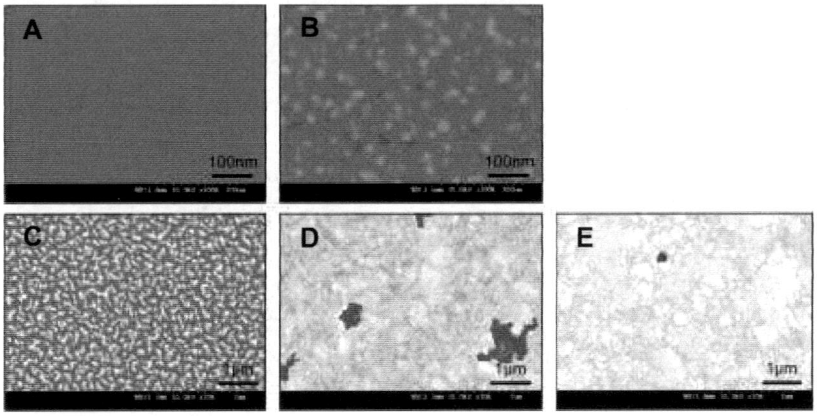

Figure 3. FE-SEM images of Au films (A) as-deposited (thickness of 15 nm) and after annealing at 350°C for 1 h with thickness values of (B) 15 nm, (C) 30 nm, (D) 60 nm, and (E) 120 nm.

thermal annealing were examined by XRD. As shown in Fig. 2, Au (111) orientation appeared to be predominant in the thermally annealed Au film, which is considered to be preferable for forming SAMs of probing molecules having thiol terminal groups.

As shown in Fig. 3, more pinholes were observed in the thinner films after thermal annealing, which might be caused by both the accelerated growth of the Au-crystalline particles at high temperatures and a weak affinity between Au and carbon. An oligonucleotide probe, HS-$(CH_2)_6$-$(dT)_{30}$, was then immobilized via thiol and Au binding. The biofunctionality of the immobilized HS-$(CH_2)_6$-$(dT)_{30}$ was confirmed by epifluorescence microscopic observation of the hybrids formed with a complementary target oligonucleotide (Fig. 4). From these results, Au coating and subsequent immobilization of thiolated biomolecules were shown to be suitable for developing new biofunctional SPM probes composed of CNFs.

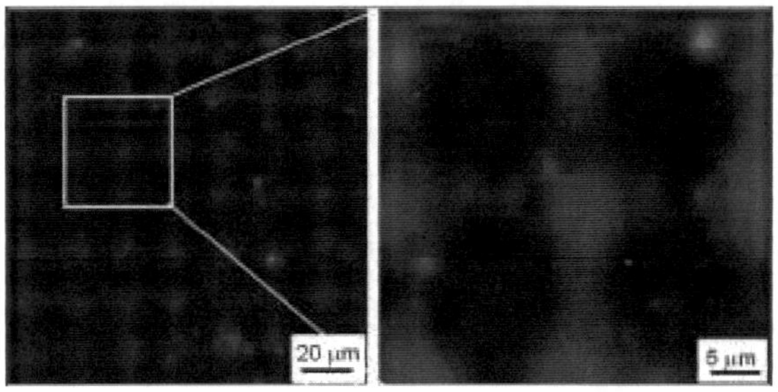

Figure 4. Epifluorescence microscopic observation of the hybrids formed between $(dT)_{30}$ and 5'-biotin-labeled A_{30} (red fluorescence). In advance to the hybridization, the immobilized HS-$(CH_2)_6$-$(dT)_{30}$ probes on the Au film were localy removed by masked VUV photooxidation.

4. Conclusions

Based on the results of Au film deposition on a carbon-coated Si wafer by ion sputtering, it was determined that deposition can occur with negligible increase in the surface roughness for Au films up to 120 nm in thickness. It was also found that thermal annealing enhances the growth of Au-crystalline particles, making Au (111) orientation predominant in the annealed Au film. Considering more pinholes were observed after thermal annealing in the thinner Au films on a carbon-coated Si wafer, it is necessary to improve the low binding affinity between Au and carbon, for example, by forming a buffer metal layer between them. The biofunctionality of immobilized HS-$(CH_2)_6$-$(dT)_{30}$ on the carbon-coated Si wafer was confirmed by epifluorescence microscopic observation of the hybrid formed with a complementary target oligonucleotide, A_{30}. Therefore, Au coating and subsequent immobilization of thiolated biomolecules were shown to be suitable for developing new biofunctional SPM probes composed of CNFs.

References

[1] Tanemura M, Okita T, Yamauchi H, Tanemura S and Morishima R 2004 *Appl. Phys. Lett.* **84**, 3831
[2] Tanemura M, Hatano H, Kitazawa M, Tanaka J, Okita T, Lau SP, Yang HY, Yu SF, Huang L, Miao L and Tanemura S 2006 *Surf. Sci.* **600,** 3663
[3] Iijima S 1991 *Nature* **354**, 56
[4] Dai H, Hafner JH, Rinzler AG, Colbert DT and Smalley RE 1996 *Nature* **384**, 147
[5] Hafner JH, Cheung CL and Lieber CM 1999 *Nature* **398**, 761
[6] Tang J, Yang G, Zhang Q, Parhat A, Maynor B, Liu J, Qin L-C and Zhou O 2005 *Nano Lett.* **5**, 11
[7] Tanemura M, Kitazawa M, Tanaka J, Okita T, Ohta R, Miao L and Tanemura S 2006 *Jpn. J. Appl. Phys* **45**, 2004
[8] Kitazawa M, Ohta R, Sugita Y, Inaba K and Tanemura M 2009 *J. Vac. Sci. Technol.* **B 27**, 975
[9] Obataya I, Nakamura C, Han SW, Nakamura N and Miyake J 2005 *Biosens. Bioelectron.* **20**, 1652
[10] Kitazawa M, Ohta R, Okita T, Tanaka J and Tanemura M 2007 *Jpn. J. Appl. Phys.* **46,** 6324
[11] Kitazawa M, Ohta R, Tanaka J and Tanemura M 2007 *Jpn. J. Appl. Phys.* **46,** 5607

Effects of electrical conductivity of substrate materials on microstructure of diamond-like carbon films prepared by bipolar-type plasma based ion implantation

S. Nakao, T. Sonoda

Materials Research Institute for Sustainable Development (MRISUS), National Institute of Advanced Industrial Science and Technology (AIST) – Chubu, 2266-98 Anagahora, Moriyama, Nagoya 463-8560, Japan

E-mail: nakao.s@aist.go.jp

Abstract. Diamond-like carbon (DLC) films are prepared by a bipolar-type plasma based ion implantation, and the structural differences between DLC films deposited on different electrical conductive substrates, i.e., conductive Si wafers and insulating glass plates are examined by Raman spectroscopy and x-ray photo emission spectroscopy (XPS). In the Raman measurements, graphite (G) and disorder (D) peaks are observed for both samples. However, the additional photo luminescence is overlapped on the spectra in the case of on-glass sample. To elucidate the structural difference, the intensity ratio of D to G peak (I(D)/I(G)), G peak position and full width at half maximum (FWHM) are obtained by curve fitting using Gaussian function and linear baseline. It is found that the I(D)/I(G) is lower, G peak position is higher and FWHM of G peak is narrower for on-glass sample than for on-Si sample. According to Robertson [1], lower I(D)/I(G) seems more sp^3 C-C bonding in amount for on-glass sample. In contrast, higher G peak position and narrower FWHM of G peak suggest less sp^3 C-C bonding in amount for on-glass sample. The results of XPS analysis with C1s spectra reveal that sp^3 ratio, i.e., the intensity ratio of $sp^3/(sp^3+sp^2)$ is smaller for on-glass sample than for on-Si sample. The inconsistency of the trend between I(D)/I(G) and other parameters (G peak position and FWHM of G peak) might be caused by the overlap of photo luminescence signal on Raman spectrum as to on-glass sample. From these results, it is considered that sp^3 C-C bonding is reduced in amount when using insulating substrate in comparison with conductive substrate.

1. Introduction

Diamond-like carbon (DLC) films have attracted much attention as a hard coating due to excellent mechanical properties such as high hardness and low friction coefficient, as reviewed in Ref. [1, 2]. Therefore, DLC films have been studied and prepared by many researchers using a lot of coating techniques, chemical and physical vapor depositions. Recently, plasma based ion implantation (PBII) has attracted much attention because of a possibility of three dimensional coating for industrial workpieces [3, 4]. In this technique, a high negative pulse voltage is applied to the workpieces so that they need electrical conductivity. On the other hand, a bipolar-type (subsequently to apply positive and negative pulse voltages) PBII system has been developed by AIST-Chubu, and it is possible to prepare the films on not only electrically conductive but also insulating materials [5]. To apply subsequent bipolar pulse voltages make the charge-up effect diminish when using the electrically insulating material. However, it is recognized that the microstructure of DLC films prepared even by the bipolar-type PBII system is slightly different depending on the electrical conductivity of the substrate materials. To make the difference clear, the microstructure of DLC films prepared on conductive and insulating substrates are examined by Raman spectroscopy and X-ray photoemission spectroscopy (XPS).

2. Experimental

DLC films were prepared on p-type Si(100) wafers (0.18 Ω/□) of 0.5 mm in thickness and quartz glasses of 1 mm in thickness by the bipolar-type PBII system. The details on the PBII system are described elsewhere [5, 6]. Toluene (C_7H_8) was used as a source gas and the flow rate was 4 sccm. Pulse frequency was 2 kHz. Positive and negative pulse voltages were +4 kV and -18 kV, respectively. The pressure during deposition was kept about 0.2 Pa. Deposition time was 15 min. Each thickness of DLC films was ~100 nm for on-Si sample and ~150 nm for on-glass sample. In the case of on-glass sample, the conductivity of the film was estimated to ~8 S/cm by the four probe method.

Micro-Raman spectroscopy (Renishaw, inVia) was performed to evaluate the microstructure of the films with 514.5 nm line of an Ar ion laser. The laser light was focused about 1 μm in diameter on the surface of the films. The laser power and typical measurement time were 10 mW and 10 s, respectively. XPS (Thermo VG Scientific, Sigma Prove) was carried out using a monochromated Al kα X-rays of 1486.6 eV.

3. Results and discussion

Figure 1 shows Raman spectra of DLC films deposited on (a) Si and (b) glass substrates. Two broad peaks are observed for both samples, as well known, disorder (D) and graphite (G) peaks. In the case of on-glass sample, photo luminescence signals are overlapped on the Raman spectrum. The intensity ratio of D to G peaks (I(D)/I(G)), the G peak position and the full width at half maximum (FWHM) of G peak are estimated by curve fitting using Gaussian function and linear baseline as shown in the figure. Figure 2 shows (a) the I(D)/I(G), (b) the G peak position and (c) the FWHM of G peak in each sample (on-Si and on-glass). In comparison with on-Si sample, on-glass sample has lower I(D)/I(G), higher G peak position and narrower FWHM of G peak. According to Robertson [1], lower I(D)/I(G) means larger sp^3 C-C bonding in amount in DLC films at stage 2. In contrast to this, higher G peak position and narrower FWHM of G peak suggest smaller sp^3 C-C bonding in amount as a consequence of the increase of aromatic ring clusters in the films. The trend of I(D)/I(G) is not consistent with that of other parameters, the G peak position and FWHM of G peak. To check sp^3 ratio in the films, XPS measurements were carried out.

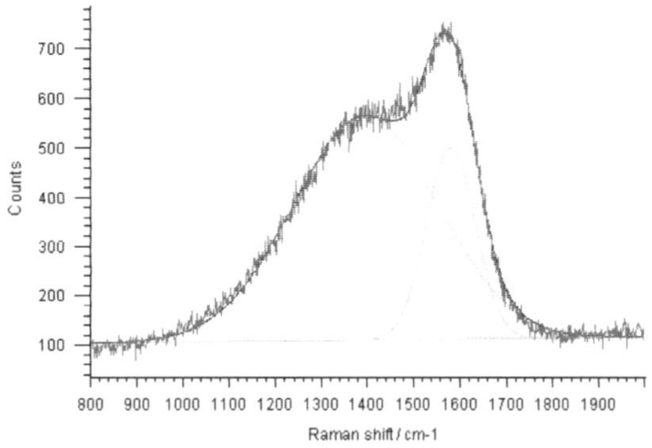

(a) Conductive Si substrate

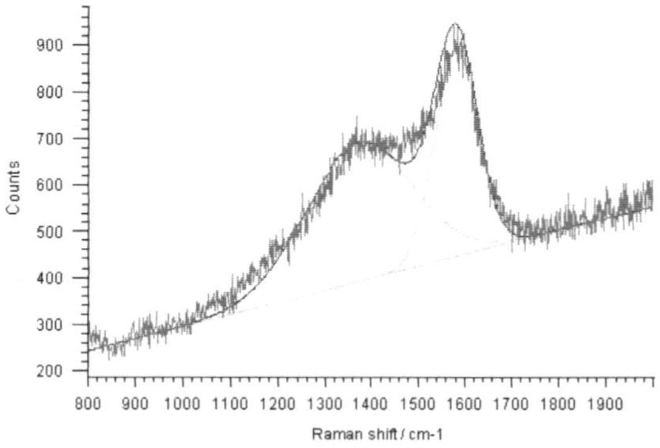

(b) Insulating glass substrate

Figure 1. Raman spectra of DLC films deposited on (a) Si and (b) glass substrates.

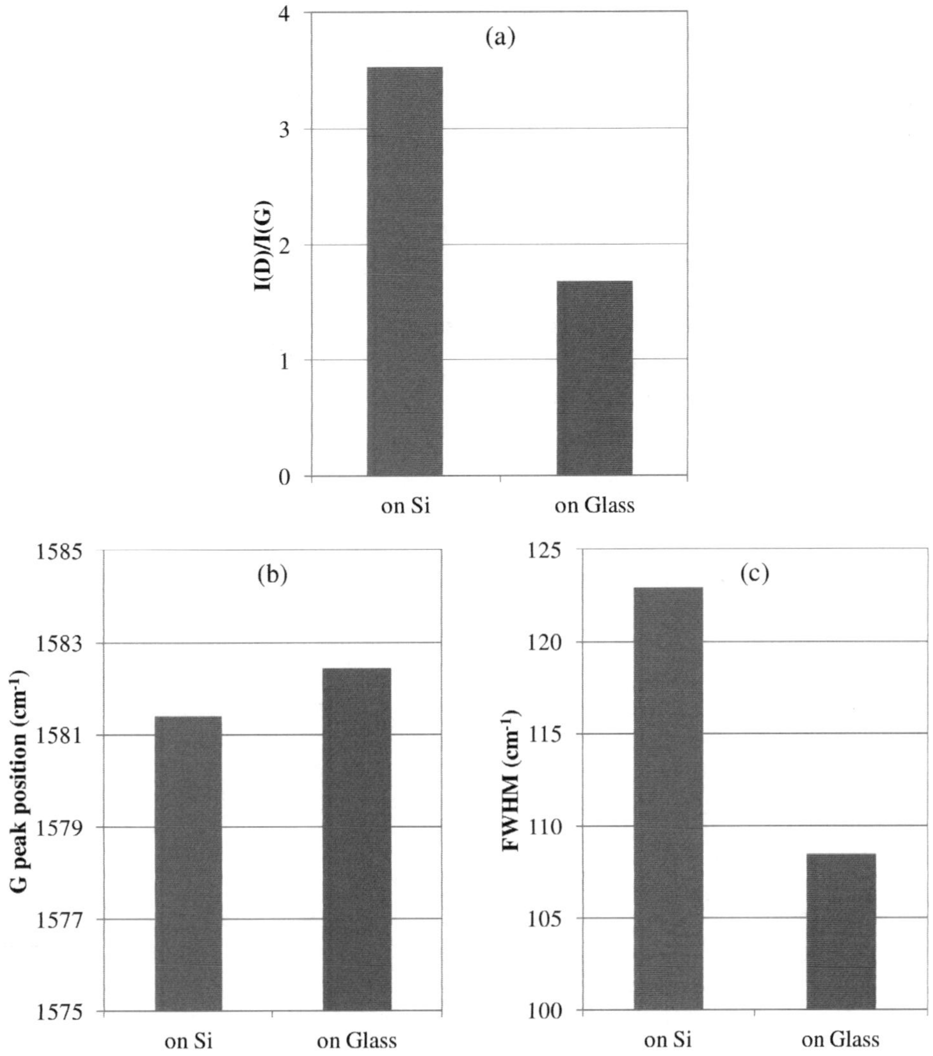

Figure 2. The parameters obtained from Raman spectra of DLC films on Si and glass substrates: (a) the intensity ratio of I(D)/I(G), (b) G peak position and (c) FWHM of G peak.

Figure 3 shows the XPS C1s spectra of (a) on-Si sample and (b) on-glass sample. The C1s core spectra is decomposed into 3 peaks of sp^2 C, sp^3 C and CO by curve fitting using Mix function and Shirley background line in order to estimate to sp^3 ratio, namely, the intensity ratio of $sp^3/(sp^2+sp^3)$. The sp^3 ratios are ~0.19 and ~0.15 for on-Si and on-glass samples, respectively. These results suggest that sp^3 C-C bonding is less amount for on-glass sample as compared with on-Si sample.

The results of XPS analysis are consistent with the trend of G peak position and FWHM of G peak, but not that of I(D)/I(G), in Raman analysis. In our opinion, the inconsistency of the trend of I(D)/I(G) should be caused by the overlap of photo luminescence signals on Raman spectrum for on-glass sample, as shown in Figure 1(b). The D peak may be buried in the wing of photo luminescence so that the I(D)/I(G) is probably underestimated. To make this point of view clear, it is necessary for another Raman measurement to avoid photo luminescence signals, e.g., UV-Raman analysis.

The presence of photo luminescence suggests the formation of polymer-like structure in the films [7]. It is noted that the thickness of the film for on-glass sample (~150 nm) is larger than that for on-Si sample (~100 nm). The fact also support to the consideration that low dense polymer-like structure is included in the case of on-glass sample. It is deduced that the charge-up effect still appears on the surface of glass substrate even in the bipolar-pulse mode, and then the suppression of the ion bombardment causes the formation of polymer-like structure. According to S. Hamaguchi [8], the ratio of the number of C atoms with each C-to-C coordination number depends on the incident energy for CH_3 and CH beam incidences in the molecular dynamics simulation. Low incident energy causes the increase of low C-to-C coordination and C atoms bonded with one or more H atoms. Therefore, it is reasonable to consider that the formation of polymer-like structure proceeds due to the charge-up effect and then the sp^3 ratio is reduced when using insulating substrate.

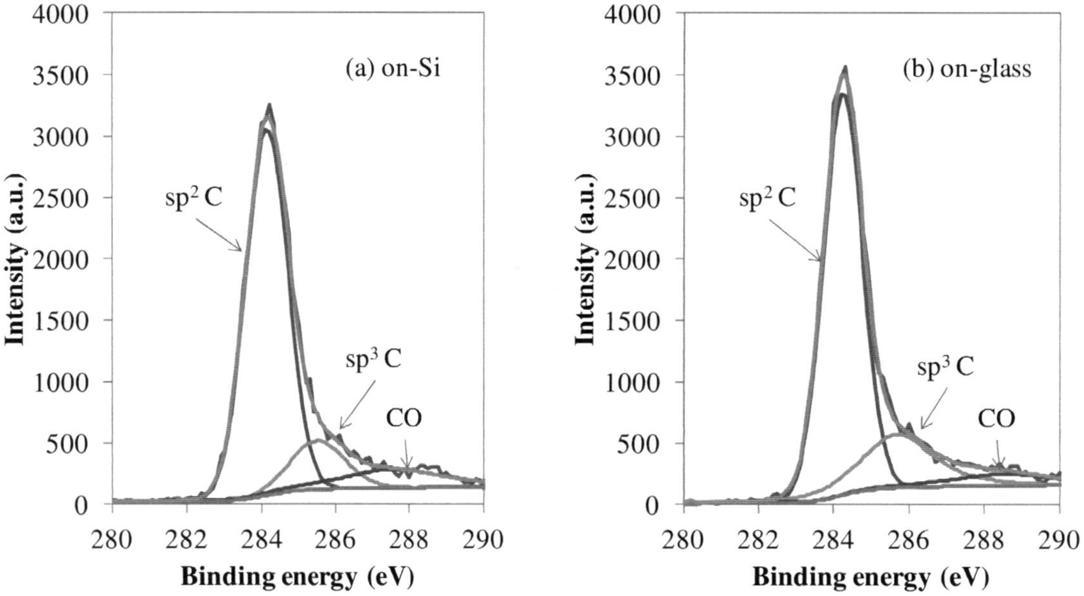

Figure 3. XPS spectra of DLC films of (a) on-Si and (b) on-glass samples.

4. Conclusion

DLC films are prepared by the bipolar-type PBII system and the structural differences in the films depending on the electrical conductivity of the substrate materials (p-type conductive Si wafer and quartz glass) are examined by Raman spectroscopy and XPS measurement. It is found that typical Raman spectra of DLC films, G and D peaks, are observed. In the case of on-glass sample, photo luminescence signals also appear, suggesting the presence of polymer-like structure. In the Raman analysis, the trend of G peak position and FWHM of G peak suggests sp^3 ratio for on-glass sample is smaller than that for on-Si sample. However, the trend of I(D)/I(G) indicates opposite. XPS measurement reveals that sp^3 ratio for on-glass sample is reduced as compared with that for on-Si sample. It is deduced that the inconsistency of the trend of I(D)/I(G) may be caused by the overlap of photo luminescence signals. From the results, it can be said that sp^3 ratio is reduced as a consequence of the formation of polymer-like structure when using insulating substrate.

References

[1] J. Robertson, 2002, Diamond-like amorphous carbon, *Mater. Sci. Eng. R* , **37**, 129-281.

[2] A. Frdemir and C. Donnet, 2006, Tribology of diamond-like carbon films: recent progress and future prospects, *J. Phys. D: Appl. Phys.* **39**, R311-R327.

[3] J. X. Liao, Z. Tian, E. Q. Li, J. Xu, L. Jin and H. G. Yang, 2007, The tribological properties of nanometre carbon films prepared by plasma-based ion implantation at various implanting voltages, *J. Phys. D: Appl. Phys.* **40**, 534-540.

[4] W. Ensinger, K. Volz, K. Baba and R. Hatada, 2007, Diamond-like carbon films formed by hydrocarbon plasma immersion ion implantation with methane/toluene mixtures, *Nucl. Instrum. Methods Phys. Res., Sect. B* **257**, 692-695.

[5] S. Miyagawa, S. Nakao, M. Ikeyama, and Y. Miyagawa, 2002, Deposition of diamond-like carbon films using plasma based ion implantation with bipolar pulses, *Surf. Coatings Technol.* **156**, 322-327.

[6] J. Choi, S. Nakao, M. Ikeyama and T. Kato, 2008, Effect of oxygen plasma treatment on the tribological properties of Si-DLC coatings, *Phys. Status Solidi. (c)*, **5**, 956-959.

[7] C. Godet and M. N. Berberan-Santos, 2001, Evidence for excitonic behavior of photoluminescence in polymer-like a-C:H films, *Diamond Relat. Mater.* **10**, 168-173.

[8] S. Hamaguchi, 2010, Plasma-surface Interactions in Material Processing, *J. Phys.: Conference Series* **257**, 1-11.

Effects of Substrate Temperature on Ti-Si-C Thin Film Deposited by Sputtering Using Elemental Targets

T Sonoda[1], S Nakao and M Ikeyama

National Institute of Advanced Industrial Science and Technology (AIST)

Email: tsutomu.sonoda@aist.go.jp

Abstract. Deposition of Ti-Si-C thin films by magnetron sputtering was examined using elemental targets of titanium, silicon and carbon, in order to investigate the effects of substrate temperature on the formation of carbide compounds such as TiC_x and/or Ti_3SiC_2, and thereby on mechanical properties and on physical properties of the thin films. It was found that Ti-Si-C thin films of both excellent electrical conductivity and high hardness could be formed by depositing at much lower temperature than the conventional depositing temperature such as 800 degree centigrade. And it was noteworthy that the Ti-Si-C film of electrical resistivity around 200μΩcm and hardness higher than 10GPa could be obtained by depositing at 500 degree centigrade.

1. Introduction

Ternary carbide compounds known as $M_{n+1}AX_n$ phases (M: early transition metal, A: A-group elements, X: C and/or N, n=1~3) have not only the benefits of ceramics such as resistance against oxidation and a high decomposition temperature but also metallic properties like ductility, electrical and thermal conductivity [1,2]. Thin films consisting of the MAX-phase family, in particular Ti_3SiC_2 phase, can be highly expected to be applied for electrical contacts, etc., thanks to their own good electrical conductivity. However the formation of Ti_3SiC_2 phase is competitive with the growth of TiC_x phase through Si segregations or Si substitutional incorporations in TiC_x [3]. Moreover thin films within the MAX phase can only be synthesized at preparation temperatures above 700 degrees centigrade and on textured or single crystalline substrates [4].

Nevertheless lower preparation temperatures are more practical in applying the thin films for protective coatings of not only metallic conductivity but also excellent resistance against wear and/or corrosion. Therefore the deposition of Ti-Si-C thin films by magnetron sputtering was examined using elemental targets of titanium, silicon and carbon, in order to investigate the effects of substrate temperature on the formation of carbide compounds such as TiC_x and/or Ti_3SiC_2, and thereby on mechanical properties such as hardness and on physical properties such as electrical conductivity of the thin films.

2. Experimental procedure

A planar magnetron sputtering system (ANELVA L-332S-FHS) with three cathodes was used. The planar targets used for the sputter-deposition of Ti-Si-C thin films were a pure titanium disk, a pure silicon disk, and a pure carbon disk of 80 mm diameter. Si(100) substrates (12×10mm^2, thickness 0.35mm) were mounted on the substrate holder at room temperature, and heated at 500 or 800 degree centigrade. The thin films were deposited on the substrates in an atmosphere of argon at a pressure of 0.4 Pa by co-sputtering of the three elemental targets with an RF electric power supply and two DC electric supplies respectively, where an independent electric power source was used for the sputtering of each target. The two DC power supplies for sputtering the titanium target and carbon one were respectively fixed at DC120W and DC150W, while the RF power supplied for sputtering the silicon target was fixed at RF50W. The deposition time was of 40min, and the substrate holder was rotated with 20r.p.m. during the sputter-deposition to obtain uniform deposition conditions for homogeneous film composition over the substrate area. The schematic illustration of the co-sputtering system for the deposition of Ti-Si-C thin films is shown in Fig.1.

Characterization of the obtained films was performed by X-ray diffractometry (XRD). And the electrical resistivity of the films was measured by the four-point-probe method. A resistivity correction factor (RCF) of 4.532 was used to calculate the resistivity. Furthermore the indentation hardness was measured using a dynamic ultra-microhardness tester (Shimadzu DUH-201S) with a Vickers indenter. Uploading tests following loading tests were performed to exclude the displacement caused by elastic deformation. Thus, the dynamic hardness was calculated from the depth of indentation as a result of plastic deformation.

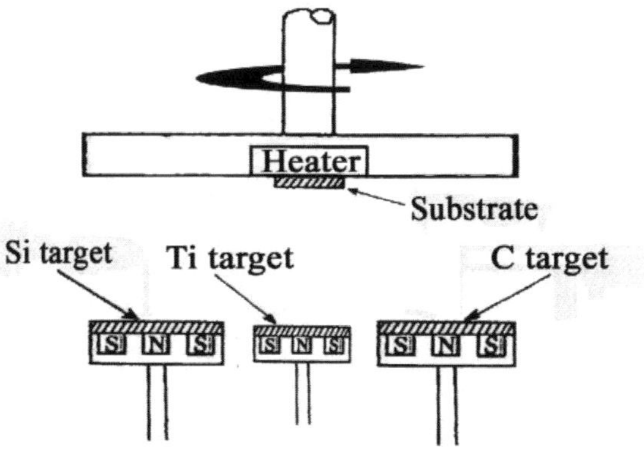

Figure 1. Schematic illustration of the co-sputtering system for the deposition of Ti-Si-C thin films onto Si(100) substrates.

3. Results and Discussion

3.1. Structural properties

The obtained Ti-Si-C thin films appeared to be uniform and adhesive, and they appeared metallic and dark gray. Figure 2 shows the X-ray diffraction patterns of the Ti-Si-C thin films deposited at room temperature, at 500 and 800 degree centigrade. It should be noted that in the figure the intensity obtained for the 500 degree centigrade is shifted by +200 cps while that for the 800 degree centigrade is shifted by +400 cps. The diffraction pattern of the film obtained at 800 degree centigrade exhibited peaks for a titanium carbide phase of TiC and a titanium silicide phase of Ti_5Si_3 as well as a MAX phase of Ti_3SiC_2. Thus it was found that the MAX phase (Ti_3SiC_2), the carbide phase (TiC) and the silicide phase (Ti_5Si_3) were formed in the Ti-Si-C film. On the other hand, any distinct diffraction peaks were not detected but a halo pattern was detected for the film obtained at 500 degree centigrade and a broad peak was detected for the film obtained at room temperature. Upon more detailed observation of the broad peak, it was guessed that the peak contained the diffraction by Ti(002) plane because of its peak-position. Thus it was assumed that fine crystals of α-titanium existed in the Ti-Si-C film. On the other hand, it was guessed that the halo pattern obtained at 500 degree centigrade reflected the formation of those MAX, carbide and silicide phase because of its pattern and its peak-position. Thus it was assumed that fine crystals of the MAX, the carbide and/or the silicide phase might also exist in the Ti-Si-C film, but which is not still clarified from the result of the XRD.

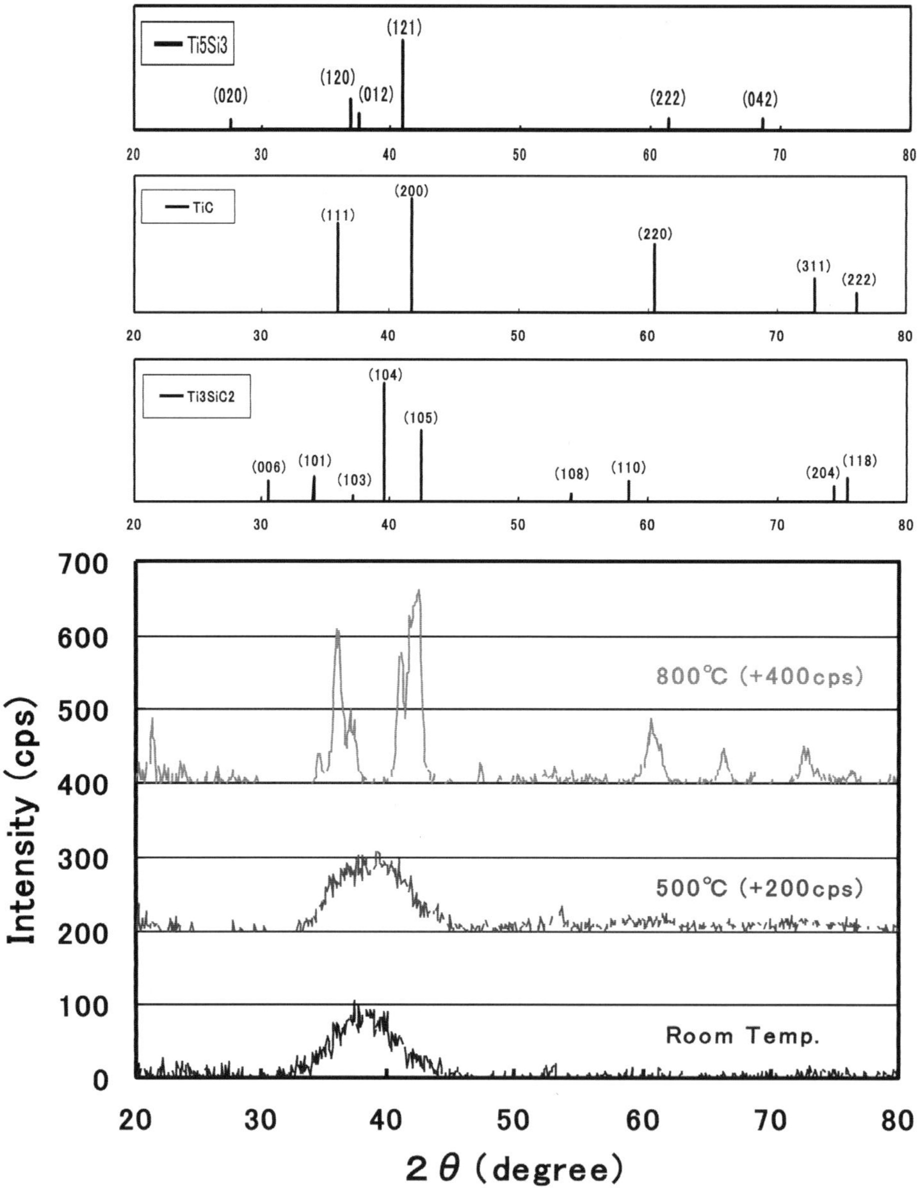

Figure 2. X-ray diffraction patterns of the Ti-Si-C thin films deposited onto Si(100) substrates, which were obtained at room temperature, 500℃ and 800℃.

3.2. Electrical resistivity

Figure 3 shows the electrical resistivity of the Ti-Si-C thin films deposited at room temperature, at 500 and 800 degree centigrade. The electrical resistivity of the thin film obtained at 800 degree centigrade was of around 160μΩcm, while the others were of around 220μΩcm, finding that they had excellent electrical conductivity, compared with the electrical resistivity (approximately 3000μΩcm) of chromium oxycarbide films having similar physical and chemical properties.[5] Then the electrical conductivity of the films obtained at 500 degree and 800 degree centigrade seemed to be due to the formation of the MAX phase (Ti_3SiC_2) and/or the carbide phase (TiC) while that obtained at room temperature seemed to be due to the formation of the α-titanium phase, considering the above-mentioned results of XRD.

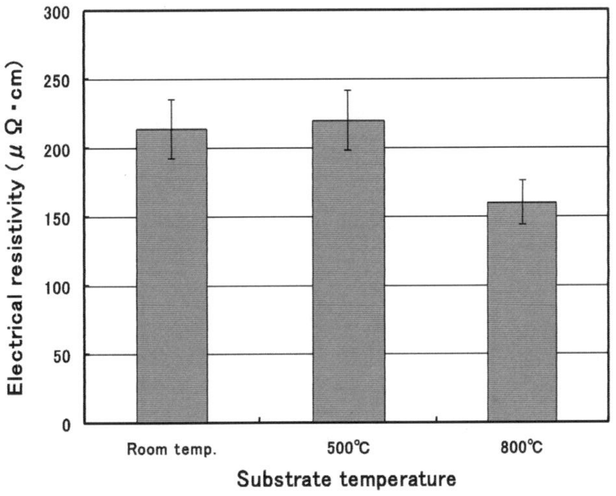

Figure 3. Electrical resistivity of of the Ti-Si-C thin films deposited onto Si(100) substrates, which were obtained at room temperature, 500℃ and 800℃.

3.3. Surface hardness

The displacement of the indenter in the loading process was caused by both elastic and plastic deformation. On the other hand, the displacement in the unloading process was caused by elastic deformation. Thus, the dynamic hardness was calculated from the displacement of the indenter caused by only plastic deformation, excluding that caused by elastic deformation.

Figure 4 shows the dynamic hardness of the Ti-Si-C thin films deposited at room temperature, at 500 degree and 800 degree centigrade normalized by that of a Si(100) wafer whose hardness is approximately 10GPa. The hardness of the film obtained at 500 degree centigrade was the highest of those under the three conditions, and the hardness was estimated to be approximately 11GPa by comparing with that of Si(100) substrates. Then the high surface hardness seemed to be due to the formation of the binary carbide TiC phase while the lower surface hardness of the film obtained at room temperature and at 800 degree centigrade seemed to be respectively due to the formation of the α-titanium phase and that of the MAX phase (Ti_3SiC_2), because the Vickers hardness of ultrafine-grained titanium phase has been reported to be approximately 4 Gpa [6], and because the ternary carbide Ti_3SiC_2 is relatively soft and its Vickers hardness has also been reported to be approximately 4GPa. [7,8]

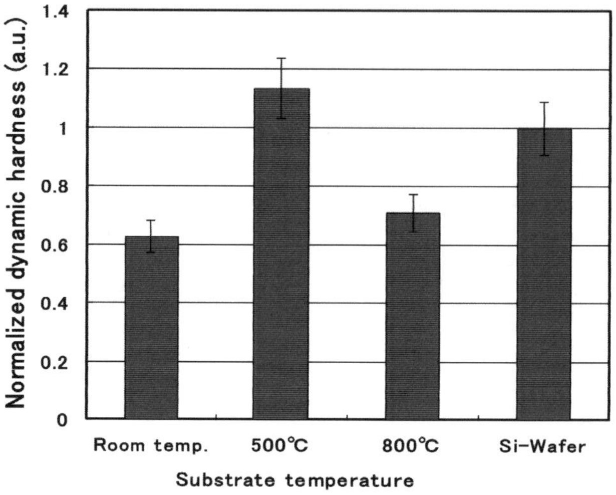

Figure 4. Dynamic hardness of the Ti-Si-C thin films deposited onto Si(100) wafer, normalized by that of Si(100) wafer whose hardness is approximately 10GPa.

4. Conclusions

Based on XRD, it was found that a titanium silicide phase of Ti_5Si_3 as well as titanium carbide phases such as Ti_3SiC_2 and TiC were formed in the film obtained at 800 degree centigrade. On the other hand, any distinct diffraction peaks were not detected, but a halo pattern was detected for the film obtained at 500 degree centigrade and a broad peak was detected for the film obtained at room temperature. Then, because of those pattern and those peak-position, it was assumed that fine crystals of α-titanium existed in the Ti-Si-C film obtained at room temperature while fine crystals of the MAX, the carbide and/or the silicide phase might also exist in the Ti-Si-C film obtained at 500 degree centigrade. The electrical resistivity of the thin film obtained at 800 degree centigrade was of around 160μΩcm, while the others were of around 220μΩcm, finding that they had excellent electrical conductivity. According to nano-indentation tests, the hardness of the film obtained at 500 degree centigrade was the highest of those under the three conditions, and was estimated to be approximately 11GPa by comparing with that of Si(100) substrates.

Therefore it was found that Ti-Si-C thin films of both excellent electrical conductivity of around 200 μΩcm and surface hardness higher than 10GPa could be formed by depositing at much lower temperature of 500 degree centigrade than the conventional depositing temperature such as 800 degree centigrade. Thus it is expected that this deposition process will be more useful for providing protective coatings for electrical friction contacts.

References

[1] K. L. Choy and E. Felix, Mater. Sci. Eng. A278, 481 (2000)
[2] J. Emmerlich, H. Hoegberg, S. Sasvari, P. O. A. Persson and L. Hultman, J. Appl. Phys. 96, 4817 (2004)
[3] A. A. Woevodim and J. S. Zabinski, Diamond Relat. Mater. 7, 463 (1998)
[4] C. Lopes, N.M.G.. Parreira, S. Carvalho, A. Cavaleiro, J.P. Riviere, E. Le Bourhis and F. Vaz, Surf. Coat. Tech. 201, 7180 (2007)

[5] T. Kado, Y. Noda, Jpn. J. Appl. Phys. 28, 1450 (1989).
[6] K. Edalati, Z. Horita, Scripta Materialia 64, 161 (2011)
[7] T. El-Raghy, M. W. Barsoum, A. Zavaliangos and S. R. Kalidindi, J. Am. Ceram. Soc. 82, 2855 (1999).
[8] M. Radovic, M. W. Barsoum, T. El-Raghy, J. Seidensticker, and S. Wiederhorn Acta Mater. 48, 453 (2000)

Journal of Physics: Conference Series **417** (2013) 012064 doi:10.1088/1742-6596/417/1/012064

Microstructure and Mechanical Properties of Nanostructured Co-Cu Alloy Films Processed by Electrodeposition

M Yuasa[1], H Nakano[2] and M Mabuchi[1]

[1] Graduate School of Energy Science, Kyoto University, Yoshidahonmachi, Sakyo-ku, Kyoto, 606-8501, Japan

[2] Cooperative Research Facility Center, Toyohashi University of Technology, 1-1 Hibarigaoka, Tempaku-cho, Toyohashi 440-8580, Japan

E-mail: yuasamotohiro@t03.mbox.media.kyoto-u.ac.jp

Abstract. Two kinds of nanostructured Co-Cu alloy films: a nanolamellar Co-Cu alloy film and an ultrafine-grained two-phase Co-Cu alloy film were processed by electrodeposition, and their microstructures and mechanical properties were investigated. These nanostructured Co-Cu alloy films showed the high hardness and the low activation volume. The mechanical properties of the nanostructured Co-Cu alloy films strongly depended on the grain boundary characteristics. Molecular dynamics simulations were performed in the two-phase Co-Cu alloy film to investigate the dislocation emission at the Co/Cu interface. The molecular dynamics simulations showed that the stacking faults, which are generated by the intense geometrical strain at the Co/Cu interface, play an important role in the dislocation emission.

1. Introduction

Ferromagnetic thin films are widely used as computer components, magnetic sensors and so on. Co alloy films are one of promising ferromagnetic thin films. For various applications, it is desirable to improve the mechanical properties of Co alloy films. Nanocrystallization can give rise to a significant enhancement of mechanical properties in metallic materials. However, nanocrystalline metals tend to be very brittle with a ductility of less than a few percent in tensile tests [1-3], as the result of the absence of dislocation activity [4]. It is accepted that nanocrystalline metals show the high hardness (high strength) and the low activation volume [5-9]. These features of nanocrystalline metals are attributed to emission of dislocations at the grain boundaries, and the grain boundaries play a critical role in deformation of nanocrystalline metals. Hence, it is required to develop nanocrystalline Co alloys with unique grain boundaries for enhancement of the mechanical properties.

In the present work, two kinds of nanostructured Co-Cu alloy films with unique grain boundaries are processed by electrodeposition, and their microstructures and mechanical properties are investigated. In addition, molecular dynamics (MD) simulations are performed in the ultrafine-grained two-phase Co-Cu alloy to investigate the dislocation emission at the Co/Cu interface.

2. Experimental

Two kinds of nanostructured Co-Cu alloy films, that is, a nanolamellar Co-Cu alloy film and an ultrafine-grained two-phase Co-Cu alloy film were processed by electrodeposition [10,11]. The electrolyte composition was $CoSO_4 \cdot 7H_2O$ (0.4-1 M) and $CuSO_4 \cdot 5H_2O$ (0.025-0.1 M). Microstructure of the Co-Cu alloy films was investigated by transmission electron microscopy (TEM). A TEM observation was carried out with a JEOL JEM-2100 at an operating voltage of 200 kV. A high-resolution TEM image was observed with a JEOL JEM-3100F operating at 300 kV with a spherical aberration of C_s=1.0 nm. Energy dispersive X-ray (EDX) analysis was carried out at 200kV using the TEM with EDX equipment (Noran Instruments Vantege) to investigate the chemical composition of the Co-Cu alloy films. Mechanical properties of the Co-Cu alloy films were investigated by the hardness tests at room temperature. The hardness tests were performed with a diamond Berkovich tip at constant loading rates of 13.24, 1.324 and 0.378 mN/s.

3. Results and discussion

3.1. Nanolamellar Co-Cu alloy film

3.1.1. Microstructures

A TEM image of a nanolamellar Co-Cu alloy film is shown in figure 1. The grain size of the Co-Cu alloy film was 110 nm. Note that most of the grains contained a high-density fine nanoscale lamellar structure. In previous studies [12,13], the nanocrystalline Cu with nanoscale twins with a spacing of tens of nanometers was fabricated by electrodeposition. On the other hand, the Co-Cu alloy developed in the present work contained nanoscale lamellar structure with a much smaller spacing of 3 nm. The EDX analysis showed that the Co content was 93 wt.% and the Cu content was 7 wt.%. The solid solubility limit of Cu in Co at room temperature is almost 0% in the equilibrium state. It is therefore suggested that the Cu is forced to dissolve into the Co because electrodeposition tends to cause the nonequilibrium state. Figure 2 shows a selected area diffraction (SAED) pattern and a high-resolution TEM image in the nanolamellar Co-Cu alloy film. The SAED pattern shows strong streaks between the reflections along the [111] directions, which means the stacking disorder along the [111] direction. The high-resolution TEM image also shows that the microstructure of the nanolamellar Co-Cu alloy film exhibits many stacking disorder composed of various stacking sequences, such as ABCABACB..., along the [111] direction of the fcc structure, as shown in figure 2.

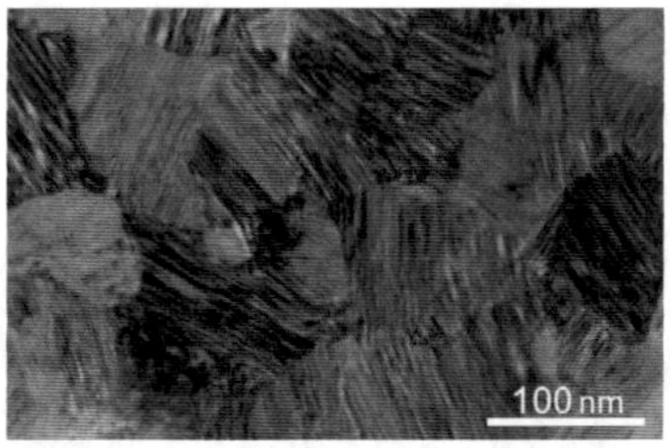

Figure 1. A TEM image of the nanolamellar Co-Cu alloy film.

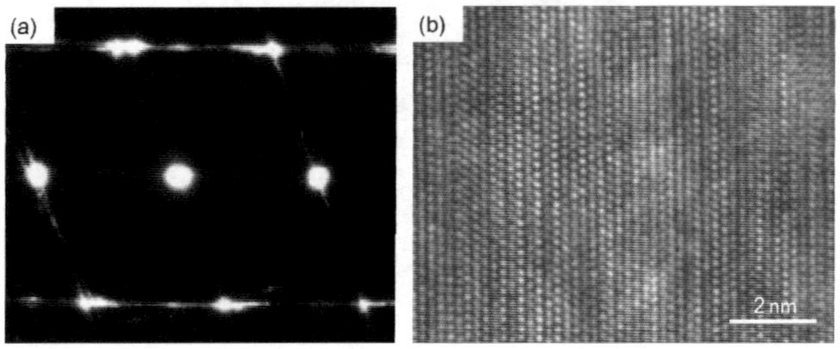

Figure 2. An SAED pattern (a) and a high-resolution TEM image (b) of the nanolamellar Co-Cu alloy film

3.1.2. Mechanical properties

Load-displacement curves obtained from the hardness tests at the three loading rates are shown in figure 3(a). The hardness of the nanolamellar Co-Cu alloy film was 4.12-5.02 GPa. As shown in figure 3(a), a higher load was required at a higher loading rate to impose the same displacement. The variation in hardness as a function of loading rate is shown in figure 3 (b). From the results in figure 3, the strain rate sensitivity and activation volume were 0.055 and $3.3b^3$ for the nanolamellar Co-Cu alloy film, respectively. It is reported that the presence of nanotwins leads to a reduction in activation volume [13,14], and the reduction is attributed to the emission of partial dislocations at twin boundaries. In the present study, the activation volume for the nanolamellar Co-Cu alloy film is much lower than those for the nanotwin Cu [14]. Clearly, the low activation volume for the nanolamellar Co-Cu alloy film is attributed to the nanoscale lamellar structure of 3 nm.

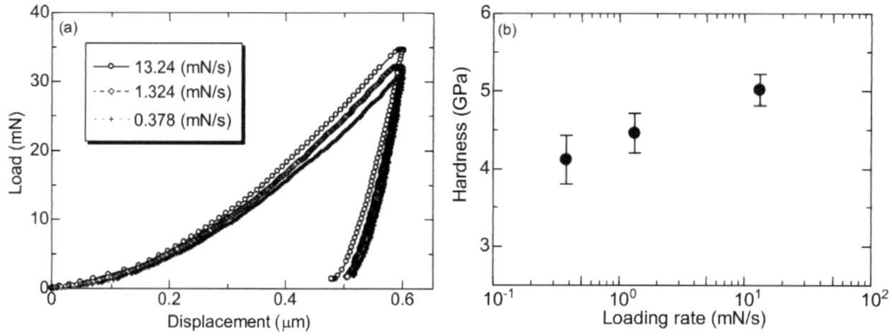

Figure 3. The results of hardness tests for the nanolamellar Co-Cu alloy film, (a) load-displacement curves at three different loading rates and (b) variation in hardness as a function of loading rate.

3.2. Ultrafine-grained two-phase Co-Cu alloy film

3.2.1. Microstructures

A TEM image of an ultrafine-grained two-phase Co-Cu alloy film is shown in figure 4. From the X-ray diffraction spectrum and EDX measurements, the Co-Cu alloy film consisted of two phases with Co grains and Cu grains. The average concentration of Cu was 38 wt.%. The grain sizes were 141 nm for Co phase and 181 nm for Cu phase, respectively. No (Co, Cu) intermetallic compounds were observed in the Co-Cu alloy.

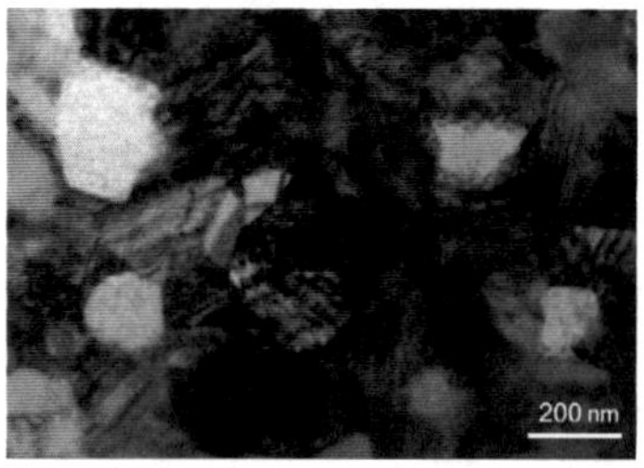

Figure 4. A TEM image of the ultrafine-grained two-phase Co-Cu alloy film.

3.2.2. Mechanical properties

The variation in hardness as a function of loading rate is shown in figure 5 for the ultrafine-grained two-phase Co-Cu alloy film, the ultrafine-grained single phase Co and the ultrafine-grained single phase Cu. From the results, the activation volume was $3.1b^3$ for the ultrafine-grained two-phase Co-Cu alloy film, $13.1b^3$ for the ultrafine-grained single phase Co and $32.8b^3$ for the ultrafine-grained single phase Cu, respectively. Note that the strain rate dependence of the two-phase Co-Cu alloys was much larger than those of single phase Co and Cu despite the almost same grain sizes. The rate controlling process of inelastic deformation in ultrafine-grained and nanocrystalline metals is the emission of dislocations from the grain boundaries [5,8,9], and the grain boundary characteristics affect the deformation behaviour [15].

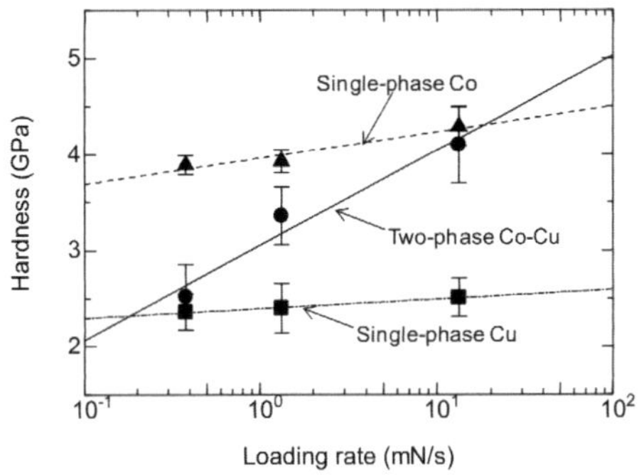

Figure 5. The variation in hardness as a function of loading rate for the ultrafine-grained two-phase Co-Cu alloy film, the ultrafine-grained single phase Co and the ultrafine-grained single phase Cu.

3.2.3. Molecular dynamics simulations

It is accepted that nanocrystalline metals exhibit the low activation volume, resulting from the emission of dislocations at the grain boundaries. In the present work, the nanostructured Co-Cu alloy films with the characteristic grain boundaries: the nanolamellar Co-Cu alloy film and the ultrafine-grained two-phase Co-Cu alloy film showed much lower activation volumes of about $3b^3$, compared with the nanocrystalline pure metals. The much lower activation volumes of the nanostructured Co-Cu alloy films with the characteristic grain boundaries cannot be explained only by the refinement of grains. Therefore, it is suggested that the unique grain boundary characteristics in the Co-Cu alloy films affect the emission of dislocations at the grain boundaries.

MD simulations were performed in the ultrafine-grained two-phase Co-Cu alloy film to investigate the dislocation emission at Co/Cu interface. Figure 6 shows atomic configuration in Co/Cu (554) grain boundary before and after dislocation emission. Note that the stacking faults were generated before the tensile testing, as shown in figure 6 (a). This results from the intense geometrical strain at the Co/Cu interface. Partial dislocations were emitted to the Cu grain from the stacking faults. Dislocations were not emitted under an applied tensile stress of 1.7 GPa, but dislocations were emitted under an applied tensile stress of 2.0 GPa, as shown in figure 6 (b). Thus, a large stress was needed to emit the dislocations at the interface.

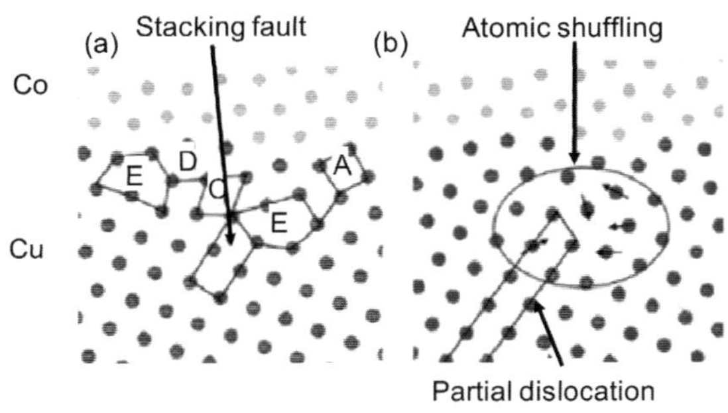

Figure 6. Atomic configuration in Co/Cu (554) grain boundary during the tensile tests under an applied stress of 2.0 GPa, (a) before dislocation emission (t=0 ps) and (b) after dislocation emission (t=5 ps).

Journal of Physics: Conference Series **417** (2013) 012064 doi:10.1088/1742-6596/417/1/012064

4. Conclusions

Two kinds of nanostructured Co-Cu alloy films: a nanolamellar Co-Cu alloy film an ultrafine-grained two-phase Co-Cu alloy film were processed by electrodeposition, and their microstructures and mechanical properties were investigated. These nanostructured Co-Cu alloy films showed the high hardness and the low activation volume. The mechanical properties of the nanostructured Co-Cu alloy films strongly depended on the grain boundary characteristics.

MD simulations revealed that the stacking faults, which are generated by the intense geometrical strain at the Co/Cu interface, play an important role in the dislocation emission at the interface.

References

[1] Wang N, Wang Z, Aust K T and Erb U 1997 *Mater. Sci. Eng. A* **237** 150
[2] Koch C C 2003 *Scripta Mater.* **49** 657
[3] Iwasaki H, Higashi K and Nieh T G 2004 *Scripta Mater.* **50** 395
[4] Schiotz J, Tolla F D D and Jacobsen K W 1998 *Nature* **391** 561
[5] Kumar K S, Swygenhoven H V and Suresh S 2003 *Acta Mater.* **51** 5743
[6] Asaro R J and Suresh S 2005 *Acta Mater.* **53** 3369
[7] Swygenhoven H V, Derlet P M and Frøseth A G 2006 *Acta Mater.* **54** 1975
[8] Wang Y M, Hamza A V and Ma E 2006 *Acta Mater.* **54** 2715.
[9] Dao M, Lu L, Asaro R J, Hosson J T M D and Ma E 2007 *Acta Mater.* **55** 4041
[10] Nakamoto Y, Yuasa M, Chen Y, Kusuda H and Mabuchi M 2008 *Scripta Mater.* **58** 731
[11] Yuasa M, Kajikawa K, Nakazawa T, Hakamada M and Mabuchi M 2010 *Scripta Mater.* **63** 132
[12] Shen Y F, Lu L, Lu Q H, Jin Z H and Lu K 2005 *Scripta Mater.* **52** 989
[13] Shen Y F, Lu L, Dao M and Suresh S 2006 *Scripta Mater.* **55** 319
[14] Lu L, Schwaiger R, Shan Z W, Dao M and Lu K 2005 *Acta Mater.***53** 2169
[15] Valiev R Z, Alexandrov I V, Zhu Y T and Lowe T C 2002 *J. Mater. Res.* **17** 5

NEXAFS Study of Air Oxidation for Mg Nanoparticle Thin Film

S Ogawa[1], S Murakami[1], K Shirai[1], K Nakanishi[2], T Ohta[2] and S Yagi[1]

[1]Graduate School of Engineering, Nagoya University, Furo-cho, Chikusa-ku, Nagoya 464-8603, Japan

[2]The SR center, Ritsumeikan University, 1-1-1 Noji-Higashi, Kusatsu, Shiga 525-8577, Japan

E-mail: ogawa.satoshi@g.mbox.nagoya-u.ac.jp

Abstract. The air oxidation reaction of Mg nanoparticle thin film has been investigated by Mg K-edge NEXAFS technique. It is revealed that MgO is formed on the Mg nanoparticle surfaces at the early stage of the air oxidation for Mg nanoparticle thin film. The simulation of NEXAFS spectrum using standard spectra indicates the existence of complex magnesium carbonates ($x(MgCO_3).yMg(OH_2).z(H_2O)$) in addition to MgO at the early stage of the air oxidation.

1. Introduction

Hydrogen energy system is expected as the alternative energy system to the current fossil fuel system because hydrogen is efficient and clean energy career. It is important to establish the hydrogen storage technique for use of hydrogen energy system. Mg is attractive material for hydrogen storage application because of its high hydrogen storage capacity of 7.6 wt% [1]. The more favorable features of Mg for the hydrogen storage are inexpensiveness and lightness compared with other materials. The problem for practical use of Mg is that hydrogenation and dehydrogenation of Mg are very slow reactions requiring an order of a few hours as the reaction time and occur only at high temperature region (≥350 °C) [1]. However, in recent years, it is reported that Mg nanoparticles of a few nanometers diameter can absorb and desorb the hydrogen at 200 °C [2, 3]. This improvement of hydrogen absorption property is due to high specific surface area of Mg nanoparticle, and it causes simultaneously easy oxidizability of Mg nanoparticle. Because Mg loses hydrogen absorption property by the air oxidation [1], it is required firstly to investigate the air oxidation reaction of Mg nanoparticles in order to prevent Mg nanoparticles from the air oxidation.

We have investigated the air oxidation reaction of Mg nanoparticle thin film by near edge X-ray absorption fine structure (NEXAFS) method because NEXAFS spectrum can be used as finger print of each chemical species formed by the air oxidation [4-6]. In our previous study, we have investigated Mg nanoparticle thin film exposed to the air for 3, 25 and 75 days by NEXAFS technique. In this work, we show NEXAFS study of the early stage of the air oxidation reaction for Mg nanoparticle thin film.

2. Experimental

Mg nanoparticles were fabricated by the gas evaporation method using He gas [4]. The rod-shaped Mg (purity: 99.98 %) was used as an evaporation source. The Mg source was evaporated by resistive heating of W basket in the He gas atmosphere (purity: 99.99995 %) and Mg atoms were condensed into the Mg nanoparticle by the collision with He atoms. The Mg nanoparticles were deposited on Si single-crystal and Ni polycrystalline substrates. The Si substrate is suitable for the estimation of the size of monodispersed Mg nanoparticles by the atomic force microscopy (AFM) observation due to its good flatness. The Ni substrate was used as the conventional substrate for NEXAFS measurements [8-10]. The deposition amount of Mg nanoparticles was controlled by deposition time. The amount of Mg nanoparticles on Si and Ni substrates were less than 1 ML and multilayer (nominal thickness estimated by AFM: 97.6±45.4 nm), respectively. The AFM observation of monodispersed Mg nanoparticles on Si substrate was carried out by tapping mode under the atmospheric condition (Figure 1). The diameter of Mg nanoparticles has been estimated by the height value of AFM observation because of high resolution in vertical direction of AFM observation compared with horizontal one [7]. The Mg nanoparticles deposited on Ni substrate were exposed to the air under the laboratory environment (30-45 %RH) at room temperature (around 20 °C) prior to NEXAFS measurements.

In order to identify the chemical state of the Mg nanoparticle thin film exposed to the air, Mg K-edge NEXAFS measurements using synchrotron radiation were carried out at BL-10 of the SR center in Ritsumeikan University [11]. A Golovchenko-type double-crystal monochromator with beryl(10-10) crystal was used for the selection of incident X-ray energy around the energy region of Mg K-edge absorption. NEXAFS spectra were obtained simultaneously by both total electron yield (TEY) method with sample drain current and fluorescence X-ray yield (FY) method with silicon drift detector (the NEXAFS spectra obtained by FY method are not shown in this paper) under high vacuum condition ($\sim 7\times10^{-6}$ Pa). The incident X-ray energy was calibrated with the first peak of the NEXAFS spectrum for MgO powder (purity: 99 %) at 1309.3 eV.

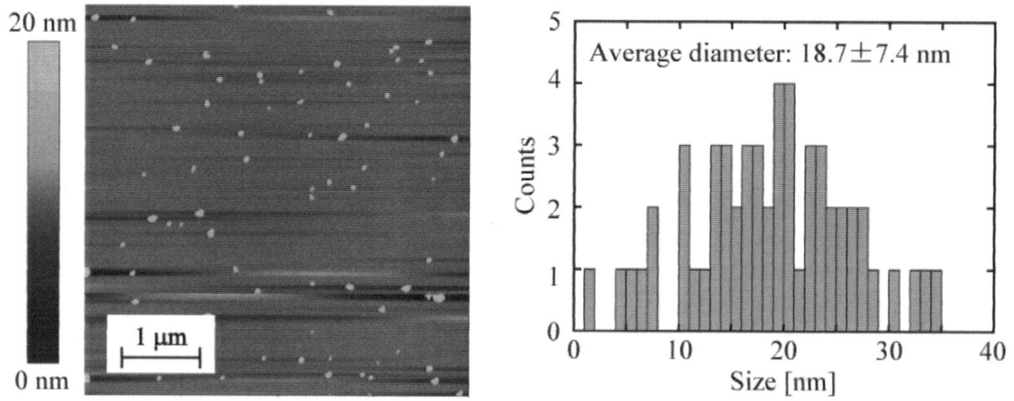

Figure 1. AFM image and the size distribution of monodispersed Mg nanoparticles on Si single-crystal substrate.

3. Results and discussion

The dotted line in Figure 2 represents Mg K-edge NEXAFS spectrum for Mg nanoparticle (average diameter: 18.7±7.4 nm) thin film exposed to the air for 3 hours. This NEXAFS spectrum were obtained by TEY method and normalized with respect to the height of the edge jump. It can be seen clearly in Figure 2 that the dotted spectrum has a shoulder structure at 1304 eV and three features in the energy range of 1307-1320 eV. The shoulder structure at 1304 eV is apparently associated with the chemical state of metallic Mg. Similarities of peak structures and peak position of the features at 1307-1320 eV with that of MgO indicates existence of the chemical state of MgO in the Mg nanoparticle thin film exposed to the air for 3 hours. The oscillation structures in the energy range of 1320-1370 eV are also consistent with that of MgO. From these results, it is obvious that MgO forms on the Mg nanoparticle surfaces by the exposure to the air for 3 hours and it is speculated that this oxidation reaction of Mg nanoparticle is the earliest stage of the air oxidation reaction of Mg nanoparticle.

In order to estimate quantitatively the fraction of chemical species, the simulation of NEXAFS spectra has been attempted by superposition of the NEXAFS spectra for metallic Mg and MgO as standards. However, it is very difficult to simulate completely the NEXAFS spectrum for Mg nanoparticle thin film exposed to the air for 3 hours. The superposition of the NEXAFS spectra for metallic Mg (62 %) and MgO (30 %), and the residual spectrum as a subtraction of superposition spectrum from dotted spectrum are shown by dashed and solid lines in Figure 2, respectively. The magnified figure of the residual spectrum is shown in Figure 3 with the Mg K-edge NEXAFS spectra for $Mg(OH)_2$ and basic magnesium carbonate ($4MgCO_3$. $Mg(OH)_2$. $4H_2O$: BMC) as standard samples. The residual spectrum has four features in the energy range of 1308-1322 eV and any features can not be simulated by standard spectra. This result indicates the existence of other chemical states in the Mg nanoparticle thin films exposed to the air. It is reasonable to assign these chemical states to the complex magnesium carbonates ($x(MgCO_3).yMg(OH_2).z(H_2O)$). The local structure around the Mg atom in the complex magnesium carbonates is MgO_6 octahedron and the crystal structure arrangements of each magnesium carbonates are different drastically, e.g. corner-sharing, edge-sharing or double-chain of edge-sharing of MgO_6 octahedron. These various structure arrangements cause many features in the Mg K-edge NEXAFS spectra of complex magnesium carbonates [12, 13]. F. Farges *et al.* have measured various minerals containing magnesium carbonates by Mg K-edge NEXAFS technique [13]. The NEXAFS spectra for magnesite ($MgCO_3$) and lansfordite ($MgCO_3.5H_2O$) [13] are shown in Figure 3 (horizontal axis for complex magnesium carbonates has been calibrated with the first peak of NEXAFS spectrum for pericrace (MgO) [13] at 1309.3 eV). Some features are consistent with that of magnesite or lansfordite, unfortunately precise assignment of these NEXAFS features has not been done yet. However, it reveals that the carbonation of Mg atoms on the surface of Mg nanoparticles occurs at the early stage of the air oxidation of Mg nanoparticles (3 hours). In our previous study, it has revealed that $Mg(OH)_2$ is formed as precursor of complex magnesium carbonates [5]. In this study, it has been indicated the existence of the direct reaction path from MgO into complex magnesium carbonates.

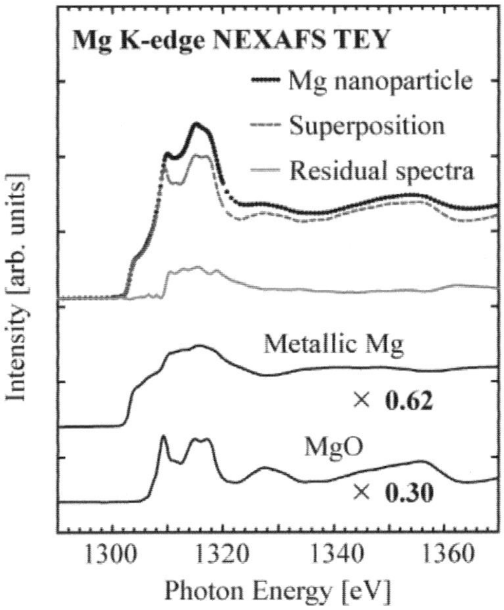

Figure 2. Mg K-edge NEXAFS spectra for Mg nanoparticle thin film exposed to the air

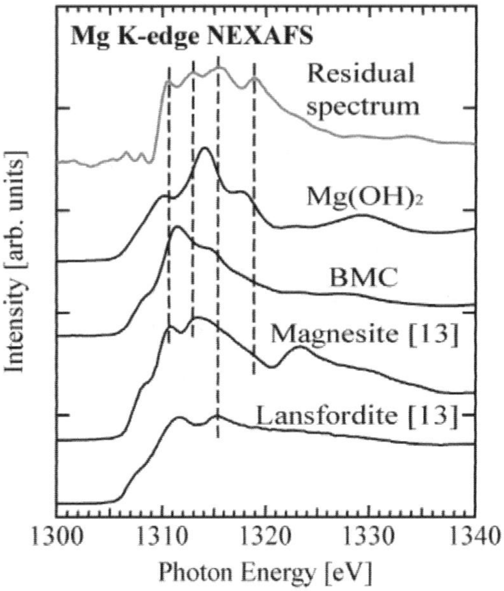

Figure 3. The magnified residual spectrum and the Mg K-edge NEXAFS spectrum

for 3 hours (dotted line). Superposition of the standard NEXAFS spectra (dashed line) and residual spectrum (solid line) are also shown.

obtained by TEY method for standard samples and minerals containing Mg carbonates [13].

4. Conclusion

We have investigated the early stage of the air oxidation for Mg nanoparticle thin film by Mg K-edge NEXAFS technique using TEY method. The NEXAFS spectrum for Mg nanoparticle thin film exposed to the air for 3 hours has some features associated with the chemical states of metallic Mg and MgO. This result indicates the formation of MgO on the Mg nanoparticle surfaces at the early stage of the air oxidation for Mg nanoparticle thin films. In order to estimate quantitatively the fraction of chemical species, the simulation of the NEXAFS spectrum for Mg nanoparticle thin film exposed to the air for 3 hours has been attempted by the superposition of the standard spectra of metallic Mg and MgO. It is very difficult to simulate the NEXAFS spectrum completely, and the residual spectrum has been obtained. The features in the residual spectrum indicates that complex magnesium carbonates ($x(MgCO_3).yMg(OH_2).z(H_2O)$) exists in addition to MgO at the early stage of the air oxidation.

References

[1] Zaluska A, Zaluski L and Ström–Olsen J.O. 1999 *J. Alloys Compd.* **288** 217
[2] Wagemans R W P, van Lenthe J H, de Jongh P E, van Dillen A J and de Jong K P 2005 *J.Am. Chem. Soc.* **127** 16675-80
[3] Jeon K J, Moon H R, Ruminski A M, Jiang B, Kisielowski C, Bardhan R and Urban J J 2011 *Nature Materials* **10** 286-90
[4] Ogawa S, Niwa H, Nomoto T and Yagi S 2010 *e-J. Surf. Sci. Nanotech.* **8** 246
[5] Ogawa S, Niwa H, Nakanishi K, Ohta T and Yagi S 2011 *J. Surf. Anal.* **17** 319-23 (in press)
[6] Ogawa S, Murakami S, Shirai K, Nakanishi K, Tsukada C, Ohta T and Yagi S 2011 *e-J. Surf. Sci. Nanotech.* **9** 315
[7] Yagi S, Sumida H, Miura K, Nomoto T, Soda K, Kutluk G, Namatame H and Taniguchi M 2006 *e-J. Surf. Sci. Nanotech.* **4** 258
[8] Ashida T, Miura K, Nomoto T, Yagi S, Sumida H, Kutluk G, Soda K, Namatame H and Taniguchi M 2007 *Surf. Sci.* **601**, 3898
[9] Yagi S, Ashida T, Nomoto T, Namatame H and Taniguchi M 2008 *J. Surf. Anal.* **14** 444
[10] Morihara M, Miura K, Nomoto T and Yagi S 2008 *J. Surf. Anal.* **14** 462
[11] Nakanishi K, Yagi S and Ohta T 2010 *AIP Conf. Proc.* **1234**, 931
[12] Li D, Peng M and Murata T 1999 *The Canadian Mineralogist* **37** 199
[13] Farges F, Meibom A, Flank A-M, Lagarde P, Janousch M and Stolarski J 2009 *J. Phys. : Conf. Ser.* **190** 012175

Acknowledgments

The authors are grateful for the financial supported of a Grant-in-Aid for Scientific Research from the Ministry of Education, Science and Culture, Japan (No. 19008872), Knowledge Cluster II in Aichi/Nagoya area and Research Fellow of the Japan Society for the Promotion of Science. This work was performed under the approval of Ritsumeikan University SR center Program Advisory Committee (Nos. S21-06, S22-07). Especially NEXAFS measurements are supported by the nanotechnology network project of the Ministry Education, Culture, Sports, Science and Technology, JAPAN (MEXT).

$Ba_2NaNb_5O_{15}$ thin film formed by electron cyclotron resonance plasma sputtering

A Watazu[1,3] and H Masumoto[2]
[1] National Institute of Advanced Industrial Science and Technology (AIST), 2266-98 Anagahora, Shimoshidami, Moriyama-ku, Nagoya 463-8560, Japan
[2] Center for Interdisciplinary Research, Tohoku University, 6-3 Aramaki-Aoba, Aoba-ku, Sendai 980-8578, Japan

E-mail: a-watazu@aist.go.jp

Abstract. $Ba_2NaNb_5O_{15}$ (BNN) thin films were formed on sapphire (0001) substrates and magnesium oxide substrates using electron cyclotron resonance plasma sputtering. The BNN films were 7.2-8.4 μm thick and had atomic ratio compositions of Ba=2.0-2.1/5Nb and Na=0.9-1.1/5Nb. The films had no cracks. Single-phase BNN and (001)-oriented BNN films on sapphire (0001) were obtained at 823 K and 923 K. An in-plane-oriented BNN film on sapphire (0001) was formed at 923 K. The films were transparent and had the ability of second harmonic generation. The films on magnesium oxide (100) were also single-phase BNN, and (001)-oriented BNN films were obtained at 823 K, 873 K, and 923 K. An in-plane-oriented BNN film on magnesium oxide (100) was formed at 823 K. The films were transparent.

1. Introduction

$Ba_2NaNb_5O_{15}$ (BNN) with a tungsten bronze structure has large nonlinear optical and Pockels coefficients [1,2,3]. Optical damage is less likely to occur in BNN compared to $LiNbO_3$ [4]. Therefore, BNN is a useful material for optical IC, piezoelectric devices, and other optical devices [5]. However, single BNN crystals tend to become cracked because of a sharp change in the coefficient of thermal expansion of the c-axis around 800-850 K during cooling from the melting temperature, and it is difficult to form single crystals with high transmittance by melting [6,7]. Application of BNN to optical IC requires the formation of thin films, preferably crystalline films. Even so, when preparing thin films by, for example, RF sputtering, cracks develop, and it becomes difficult to form BNN thin films with high transmittance or high levels of other optical abilities.

BNN thin films were prepared without cracking on sapphire ($11\bar{2}0$) and (012) substrates using electron cyclotron resonance (ECR) plasma sputtering (ES037; Sumitomo Metal Industries, Ltd.) in our previous studies [7-9]. This system generates ECR plasma with high electron density, so the substrate temperature is low during film formation and compositional deviation is not likely [10]. There is compositional deviation in BNN film deposition because Na has high vapor pressure; we reported on the amount of deviation and a method for controlling deviation [7,8]. In this study, we report on various BNN films formed by ECR plasma sputtering on sapphire (0001) of hexagonal crystal structure and magnesium oxide (MgO) single-crystal substrates of cubic crystal structure, crystal

[3] Author to whom any correspondence should be addressed.

orientations of resulting BNN films. In addition, second harmonic generation (SHG) waves were measured for the BNN films on sapphire (0001).

2. Experimental Procedures

In ECR plasma sputtering, microwaves (2.45 GHz, 800 W) were introduced into a plasma chamber through a rectangular waveguide, and a magnetic field (87.5 mT) for ECR was applied to generate ECR plasma, as shown in Figure 1. Target raw materials were $BaCO_3$ (99% pure, Wako Pure Chemical Industries, Ltd.), Na_2CO_3 (JIS primary standard, Wako), and Nb_2O_5 (99.9% pure, Wako). The materials were mixed in an agate mortar, calcined for 12 h at 1073 K in the air, mixed again, and then sintered for 12 h at 1373 K [7,8]. Then, 10 mm×10 mm substrates with Pt film or no Pt film (for transmittance measurement) were heated with an infrared lamp to 773-923 K. After the film formation chamber was evacuated to 1×10^{-4} Pa, 2.5×10^{-5} m^3/min argon gas and 2.5×10^{-5} m^3/min oxygen gas were introduced into the plasma chamber to form film at 0.1 Pa. In the case of the formation of multimodal films, the experiment was halted every 2 h to cool the system, dividing the experiment into 13 periods. The BNN films obtained were several microns thick. To avoid film discontinuity caused by dividing the experiment, we pre-sputtered the films for 5 min before each cycle for activation during film growth. The consistency of the film's thickness was evaluated based on the film's composition and texture. The BNN films were 7.2-8.4 μm thick and had atomic ratio compositions of Ba=2.0-2.1/5Nb and Na=0.9-1.1/5Nb. The samples were checked by eye and by scanning electron microscopy (SEM). Crystal structures and orientations of the films were measured by X-ray diffraction (XRD, RAD-C, Rigaku). In-plane orientations of the films were measured by pole figure (RAD-C, Rigaku) using BNN (402) peaks. Transmittances of the films were measured by a spectrophotometer (UV-31010PC, Shimadzu). SHG waves were measured using a Nd^{3+}: YAG laser of 1.06 μm wavelength. The laser was applied at a 45° angle with respect to the BNN film surface, and the spectrometer was located in another 45° direction.

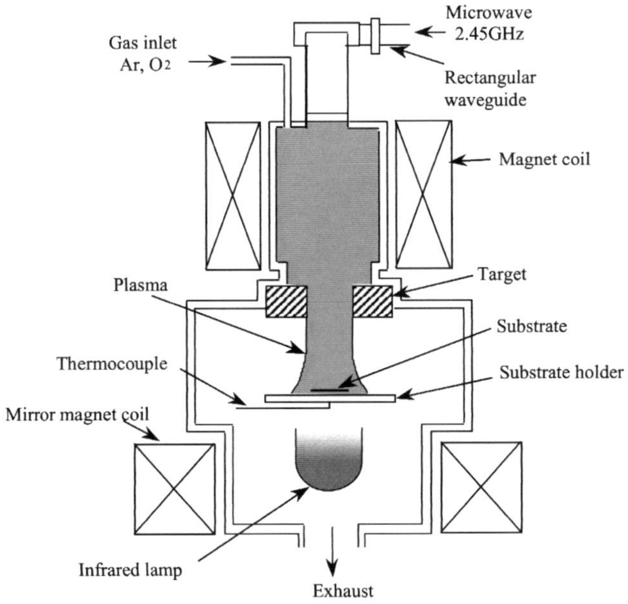

Fig. 1 Schematic diagram of electron cyclotron resonance plasma sputtering.

3. Results and Discussion

3.1 $Ba_2NaNb_5O_{15}$ thin films on sapphire (0001) plane substrates

None of the surfaces of the films formed at the substrate temperature of 823 K, 873 K, or 923 K had any cracks, and all were transparent by visual observation. As shown in the SEM photograph in Fig. 2, the films had no cracks in their surfaces or their interiors. Formation temperatures of the films were lower than those of typical melting processes, such as BNN single crystal formation by the Czochralski method [6,7], and therefore no cracks were formed on the BNN films during the decrease of temperature from the deposition temperature to room temperature. Though the formation of each film needed 13 periods of deposition by ECR plasma sputtering, discontinuous interaction was not observed in the cross sections of the films and the films were formed uniformly, as occurred in our previous study [8]. A BNN thin film formed at 773 K had a smooth surface in our previous study [8], but angular shapes were observed on the surface of the BNN films in the present study. The XRD patterns of the films had peaks only of BNN, as shown in Fig. 3. This indicated that single-phase BNN was deposited on sapphire (0001) without cracking using the low-temperature process of the ECR plasma sputtering method. The film formed at 873 K had lower (00l)-orientation, which might have

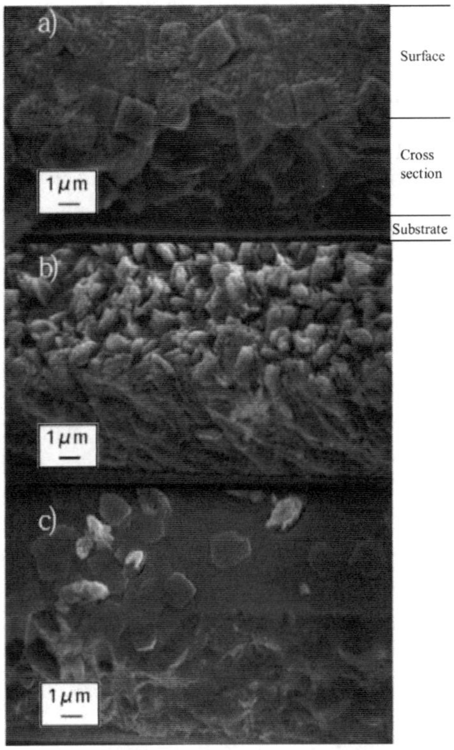

Fig. 2 SEM photographs of BNN films formed on sapphire (0001) substrates at (a) 923 K, (b) 873 K, (c) 823 K.

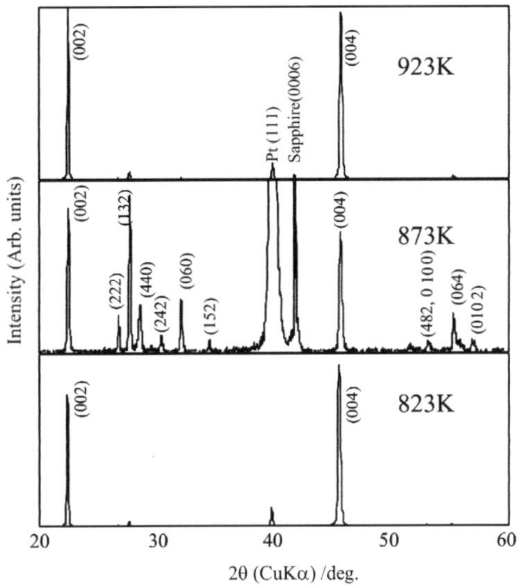

Fig. 3 XRD patterns of BNN films formed on sapphire (0001) substrates at 823-923 K. (Index numbers are BNN.)

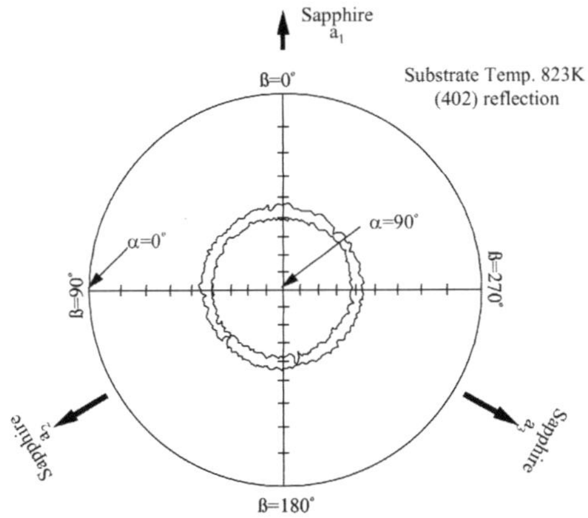

Fig. 4 Pole figure of BNN films formed on sapphire (0001) substrates at 823 K.

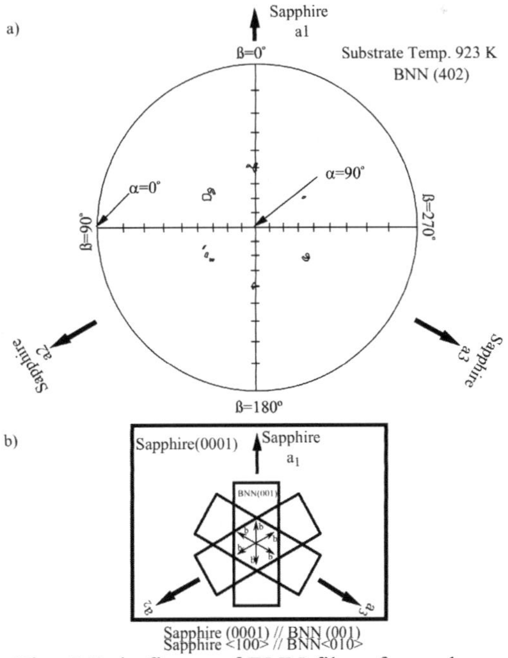

Fig. 5 Pole figure of BNN films formed on sapphire (0001) substrates at 923 K.

caused the formation of the jagged structure of the surface of the film. The BNN thin films formed at 823 K and 923 K were (00l)-oriented films. Pole figures of BNN films are shown in Figs. 4 and 5. The BNN film formed at 823 K had no in-plane orientation, but that formed at 923 K did have in-plane orientation of sapphire <100> // BNN <010>, sapphire (0001) // BNN (001). BNN single crystal had a sharp change in the coefficient of thermal expansion of the c-axis around 800-850 K [6,7]; thus, the BNN film formed at 823 K might not have in-plane orientation. A cause of the 3-direction in-plane orientation might have been that sapphire has a hexagonal crystal structure and 11 times the A-O distance of the a-axis in sapphire (0001) is an 0.8% mismatch of the BNN b-axis. Transmittances of the BNN films are shown in Fig. 6. The BNN films had lower transmittances than those formed at 823 K, but the BNN films in this work had high transmittances at high wavelengths. The transmittance of the polycrystalline BNN film formed at 823 K was the lowest. SHG measurement results of BNN films are shown in Fig. 7. All BNN films formed at 823-923 K showed 532-nm SHG waves, and the BNN film with low transmittance had a high SHG peak. The BNN film with the highest SHG ability was formed at 873 K and was polycrystalline. A laser with one direction might form SHG waves with the polycrystalline BNN film effectively because the nonlinear optical coefficients of BNN are anisotropic; d_{31}=-13.2×10^{-12}, d_{32}=-13.22×10^{-12}, and d_{33}=-18.2×10^{-12} m/V [4].

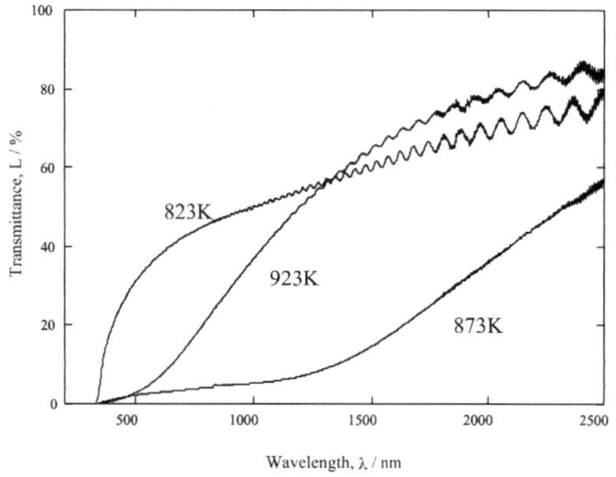

Fig. 6 Transmittances of BNN films formed on sapphire (0001) substrates at 823-923 K.

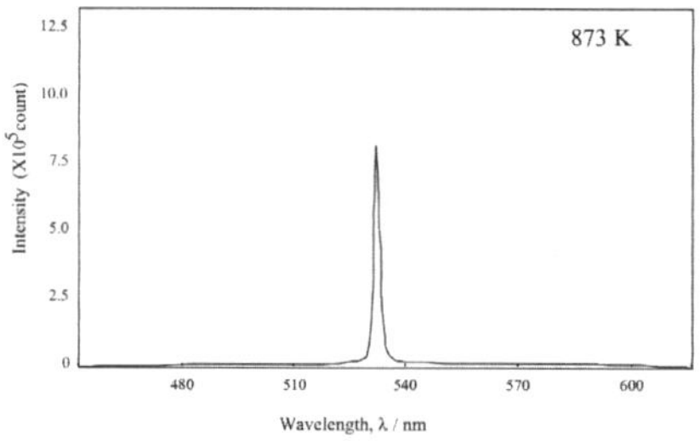

Deposition temperature	Intensity (X10^{5}count)
823K	1.1
873K	8.1
923K	4.8

Fig. 7 SHG using a Nd^{3+}: YAG laser (1.06 μm wavelength) and BNN films formed on sapphire (0001) substrates at 873 K.

3.2 $Ba_2NaNb_5O_{15}$ thin films on sapphire MgO substrates

The XRD patterns of the samples had only peaks of BNN, as shown in Fig. 8. Single-phase BNN was also deposited on MgO substrates. The BNN thin films formed at 823 K, 873 K, and 923 K were (00l)-oriented films. A pole figure of the BNN film formed at 823 K is shown in Fig. 9. The BNN thin film had in-plane orientation at a lower substrate temperature than the deposition temperature of in-plane-oriented BNN on sapphire (0001). The BNN single crystal had a sharp change in the coefficient of thermal expansion of the c-axis around 800-850 K [6,7], and the BNN film on sapphire (012) was also not in-plane oriented [8]. MgO might be more congenial to BNN than sapphire at 823 K. The films formed at 823 K on MgO (100) substrate had 2 patterns of 4-direction in-plane orientation, and the direction pattern of in-plane orientation was different from the 3-direction in-plane orientations of the films on sapphire (0001). Though matching positions of atoms between BNN and MgO were not clear, the cause of 4-direction in-plane orientations might be that MgO had a cubic crystal structure of 4-times rotational symmetry.

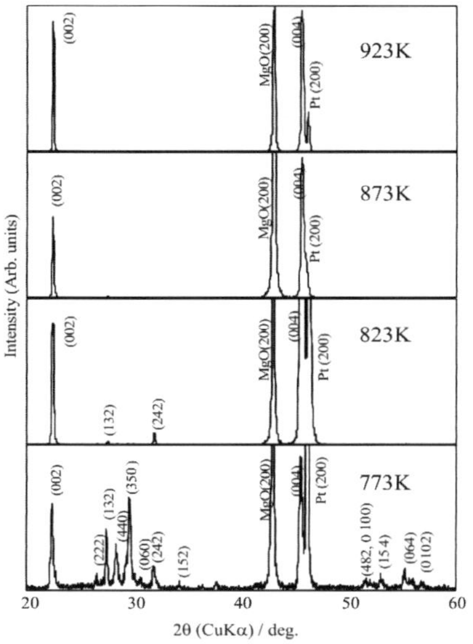

Fig. 8 XRD patterns of BNN films formed on MgO substrates at 773-923 K. (Index numbers are BNN.)

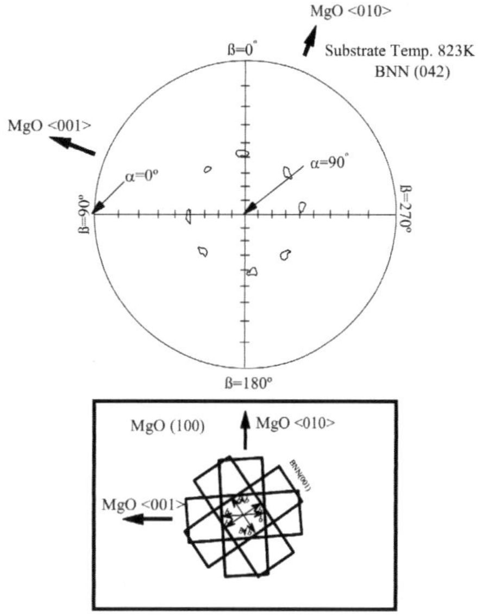

Fig. 9 Pole figure of BNN films formed on MgO substrates at 823 K.

4. Summary

By the ECR plasma sputtering method, we formed BNN single-phase thin films without any cracks on sapphire (0001) substrates and MgO (100) substrates. The crystal orientations of the films on sapphire (0001) were different from those of the films on MgO (100) substrates. The in-plane-oriented BNN thin film was formed at 923 K. The films had second harmonic generation ability. BNN single-phase thin films were also formed on MgO (100) substrates, in which case the in-plane-oriented BNN thin film was formed at 823 K. SHG waves were observed by the films on sapphire (0001) using a Nd^{3+}: YAG laser. These results indicate that the ECR plasma sputtering method is useful for forming single-mode thin BNN film and multimodal BNN film, and this method is expected to be useful for preparing optical BNN films for optical and piezoelectric IC.

Acknowledgments

Authors thank Dr. T. Hirai and Dr. T. Goto for their advices. Authors thank Dr. H. Itou for his helping of SHG wave measurement.

References

[1] Quandt H D and Rothrock L R 1977 *J. Cryst. Growth* **42** 435
[2] Jamieson P B, Abrahams S C and Bernstein J L 1969 *J. Chem. Phys.* **50, 69J1** 4352
[3] Scott B A, Giess E A and O'Kane D F 1969 *Mat. Res. Bull.* **4, 69S5** 107
[4] Shimazu M, Tsukioka M, Kuroiwa S, Tanokura Y and Tsutsumi S 1991 *Jpn. J. Appl. Phys.* **30** 2002
[5] Masuda Y, Masumoto H, Kodachi Y, Watazu A, Baba A, Goto T and Hirai T 1995 *Jpn. J. Appl. Phys.* **34** 5124
[6] Ballman A A, Carruthers J R and O'Brown Jr H M 1970 *J. Cryst. Growth* **6** 184
[7] Watazu A, Masumoto H, Masuda Y, Baba A, Goto T and Hirai T 1997 *J. Jpn. Soc. Powder Powder Metal.* **44-1** 86
[8] Watazu A, Usui F, Masumoto H, Masuda Y, Baba A, Goto T and Hirai T 1999 *J. Jpn. Soc. Powder Powder Metal.* **46-9** 909
[9] Watazu A, Masumoto H, Masuda Y and Hirai T 1998 *J. Ceram. Soc. Japan* **105** 740
[10] Matsuoka M and Ono K 1998 *Oyo Buturi* **57** 1301

Adhesion of DLC Film Prepared on WC-FeAl Substrate by Sputtering

A Matsumoto, K Kobayashi and S Nakao

Materials Research Institute for Sustainable Development, National Institute of Advanced Industrial Science and Technology, 2266-98 Anagahora Shimoshidami, Moriyama, Nagoya, 463-8560, Japan

E-mail: akihiro-matsumoto@aist.go.jp

Abstract. DLC films were prepared on WC-25vol%FeAl and WC-25vol%Co substrates using r.f. magnetron sputtering at substrate temperature range between room temperature and 773K. The Raman spectra was analysed using $I_{G2}/I_G(I_G=I_{G1}+I_{G2})$. The critical loads of sputtered WC-25vol%FeAl specimen were higher than those of sputtered WC-25vol%Co specimen over a wide range of temperatures between room temperature and 573K. TEM observation revealed the existence of an aluminium oxide between the DLC film and FeAl binder.

1. Introduction

It is well known that WC-Co hard metals are used in a variety of scientific and industrial applications. Over the last few decades significant research was carried out to replace the cobalt binder by alternative binder systems to overcome some of its shortcomings, such as, poor corrosion resistance, high cost, its scarcity of supply and environmental toxicity. FeAl intermetallic compound might be one of the most promising candidate, because of good oxidation resistance, being much less expensive and also less toxic. An attempt was made by Subramanian *et al.* using melt infiltration technique [1] and by Schneibel *et al.* using liquid phase sintering of mixed powders technique [2] to prepare WC-FeAl composites. However, insufficient mechanical properties were obtained in both investigations for practical use. Recently, we have successfully improved mechanical properties using pulse current sintering method under pressure application [3, 4].

Diamond-like carbon (DLC) films have a range of technological applications, because of their optical, electrical and mechanical properties [5]. Particularly, they can be used as a coating film for cutting tools and dies because of its own high hardness, low friction coefficient and excellent smooth surface. However, in many cases, an additional metallic interlayer, such as tungsten, titanium, silicon and/or chromium, should be prepared owing to improve the interface adherence between DLC films and cemented carbide substrates. In this paper, the adhesion of the DLC film prepared on WC-FeAl substrate by r.f. magnetron sputtering was investigated using the scratch test and transmission electron microscopy.

2. Experimental

Fe (99.5mass%, 5μ m), Al (99.9mass%, 10μ m), Co (99.9%, 5μ m) and WC powders (2.5 μ m) with the composition of WC-25vol%FeAl (Fe-40at%Al) and WC- 25vol%Co were dry blended using an automatic agitator for 3.6ks. The blended powder mixtures were poured in a graphite die with

15mm in inner diameter and subsequently consolidated under a pressure of 50MPa in vacuum using pulse current sintering method. The heating rate, sintering temperature and holding temperature were 1.67K/sec, 1373K and 180 sec, respectively. The obtained compacts were polished using a planar grinding machine with a grinding wheel of 240 mesh roughness and subsequently to a mirror gloss with diamond abrasive finishing. Vickers hardness tests were performed under the load of 10kgf. The microstructures of the compacts were investigated using a scanning electron microscope (SEM) and a field emission transmission electron microscope (FE-TEM) equipped with energy dispersive X-ray spectral analysis system.

DLC films were deposited on WC-25vol%FeAl and WC-25vol%Co substrates using r.f. magnetron sputtering system with a graphite disk target (50mm in diameter), mounted on the cathode at a distance of ~100 mm from a substrate table. The base pressure of the deposition chamber was around 1.0×10^{-3} Pa. Pure argon (99.999% purity) was used as inert gas in the chamber. The sputtering target was pre-sputtered for about 5 min before deposition to remove any surface impurities that may have been present on the target. The films were deposited at temperatures of room temperature, 373, 473, 573, 673 and 773K, using a resistance heater located behind the substrate holder. The process pressure was 1 Pa with r.f. power of 300W. WC-25vol%FeAl and WC-25vol%Co compacts were used as substrate materials and a standard chemical cleaning procedure was maintained before deposition. Raman spectra of the deposited films were obtained using a Raman spectrometer (Renishaw, inVia) with a green laser of argon ion laser 514.5 nm. The spectra were recorded from 800 to 2000 cm^{-1}.

Thickness values of the films were determined from an atomic force microscope (AFM) line analysis. Scratch tests were conducted using a commercialized scratch tester (RHESCA, CSR-2000). A cartridge with a diamond stylus having a radius of 5 μ m was swung parallel to the surface of a DLC film sample on a X-Y translator with a tilt-table and moved horizontally, normal to the swing. As the sample moved, the stylus was gradually pressed down upon the sample until the friction force abruptly increased. The load which leads to the abrupt increase of friction coefficient is defined as the critical load.

3. Results and discussion

Figure 1 shows SEM micrographs of WC-25vol%FeAl and WC-25vol%Co compacts. FeAl phase was formed through the combustion synthesis reaction between Fe and Al powders, and confirmed by X-ray diffraction measurement. Both of the compacts consist of WC particles (white color) and binder (gray color) lying among them. Binder pools with about 10 μ m were dispersed in places in a similar way. Vickers hardness of the compacts was 1200 and 1150, respectively.

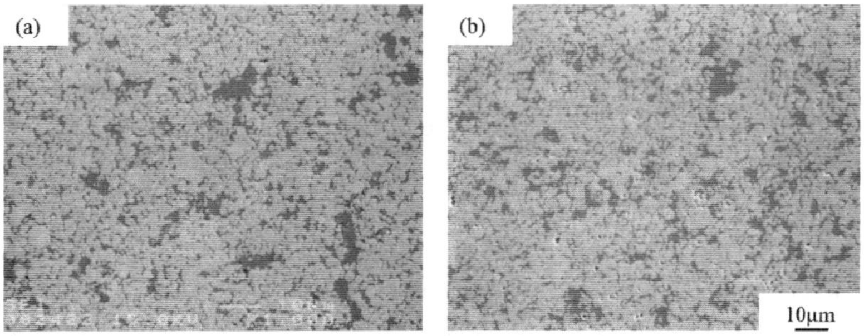

Figure 1. SEM micrographs of (a) WC-25vol%FeAl compact and (b) WC-25vol%Co compact.

Figure 2 shows the Raman spectrum of DLC films deposited on WC-25vol%FeAl substrates at different temperatures. The thickness of the obtained DLC film was about 800nm and slightly tended to increase with increasing the substrate temperature. In this study, 4 peak curve fitting technique is adopted to unify the curve fitting way for all spectra. The analysis of Raman spectra using 4 peaks was carried out by several researchers [6, 7]. Iwaki *et al.* pointed out at least 4 peaks are necessary to obtain the best curve fitting [8]. Referring to their studies, an assumption is made as follows: (1) the spectrum is composed of 4 peaks at 1190 cm^{-1} (D1 peak), 1380 cm^{-1} (D2 peak), 1500-1560 cm^{-1} (G1 peak) and 1590 cm^{-1} (G2 peak) (2) the peak is fitted by Gaussian line shape, (3) D1, D2 and G2 peaks do not shift, i.e., 3 peak positions are fixed except for G1 peak. The origins of G1 and G2 peaks should be assigned to disordered graphite and ordered graphite. The peak summation of G1 and G2 peaks is equal to the intensity of G peak. As well known, G band is Raman active E_{2g} mode in graphite. Therefore, Nakao *et al.* pointed out the intensity ratio of G2 to G peak (I_{G2}/I_G) may also be useful for the evaluation of the magnitude of ordered graphite structure [9]. Figure 3 shows the relationship between I_{G2}/I_G and the substrate temperature for DLC films prepared on WC-25vol%FeAl compact. The value of I_{G2}/I_G increases at the substrate temperature 673K sharply. Some work on the thermal stability of DLC films have been reported. Tallant *et al.* showed the significant conversion of DLC films to nano-crystalline graphite on heating at temperatures above 573K [10]. Our study is well coincident with the result above.

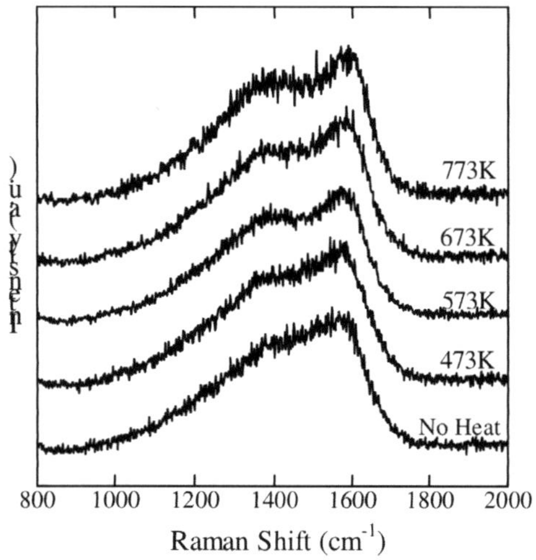

Figure 2. Raman spectra of sputtered films prepaered on WC-25vol%FeAl substrates at different temperatures.

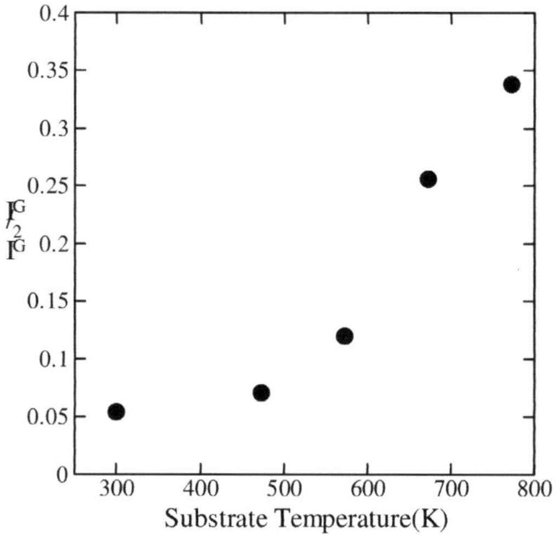

Figure 3. The relationship between I_{G2}/I_G and the substrate temperature for sputtered films prepared on WC-25vol%FeAl substrates.

Figure 4 shows the relationship between the critical load and the substrate temperature for WC-25vol%FeAl and WC-25vol%Co compacts. With increasing the substrate temperature up to 573K, the critical loads of both of specimen increased. The drop of the critical loads at 673K is due to graphitization of the DLC films. It should be noted that the critical loads for WC-25vol%FeAl substrate are 10 ~ 30% higher than those for WC-25vol%Co substrate. Figure 5 shows a bright field image of the interface between the DLC film and WC-25vol%FeAl substrate. The DLC film shows a typical hallo pattern. It is noticeable that there exists a thin film with ~3nm between the DLC film and FeAl binder. As shown in fig.5 (b), oxygen was detected clearly in EDX spectra. Graupner *et al.* have reported that Al_2O_3 films were easily formed at room temperature even under ultra high vacuum condition [11]. Therefore, the thin film observed at the interface between the DLC film and FeAl binder is considered as an aluminum oxide. On the other hand, no film was observed at the interface between DLC film and Co binder for WC-25vol%Co substrate. Nie *et al.* prepared duplex Al_2O_3/DLC films on Al alloys using a combined micro-arc oxidation and plasma-immersion ion implantation technique and investigated the influence of C_2H_2/Ar ratio on the interfacial adhesion between the DLC and alumina layers [12]. They clarified an adequate condition led to an excellent adhesion. Therefore, the excellent adhesion between the DLC film and WC-25vol%FeAl substrate might be attributed to the existence of thin alumina film.

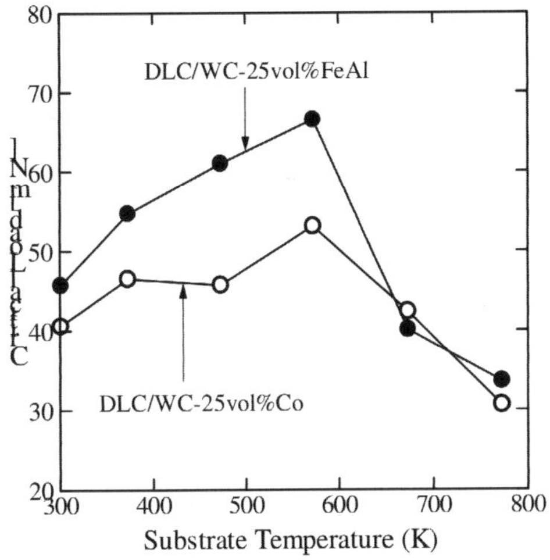

Figure 4. The Relationship between the critical loads and substrate temperatures for DLC/WC-25vol%FeAl and DLC/WC-25vol%Co specimen..

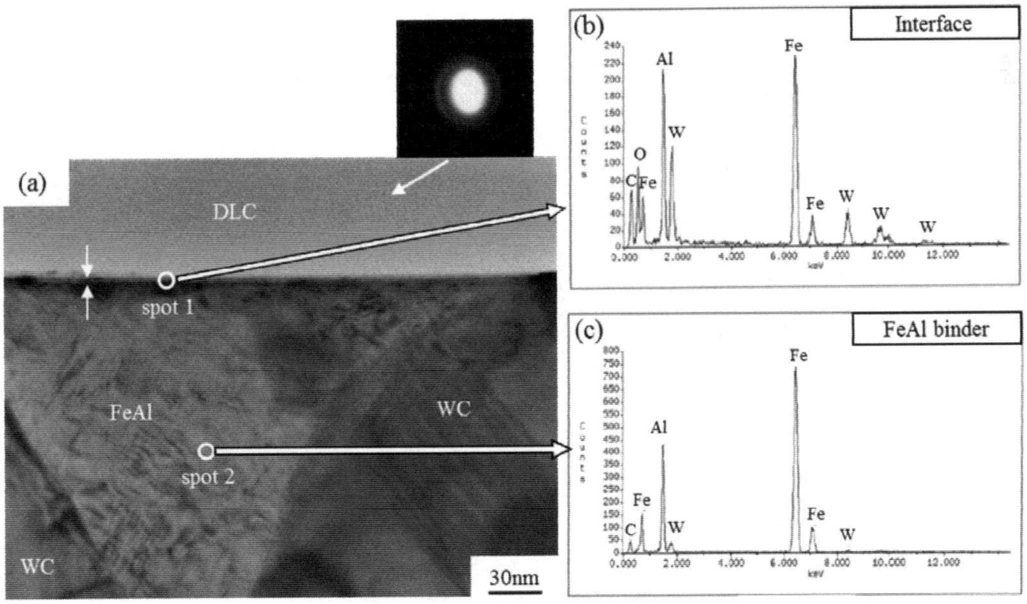

Figure 5. (a) The bright field image of the interface between the DLC film and WC-25vol%FeAl substrate. (b) Energy dispersive spectra at the interface. (c) Energy dispersive spectra in FeAl binder.

4. Conclusion

DLC films were successfully prepared on WC-25vol%FeAl substrate using r.f. magnetron sputtering. The critical loads of sputtered WC-25vol%FeAl specimen were higher than those of sputtered WC-25vol%Co specimen over a wide range of temperatures between room temperature and 573K.

References

[1] R Subramanian and J H Schneibel, *Intermetallics* **5** 23 (1997).
[2] J H Schneibel, C A Carmichael, E. D. Specht and R. Subramanian, *Intermetallics* **5**, 61 (1997).
[3] A Matsumoto, K Kobayashi, T Nishio and K Ozaki, *Proc. PLANSEE Seminar 2005* **2**, 287 (2005).
[4] A Matsumoto, K Kobayashi, T Nishio, K Ozaki and S Tada, *Euro PM2004 Conference Proc.***3**, 641 (2004).
[5] R E Clausing, L L Horton, J C Angus and P Koidl, *Proc. NATO ASI Series* **266**, 911 (1990).
[6] B Oral, R Hauert, U Muller and K H Emst, *Diamond Rel. Mater.* **4**, 482 (1995).
[7] J Schwan, S Ulrich, V Batori, H Ehrhardt and S R Silva, *J. Appl. Phys.* **80**, 440 (1996).
[8] M Iwaki and H Watanabe, *Nucl. Instr. Methods.* **B 141**, 206 (1998).
[9] S Nakao, J Kim, J Choi, S Miyagawa, Y Miyagawa and M Ikeyama, *Trans. Mater. Res. Soc. Japan* **31**, 685 (2006).
[10] D R Tallant, J E Parmeter, M P Siegal and R L Simpson, *Diamond Rel. Mater.* **4**, 191 (1995).
[11] H Graupner, L Hammer, K Heinz and D M Zehner, *Surf. Sci.* **380**, 335 (1997).
[12] X Nie, A Wilson, A Leyland and A Matthews, *Surf. Coat. Technol.* **121**, 506 (2000).

Studies on ultrathin films of tricycloquinazoline (TCQ) based discotic liquid crystal molecules

Raj Kumar Gupta[1*], Manjuladevi V.[1], C. Karthik[1], Sandeep Kumar[2]

E-mail: *raj@bits-pilani.ac.in
[1]Department of Physics, Birla Institute of Technology and Science, Pilani, Rajasthan - 333031, India
[2]Raman Research Institute, Sadashivanagar, Bangalore - 560080, India

Abstract. The assembly of disk-shaped molecules on surfaces has drawn considerable attention because of their unique electro-optical properties. We have studied the monolayer of the tricycloquinazoline based disk-shaped (TCQCB) molecules at air-water and air-solid interfaces. The TCQCB molecules form a stable Langmuir monolayer at the air-water interface. The monolayer exhibits gas, low density liquid (L_1), and high density liquid (L_2) phases. In L_1 and L_2 phases, the molecules prefer edge-on conformation. The atomic force microscope study on the Langmuir-Blodgett films of the molecules reveals a structural transformation. The elongated domains in the film deposited in L_1 phase transformed to a compact grainy texture in the film deposited in the L_2 phase.

1. Introduction

The field of disk-shaped (discotic) liquid crystals has grown enormously because of the interesting electrical and optical properties of the molecules. Depending on the temperature, the discotic molecules assemble itself onto solid substrates yielding various columnar phases [1]. The charge mobility along the length of the column has been found to be enormously high when compared to its perpendicular direction. This makes such columnar system to act as a semiconducting material with a very high value of anisotropy in the conducting property. The discotic liquid crystals find potential industrial application in the fields of opto-electronics and photovoltaic devices such as light emitting diodes [2, 3], field-effect transistors [4], solar cells [5] and sensors [6]. Mostly, the cores of commonly studied discotic molecules are rich in π-electrons, and such molecules are known to behave as a p-type semiconductor with appropriate doping [7, 8]. The tricycloquinazoline (TCQ) based molecules are deficient in π-electrons and are reported to behave as a n-doped semiconducting material [9, 11, 12]. The nature of aggregation of the molecules on the surface is important for the device applications. The surface can induce an ordering in the bulk, and hence the material properties can be varied by varying the structure of aggregates on the surface [13, 14]. The surface behavior of such molecules are relevant and demands a systematic study. We have studied the behavior of TCQ core substituted with cyano-alkyloxybiphenyl (TCQCB) molecules on the surfaces like water and glass. A stable single layer of molecules at an air-water (A-W) interface is known as Langmuir monolayer [15]. The Langmuir monolayer at different phases can be transferred to a variety of solid substrates in a highly controlled manner by vertical dipping mechanism known as Langmuir-Blodgett (LB) technique [15, 16]. There are numerous studies on the Langmuir monolayer of the disk-shaped molecules at the A-W interface [17, 18, 19, 20, 21]. Such systems are known to exhibit monolayer phases where the disk-plane of the molecules are either planar to the water surface (face-on conformation) or it is perpendicular to the water surface (edge-on conformation).

In a study on ethylenoxy substituted side chains of TCQ core, we have found a stable Langmuir monoayer at the A-W interface. Our studies have revealed a face-on molecular conformation in a low density liquid phase. Interestingly, the molecular conformation in a high density liquid phase were found to be both face-on and edge-on [22]. In this paper, we report our study on the Langmuir monolayer of TCQCB molecule. The Langmuir monolayer of the molecule at A-W interface exhibits a low density and a high density liquid phases where the molecules prefer edge-on conformation. The AFM images reveal a structural transformation from elongated domains to very compact grainy texture for the LB films deposited in the low and high density liquid phases, respectively.

2. Experimental

The cyano-alkyloxybiphenyl substituted TCQ based amphiphilic (TCQCB) molecule was synthesized in the laboratory [10, 11, 23, 24]. The chemical structure of the molecule is shown in Figure 1. The molecule possesses six polar cyano-alkyloxybiphenyl side-group symmetrically

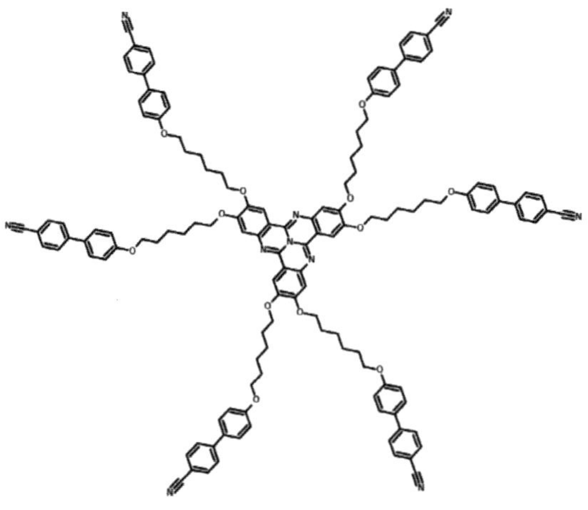

Figure 1. The chemical structure of the TCQCB molecule.

attached to the TCQ core so that it retains its C_3 fold symmetry. Such a novel covalently linked rod-disk oligomer exhibited a nanophase segregated liquid crystalline phase wherein the rods and disks are segregated into separate layers to form a smectic phase (SmA). The TCQCB molecules exhibit liquid crystalline phase while cooling from isotropic phase (I): I 173.4 °C SmA 169.5 °C cubic 156.4 °C crystal. The diameter of the TCQCB molecule is found to be around 5.3 nm. A solution of the sample was prepared in HPLC grade chloroform having a concentration of 1.7×10^{-5} M. The monolayer at the A-W interface was formed by spreading the solution onto the ultrapure ion-free (Millipore, DQ5) water having the resistivity greater than 18 MΩ-cm. The surface pressure (π) - area per molecule (A_m) isotherms were obtained by compressing the monolayer at the A-W interface in a Langmuir trough (LB2004C, Apex Instruments). The compression speed was maintained at 0.5 (nm^2/molecule)/min. A high resolution microscopy was done by transferring the monolayer onto very clean solid (glass coverslip) substrates, and obtaining the topographic images of the films using atomic force microscope (AFM). Prior to deposition, the glass coverslip substrates were treated hydrophilically by boiling them in piranha solution (3:1, conc H_2SO_4:H_2O_2) for 1 minute, and then rinsed successively with ion-free water, absolute alcohol and acetone solvents. The substrates were then dried by blowing hot air at 70 °C. The average root mean square roughness of glass coverslip substrates were found to be ~ 0.1 nm for a scan range of 1.0×1.0 μm^2. This indicate a reasonably smooth surface of the glass coverslip [25]. We deposited a single layer of LB films on these substrates at different surface pressures by a single upstroke of the dipper. The film deposited substrate was transferred to the scanning stage of AFM, (Solver P47-Pro, NT-MDT) for imaging. The AFM imaging were done in semicontact mode in ambient. The spring constants and the resonance frequencies of the Si tips were in the ranges of 10-20 N/m and 180-270 KHz, respectively. The experiments were carried out at the room temperature (~22 °C).

3. Results and Discussion

The surface pressure (π) - area per molecule (A_m) isotherm of TCQCB molecules is shown in Figure 2. The isotherm shows zero surface pressure at very large A_m. It shows a lift-off area

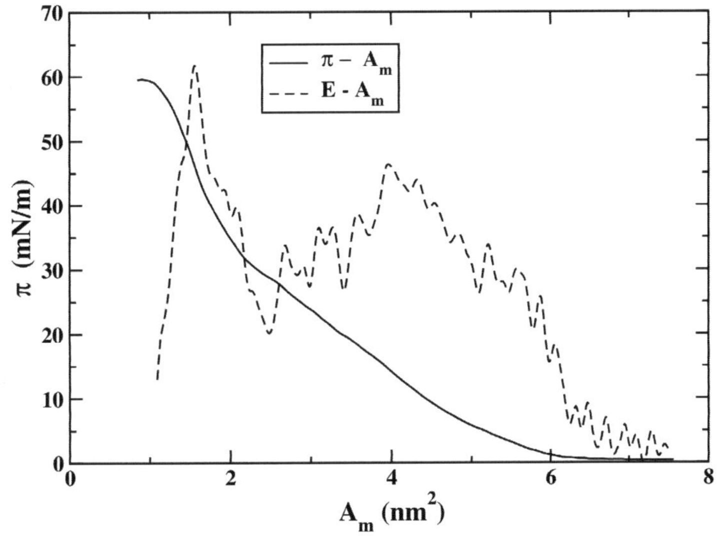

Figure 2. Surface pressure (π) - area per molecule (A_m) isotherm (solid line) and inplane elastic modulus (E) - $A_m$of the TCQCB molecule (dashed line).

per molecule (A_i) at around 7 nm^2. On decreasing the A_m, the isotherm indicates a change in the slope and then followed by a gradual rise in surface pressure. This region of the isotherm may corresponds to a low density liquid (L_1) phase. On further decreasing the A_m, another slope change is obtained at 30 mN/m. The surface pressure rises rapidly till the monolayer collapses at around 58 mN/m. This region of the isotherm may corresponds to a high density liquid (L_2) phase. The average molecular area in a phase can be estimated by extrapolating the corresponding region of the isotherm at the zero surface pressure [15]. The extrapolation of the region of the isotherm between 6 to 2.2 nm^2 to zero surface pressure yields an A_m value of around 5.6 nm^2 (=A_o^1). The extrapolation to the zero surface pressure of the rapidly increasing surface pressure region of the isotherm yields a limiting area per molecule of around 3.0 nm^2 (=A_o). The isothermal in-plane elastic modulus (E) is an appropriate quantity for distinguishing very weak phase transitions [26]. It is defined as $E = -A_m \frac{d\pi}{dA_m}$. The maximum value of E in the L_1 and L_2 phases were obtained as 46 and 61 mN/m, respectively. These are relatively smaller as compared to a commonly studied molecule, octadecanoic acid. The maximum values of E of octadecanoic acid in the liquid expanded and liquid condensed phases were found to be 100 and 400 mN/m, respectively. This indicates that the observed L_1 and L_2 phases are more fluidic as compared to that of liquid expanded phase of octadecanoic acid. For a face-on conformation of the TCQCB molecules on the water surface, the average molecular area should be around 22 nm^2. However, the π - A_m isotherm of the TCQCB molecule indicates the average molecular areas in L_1 and L_2 phases to be around 5.6 and 3.0 nm^2, respectively. Such smaller average molecular areas in the different phases can be accounted for by the edge-on conformation of the molecules. Therefore, the L_1 and L_2 phases are low density and high density liquid phases wherein the TCQCB molecules exhibits edge-on conformation.

The stability of the Langmuir monolayer of TCQCB molecules on water subphase is studied by repeatedly compressing and expanding the monolayer and recording the corresponding surface pressures. The results from such isocycles are shown in Figure 3. The isocycle studies of

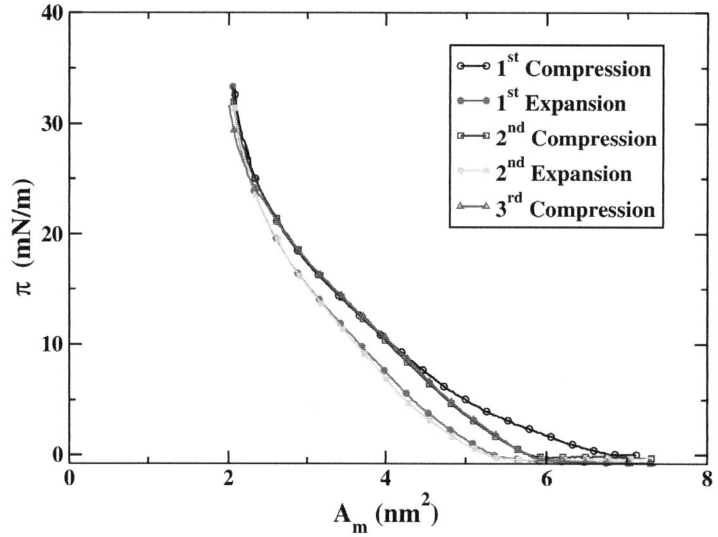

Figure 3. Isocycles obtained during repeated compression and expansion of the TCQCB monolayer at A-W interface.

TCQCB indicate that after the first compression, the expansion and compression curves remain unchanged at higher cycles. The isocycles also show negligible loss due to either dissolution or 3D aggregation. This indicates a better stability of TCQCB monolayer at the A-W interface.

The LB films of the TCQCB molecules are deposited on glass coverslip at different target surface pressure (π_t), and scanned using an AFM. The LB films are deposited in the L_1 and L_2 phases. These are shown in Figure 4. The AFM image of the LB films deposited at 10 mN/m (Figure 4A) shows the bright domains grown over a less bright background. The average height of such bright domains is 5.2±0.3 nm. This height corresponds to a vertical orientation of TCQCB molecules. The average height of the less bright region is around 3 nm. This can be accounted for by a tilt of the molecules. The bright domains do not show any preferential growth direction, and are found to be grown randomly. The morphology of the domains are found to be changed with the increase in π_t. The less bright region appear to vanish whereas the randomly grown bright domains assemble to yield interesting pattern. The AFM image of LB films of TCQCB molecules at 20 mN/m (Figure 4B) shows the domains to be elongated and oriented in a preferential direction. The height variation data also reveal that the average height of such elongated domains to be around 5.5 nm. With the further increase in π_t (= 30 mN/m), the elongated domains appear to be vanishing. This can be observed from the Figure 4C. The

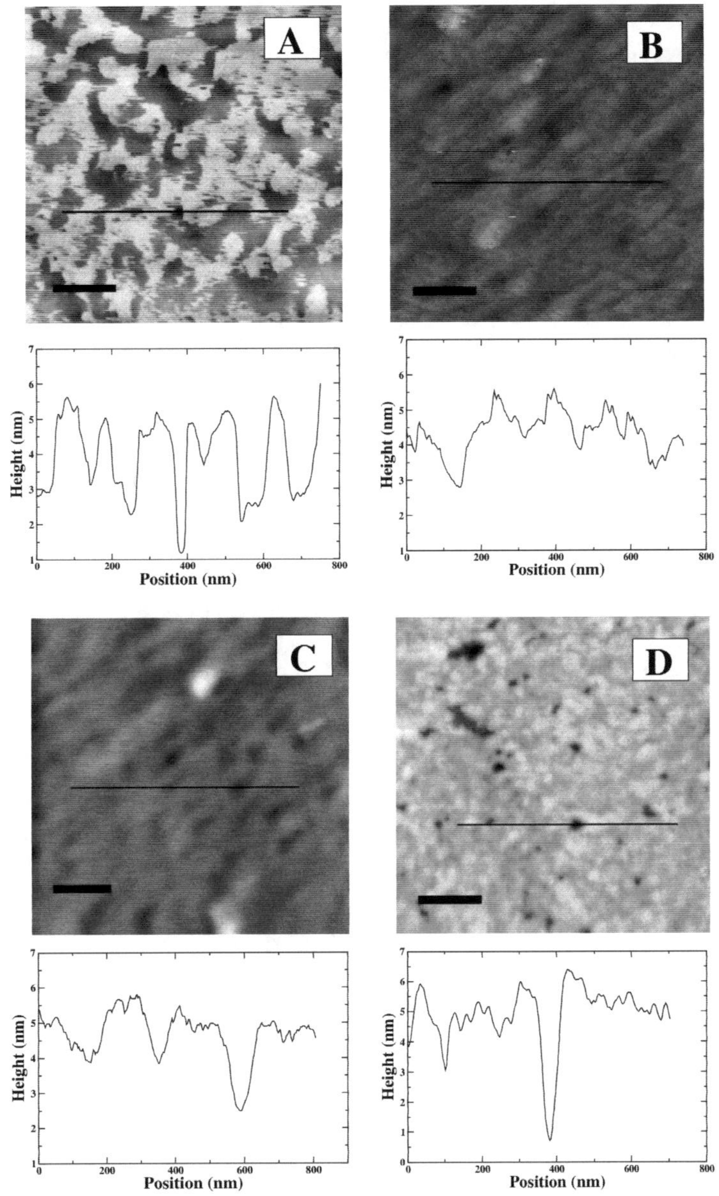

Figure 4. AFM images of LB films of TCQCB molecule. The target surface pressure (π_t) for LB films deposition corresponding to the images A, B, C and D are 10, 20, 30 and 35 mN/m, respectively. The height profile along the dark line drawn on the images are shown below the respective images. The scale bar represent a length of 200 nm.

AFM image at 35 mN/m (Figure 4D) reveal grainy texture wherein the elongated domains as observed in Figures 4(B - C) have completely transformed to isotropic small grain-like domains. The π_t = 30 mN/m is close to the L_1-L_2 phase transition. Therefore, the elongated domains as observed in the L_1 phase undergo a structural transformation to very compact grainy and isotropic texture in the L_2 phase.

The π-$A_m$isotherm of TCQCB molecules has indicated the low density L_1 phase where the molecules exhibit edge-on conformation. Because of the low density of L_1 phase, when the LB films were deposited at the low surface pressure ($\leq$10 mN/m), two regions were obtained: region with vertically oriented molecules and region with tilted molecules. With the increasing π_t, the titled region vanishes and interesting domains with vertically oriented molecules were obtained. The LB films deposited in L_1 phase at a higher π_t (>10 mN/m) exhibited elongated domains. Such domains undergo a structural transformation to a compact grainy texture in the LB films deposited in L_2 phase.

4. Conclusions

We found a stable Langmuir monolayer of TCQCB molecules. The monolayer exhibits gas phase, a low density liquid (L_1), and a high density liquid (L_2) phases. The molecular orientation in L_1 and L_2 phases were found to be edge-on. The AFM images of the LB films deposited at $\pi_t \leq$ 10mN/m reveal the domains with vertical and tilted orientation of molecules. With increasing π_t, the elongated domains were observed. The elongated domains as observed in L_1 phase undergoes a structural transformation to form a dense grainy texture in L_2 phase.

5. Acknowledgments

The authors (BITS, Pilani) are thankful to University Grants Commission of India for their support under DRS SAP programme (Grant No. F,530/3/DRS/2009 SAP-I).

[1] S. Chandrasekhar and G. S. Ranganath, Rep. Prog. Phys. **53**, 57 (1990).
[2] G. Lüssem and J. H. Wendroff, Polym. Adv. Technol. **9**, 443 (1998).
[3] I. Seguy, P. Jolinat, P. Destruel, J. Farenc, R. Mamy, H. Bock, J. Ip, and T. P. Nguyen, J. Appl. Phys. **89**, 5442 (2001).
[4] N. Stutzmann, R. H. Friend, and H. Sirringhaus, Science **299**, 1881 (2003).
[5] L. Schmidt-Mende, A. Fechtenkötter, K. Müllen, E. Moons, R. H. Friend, and J. D. MacKenzie, Science **293**, 1119 (2001).
[6] N. Boden, R. J. Bushby, J. Clements, and B. Movaghar, J. Mater. Chem. **9**, 2081 (1999).
[7] L. Y. Chiang, J. P. Stokes, C. R. Safinya, and A. N. Bloch, Mol. Cryst. Liq. Cryst. **125**, 279 (1985).
[8] P. G. Schouten, J. M. Warman, M. P. de Haas, M. A. Fox, and H. -L. Pan, Nature **353**, 736 (1991).
[9] N. Boden, R. C. Borner, R. J. Bushby, and J. Clements, J. Am. Chem. Soc. **116**, 10807 (1994).
[10] H. K. Bisoyi, V. A. Raghunathan, and S. Kumar, Chem. Comm., 7003 (2009).
[11] S. Kumar, D. S. S. Rao, and S. K. Prasad, J. Mater. Chem. **9**, 2751 (1999).
[12] R. Hiesgen, H. Schönherr, S. Kumar, H. Ringsdorf, and D. Meissner, Thin Solid Films **358**, 241 (2000).
[13] P. Ruffieux, O. Gröning, M. Bielmann, C. Simpson, K. Müllen, L. Schlapbach, and P. Gröning, Phys. Rev. B **66**, 073409 (2002).
[14] R. Friedlein, X. Crispin, C. D. Simpson, M. D. Watson, F. Jäckel, W. Osikowicz, S. Marciniak, M. P. de Jong, P. Samori, S. K. M. Jönsson, M. Fahlman, K. Müllen, J. P. Rabe, and W. R. Salaneck, Phys. Rev. B **68**, 195414 (2003).
[15] G. L. Gaines, Jr., *Insoluble Monolayers at Liquid-Gas Interfaces* (Wiley-Interscience, New York, 1966).
[16] G. G. Roberts, *Langmuir-Blodgett Films* (Plenum Press, New York, 1990).
[17] A. Albrecht, W. Cumming, W. Kreuder, A. Laschewsky, and H. Ringsdorf, Col. Polym. Sci. **264**, 659 (1986).
[18] N. C. Maliszewskyj, P. A. Heiney, J. K. Blasie, J. P. McCauley Jr., and A. B. Smith III, J. Phys. II (France) **2**, 75 (1992).
[19] D. Gidalevitz, O. Y. Mindyuk, P. A. Heiney, B. M. Ocko, P. Henderson, H. Ringsdorf, N. Boden, R. J. Bushby, P. S. Martin, J. Strzalka, J. P. McCauley Jr., and A. B. Smith III, J. Phys. Chem. B **101**, 10870 (1997).
[20] N. C. Maliszewskyj, P. A. Heiney, J. Y.Josefowicz, J. P. McCauley Jr., and A. B. Smith III, Science **264**, 77 (1994).
[21] A. Nayak, K. A. Suresh, S. K. Pal, and S. Kumar, J. Phys. Chem. B **11**, 11157 (2007).
[22] *Submitted for publication.*
[23] S. Kumar, *Chemistry of discotic liquid crystals: From monomers to polymers* (CRC Press: New York, 2010).
[24] TCQCB compound was synthesized by Prof. Sandeep Kumar.
[25] R. K. Gupta and Manjuladevi V., Int. J. Nanoscience and Nanotechnology **101**, 171 (2011).
[26] H. Mohwald, Handbook of Biological Physics, R. Lipowski and E. Sackmann (editors), Elsevier Science, Amsterdam, Vol. 1, Chapter 4.

Height-Height Correlation Function to Determine Grain Size in Iron Phthalocyanine Thin Films

Thomas Gredig, Evan A. Silverstein, Matthew P. Byrne

Department of Physics and Astronomy, California State University Long Beach, 1250 Bellflower Blvd., Long Beach, CA 90840, U.S.A.

E-mail: tgredig@csulb.edu

Abstract. Quasi one-dimensional iron chains are formed in thermally evaporated iron phthalocyanine ($FeC_{32}N_8H_{16}$) thin films on silicon substrates. The chain length is modified by the deposition temperature during growth. Atomic force microscopy images show spherical grains at low deposition temperatures that become highly elongated at high deposition temperatures due to diffusion. The grain distributions are quantitatively characterized with a watershed-based segmentation algorithm and a height-height correlation function. The grain size distributions are found to be characteristically distinct for the α-phase and β-phase samples. The average effective grain size from the distribution is proportional to the correlation length found from the height-height correlation function and grows exponentially with deposition temperature. The long-range roughness and Hurst parameter increase only slightly with the deposition temperature.

1. Introduction

The surface morphology of organic thin films plays an important role in device fabrication. The grain boundaries and grain size of polycrystalline thin films can affect the performance of gas sensors, field-effect transistors, and photovoltaic devices.[1, 2, 3] Understanding the growth mechanism and the surface structure at the nanometer length scale would allow fabrication of devices with optimized performance characteristics. The growth of small molecules is different due to weaker van-der-Waals interactions and generally more complex as compared to inorganic atomic growth. In order to understand the growth mechanism, the polycrystalline surface of many materials is described with its grain size distribution, long-range roughness, the average grain size, and the short-range roughness. These parameters can also be modeled and so compared with theory.[4, 5]

Here, we extract the quantitative surface structure data to understand the growth process in metallo-organic thin films. The metallo-phthalocyanine molecule ($M\text{-}C_{32}N_8H_{16}$) is a planar molecule with four-fold symmetry and a metal ion at its center, see inset in Fig. 1(b). While iron phthalocyanine (FePc) grows similar to other low-Z metallo-phthalocyanine molecules, such as copper phthalocyanine, it is used here due to its ability to form quasi one-dimensional magnetic iron chains.[3, 6] The iron chain length effectively corresponds to the crystalline or grain size and can be controlled by deposition parameters including the substrate temperature. The surface parameters will be extracted from a watershed-based segmentation algorithm and a correlation function based on the atomic force microscopy (AFM) data. By measuring the AFM correlation length the average iron chain length is inferred.

2. Experiment

Thin films of iron phthalocyanine (FePc) were prepared via thermal evaporation. Silicon (100) substrates are used, because they are known to weakly interact with the molecule and lead to growth where the molecular plane is perpendicular to the substrate (edge-on growth). The substrates were cleaned via suspension in ultrasonic baths of acetone and methanol. Two substrates were mounted in the thermal evaporator for each deposition on a temperature-controlled holder that is allowed to rotate during deposition. The chamber's base pressure was $2 \cdot 10^{-6}$ torr and the substrates are preheated to about 200°C to remove water from the surface, while the organic material was outgased for 12 - 24 hours before deposition. Commercially available FePc was pre-purified by the thermal gradient method. For the deposition FePc was heated to roughly 350°C for continuous sublimation. A steady sublimation rate of 2 nm/min was used for all depositions for a total film thickness of 80 ± 4 nm as monitored with an in-situ quartz crystal monitor. While the film thickness is kept constant, the deposition temperature is varied from room temperature to 230°C. Additional samples were used for the watershed-based segmentation algorithm analysis and the sample preparation procedure is discussed elsewhere.[7]

The FePc thickness was calibrated from Kiessing fringes in the Cu K_α x-ray diffraction spectrum. The spectrum shows a prominent (200) FePc peak at the 2θ angle of 6.8°.[7, 8] This angle corresponds to a lattice spacing of 1.3 nm and confirms that the molecules crystallize and tend to align edge-on with respect to the substrate. Since the x-ray peak width is only suggestive of the correlation length, AFM images are used instead.

The morphology of FePc thin films was studied utilizing AFM with a Digital Instruments Nanoscope MultiMode scanning probe microscope operated in tapping mode. A total of twelve samples were studied and for each sample two image sizes were taken: one 4μm x 4μm image and one 1μm x 1μm image. The AFM images were converted to a linear gray scale using WSxM software[9] and then processed with a Matlab code to generate the correlation functions, see Fig. 1. The statistics analysis was performed using a watershed-based segmentation algorithm.[10]

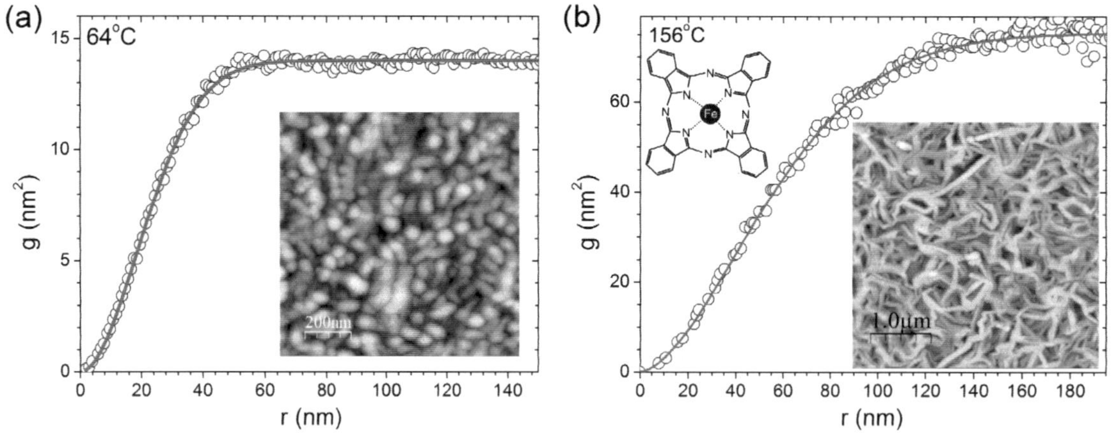

Figure 1. (Color online) The calculated height-height correlation functions (circles) according to Eq. 1 from their respective atomic force microscopy images for two iron phthalocyanine thin films. The continuous lines are best fits according to Eq. 2. The sample was deposited at (a) 64°C and (b) 156°C with equal thickness of 80 nm. The corresponding AFM image shown in the inset has a lateral size of (a) 1μm and (b) 4μm, respectively. The inset shows the chemical structure of the iron phthalocyanine molecule.

3. Analysis and Results

The surface morphology of the iron phthalocyanine thin films is quantified with two methods. In the first method, a correlation function is used to determine the average grain size, film roughness, and short-range roughness. In the second method the grain size distribution is measured from several AFM images and the statistics are analyzed. The former characteristics are obtained from the Height-Height Correlation Function (HHCF)

$$g(r) = \langle [h(\vec{x}) - h(\vec{x} - \vec{r})]^2 \rangle, \tag{1}$$

where $\vec{x}$ is a any specific point in the image and $\vec{r}$ is a displacement vector. The average height difference between any two points separated by a distance r are described by the function $g(r)$. Phenomenologically the HHCF can be interpreted by

$$g'(r) = 2\sigma^2 \left[1 - e^{-\left(\frac{r}{\xi}\right)^{2\alpha}}\right], \tag{2}$$

where σ is the surface roughness, α is the Hurst parameter, and ξ is the correlation length.[11, 12] The Hurst parameter measures the short-range roughness; i.e. it describes the grain surface structure.[13] In order to fit the function $g'(r)$ to the HHCF data, only small values of r (<250 nm) are used due to the fact that the HHCF quickly reaches a saturation value, see Fig. 1. The saturation value corresponds to the long-range film roughness $g'(r) \sim 2\sigma^2$ with $r \gg \xi$. On the other hand, for values of $r \ll \xi$, Sinha et al.[14] found that Eq. 2 is not a good fit to the data and will overestimate the Hurst parameter. Instead a linear fit on a log-log plot is used to fit α, since $g'(r) \simeq Ar^{2\alpha}$, where A is a constant.

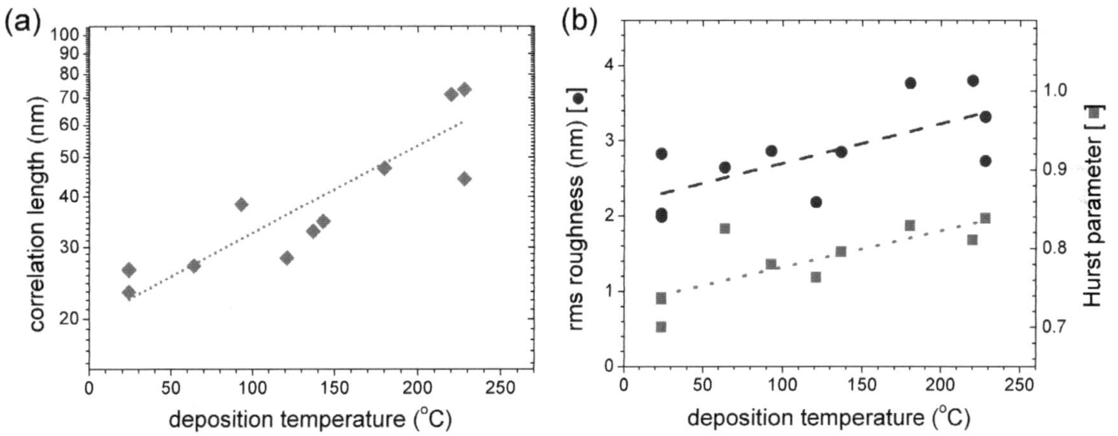

Figure 2. (Color online) (a) The correlation length ξ increases with the deposition temperature due to the increased diffusion. The dashed line is a fit to an exponential growth. (b) The rms roughness σ (circles) and the Hurst parameter α (squares) depend only weakly on the deposition temperature. Linear fits are drawn as guides to the eye.

The most common phases of FePc growth are the α- and β-phase.[15] Both have a characteristic herringbone structure but the phases differ in the angles made to adjacent molecular chains. When deposited onto substrates at temperatures lower than 200°C, FePc will assemble in the α-phase and otherwise in the β-phase. Moreover, as the substrate temperature is increased during deposition, the molecules tend to diffuse over longer distances and the grain size increases. For thin films deposited at room temperature, the correlation length ξ is 25 nm and increases to over 70 nm for samples deposited above 200°C, see Fig. 2(a). An exponential fit illustrates the rapid increase. While the diffusion leads to strong growth of the grain size, it changes the long-range roughness σ only slightly, see Fig. 2(b). Using the smaller AFM images, the rms roughness σ increases from $2.2 - 3.2$ nm. Equally, the Hurst parameter α, which characterizes the short-range roughness, increases only from $0.74 - 0.84$. These three parameters describe the grain growth as a function of the deposition temperature.

Concurrent with the increase of the correlation length, the grain shape changes from circular to strongly elongated. The surface area of each grain can be approximated as an ellipsis with a defined minor and major axis with an area equal to that of the actual grain.[7] The circularity (axes ratio) decreases from 0.75 to 0.50 with deposition temperature. Since the grains are elongated, an effective grain diameter d_0 is defined by the total grain area $A = \pi(d_0/2)^2$. The comparison of the watershed-based segmentation algorithm that measures individual grains to the HHCF shows that there is a linear correlation between the average effective grain size diameter d_0 and the correlation length ξ. The proportionality factor is determined from a best fit: $d_0/\xi = 3.4$. Similarly, a factor of 2.8 was previously found for inorganic materials.[12]

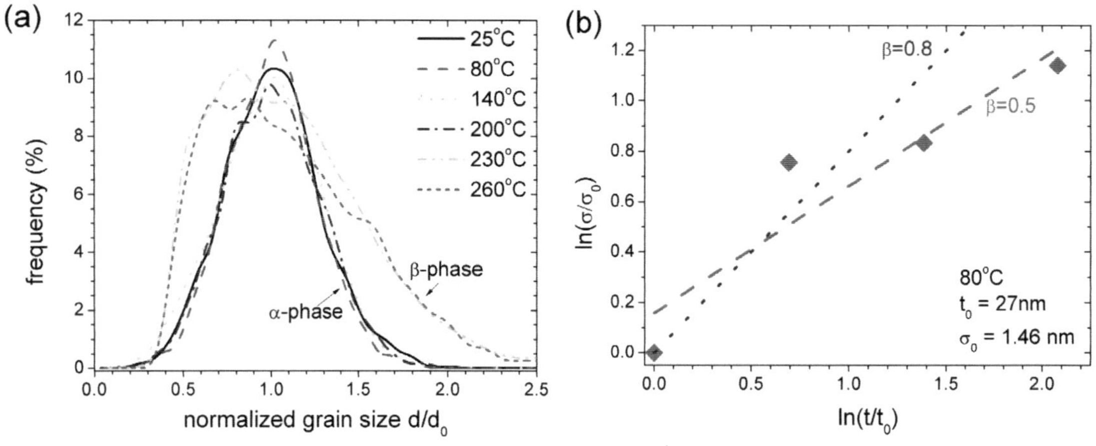

Figure 3. (Color online) (a) The effective grain diameter is determined from elliptical fits to individual grains of many AFM images using a watershed-based segmentation algorithm. The average effective grain diameter d_0 at a given deposition temperature is used as a scaling parameter. Samples with different deposition temperatures show two fundamentally distinct distributions. The α-phase FePc ($< 200°C$) samples fit better to a Gaussian distribution, whereas the β-phase FePc ($> 200°C$) samples fit better to a log-normal distribution. (b) From the sample thickness dependence t at fixed deposition temperature (80 °C), the scaling exponent $\beta = 0.51 \pm 0.14$ is determined using a best fit (long dashes). As a guide to the eye, $\beta = 0.8$ is drawn (short dashes). The thickness is normalized by t_0=27 nm and the roughness by σ_0=1.46 nm corresponding to the parameters of the thinnest sample.

The watershed-based segmentation algorithm from many AFM images results in explicit grain size distribution functions shown in Fig. 3(a). Despite the strong growth of the average grain size, once the distribution is scaled by the average grain size d_0, the curves collapse to two distinct distributions. At low deposition temperatures ($< 200°C$) the distributions are similar to a Gaussian distribution and at high deposition temperatures ($> 200°C$) the distribution can be easily fit to a lognormal distribution. These distinct differences in grain size distributions overlap with the documented $\alpha-$ to β-phase transition that occurs during the growth of the phthalocyanine molecule.[7, 16] Additionally, we observe round mounds at low temperatures and pin holes at high deposition temperatures. Rapid roughening and pin holes are generally attributed to energy barriers that appear at molecular steps on the surface. They are also referred to as Ehrlich-Schwoebel energy barriers.[17, 18] Recently, evidence for such an energy barrier was found in organic thin films.[19] Rapid roughening is referred to the anomalous growth exponent in the power law of roughness σ versus thickness t, where $\sigma \sim t^{\beta}$. For a simple random deposition $\beta = 0.5$.[20] Specifically in samples of di-indenoperylene thin films good agreement between the experiment and the Ehrlich-Schwoebel growth model was found. In support of this, a growth scaling exponent of $\beta = 0.8$ was measured.[21] In the FePc thin films discussed here, the scaling exponent β is fit to four samples deposited at 80°C and found to be 0.51 ± 0.14, see Fig. 3(b). As a comparison $\beta = 0.8$ is plotted in the same graph suggesting no or minimal rapid roughening occurs in the measured FePc samples.

4. Conclusion

Thermally evaporated iron phthalocyanine thin films were quantitatively characterized with a watershed-based segmentation algorithm and a correlation function. The grain size distributions revealed two distinct distributions coinciding with the α- and β-phase. This suggests that the grain size distribution may be fundamentally impacted by the underlying unit cell. An atomic force microscopy study of samples deposited at different temperatures allowing for differences in the diffusion during growth shows that while the grain size increases exponentially with deposition temperature, both long-range and short-range roughnesses grow only slowly. The formation of mounds, pin holes, and asymmetric grains suggests that there is an Ehrlich-Schwoebel-like energy barrier present that breaks the growth symmetry. Although the grains are strongly asymmetric, the measured growth scaling exponent was found to be close to the stochastic limit explained by a random deposition.

Acknowledgments

This research was supported by the National Science Foundation (NSF) grant DMR-0847552.

References

[1] R. D. Yang, T. Gredig, C. N. Colesniuc, J. Park, I. K. Schuller, W. C. Trogler, and A. C. Kummel, Applied Physics Letters **90**, 263506 (2007).
[2] A. Shehu, S. D. Quiroga, P. DAngelo, C. Albonetti, F. Borgatti, M. Murgia, A. Scorzoni, P. Stoliar, and F. Biscarini, Physical Review Letters **104**, 246602 (2010).
[3] T. Gredig, K. P. Gentry, C. N. Colesniuc, and I. K. Schuller, Journal of Materials Science **45**, 5032 (2010).
[4] R. L. Schwoebel, Journal of Applied Physics **40**, 614 (1969).
[5] A. V. Teran, A. Bill, and R. B. Bergmann, Physical Review B **81**, 075319 (2010).
[6] J. Bartolomé, F. Bartolomé, L. M. García, G. Filoti, T. Gredig, C. N. Colesniuc, I. K. Schuller, and J. C. Cezar, Physical Review B **81**, 195405 (2010).
[7] K. P. Gentry, T. Gredig, and I. K. Schuller, Physical Review B **80**, 174118 (2009).
[8] G. Liu, T. Gredig, and I. K. Schuller, Europhysics Letters **83**, 56001 (2008).
[9] I. Horcas, R. Fernandez, J. Gomez-Rodriguez, J. Colchero, J. Gomez-Herrero, and A. M. Baro, Review of Scientific Instruments **78**, 013705 (2007).
[10] M. Tsukahara *et al.*, Physical Review E **77**, 061306 (2008).
[11] V. W. Stone, A. M. Jonas, B. Nysten, and R. Legras, Physical Review B **60**, 5883 (1999).
[12] S. Labat, C. Guichet, O. Thomas, B. Gilles, and A. Marty, Applied Surface Science **188**, 182 (2002).
[13] J. Krim and J. O. Indekeu, Physical Review E **48**, 1576 (1993).
[14] S. K. Sinha, E. B. Sirota, S. Garoff, and H. B. Stanley, Physical Review B **38**, 2297 (1988).
[15] G. Liu, T. Gredig, and I. K. Schuller, Europhys. Lett. **83**, 56001 (2008).
[16] M. T. Robinson and G. E. Klein, Journal of the American Chemical Society **74**, 6294 (1952).
[17] G. Ehrlich, The Journal of Chemical Physics **44**, 1039 (1966).
[18] R. L. Schwoebel, Journal of Applied Physics **37**, 3682 (1966).
[19] R. Ruiz *et al.*, Chemistry of Materials **16**, 4497 (2004).
[20] G. W. Collins, S. A. Letts, E. M. Fearon, R. L. McEachern, and T. P. Bernat, Physical Review Letters **73**, 708 (1994).
[21] X. Zhang, E. Barrena, D. Goswami, D. G. de Oteyza, C. Weis, and H. Dosch, Physical Review Letters **103**, 136101 (2009).

Exact Analytical Solutions of Charged Monomer and Dimer Deposition Models in One and Two Dimensions

D A Mazilu, I Mazilu and H T Williams

Department of Physics and Engineering, Washington and Lee University, Lexington, VA, 24450, USA

E-mail: mazilud@wlu.edu

Abstract. Our work is motivated by the manufacturing process of self-assembled antireflective coatings using silica and titania nanoparticles. During the manufacturing process, it is highly desirable to know the time dependent analytical relationship between the index of refraction of the coating and the particle density of the jammed state. We present exact analytical results of one-dimensional cooperative sequential adsorption (CSA) models for two cases: monomer and dimer deposition of charged particles. We extend our study to two-dimensional deposition models on Cayley trees for charged monomers and dimers. Using standard linear algebra techniques, we introduce the general methodology for finding the exact analytical solutions for the monomer and dimer deposition of charged particles and discuss the time dependence of particle density (surface coverage) for some special cases of particle-particle interaction.

1. Introduction

The use of statistical physics methods in analyzing the cooperative behavior of multi-particle systems is ubiquitous in many fields, from physics, chemistry and engineering to modern biology and social sciences. Simple lattice models play a fundamental role in this quest to understand complexity, and were at the core of the development of equilibrium statistical physics. They have also been employed successfully in the field of non-equilibrium statistical physics, for modeling surface deposition, polymer chains dynamics and chemisorption on crystal surfaces [1], [2].

One-dimensional sequential adsorption models have been studied thoroughly in different physics contexts [3], [4]. The two basic one-dimensional models are known in literature as RSA (random sequential adsorption) when the adsorption sites are chosen randomly, and CSA (cooperative sequential adsorption) when the adsorption mechanism is influenced by the local environment. Despite the large number of studies of one-dimensional models, deposition in two dimensions is not as well understood. There are quite a few computational studies [3] on the matter, but no exact solutions exist for the general two-dimensional case. Recently, some progress was made for the case of a Cayley tree lattice structure [5], [6], [7].

Our paper was motivated by the manufacturing process of self-assembled antireflective coatings using silica and titania nanoparticles. The index of refraction of the coating depends on the surface coverage of the substrate. During the manufacturing process, it is highly desirable to know the analytical relationship between the index of refraction and the particle density of the jammed state. We study the ionic self-assembly deposition from the point of view of stochastic cooperative sequential adsorption models in one dimension and for a two-dimensional Cayley tree lattice. For the purpose of building a theoretical model for this process, we emphasize some of its essential elements: i) The deposition process is stochastic, so a random sequential adsorption model is justified; ii) The nanoparticles being deposited are charged particles, so a cooperative sequential adsorption model with rates dependent on the nearest neighbor occupation is appropriate; iii) The surface on which particles are being deposited is covered with a polyelectrolyte. In a sense, this is a pre-patterned lattice, because the negatively-charged nanoparticles will only stick to the positively charged monomers of the polyelectrolyte. Therefore, we consider appropriate a Cayley tree model of the surface.

2. Cooperative sequential adsorption models in one dimension

From the broad category of lattice gas models [1], we focus on one-dimensional sequential adsorption models. For this set of models, a linear string of lattice sites changes as each site switches irreversibly from empty to filled. In particular, we consider cooperative sequential adsorption (CSA) in which filling rates are influenced by the local environment (Fig. 1). If the incoming particles can each fill a single site (monomers), the density of filled sites increases with time, and the number of vacancies shrinks at an exponential rate. The problem becomes more complex when the incident particles are dimers, for example. In this case, the system evolves to a jammed state with vacancies remaining. Here, we recast such models into an eigenvalue/eigenvector problem.

2.1. General approach to monomer deposition

We assume N sites, initially empty, on a linear chain. A single entity can attach to an empty site with rates (attachments/time/empty site) α' when the two sites adjacent to the site to be filled are empty; β' if only one adjacent site is empty; and γ' when both adjacent sites are filled. Walls at $i = 0$ and $i = N + 1$ are treated as initially filled cells.

We define Q_i to be the probability that, through random deposition of monomers into empty cells, there exists a subchain of i empty cells bounded by filled cells. For example, Q_1 decreases each time a particle is deposited in the single empty site (rate $-\gamma' Q_1$); increases each time a string of two empties has a deposit in either of the empty sites (rate $2\beta' Q_2$); increases by two each time a string of length three has a deposit into its middle empty spot (rate $2\alpha' Q_1$); and increases each time a string of length four or more has a deposit made into a spot, leaving a single empty at either end of the original string (rate $2\alpha' Q_{i>3}$). From the master equation follows the rate equation for Q_i:

$$\frac{dQ_1}{dt} = -\gamma' Q_1 + 2\beta' Q_2 + 2\alpha' \sum_{j=3}^{N} Q_j. \tag{1}$$

At this point and henceforth we will simplify the rate equations by redefining our time scale $t \to \alpha' t$, and defining relative rates $\beta \equiv \beta'/\alpha'$ and $\gamma \equiv \gamma'/\alpha'$. With this new notation, the deposition probability at a site adjacent to two empty sites becomes $\alpha'/\alpha' = 1$. This renders the previous rate equation as

$$\frac{dQ_1}{dt} = -\gamma Q_1 + 2\beta Q_2 + 2\sum_{j=3}^{N} Q_j.$$

By similar logic, the general form for the rate equation for all Q_k, $k > 1$, is seen to be

$$\frac{dQ_k}{dt} = -(2\beta + (k-2)\alpha) Q_k + 2\beta Q_{k+1} + 2\sum_{j=k+2}^{N} Q_j.$$

A useful simplification comes from the change of variables $R_j = \sum_{n=j}^{N}(n+1-j)Q_n$, where R_j represent the probability of finding a substring of j empties that may have particles or empty cells on either end. The rate equations now become (for $k > 1$)

$$\frac{dR_k}{dt} = -(2\beta + k - 2)R_k - (2 - 2\beta)R_{k+1};$$

and

$$\frac{dR_1}{dt} = -\gamma R_1 - 2(\beta - \gamma)R_2 - (1 - 2\beta + \gamma)R_3,$$

in agreement with [4].

2.2. Matrix solution to rate equations

We now introduce a row vector $\vec{R}$ with components R_1, R_2, , R_3, ... and a matrix M so that the rate equations become

$$\frac{d\vec{R}}{dt} = M\vec{R},$$

with

$$M = \begin{bmatrix} -\gamma & -2(\beta-\gamma) & -(1-2\beta+\gamma) & 0 & 0 & 0 & \cdots \\ 0 & -2\beta & -2(1-\beta) & 0 & 0 & 0 & \cdots \\ 0 & 0 & -(2\beta+1) & -2(1-\beta) & 0 & 0 & \cdots \\ 0 & 0 & 0 & -(2\beta+2) & -2(1-\beta) & 0 & \cdots \\ 0 & 0 & 0 & 0 & -(2\beta+3) & -2(1-\beta) & \cdots \\ 0 & 0 & 0 & 0 & 0 & -(2\beta+4) & \cdots \\ \vdots & \vdots & \vdots & \vdots & \vdots & \vdots & \vdots \end{bmatrix}. \quad (2)$$

The general solution to this system of equations is

$$\vec{R}(t) = \sum_{j=1}^{N} c_j \vec{V}_j \exp(\lambda_j t), \quad (3)$$

where λ_j is the j-th eigenvalue of M, V_j is the corresponding eigenvector of M, and the c_j's are to be determined from the initial conditions, which for an initially empty chain are $R_i(t=0) = N + 1 - i$. Eigenvalues of a triangular matrix are its diagonal elements [9], thus $\lambda_j = -\gamma, -2\beta, -2\beta-1, \ldots, -2\beta-N+2$ for $j = 1, 2, 3, \ldots, N$. We denote the k-th element of the j-th eigenvector as $v_{k,j}$. The eigenvectors can be determined to within a constant multiplicative factor from

$$(M - \lambda_j I)V^j \equiv \mathcal{M}_j V^j = 0. \quad (4)$$

$\mathcal{M}_j$ is also upper triangular, with a zero in its j-th diagonal slot. Thus, solutions exist for V^j with $v_{k,j} = 0$ for $k > j$. These are the appropriate physical solutions for which long substrings of empty cells are quickly eliminated in the deposition process. Without loss of generality, we assume $v_{j,j} = 1$. Proceeding along the diagonal of $\mathcal{M}_j$ up and to the left of the zero value is the sequence of values: $1, 2, \ldots, j-2, -\gamma + 2\beta + j - 2$. The two-term recursion relation for the eigenvalue elements resulting from Eq. (4) produces

$$v_{k,j} = \begin{cases} 0 & \text{if } k > j \\ \frac{(2(1-\beta))^{j-k}}{(j-k)!} & \text{if } 2 \le k \le j \\ \frac{(2(1-\beta))^{j-3}(4(1-\beta)(\beta-\gamma)-(j-2)(2\beta-\gamma-1))}{(2\beta-\gamma+j-2)(j-2)!} & \text{if } k = 1 \end{cases}. \quad (5)$$

Eq. (3), at $t = 0$ can be cast as $\vec{R}(0) = \mathcal{V}\vec{c}$, where $\mathcal{V}$ is a matrix with elements $v_{i,j}$. Expansion coefficients c_i can be evaluated in terms of $R_i(0)$ through the inverse matrix $\mathcal{V}^{-1}$.

From Eq. (5) $\mathcal{V}$ is seen to be an upper-triangular matrix. The inverse of an upper-triangular matrix is itself upper-triangular [9]; also the upper left $n \times n$ values of $\mathcal{V}^{-1}$ are the inverse of the upper left $n \times n$ values of $\mathcal{V}$; and the diagonal elements of $\mathcal{V}^{-1}$ are the numerical inverses of corresponding diagonal elements of $\mathcal{V}$. We can sequentially find values for off-diagonal elements of $\mathcal{V}^{-1}$ utilizing the expressions that follow from the explicit relationship $\mathcal{V}\mathcal{V}^{-1} = I$

$$\sum_{k=i}^{j} \mathcal{V}_{i,k} v_{k,j} = v_{i,j} + \sum_{k=i+1}^{j-1} \mathcal{V}_{i,k} v_{k,j} + \mathcal{V}_{i,j} = \delta_{i,j} \tag{6}$$

noting that $v_{i,i} = \mathcal{V}_{j,j} = 1$. We apply Eq. (6) sequentially to find all off-diagonal elements of $\mathcal{V}^{-1}$: for $j = i+1$ it yields $v_{i,i+1} = -\mathcal{V}_{i,i+1}$; for $j = i+2, i+3, \ldots$ we use $v_{i,j}$ values determined in previous applications. This process is an efficient way to numerically determine all elements of a reasonable-sized upper left rectangle of the inverse of $\mathcal{V}$ of arbitrarily large finite dimensions.

The determination of the expansion coefficients c_i makes use of the fact that for physical cases of interest, the size of the systems (N) is asymptotically large, so that for such cases we can approximate the initial state values $R_i(t = 0) = N + 1 - i \approx N$ for $i << N$. This relates the expansion coefficients to row sums of the matrix $\mathcal{V}^{-1}$: $c_i = N \sum_{j=i}^{N} v_{i,j}$. In our particular case all row sums after the first are identical, so we need only evaluate c_1 and c_2. The fact that these row sums converge quickly compared to the size of N makes our approximation of constant values for $R_i(t = 0)$ quite good.

Results obtained by this method are in complete agreement with previously obtained analytical results for this case [3]. The value of the methodology comes in the more challenging case of deposition of dimers.

2.3. Dimer deposition

Deposition of entities that occupy two neighboring cells, dimers, can be treated with the same formalism. Defining the R_i's as before, their time behavior obeys

$$\begin{aligned}
\frac{dR_1}{dt} &= -2\gamma R_2 - 4(\beta - \gamma)R_3 + 2(2\beta - \gamma - 1)R_4; \\
\frac{dR_2}{dt} &= -\gamma R_2 - 2(2\beta - \gamma)R_3 + (4\beta - \gamma - 3)R_4; \\
\frac{dR_k}{dt} &= (3 - 2\beta - k)R_k - 2R_{k+1} + (2\beta - 2)R_{k+2}.
\end{aligned} \tag{7}$$

Rewriting as before, in terms of vectors and matrices, $\frac{d\vec{R}}{dt} = M_d \vec{R}$, with

$$M_d = \begin{bmatrix}
0 & -2\gamma & -4(\beta-\gamma) & 2(2\beta-\gamma-1) & 0 & 0 & 0 & \cdots \\
0 & -\gamma & -2(2\beta-\gamma) & 4\beta-\gamma-3 & 0 & 0 & 0 & \cdots \\
0 & 0 & -2\beta & -2 & 2\beta-2 & 0 & 0 & \cdots \\
0 & 0 & 0 & -(2\beta+1) & -2 & 2\beta-2 & 0 & \cdots \\
0 & 0 & 0 & 0 & -(2\beta+2) & -2 & 2\beta-2 & \cdots \\
0 & 0 & 0 & 0 & 0 & -(2\beta+3) & -2 & \cdots \\
\vdots & \vdots & \vdots & \vdots & \vdots & \vdots & \vdots &
\end{bmatrix}$$

The time behavior of the system is as seen earlier, $\vec{R}(t) = \sum_{j=1}^{N} d_j \vec{V}_j \exp(\lambda_j t)$, in terms of eigenvalues λ_j and eigenvectors $\vec{V}$ of M_d.

The eigenvalues of this upper triangular matrix are its diagonal elements, all negative and in decreasing order as long as $2\beta > \gamma$. The zero eigenvalue corresponds to the asymptotic time behavior with some equilibrium number of empty slots of width 1. The k-th eigenvector has only its first k elements non-zero, and it is convenient to pick the k-th element to be 1. The eigenvector equation for this eigenvector gives recursion relations for remaining nonzero eigenvector components.

The inverse of the upper-triangular matrix of eigenvectors can be evaluated as was done for the monomer case, and it is also upper-triangular. Row-sums of the inverse matrix converge quickly, allowing evaluation of expansion coefficients d_i in terms of the initial state characterized by $R_i(0)$. The agenda for resolution of the dimer deposition problem is thus also set, and straightforward computer code can be written for families of parameter sets worthy of examination.

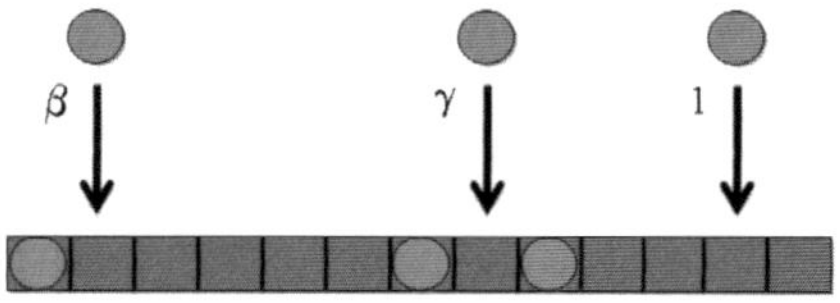

Figure 1. Cooperative sequential adsorption (CSA) of monomers on a one-dimensional lattice.

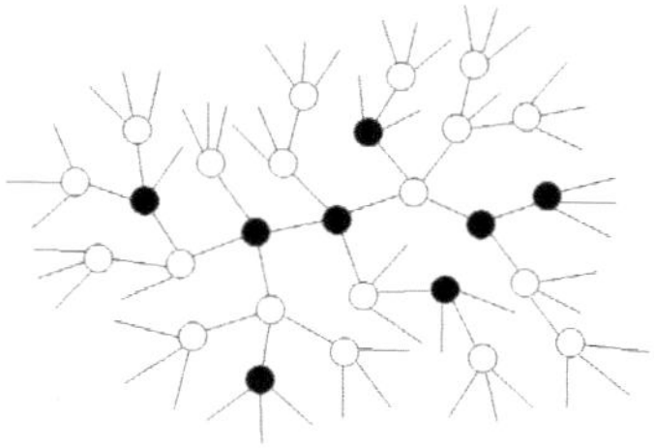

Figure 2. Cooperative sequential adsorption (CSA) of monomers and dimers on a Cayley tree with $z = 4$.

3. Cooperative sequential adsorption of monomers and dimers on a Cayley tree

Adsorption of particles is exactly solvable in higher dimensions only for a special case, a tree-like structure called the Cayley tree (or a Bethe lattice). A Cayley tree is a connected, cycle-free graph, where each node is connected to z neighbors, where z is called the coordination number. For $z = 2$ we recover the one dimensional case. A fully two-dimensional model can be approximated by a Cayley tree with a coordination number $z = 4$, as shown in Fig. 2.

It was proven that a Cayley tree structure gives a very good approximation of a regular lattice. For example, for the coverage of a two-dimensional lattice with dimers, the Cayley tree approximation gives $8/9 = 0.889$, while the regular lattice gives 0.908 (result obtained from computer simulations [3]). The use of a Cayley tree lattice is appropriate for the ionic self-assembly process, because we can approximate the polyelectrolyte as a web with the attractive sites being the positive monomers of the polyelectrolyte. The incoming particles are going to land only at the attractive sites, or in close proximity to these sites.

The problem was solved for the deposition of a mixture of monomers and dimers on a Bethe lattice by Cadilhe and Privman [6] for the case of random sequential adsorption of uncharged particles. Cadilhe and Privman found exact analytical expressions for the time evolution of clusters of connected empty sites. In the context of reaction-diffusion models on a Cayley tree, it was shown in [5] and [7] that the only reactions possible (in the sense of getting a closed system of equations for clusters of empty connected sites) are:

$$\begin{aligned} AB &\rightarrow AA \text{ with rate } r_1 \\ BB &\rightarrow BA \text{ with rate } r_2 \\ BB &\rightarrow AA \text{ with rate } r_3 \end{aligned}$$

where A are filled sites and B are empty sites. The rates are assumed to be non-zero.

The first two processes can be seen as monomer deposition processes, with different rates depending on the nearest neighbor occupation. The third process is a dimer deposition process. In this model, it is assumed that each branch of the Cayley tree is independent of the rest, and only nearest neighbors are being considered. As reported in [5], the evolution equations for the set of k connected empty sites are:

$$\frac{dP_k}{dt} = N_k[-r_1(P_k - P_{k+1}) - (r_2 + r_3)P_{k+1}] - (k-1)(2r_2 + r_3)P_k \tag{8}$$

P_k is the probability of finding a cluster of k connected empty sites, $N_k = k(z-2) + 2$ is the number of sites adjacent to a collection of k connected sites, and z is the coordination number. The matrix approach used for the one-dimensional case can also be applied here. We re-write Eq. 8 as:

$$\frac{dP_k}{dt} = [-(k-1)(2r_2 + r_3) + r_1 N_k]P_k - (r_2 + r_3 - r_1)N_k P_{k+1} \tag{9}$$

with the associated upper triangular matrix C with the diagonal elements $a_{k,k} = -[(k-1)(2r_2 + r_3) + r_1 N_k]$, and the off diagonal elements $a_{k,k+1} = -(r_2 + r_3 - r_1)N_k$.

Just like for the one-dimensional CSA, the eigenvalues of the matrix are the diagonal elements $a_{k,k}$, and the eigenvectors can be easily found using two term recursion relationships similar to the ones for matrix M (for deposition of monomers in one dimension). The solution for initially randomly occupied sites with occupation probability ρ is found to be:

$$P_k(t) = P_k(0)e^{-(zr_1 + (k-1)\beta)t}\{\frac{1}{1 + \alpha b(0)[1 - e^{-\beta t}]}\}^{\frac{z}{z-2} + k - 1} \tag{10}$$

where $\beta = (z-2)r_1 + 2r_2 + r_3$, $\alpha = \frac{(z-2)(r_2 + r_3 - r_1)}{(z-2)r_1 + 2r_2 + r_3}$, and $b(0)$ is the initial number of empty spaces, consistent with a pre-patterned surface. For easy comparison, we use the same notation as in [5]. By looking at Eq. 10, we see that the number of empty sites decreases exponentially with time, just like for the one-dimensional case. However, in the context of layer-by-layer self-assembly experiments, the dipping process is finite, so knowing exactly the time dependence of the number of empty sites can give us a good prediction for the overall particle density after a certain amount of time.

The analytical results presented here are quite general, and can be tailored to particular experimental scenarios, by tuning the values of the deposition rates in accordance with the experimental parameters. These deposition models can also be used successfully in other areas of study, such as epidemics or voter models.

4. References

[1] Schmittmann B and Zia R.K.P. *Phase transitions and critical phenomena, volume 17* Academic Press, London (1995.)
[2] Liggett T M *Stochastic Interacting Systems: Contact, Voter, and Exclusion Processes* New York: Springer (1999.)
[3] Evans J W, *Rev. Mod. Phys.*, **65**, 1281 (1993.)
[4] Privman V*Nonequilibrium statistical mechanics in one dimension*, Cambridge University Press, Cambridge, England (1997.)
[5] Matin L F, Aghamohammadi A and Khorrami M, *Eur. Phys. J. B*, **56**, 243 (2007.)
[6] Cadilhe A, Araujo N A M, and Privman V, *J. Phys. Cond. Mat.* **19** (2007.)
[7] Alimohammadi A, and Olanj N, *Physica A*, **389**, 549 (2010).
[8] Schütz G M, in *Phase Transitions and Critical Phenomena, Vol. 19*, ed C. Domb and L. Lebowitz Academic Press, London (2001.)
[9] Bernstein D *Matrix Mathematics.* Princeton University Press, Princeton, (2005.)
[10] Gouet R and Sudbury A, *J. Stat. Phys.*, **130**, 935 (2008.)

Analysis of Hybrid Type Boron-Doped Carbon Stripper Foils in J-PARC RCS

Y Yamazaki[1], M Yoshimoto[1], O Takeda[1], M Kinsho[1], T Taguchi[2], S Yamamoto[3], T Kurihara[4], I Sugai[4]

[1] J-PARC Center, 2-4 Shirane Shirakata, Tokai-mura, Ibaraki 319-1195, Japan
[2] Quantum Beam Science Directorate, Nuclear Science Research Institute, Japan Atomic Energy Agency, 2-4 Shirane Shirakata, Tokai-mura, Ibaraki 319-1195, Japan
[3] Environment and Industrial Materials Research Division, Takasaki Advanced Radiation Research Institute, Japan Atomic Energy Agency, 1233 Watanuki-machi, Takasaki, Gunma 370-1292, Japan
[4] High Energy Accelerator Research Organization (KEK), 1-1 Oho, Tsukuba, Ibaraki 305-0801 Japan

Email: yamazaki.yoshio@jaea.go.jp

Abstract. J-PARC (Japan-Proton Accelerator Research Complex) requires a carbon stripper foil to strip electrons from the H^- beam supplied by the linac before injection into the Rapid Cycling Synchrotron (RCS) [1]. The foil thickness is about 1μm (200μg/cm^2) corresponding to conversion efficiency of 99.7% from the primary H^- beams of 181MeV energy to H^+. We have successfully developed the Hybrid type thick Boron-doped Carbon (HBC) stripper foil, which showed a drastic improvement the lifetime without thickness reduction and shrinkage at the irradiated area. We started to study carbon stripper foils microscopically why carbon foils have considerable endurance for the beam impact by boron-doped. At first step, we made a comparison of ion irradiation effect between normal carbon and HBC by the electric microscope, ion-induced analysis. In particular, it seems that grain size of boron-rich area became much larger by irradiation for HBC. It was also observed that the boron-rich grain grew up by taking around material and generated pinholes more than 100 nm near itself consequently.

1. Introduction

A charge stripper foil is one of key technology for multi-turn injection of high power proton synchrotron. Typically carbon-based material is used to strip two electrons off the incident H^- beams. Recently injection of high power ring accelerators such as SNS [2] and J-PARC demands much longer lifetime for the stripping foils to reduce beam dead time. J-PARC RCS applied HBC foil which is developed with the arc-discharge method by Sugai [3]. HBC foil is much better than a normal carbon foil for uniformity and dynamic strength of large area foil and durability against irradiation of ions. The foil thickness is about 1μm (200μg/cm^2) corresponding to conversion efficiency of 99.7%. The usual foil is composed two pieces of foil because the double foil can endure against beam irradiation much better than the single one [4]. The foil size is 40mm x 110mm, but the area hit by the beam is about 10mm diameter around the end. Figure 1 shows a picture of our foil's holder which is attached by SiC wires of 10μm diameter. In order to guarantee not to disturb beam demand time by foil damage, we can exchange a new foil immediately without vacuum purge because of 15 foil's holders in the stock chamber. The Figure 1 also shows "the magazine rack" for the foil's holders. At the present time, HBC foil endures our usage of neutron's generation by RCS output beam power of 200kW continuously. However, the final goal of J-PARC RCS is to provide beam output power of 1MW.

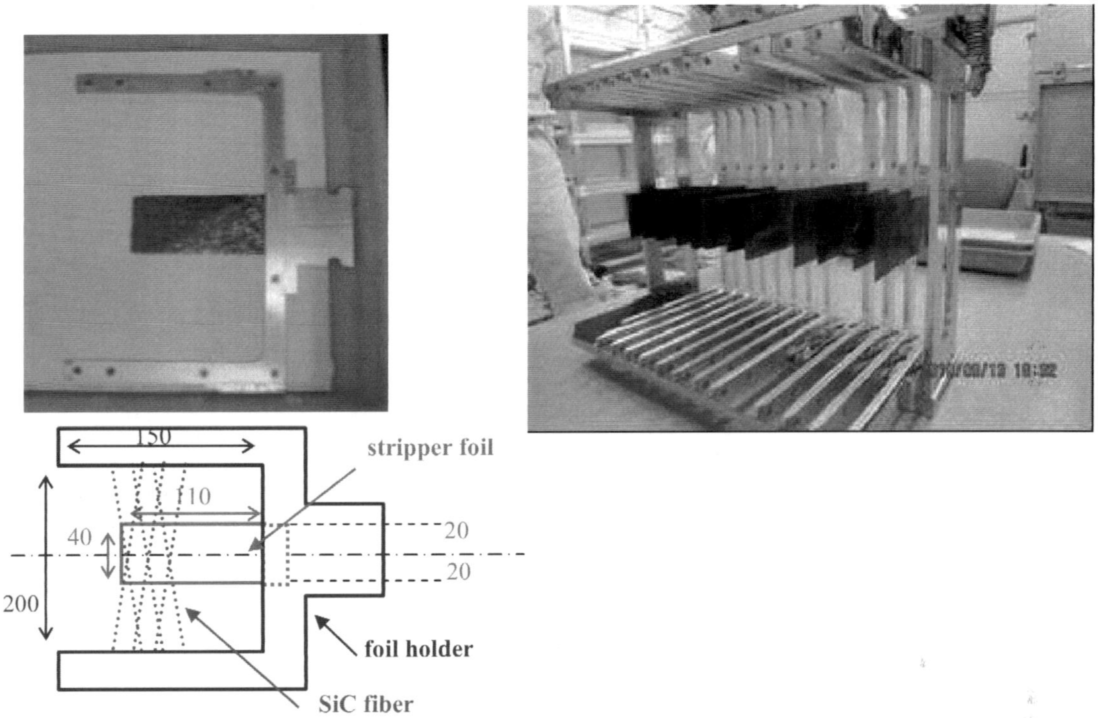

Figure 1 HBC foil's holder, the schematic diagram and "the magazine rack" which can stock 15 holders

We began to study why HBC foil has the much more durability against beam irradiation than a commercial available carbon foil, and how foils are broken by beam damage. In particular, it is important to investigate destruction mechanism by microscopic methods. In order to reveal the boron-doped effect for a carbon foil made by arc-discharge method against ion beam irradiation, ion accelerators of TIARA (Takasaki Ion Accelerators for Advanced Radiation Application) was used for ion-induced analysis. TIARA belongs in JAEA and has four accelerators, an AVF cyclotron, a 3-MV tandem accelerator, a 3-MV single-ended accelerator and a 400- kV ion implanter. This time we used ion implanter for ion irradiation and 3-MV single ended accelerator for RBS (Rutherford Backscattering Spectrometry). Furthermore, TEM (Transmission Electron Microscope) observation were tried to compare between normal carbon (NC) and HBC foil before and after ion irradiation.

2. Sample Preparation

Although typical foil thickness is 200μg/cm^2 for the practical usage, we used foil thickness of 15 μg/cm^2 because of ease for TEM observation. We prepared several foils each of NC and HBC foils of about 15 μg/cm^2. Ar^+ of 300keV, which we use as irradiated ion from ion implanter in TIARA this time, can deliver the energy deposition by the ion beam into this thickness foil without ion implantation. The preparation method of HBC foils is the controlled DC arc discharge process. The cathode(-) was a boron-doped (25%) carbon rod of 10mm diameter while the opposite electrode(+)

was a pure graphite rod of 15mm diameter. The distance between the evaporation source of the carbon rods and the substrate on the flat glass was 250mm. NC and HBC foils were analyzed the major atomic composition with RBS. The probe beam was H^+ of 3MeV from 3-MV single ended accelerator, in which the size was 1mm diameter and the current was about 15 nA on the foil surface. It is possible to obtain the boron to carbon ratio and the foil thickness information as shown in Figure 2. In HBC foil, the boron to carbon ratio is close to the rod's component.

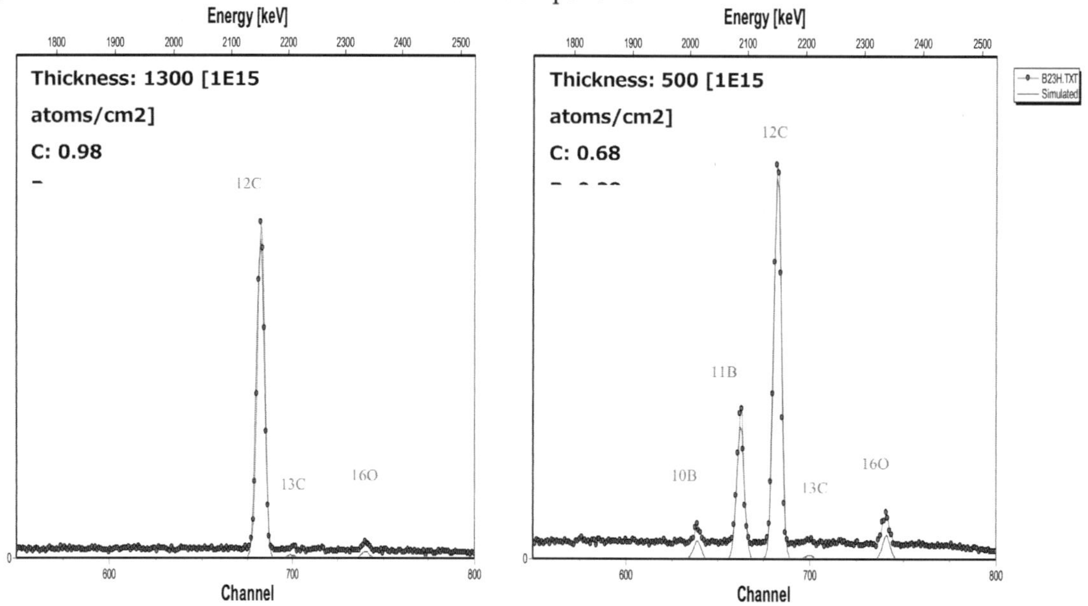

Figure 2 RBS results for NC (normal carbon): left side and HBC: right side

3. Analysis Results

3.1 Macroscopic observation for foils by irradiated ion beam

The ion beam size was about 5mm diameter and much smaller than the target sample foils. Figure 3 shows variance of the surface appearances for each of foils irradiated by Ar^+ of 300keV, 300nA. It seems that surface of HBC flattened out much slower than NC by the irradiation, and HBC was not broken after smoothed out. The visual light from the foil by the energy deposition wasn't observed. The foil surface temperature was lower than four hundred degrees C. We need to evaluate dynamic strength quantitatively before and after irradiated ion beam.

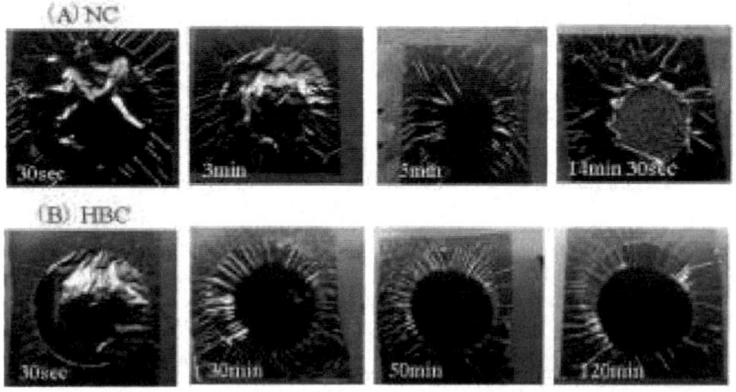

Figure 3 Time transition of sample surface for (A) NC (normal carbon) and (B) HBC irradiated with 300keV, 300nA Ar^+ ions. Aperture of the sample holder is diameter of 8mm.

3.2 Microscopic observation (TEM) for foils by irradiated ion beam

We have great interest how the irradiated area of the foil changes microscopically. In TIARA we have TEM (JOEL JEM-4000F:400keV) system which can inject ion beams from the ion implanter. We tried to observe the time variation by Ar ion irradiation (300kV, 20nA, <3mmϕ) for these foils in the TEM system. Both NC and HBC had generally amorphous phase, but they involved a lot of micro graphite grains of a few nanometer before irradiation of ions. In the NC foil case, though some graphite grains grew up, drastic change was not occurred. On the other hand, in the HBC case there were many boron-rich micro grains before and after ion irradiation. In particular, it seems that grain size of boron-rich area became much larger by irradiation in shown by the Figure 4. In the Figure 5, it was also observed that the boron-rich grain grew up by taking around material and generated pinholes more than 100 nm near itself consequently. We guess that this phenomenon is one of the destructive mechanisms of the HBC foil. We should detect atomic component and crystal structure of the boron-rich grain as shown in the black contrast.

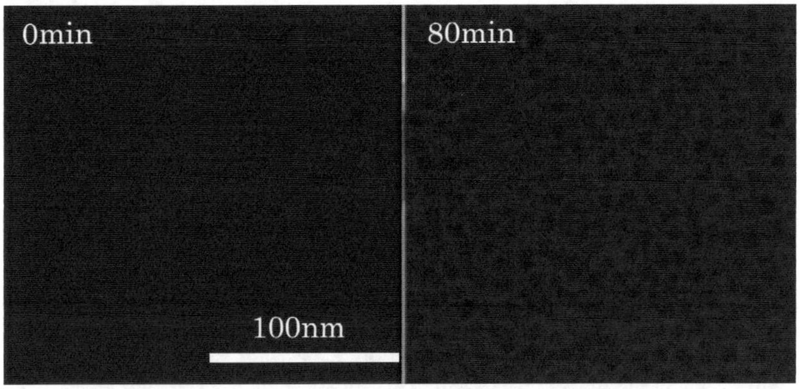

Figure 4 The time variation by Ar ion irradiation (300kV, 20nA) for the boron-rich area of HBC foil in the TEM system. Left side is before irradiation, and Right is after irradiation for 80min.

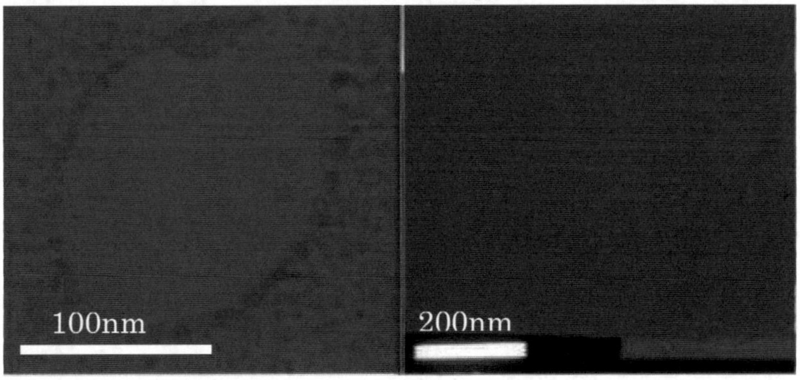

Figure 5 Some large pinholes by Ar ion irradiation of HBC foil in the TEM system

4. Summary

At first step, some results with microscopic analysis methods, such as RBS and TEM, were demonstrated for NC and HBC foils before and after Ar+ ion irradiation. This time we could obtain effective microscopic techniques adapted to analyze charge stripper foils, and got some remarkable results for transition by ion beam irradiation. In the future, we will try to evaluate atomic component and structure for boron-rich area in HBC foil, and reveal relation between boron and carbon in the foil.

References

[1]"The Joint Project for High-Intensity Proton Accelerators",JAERI-Tech 99-056, KEK Report 99-4, and JHF-99-3(1999).
[2] R.W. Shaw et al., PAC2007,AQ, NM, p.620.
[3] I. Sugai et al., Nucl. Instr. and Meth. A561(2006)16.
[4] I. Sugai et al., Nucl. Instr. and Meth. A613(2010)457.

Robust operation and performance of integrated carbon nanotubes atomic force microscopy probes

G Rius, I T Clark and M Yoshimura
Toyota Technological Institute, 2-12-1 Hisakata, Tempaku, 468-8511, Nagoya, Japan

E-mail: rius@toyota-ti.ac.jp, yoshi@toyota-ti.ac.jp

Abstract. We present a complete characterization of carbon nanotubes-atomic force microscopy (CNT-AFM) probes to evaluate the cantilever operation and advanced properties originating from the CNTs. The fabrication consists of silicon probes tip-functionalized with multiwalled CNTs by microwave plasma enhanced chemical vapor deposition. A dedicated methodology has been defined to evaluate the effect of CNT integration into the Si cantilevers. The presence of the CNTs provides enhanced capability for sensing and durability, as demonstrated using dynamic and static modes, e.g. imaging, indentation and force/current characterization.

Since the discovery of carbon nanotubes (CNT) [1] there have been many attempts to benefit from their entirely unique properties, motivating a revolution in the, at that time, emerging field of nanoscience and nanotechnology perhaps only comparable to the invention of atomic force microscope (AFM) [2] and scanning tunneling microscope (STM). [3] One of the most sought-after applications for CNTs is their use as scanning probe microscope (SPM) tips, taking advantage of unique dimensional high aspect ratio and ideal mechanical and electronic properties. The fabrication approaches vary as much as the number of combinations that can be implemented for the catalyst materials and preparation methods, synthesis techniques, attachment, etc. [4, 5, 6, 7] as do the resulting CNTs and the probes shape and performance characteristics. [8, 9]

We present a complete characterization of carbon nanotube-based dynamic mode AFM probes, both from the point of view of the operation of the cantilevers and from their advanced properties for sensing/microscopy due to the presence of the CNTs. The well-established technological approach consists of Si AFM probes functionalized near the tip with multiwalled CNTs grown by catalytic microwave plasma-enhanced chemical vapor deposition (MPECVD). [10, 11] First, a dedicated methodology has been defined to evaluate the effect of CNT integrative processing into the Si cantilevers. This information becomes essential for proper evaluation of imaging potential uses. [12] Then, we show how the presence of the CNTs provides enhanced capability for sensing with remarkable apex durability, as demonstrated by exemplary measurements in both dynamic and static mode. In particular, the specifications of CNT-interfaced topographical imaging, indentation, electrical conduction, mechanical response and adhesion are provided, while other capabilities for similar in-house fabricated devices have already been reported. [13]

The technological choice for the production of CNT functionalized tips is based on the integration of CNTs in commercial AFM probes using on-device synthesis of CNTs. Commonly, Si cantilevers with Al reflection coating are used as starting device (Budget Sensors Inc. Model: TAP300AL, 300±100 kHz, 40 N/m). After removal of native oxide in HF solution, locally controlled deposition of thin, polycrystalline Pd layer is applied, as described in [14]. Subsequently, MPECVD is used for the growth of a few thousands vertically oriented multiwalled CNTs (MWCNT), ordered and arranged as a cover of tip's original shape and dimensions. [15] Obviously, this may not seem the ideal product for undertaking high spatial imaging resolution in high aspect ratio sample structures, as it is envisioned for the use of long (small diameter) SWCNTs [16], but actually our devices have demonstrated, a number of interesting properties that might balance this fact (*vida infra*). The collective nature of the resulting number of CNTs comes directly from the dimensions of the area of Pd electroplating, whereas the multiwalled structure derives from the effective particle size during PECVD, and probably additional contribution coming from the use of microwave field. As reported in, [17] different metal particles do present a specific heating efficiency to absorb microwave radiation, which can be crucial for the growth process since it determines the particle state as a site for nucleation or formation of CNT. Although the exact mechanism is not yet clear, in both cases, [18] the combination of the catalyst and microwave CVD determines the CNT characteristics. Comparing the chemical products of the synthesis for both approaches, the use of methane as the source of carbon may be regarded as more hazardous, however, it is also noteworthy that methane has indeed been reported for its remarkable capability to integrate SWCNTs for the fabrication of CNT field effect transistors (CNT-FET) devices at wafer scale [19] or the development of MWCNT vertically aligned forests without external field assistance for functional CNT-based electrodes. [20] In addition to a morphological inspection by scanning electron microscope (SEM), recently we apply high resolution transmission electron microscopy (HRTEM) and energy dispersive X-ray spectroscopy (EDS) analysis for a more precise determination of the structural and chemical properties of CNT-decorated tips, which is important to support study of CNT synthesis process, and indispensable for the objective of present investigation of CNT-probes for AFM sensing (Figure 1). Therefore, the knowledge of catalyst position and element, the quality of outer amorphous carbon layer and its interface to the metal, etc is applied to the discussion and understanding of force, electric and mechanical interactions and imaging experimental results. For any kind of catalyst, the CNTs consist of a few tens of layers, MWCNT, being indicative of the growth technique, the MPECVD. In the case of Pd, we systematically observe that inner structure is bamboo-like. Using EDS we confirm the purity of catalyst particles.

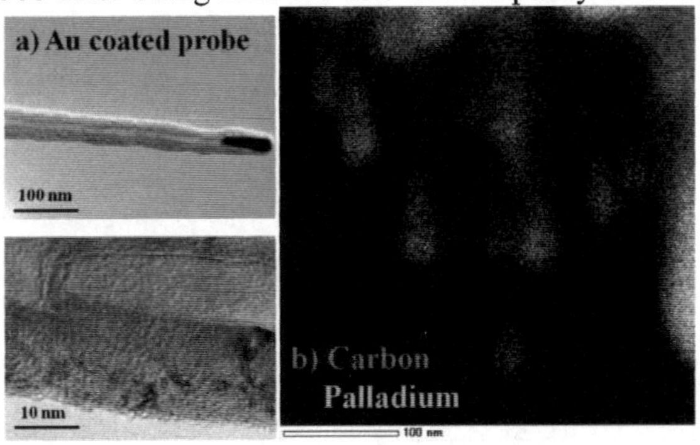

Figure 1. a) HRTEM is used to determine the structure of the CNTs. MWCNT images obtained from as-purchased Au coated probes. b) EDS provides chemical determination of catalyst particles.

Although our basic probe device is based on the specific dynamic mode probes related above, other cantilevers dimensions and starting materials have also been used and similarly studied for the production of particular functionality CNT-integrated probes. The same procedure has been used when testing local electrodepositing of Pd and MPECVD in as-purchased metal-coated softer probes (e.g. model SI-DF3-A, Rh or Au coated, 1.5 N/m, from SSI Nanotechnology Inc, Japan) as a test of the generality of the method. Provided that they could be more sensitive to the harsh fabrication conditions due to their increased length, and hence, mechanical fragility [21], they actually showed similar guarantee for the cantilever mechanical resonating properties after the CNT on tip synthesis. Results also indicate the strong dependence of the catalyst material nature in the obtained CNTs in terms of multilayer arrangement, probably linked to particle issues, such as diameter, etc (further study is ongoing). After the in-depth study of a few tens of CNT-AFM probes in terms of resonating mechanical operation of the cantilevers, morphological, chemical and structural determination, and test of performance and analysis for AFM applications, we describe and summarize the properties of our devices as follows.

For the quality control of the whole CNT-integrative fabrication process, we establish a procedure for monitoring the resonance figures of merit, so that we have access to the state of intrinsic properties of the cantilever. The resonance calibration is the fundamental step for the actual use of the probes by DFM. For doing this, we apply the automated resonance calibration routine of our SPM system (Seiko Instruments NanoNavi II E-Sweep) after, i) HF etching, ii) Pd electrodepositing, and iii) MPECVD (CNT growth), and register resonance frequency peak, together with related characteristics, such as quality factor or excitation voltage. The detection of the peak of oscillation is determined at resonance (instead of slightly shifted to, about 3 kHz lower, as commonly used for dynamic mode operation), for a target amplitude of 1 V. The calibration measurement is executed first, in air and normal pressure, then, in a vacuum ($\sim 10^{-3}$ Pa), and finally, at atmospheric pressure in a N_2 ambient, for the study of pressure, humidity and vibration dissipation, which will strongly determine the CNT-probe performance, thanks to the SPM chamber system for relative humidity (i.e. including pumping).

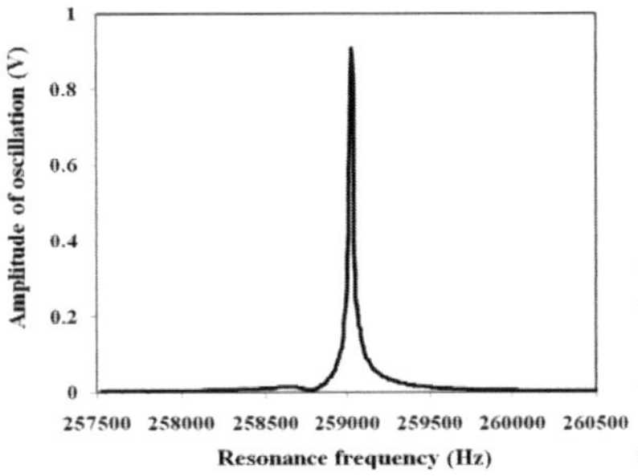

a)

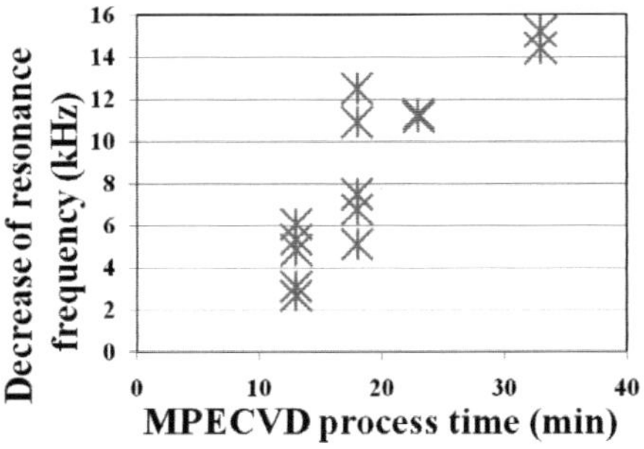

b)

Figure 2. a) Example of resonance frequency calibration obtained in a vacuum. Peak width is below 30 Hz. Perfect coupling between CNTs and tip supports a proper performance during dynamic mode imaging. b) Tendency of resonance frequency as a function of MPECVD process time. The decrease correlates with the mass loading on the tip due to the formation of CNTs.

The summary of the resonance frequency peak monitoring demonstrates that the calibration of the oscillation can be uniquely determined, for the user fixed target parameters, and for any of the tested environmental conditions with no unexpected divergences. The resonance frequency peak in a vacuum is sharp and precisely reliable, with a width of no more than a few tens of Hz (Figure 2, a)). In the case of normal pressure determination, we observe the usual broadening of the peak (FWHW about 0.5 kHz), within the specifications of as-purchased probes, and, importantly, without major difference between air versus N_2, which will be convenient for demonstration of universal DFM imaging.

Closely related to the frequency determination, the quality factor is a measure of the oscillation attenuation due to a series of factors. [22] It is indispensable to have a minimum value for normal conditions of operation [23], while it should be adjusted in absence of damping for a controllable actuation and detection. In brief, our results register no degradation of the cantilever quality factor (complementary description in [15]). In the case of the vibration voltage, the conditions of the technological approach results in no clear tendency, but remarkably (almost) always the target amplitude value could be reached within the voltage limits of the instrument (10 V). In fact, the determination of the vibration voltage for the target actuation is intrinsically less precise or more unpredictable. Even when no modification or technological process is applied to the probes, it contains an additional source of error as it can depend on slight variations of, for instance, photo detector alignment and signal.

This quality control strategy has been applied to the intermediate steps, for completeness, while strictly, the main objective as been the determination of the product specifications for the final CNT-probe devices. The results indicate that the probes intrinsic properties do not suffer any modification that could make unfeasible their application for imaging, nor additional degraded characteristics after CNT integration. The sharpness of the resonance peak in vacuum suggests that high resolution AFM could be tested using proper system (frequency modulated detection [24, 25]). In this case, probably, the CNT diameter characteristic of our MWCNTs would be the limiting factor for atomic resolution. The good resonance frequency determination is rooted in the fact that no dangling CNTs conform the CNT-modified tip (as it could be the case of a single long SWCNT), but collective arrangement coverage of the tip cone is present. Therefore, CNTs support each other and show perfect coupling to the Si original device. This point is a strict requirement for a reliable linear response when executing topographical imaging in order to guarantee a precise dimensional calibration. Partially summarizing, the possibility to be properly operated in a vacuum in terms of excitation, detection and linear response indicates that present CNT-based probes may be tested in atomic resolution SPM systems to test their ultimate properties for force or conductivity probing techniques, AFM or STM.

As stated in [15], the resonance frequency shift after the CNT synthesis indicates and quantifies the carbon mass loading due to the MPECVD process. Effectively, it correlates with the local accretion of mass as a consequence of CNT formation and allows study of the synthesis growth rate. As shown in Figure 2, b), the decrease of resonance frequency [26, 27] (due to mass loading) as a function of MPECVD process time shows a linear dependence tending to the saturation for synthesis longer than 30 minutes. Certain deviation from linearity is attributed mainly to the variability of the MPECVD and the use of more or less concentrated catalyst solution together with the concise resulting electrodeposited area of Pd for each tip. The extrapolated amount of MWCNTs (~3000) formed is in agreement with the SEM observed MWCNT structures and density. Although a fine fitting of present case is hard and inherently imprecise to be established, this procedure, together with an even more local CNT-functionalized nanomechanical device can be envisioned as an ultimate resolution mass sensor, for example, if properly integrated for optimal operation (CNTs are reduced) and signal conditioning, as in [28, 29].

After comprehensive determination of CNT role and performance on AFM probes subsequent to CNT integration in terms of technological approach, examples of their actual function as a tool for AFM in several of the sensing modes are presented. The measurements include both non-contact and contact sensing and demonstrate the wide-range of applications capability of our CNT-AFM probes.

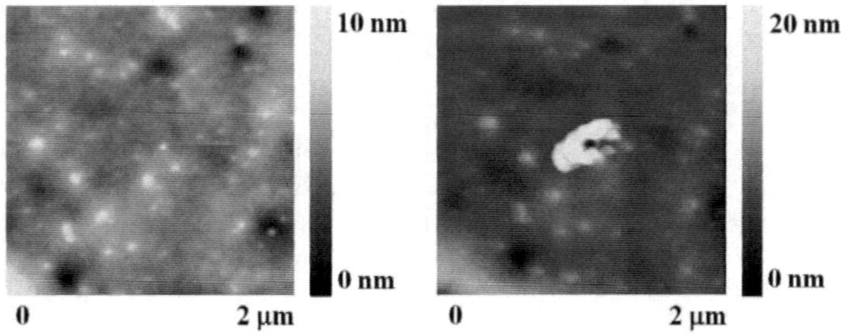

Figure 3. Tapping mode imaging using CNT-probe on an epoxy resist containing Au nanoparticles. a) The quality of the image corroborates the proper performance of the integrated CNT-tip. b) Same location inspected after indentation of the polymer with the CNT-probe, while consecutive image is also acquired by the same CNT-tip.

The proper imaging performance is easily demonstrated in Figure 3. Topography image (Figure 3 a)) has been acquired in dynamic mode on a thin layer (about 10 nm) of epoxy resist deposited on a Si substrate. The resist contains embedded Au nanoparticles (commercially available, with an average size of 6.5 nm). It is clear from the image quality and the dimensions of the observed particles inside the resist matrix that they are absolutely correct and well resolved. More remarkably, following the topography image an indentation test using CNT tip is executed, and consecutive imaging of the resist carving (bump in the center of the image) is acquired also with the same CNT-AFM probe (Figure 3 b)). As the tip has no acute cone shape, but MWCNTs 'grass-like' conformal coverage of the very same tip shape and 20-30 nm in diameter (average size of present Pd-based MWCNTs), the tip is not able to completely resolve inside the indentation hole, however, the rest of the image area is not significantly altered in terms of quality of resolution, in-plane (xy) nor vertical dimension (z). As it will be further demonstrated next, the durability of the tip apex in our CNT-AFM probes represents a truly advantage for the tip technology.

Similarly to the results presented in [18], one of the promising applications of CNT-functionalized tips is their potential suitability for measurements in contact, in particular, for force spectroscopy or electrical conductivity. In this case, the potential is a two-fold one, since CNT-probes can be applied to the determination of the sample substrate properties, or the study of very same CNT characteristics. Figure 4, a) shows a retraction force curve executed on mica, which is a standard hard substrate for analysis or calibration of cantilever elastic properties. The mechanical response is modulated (milder contact) due to the presence of the CNTs with respect to the one obtained for bare tip on same conditions. Particular adhesion interactions attributable to the CNT-interfacing are under study. In figure 4, b), an example of relatively high voltage electrical conduction measurement is presented (sample-tip approach). The simultaneous force/current data acquisition provides deeper insight on the events of mechanical and electrical contact: stable electrical current is established at very low loading forces of about 7 nN and not abruptly. In both cases, further measurements could go behind the measurement tests reported in the figures, thanks to the unaltered tip apex and the MWCNTs completely elastic deformation. More detailed communications about the performance of CNT-probes, specifically, for mechanical and electrical conduction properties will be soon published in [30].

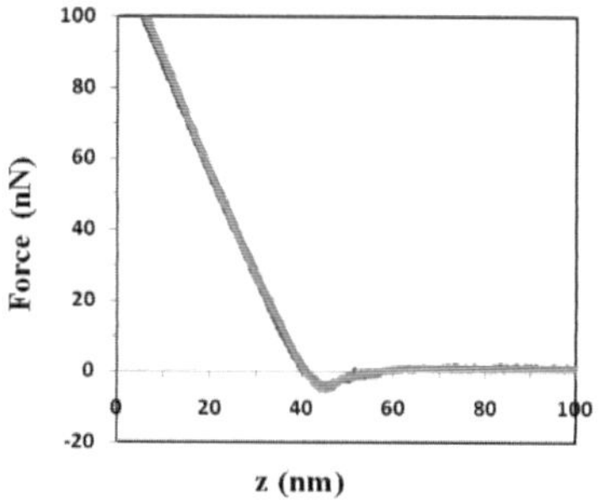

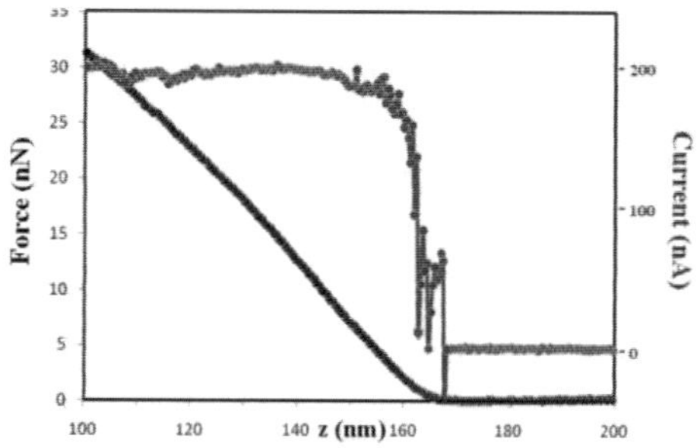

Figure 4. a) Iterative retraction curves performed with CNT-probe on a mica substrate. Controlled RH is set to 5%, whereas forces related to tip-sample adhesion interaction are still observed. b) Simultaneous acquisition of approach force curve and electrical current at tip-sample bias 7 V, as a function of z, stable current is 200 nA. Establishment of correlation between mechanical and electrical contact can be studied (RH 35 %). Sampling substrate is Si covered by a thin Pt layer.

In summary, the present communication shows the complete and comprehensive description of integrated MWCNTs into AFM silicon probes. After a detailed description of the operational performance of the CNT-probes from the point of view of devices for dynamic mode AFM application, representative results of their actual use for different sensing modes is shown. A procedure for the quality control of the fabrication process has been established, enlightening some aspects of the CNT synthesis process and framing the technology robustness for its exploitation for probing. The utility of the CNT-probes is demonstrated for both contact and dynamic modes of operation, with outstanding force sensing sensitivity of (probably) any kind of interaction. By AFM, conducting, semiconducting or insulating materials can be investigated and further exploitation of the CNT-AFM probes is envisioned for phase imaging, electric modes (electric force microscopy, Kelvin probe force microscopy, magnetic force microscopy, etc), force spectroscopy, electrical conductivity measurements, surface manipulation or nanolithography. The durability of the tip apex represents remarkable robustness, combined with the convenient CNT elasticity, for probing all sorts of substrates (either hard or soft materials).

Acknowledgments
G. Rius is grateful for the granted financial support of Strategic Research Foundation Grant-aided project for Private Universities from MEXT, Japan. The funding for JST-Sentan Project: MultiProbe Microscopy System is also acknowledged.

References

[1] Iijima S 1991 *Nature* **354**, 6348
[2] Binnig G, Quate C F and Gerber C 1986 *Phys. Rev. Lett.* **56** 930
[3] Behm R J, Garcia N and Rohrer H 1989 *Proceedings: NATO ASI* Drordrecht, Holland: Kluver 184
[4] Hafner J, Cheung C and Lieber C 1999 *Nature* **398** 761
[5] Stevens R, Nguyen C, Cassell A, Delzeit L, Meyyappan M and Han J 2000 *Appl. Phys. Lett.* **77** 3453
[6] Akita S, Nishijima H, Nakayama Y, Tokumasu F and Takeyasu K 1999 *J. Phys. D: Appl. Phys.* **32** 1044
[7] Chin S-C, Chang Y-C and Chang C-S 2009 *Nanotechnology* **20** 2853071
[8] Jia Y, Li H, Zhang B, Wei X, Zhang Z, Liu Z and Xu, Y 2010 *J. Phys.: Condes. Matter* **22** 334218
[9] Kis A and Zettl A 2008 *Phil. Trans. R. Soc. A* **366** 1591
[10] Tung F K, Yoshimura M and Ueda K 2009 *J. Nanomaterials* 612549
[11] Sakamoto T, Chiu C-C, Tanaka K, Yoshimura M and Ueda K 2009 *J. Nanomaterials* 851290
[12] Verdaguer A, Sacha G, Bluhm H and Salmeron M 2006 *Chem. Rev.* **106** 1478
[13] Tanaka K, Yoshimura M and Ueda, K. 2009 *J. Nanomaterials* 147204
[14] International patent in preparation.
[15] Thomas I T, Rius G, Matsuoka Y and Yoshimura M 2010 *J. Vac. Sci. Technol. B* **28** 1148
[16] Xu D H, Wang Z-G and Douglas J F 2008 *Macromolecules* **41** 815
[17] Druzhinina T S, Weltjens W, Hoeppener S and Schubert U 2009 *Adv. Func. Mater.***19** 1287
[18] Druzhinina T S, Hoeppener S and Schubert U 2010 *NanoLetters*, DOI: 10.1021/nl101934j
[19] Martin-Fernandez I, Sansa M, Esplandiu M J, Lora-Tamayo E, Perez-Murano F and Godignon P 2010 *Microelectronic Engineering* **87** 1554
[20] Martin-Fernandez I, Gabriel G, Rius G, Godignon P and Perez-Murano, F 2009 *Microelectronic Engineering* **86** 806
[21] Jacobs H O, Knapp H F and Stemmer A 1999 *Rev. Scient. Inst.* **70** 3, 1756
[22] Richter AM, Sengupta D and Hines M A 2008 *J. Phys. Chem. C* **112** 1473
[23] Sulchek T, Hsieh R, Adams J D, Yaralioglu G G, Minne S C, Quate C F, Cleveland J P, Atalar A and Adderton D M 2000 *Appl. Phys. Lett.* **76** 1476
[24] Oyabu N, Sugimoto Y, Abe M, Custance O and Morita S 2005 *Nanotechnology* **16** S112
[25] Morita K, Sugimoto Y, Sasagawa Y, Abe M and Morita S 2010 *Nanotechnology* **21** 305704
[26] Ekinci K L and Roukes M L 2005 *Rev. of Scient. Inst.* **76** 6
[27] Ekinci K L,Yang Y T and Roukes M L 2004 *J. Appl. Phys.* **95** 5, 2682
[28] Arcamone J, Rius G, Abadal G and Pérez-Murano F 2006 *Microelectronic Engineering* **83** 1216
[29] Arcamone J, Rius G, Llobet J, Borrisé X and Pérez-Murano F 2008 *J. Phys.: Conf. Ser.* **100** 052031
[30] Rius G and Yoshimura M 2012 *Micro and Nano Letters* (Accepted for publication at the Special Issue on Micro and Nano Technologies for Probe-Based Microscopy).

HBC Foil Beam Study and Long-term Observation at the 3-GeV RCS in J-PARC

M. Yoshimoto[1], P.K. Saha[1], Y. Yamazaki[1], M. Kawase[1], R. Saeki[1], N, Hayashi[1], K. Yamamoto[1], H. Hotchi[1], T. Ishiyama[1], M. Kinsho[1], I. Sugai[2] and Y. Irie[2]

[1] JAEA, Japan Atomoic Energy Agency, Tokai-Mura, Naka-Gun, Ibaraki-Ken, JAPAN

[2] KEK, High Energy Accelerator Reserch Organication, Oho, Tsukuba, Ibaraki-Ken, JAPAN

Yoshimoto.masahiro@jaea.go.jp

Abstract. The HBC foil are installed in the J-PARC RCS for the charge-exchange H^- beam injection. In order to examine the characteristics of the HBC foils, beam studies for the HBC foil were carried out from the viewpoint of foil's life time and the beam survival rate due to foils. According to the compromise between charge-exchange efficiency and foil scattering beam loss, we selected the foil thickness of about 200μg/cm^2 for user operation. The measurement of the charge-exchange efficiencies of the HBC foils is found to be consistent with expectation. The long-term observation during the user operation as well as the HBC foil endurance test for its life time evaluation was carried out. In the beginning of the beam irradiation the voluminous outgassing from the foil occurred and the foil was found to deform drastically but after that the outgassing decrease gradually and foil deformation didn't become worse. There was no noticeable deterioration of the stripping foils even after one year user operation, which corresponds to about 1.41×10^{21} particles irradiation.

1. Introduction

Japan Proton Accelerator Research Complex (J-PARC) is a multi-purpose research facility for materials and life sciences, nuclear and particle physics, and nuclear engineering with MW class proton beam at 3GeV and at several 10GeV. It is a joint project between Japan Atomic Energy Agency (JAEA) and High energy Accelerator Research Organization (KEK), and is located in Tokai-Mura in Japan. The accelerator complex consists of a 400-MeV linear accelerator (LINAC), a 3-GeV Rapid Cycling Synchrotron (RCS), and a 50-GeV Main Ring Synchrotron (MR) [1]. The RCS is designed to accelerate the H^+ beam from 400MeV to 3GeV and with 8.3×10^{13} protons per pulse at a repetition rate of 25Hz, which corresponds to a beam power of 1MW. At the first stage, the beam energy injected from the LINAC to the RCS is 181MeV but the extraction energy is same as 3GeV. Therefore the RCS aims to achieve the beam power of 0.6MW [2]. In the RCS, the charge-exchange H^- beam injection scheme with stripping foils is adopted to realize the MW class high-power proton beam [3]. In accelerator, multi bunched beams are injected from an injector repeatedly and are stacked in a circulating accelerator to increase the beam intensity. But Liouville's theorem precludes that identical particles are injected over the phase space volume occupied already by other particles. The charge-exchange injection scheme is one of the beating Liouville's theorem techniques. Figure 1 shows the schematic diagram of the charge-exchange H^- beam injection scheme at the RCS. The H^- particles enriched with electrons are produced in the ion source and are accelerated to 181MeV in the LINAC. In RCS, horizontal bump magnets are excited at the injection period and thus the injected H^-

particles are bent onto the circulating beam orbit. The H^- particles are then encountered a thin foil, which stripped two electrons from the H^-. Thus H^+ particles exited from the foil and were placed in the same phase space volume as an existing bunch of protons already circulating in the ring. On the other hand, there are partially stripped H^0 particles and un-stripped H^- particles slightly. In order to control these H^0 and H^- beams, there are two more stripping foils named 2nd and 3rd foil and are installed downstream of the 1st foil. The partially stripped H^0 and the un-stripped H^- beams are further stripped to H^+ by the 2nd and 3rd foils respectively and are transported to the injection beam dump called H0-beam dump (H0-BD).

The important key issues for the charge-exchange H^- injection with stripping foil are foil's life time and beam survival rate due to foils. Thus various beam studies for stripping foil were carried out. In this paper, we report the beam studies results concerned with foil's life time and beam survival rate due to foils.

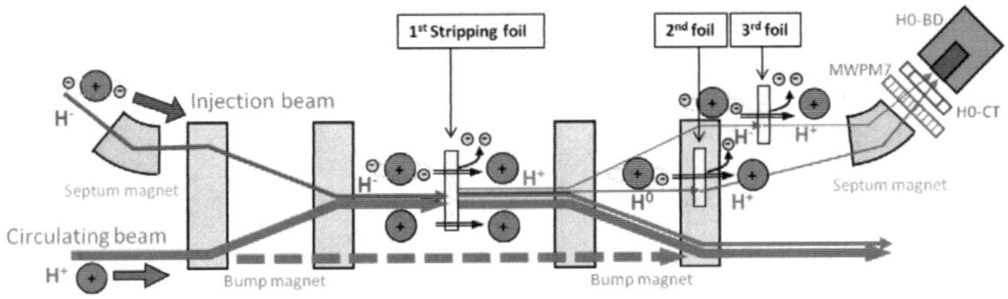

Figure 1. Schematic diagram of the charge-exchange H^- beam injection scheme at 3GeV RCS in J-PARC.

2. Experimental Procedures and Results

The charge-exchange H^- injection technique is already utilized in various accelerator laboratories around the world, and several kinds of the foils are used in respective laboratories. We adopt the Hybrid type thick Boron-doped Carbon (HBC) stripping foils in the J-PARC RCS. The HBC foils are developed by Sugai group in KEK, which improved the lifetime drastically [4]. So far the HBC foil doesn't have certain achievements in the high power user operation, although it has the lifetime's advantage in the beam irradiation test. Aiming the higher power beam operation, the HBC foil has been installed in the RCS ring on September 2008, and various beam studies in the RCS are carried out to examine the characteristics of the HBC foils [5].

2.1. Beam survival rate due to stripping foil

The thickness of the thin foil is one of the key parameters. Charge-exchange efficiency depends on the stripping foil thickness. The higher the thickness of the stripping foil, the more charge-exchange efficiency increases but the beam loss due to the foil scattering due to foil beam interaction increases too. Thus the optimum foil thickness is a compromise between charge-exchange efficiency and beam loss due to foil scattering.

To optimize the foil thickness, the beam survival and the stripping efficiency were measured using various thicknesses HBC foils from 100μg/cm^2 to 600 μg/cm^2 (nominal values; 92, 196, 288, 392, 487, 634 μg/cm^2). In the user operation, the bunched beams of 250turns are injected into the RCS ring. But in this beam study, the number of the injecting beam bunches are limited to 50turns and a long flattop pattern of the bump magnets was adopted and the beam injection timing is set at the start of the flattop, so foil hitting probability increases, to clear the foil scattering beam loss. Circulating beam intensity was measured by a Wall Current Monitor (WCM) and is shown in figure 2. Injection beam is stacked in the ring and then the beam intensity increases. After the injection, the beam intensity seems to remain constant. Then beam intensities at the various timing are compared as shown figure 3. In figure 3, red circles are obtained just after the injection (50 turns) has finished. The higher the thickness of the stripping foil, the more circulating beam intensity increased, but the beam intensity was saturated with the foil thickness more than 300μg/cm^2. Thus the red circles are similar to the charge-exchange efficiency. The other marks are obtained after the bump magnets are turned off in other words the circulating beam doesn't hit the scattering foil thus there were no observable reduction. The difference between red circles and other marks means the foil scattering beam loss and it increases depending on the foil thickness. Figure 4 shows the analysis results of the beam survival rate due to foil scattering beam loss as a function of the foil thickness. The foil scattering beam loss is directly proportional to the foil thickness. According to the compromise between charge-exchange efficiency and foil scattering beam loss, we selected the foil thickness of about 200μg/cm^2 for user operation.

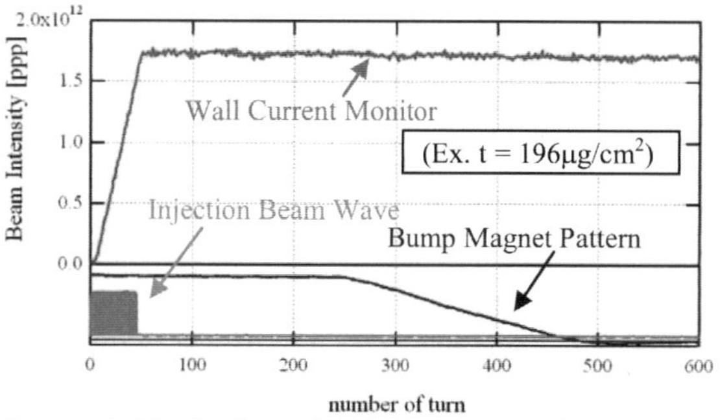

Figure 2. Typical result of the measured circulating beam intensity with WCM, wave patterns of bump magnet and injection beam. (In this case, foil thickness of 196μg/cm^2 were used.)

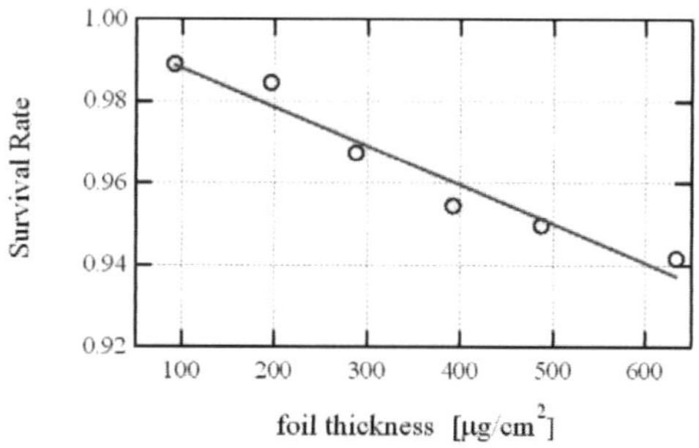

Figure 4. Analysis results of the beam survival rate due to foil scattering beam loss as a function of the foil thickness.

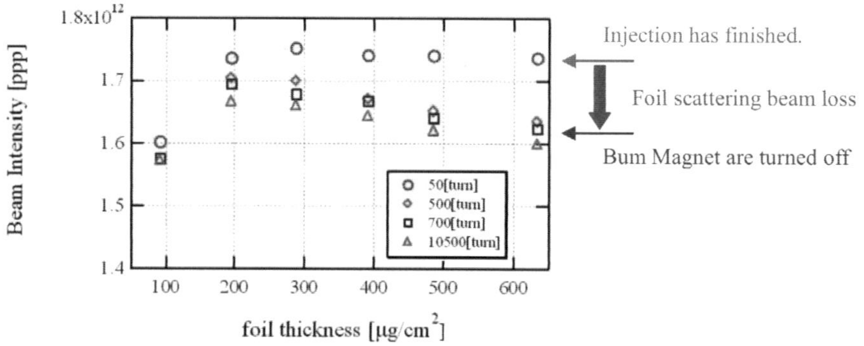

Figure 3. Beam intensities measured by WCM as a function of the foil thickness. Red circles are obtained just after the injection has finished. The higher the thickness of the stripping foil, the more circulating beam intensity increased, but the beam intensity was saturated with the foil thickness more than 300μg/cm^2. Thus the red circles are similar to the charge-exchange efficiency. The other marks are obtained after the bump magnets are turned off in other words the circulating beam doesn't hit the scattering foil thus there were no observable reduction. The difference between red circles and other marks means the foil scattering beam loss and it increases depending on the foil thickness.

In order to obtain the HBC foil's charge-exchange efficiency directly, partially stripped (H^0) beam component and un-stripped (H^-) beam component were measured respectively at the two beam monitors installed at the front of the H0-BD. One is the multi wire profile monitor 7 (MWPM7), and the other is a H0-current transformer (H0-CT). In this experiment, several thinner foils whose thickness was less than 200 μg/cm^2 (nominal values; 52, 78, 158 μg/cm^2) were used, because these two component signals were increased clearly [6]. Figure 5 shows the measurement result with MWPM7. At the front of the H0-BD, H^0 and H^- components were separated and then two beam profiles can be measured. When the 1st stripping foil was removed from beam orbit, the entire injected H^- beam was converted to H^+ at the 3rd stripping foil and the H^+ beam was transported to H0-BD because the 3rd foil thickness is 500 μg/cm^2 and has charge-exchange efficiency of almost 100%. Thus the each charge-exchange efficiency from H^- to H^+ or H^0 can be obtained by comparing between the peak signals of MWPM7 with or without the 1st stripping foil, in principle. However as seen in figure

5, the MWPM7 signal doesn't have a good resolution and thus small peak signal is hard to separate from the background noise. In the case of H0-CT measurement, it has very high resolution with Fast Fourier Transform (FFT) data analysis techniques and even very small signal can be picked up and separated from the background noise [6]. However measurement signal with H0-CT with nominal setting of all stripping foils means a total beam current transported to the H0-BD and thus H^0 and H^- components can't be separated. But as seen in figure 1, not only the 1st stripping foil but also the 2nd or 3rd stripping foil can be removed from the beam orbit independently in the J-PARC RCS. When the 1st foil is installed on the beam orbit, we can separately measure the partially stripped H^0 beam component and un-stripping H^- beam component with the H0-CT, by removing the 2nd and 3rd stripping foil, respectively. And total injected H^- beam can be measured with the H0-CT by removing the only 1st stripping foil. In Figure 6, the measured charge-exchange efficiencies from H^- to H^+ and H0 are summarized. Open circles are measurement results and solid lines are calculation results with cross sections given in the previous experiment with normal Carbon [7]. The measured charge-exchange efficiencies of the HBC foils are consistent with expectation at the present. As there is only 20% of Boron rest of 80 Carbon in the HBC foil, the charge-exchange efficiency is believed to be almost same as Carbon foil.

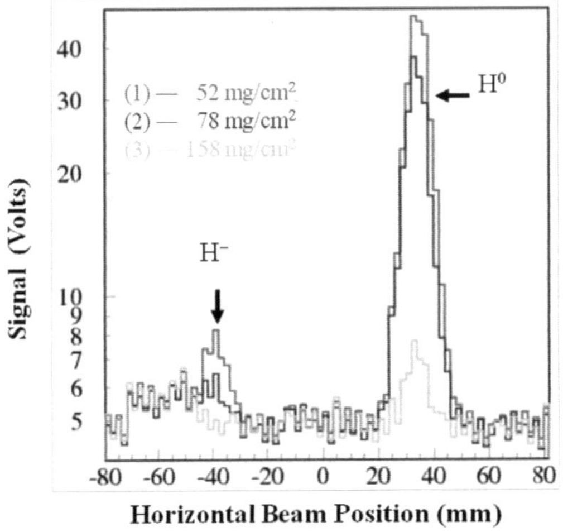

Figure 5. Measurement results of H^0 and H^- beam profile with MWPM7 with thinner stripping foils.

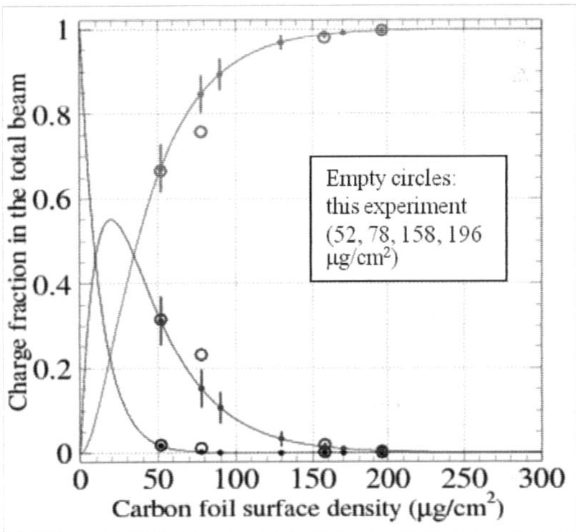

Figure 6. Measurement result of the charge-exchange efficiencies from H^- to H^+ and H^0 with H0-CT. Lines are calculated from the previous experimental cross section shown by solid dots.

2.2. Life time of the stripping foil

A life time of the stripping foil is one of the most important key issues, because it respects tightly an available beam operation time in the J-PARC. Progress in the foil's life time can be expected by adopting the HBC stripping foil. We tried the HBC foil endurance test. The outgassing from the HBC foil and the foil deformations during the beam irradiation were observed. The beam irradiation on the new stripping foil with thickness of 196μg/cm^2 was started since RUN26 in October 2009. The outgassing from the HBC foil and the foil deformations during the beam irradiation are summarized in figure 7. The upper graph in figure 7 shows the trend of the vacuum pressure measured by a Cold-Cathode ionization Gauge (CCG) installed in the neighborhood of the HBC foil and total irradiation particle in every run cycle. At the start of the beam irradiation, the voluminous outgassing from the foil occurred. After that, the outgassing from the HBC foil was decreased and it seems to be stable after the RUN29. Since Run33 started, the beam power at user operation increased from 120kW to 200kW, but the outgassing did not to increase. The lower photos in figure 7 show the deformations of the HBC foils during the beam irradiation. A surface condition of the HBC foil including a form and a texture was varied quickly until RUN28, for which about 2.5x10[20] particles were irradiated on foil. But afterward, it seems the process of the foil deformation has stopped. During the one year of the user operation (a total of about 1.41x10[21] particles irradiation), there was no noticeable deterioration of the stripping foil.

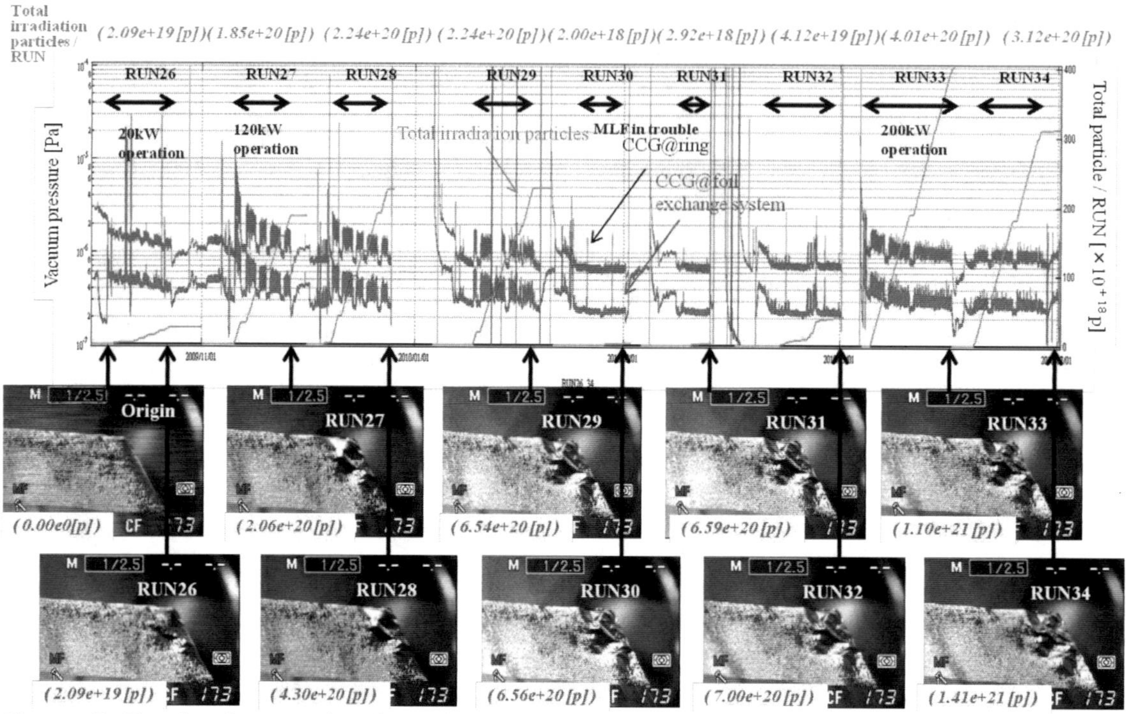

Figure 7. Trends of outgassing from the foil and the deformations on the foils during the beam irradiation. The upper graph plots the vacuum pressure measured by the CCG around the stripping foil and the total irradiated particle every run cycle. The lower photos show the deformations of the HBC foils during the beam irradiation.

3. Summary

In the J-PARC RCS, the HBC stripping foil was adopted for the charge-exchange H^- beam injection. The important key issues for the charge-exchange injection with the stripping foil are foil's life time and beam survival rate due to foils. Thus the foil beam study and long-term observation are carried out from the viewpoint of two key issues.

In order to evaluate the beam survival rate due to HBC stripping foil, circulating beam intensity was measured with various thickness stripping foils. According to the compromise between charge-exchange efficiency and foil scattering beam loss, we selected the foil thickness of about 200μg/cm^2 for user operation. Moreover the charge-exchange efficiencies in the HBC foil were measured with MWPM7 and H0-CT directly. The measured charge-exchange efficiencies of the HBC foil are consistent with the expectation at the present. In our future plan, not only HBC foils but also normal Carbon foils of various thicknesses will be produced and a difference between their charge-exchange efficiencies will be examined in the J-PARC RCS.

The long-term observation during the user operation as well as the HBC foil endurance test for its life time evaluation was carried out. The outgassing from the HBC foil and the foil deformations during the beam irradiation were observed. At the start of the beam irradiation, the voluminous outgassing from the foil occurred and the foil was deformed drastically. Even after about 2.5x10[20] particles irradiation, the outgassing rate from the foil was stable and the foil deformation did not become worse, and also no noticeable deterioration of the stripping foils could be seen. During the one year of the user operation (a total of about 1.41×10^{21} particles irradiation), there was no noticeable deterioration of the stripping foils. Until now, the causal relation between the outgassing from the foil and foil's deformation during the beam irradiation is still unclear. Thus we will install a quadruple mass spectrometry near the stripping foil, and detailed analysis of the outgassing from the stripping foil will be done.

References

[1] JAEA/KEK Joint Project Team, KEK Report 2002-13, Jaeri-Tech 2003-044, March 2003.
[2] H. Hotchi, et. al., Phys. Rev. ST Accel. Beams 12, 040402 (2009).
[3] M. Yoshimoto, et. al., Proceedings of EPAC06 (2006), pp.1765-1797
[4] I.Sugai, et. al., Nucl. Inst. and Meth. A561(2006), pp.16-23.
[5] M. Yoshimoto, et. al., Proceedings of IPAC2011(2011), pp.3927-3929
[6] P.K. Saha, et. al., Proceedings of IPAC2011(2011), pp.3511-3513
[7] R.C.Webber and C.Hojvat, IEEE Trans. NS26 (1979) , pp.4012-4014

AUTHOR INDEX

AUTHOR INDEX

AUTHOR INDEX